MY BIG TOE

AWAKENING • DISCOVERY • INNER WORKINGS

**A TRILOGY
UNIFYING PHILOSOPHY,
PHYSICS, AND METAPHYSICS**

Thomas Campbell

http://www.My-Big-TOE.com
http://www.lightningstrikebooks.com

The *My Big TOE* reality model will help you understand your life, your purpose, the totality of the reality you experience, how that reality works, and how you might interact most profitably with it.

A half dozen independent test readers of various backgrounds were asked to evaluate the My Big TOE *trilogy and record their impressions of it. This is what they said:*

■ ■ ■

"Eureka! A Theory Of Everything that actually lives up to its name! *My Big TOE* not only unifies physics, but unifies philosophy and theology as well. You will be amazed!"
— PAMELA KNIGHT, PHYSICIST

"Reading *My Big TOE* has challenged my mind and widened my horizon. Expect your worldview to radically expand and your perspective to reach a new level of understanding."
— INA KUZMAN

"*My Big TOE* is utterly original, pioneering and bold. Campbell writes with clarity and humor as he explores and answers the hard questions in this comprehensive work about the ultimate nature of reality and consciousness. Full of fresh and profound ideas, you may be astonished to find that learning how reality works actually improves the quality of your life."
— LYLE FULLER, POWER ENGINEER

"The *My Big TOE* trilogy roared through my comfy no-brainer world like a category F5 tornado that makes you laugh ... when the dust finally settled, I was left with an incredibly clear view of how and why things are as they are."
— PEG ROCHINE, FOUNDER AND CEO, CLINICAL RESEARCH

"This trilogy will profoundly change you .... you will never look at your world in the same way again."
— INA KUZMAN

"Thoroughly challenging, engaging ... a transforming experience. *My Big TOE* marks the end of humanity's childhood."
— LYLE FULLER, POWER ENGINEER

"Unique, profound, and enriching are the words that most easily come to my mind to describe *My Big TOE.*"
— TREVOR GOLDSTEIN, PHILOSOPHER AND FUTURIST

"If you have ever asked the questions: Is this all there is?, What's the purpose?, How am I related to the whole?, *My Big TOE's* logic, grounded in science, provides unequivocal answers that make you think. A profoundly fascinating read!"
— INA KUZMAN

MY BIG TOE: A TRILOGY UNIFYING PHILOSOPHY, PHYSICS, AND METAPHYSICS

AWAKENING • DISCOVERY • INNER WORKINGS

Lightning Strike Books
http://www.LightningStrikeBooks.com

Printed in the United States of America
First printing: October 2007
Second printing: May 2010

Book design by Michele DeFilippo, 1106 Design, LLC
Cover illustration by Frank Foster

Publishers Catalogue-in-Publication Data
Campbell, Thomas
My Big TOE: Awakening – Discovery – Inner Workings, A trilogy unifying philosophy, science, and metaphysics / Thomas Campbell
1. Science. 2. Philosophy. 3. Metaphysics. 4. Reality, model of.
5. Consciousness, theory of. 6. Spirituality. 7. Paranormal, theory of.
8. Theory of Everything. 9. TOE. 10. Theology (religion) and science.

ISBN 978-0-9725094-6-6 (Softcover)
ISBN 978-0-9725094-7-3 (Hardcover)

To Chris

To Bob & Nancy

To Dennis & Nancy Lea

To Todd, Lyle, Ina, and Trevor,
whose encouragement
was key to success

To those in need of
a new perspective

To all seekers of Big Truth

To Pamela, The One

To love within
A joyous
Heart

Synopsis
My Big Picture
Theory of Everything

**My Big TOE – A trilogy unifying
philosophy, physics, and metaphysics**

Book 1: Awakening

Section 1 provides a partial biography of the author that is pertinent to the subsequent creation of this trilogy. This brief look at the author's unique experience and credentials sheds some light upon the origins of this extraordinary work. The unusual associations, circumstances, training and initial research that eventually led to the creation of the *My Big TOE* trilogy are described to provide a more accurate perspective of the whole.

Section 2 lays out, logically justifies, and defines the basic conceptual building blocks needed to construct *My Big TOE*'s conceptual foundation. It discusses the cultural beliefs that trap our thinking into a narrow and limited conceptualization of reality, defines the fundamentals of Big Picture epistemology and ontology, as well as examines the inner-workings and practice of meditation. Most importantly, Section 2 defines and develops the two basic assumptions upon which this trilogy is based. From these two assumptions, time, space, consciousness, and the basic properties, purpose, and mechanics of our reality are logically inferred.

Book 2: Discovery

Section 3 develops the interface and interaction between "we the people" and our digital consciousness reality. It derives and explains the characteristics, origins, dynamics, and function of ego, love, free will, and our larger purpose. Finally, Section 3 develops the psi uncertainty principle as it explains and interrelates psi phenomena, free will, love, consciousness evolution, physics, reality, human purpose, digital computation, and entropy.

Section 4 describes an operational and functional model of consciousness that further develops the results of Section 3 and supports the conclusions of Section 5. The origins and nature of digital consciousness are described along with how artificial intelligence (AI), as embodied in AI Guy, leads to artificial consciousness, which leads to actual consciousness and to us. Section 4 derives our physical universe, our science, and our perception of a physical reality. The mind-matter dichotomy is solved as physical reality is directly derived from the nature of digital consciousness.

Book 3: Inner Workings

Section 5 pulls together Sections 2, 3, and 4 into a more formal model of reality that describes how an apparent nonphysical reality works, interacts, and interrelates with our experience of physical reality. Probable realities, predicting and modifying the future, teleportation, telepathy, multiple physical and nonphysical bodies, and the fractal nature of an evolving digital consciousness reality are explained and described in detail.

Section 6 is the wrap-up that puts everything discussed in Sections 2, 3, 4, and 5 into an easily understood personal perspective. Additionally, Section 6 points out *My Big TOE's* relationship with contemporary science and philosophy. By demonstrating a close conceptual relationship between this TOE and some of the establishment's biggest intellectual guns, Section 6 solidly integrates *My Big TOE* into traditional Western scientific and philosophical thought.

Contents

BOOK 3: Inner Workings

SECTION 5
Inner Space, the Final Frontier:

Acknowledgements

The One. In a category all to herself, I wish to acknowledge the immeasurable contribution, in all possible forms, given by my most constant, consistent, and challenging teacher: Pamela – The One.

Fellow travelers. First and foremost, acknowledgement goes to Bob Monroe and his wife Nancy who enabled my exploration of the path that eventually led to the *My Big TOE* trilogy. Next, to Dennis Mennerich, my fellow explorer and traveling companion. We pulled each other along when neither of us knew much about where we were going or how we were going to get there. Then, to Nancy Lea McMoneagle who was not only a fellow traveler but also the primary enabler of Monroe's success. All gems, every one – I could not possibly have set out on this strange journey with a better collection of friends and mentors. Finally, to the un-named many who provided me with the opportunities that enabled me to be what and who I have become. I wish I could have made more of the opportunities you offered. In the end it is these tens, these hundreds, these thousands, who made this trilogy possible. Thank you all.

Major contributors. In a more direct and immediate vein, there are a few readers of indomitable fortitude to whom I am eternally grateful. The time and effort volunteered by these remarkable people made all the difference in the world. Together we have tried hard to make all three books as clear and understandable as possible.

Special thanks go to Lyle Fuller, Todd Phillips, Ina Kuzman, and Caroline Lampert for their effort to improve the readability and clarity of *My Big TOE*. All four were quick to point out where I had left stumbling blocks lying on the path to Big TOE understanding. Additionally, Todd's and Ina's questions served as a catalyst to ferret out much interesting material. Many thanks to Chris Nelson who started me writing in the first place. Without their selfless generosity and dedication beyond all reason, this trilogy would be a poor shadow of what you have before you.

In addition, I thank Nancy Lea McMoneagle and Dennis Mennerich for aiding and corroborating the accuracy of my memories of the early years at Whistlefield. Also heartfelt thanks to Lyle Fuller, Joel Dobrzelewski, Trevor Goldstein, and Eric Campbell for their encouragement and good questions. Special thanks to Steve Tragesser for asking questions that became the catalyst for much of Chapter 90, Book 3. Likewise, to Lyle Fuller for doggedly pursuing questions that eventually produced the discussion of free will found in Chapter 45, Book 2 and that added clarity to my exposition of the psi uncertainty principle. Similarly, to Trevor Goldstein whose experience and questions precipitated the discussion of mind tectonics in Chapter 40, Book 2; and to Ina Kuzman for initiating the discussion found in Chapter 23, Book 1 about the nature and practice of meditation. Also, thanks goes to Eric Campbell for precipitating a discussion about the natural constraints of a finite consciousness system. Credit goes to Bryan Mott, Ted Vollers, Tom Hand, Zane Young, Rhonda Ganz, Kristopher Campbell, and Lynda Sterling for finding errors, offering useful comments and asking good questions. Finally, I wish to thank Steve Kaufman for being in the right place at the right time with his book, *Unified Reality Theory: The Evolution of Existence Into Experience*. I love it when a plan comes together.

Hired Help. Two ladies of great integrity and competency enabled *My Big TOE* to make the transition from an amateur creation to a professional product. Kate von Seeburg, owner of *K8 & Company*, edited the manuscript while Michele DeFilippo, owner of *1106 Design*, produced the interior and cover designs.

Family. Great appreciation goes to my wife and children, who patiently and cheerfully allowed me to work on "the book" when I should have been paying attention to them. I hope the final result will prove itself worthy of our collective sacrifice.

Non-contributors. Last, and certainly least, I wish to barely mention Kathy Cyphert and Peggy Rochine, who, along with many others too numerous to name, contributed absolutely nothing to this effort but wanted to see their names mentioned in it just the same. Additionally, Boldar, Kiana, Onyx, Joe, Nikki, Chico, Mr. Pickle, Sid, Moe, Sir Maximus, Snuffy, Sir Minimus, Kia, Gabrielle, Isabel, and Kuga-Bear also deserve honorable mentions as outstanding non-contributors.

—Tom Campbell,
Dec. 9, 2002

Preface: Author's Note to the Reader

Yes, you should read this preface.

I understand that many readers have little interest in, or patience for, lengthy prefaces or forewords. The first question is always: Should I take the time to read this ancillary text, or can I skip it without missing anything important?

Most of us are eager to zip past the preliminaries and immediately sink our teeth into the meat of the main text. Anticipation and expectation push us to get on with the real thing. We of Western culture are an impatient goal oriented people driven toward endpoints. In our rush to the finish line, we take little notice of the journey that gets us there. Such a misappropriation of emphasis often squanders our opportunities because, more often than not, the tastiest and most nourishing part of life lies in experiencing the process, not in attaining the goal.

By the end of Section 6, you will no doubt agree that these books are ... well ...different. As such, they require a different approach. The preface and foreword of the *My Big TOE* trilogy **are** integral parts of the story. Because this trilogy blazes an original trail far off the beaten path, it is essential to include introductory material that can help prepare you for what lies ahead. I know you are eager to get on with it and discover if this trilogy delivers the goods, but rushing off toward that goal too quickly actually reduces the likelihood that you will get there at all.

The function of the preface and the foreword is to maximize the return on your reading investment. The preface provides an overview of the tone, structure, process, and mechanics of the *My Big TOE* trilogy. The foreword establishes a broad view of the trilogy's content and lays out a rough map of where you will be going on this unusual journey. It provides the context and focus wherein the trilogy's content is most easily understood. The foreword and preface together improve comprehension and minimize frustration by providing a global view of the forest before you begin your descent into the trees.

I strongly suggest that you adopt an attitude of patience toward gaining an understanding of the profound mysteries and ancient secrets that are logically unraveled by this new physics. My Big Picture Theory Of Everything (*My Big TOE*) will take you to both the beginning and to the end of time. It will dive deeply into the human heart as well as probe the limits of the human mind. It will define the significance of you, and provide new meaning to your existence. It will help you realize and optimize your potential. It will develop a wholly new scientific understanding of both your inside and outside world.

You may find it more productive to pace yourself by depth of comprehension than by percent of completion. Avoid rushing from concept to concept the way children pursue presents on Christmas day. Take your time. A feast for heart, head, and soul is best ingested little by little, bite by bite, with many thoughtful pauses and much careful cogitation to aid digestion. Genuine breakthroughs must be absorbed slowly as existing paradigms grudgingly dissolve. Familiar paradigms, like a favorite teddy bear, can be extremely difficult to let go.

Every successful journey, regardless of how long or difficult, begins with a single step that is animated by gumption, directed by goals, and repeated as often as necessary by dogged perseverance. On this particular journey, the preface is located at step one, the foreword at step two, followed by the three books: *Awakening*, *Discovery*, and *Inner Workings*.

I have carefully aimed the content of this scientific and philosophic exposition at a general audience of varied background. You do not need a scientific, philosophical, or metaphysical background to understand the content of the My Big Picture Theory Of Everything trilogy. No leaps of faith or beliefs are required to get to where these books will take you. A determined and tenacious truth seeker – a sturdy, independent intellect that is by nature open minded and skeptical – constitutes the optimal reader. There are no prerequisites. If you have a logical, open, and inquisitive mind – an attitude of scientific pragmatism that appreciates the elegance of fundamental truth and the thrill of breakthrough – you will enjoy this journey of personal and scientific discovery.

Under the best of circumstances the successful communication of this trilogy's content will require much from both of us. This work presents many unique and daunting challenges to the effective communication between author and reader. Worldviews are not casually picked, like fruit, from a vendor's cart: To make the necessary connections, we must dive deeply.

Far beneath the foundation of your intellect, your culture lays out the template for your worldview upon the core belief systems that define your perception of existence. The basic assumptions that support your notion of reality are not seen by you as assumptions at all – they are accepted, without question, as the most solid of all facts. That is simply the nature of culture – belief at the bone and sinew level of awareness. The point is: The concepts presented by this trilogy are likely to challenge the belief systems of your culture – regardless of what culture you come from.

Material within *My Big TOE* may challenge your familiar assumptions, beliefs, and paradigms to the point of serious discomfort. If that discomfort leads to a profitable resolution, I am pleased; if it does not, I am saddened. My goal is to be informative and helpful. I encourage you to take what you can profitably use and leave the rest.

There are enough new concepts and unusual perspectives presented here to support and generate a multitude of books. I have purposely left much unsaid at the periphery in order to stay focused on the central idea of developing a Big TOE. Though the trilogy remains, from beginning to end, tightly focused on its primary objective, I will occasionally take short side trips in the form of asides to add color, explore related connections, and insert topics of special interest and practical value. Hopefully, you will find these side trips so interesting and informative that you will gladly excuse their interruption. Some effort will be required on your part to bridge these asides in order to maintain the logical continuity of the larger discussion. To make sure that you are never confused about whether you are reading an aside or the main text, asides are indented, have their own special font, and are clearly marked (at the beginning and end) with dingbats that look like this: ▶. If a secondary aside resides within a primary aside, it is indented yet again and marked with double dingbats ▶▶. When an aside fills an entire page, it is difficult to judge how much the text is indented, consequently, when this condition occurs, dingbats are placed in the header to let you know that the text is part of an aside. Thus, a casual glance is all that is required to determine if the text that you are reading is part of an aside, and if so, at what level.

You may find the text to be challenging in some places and obvious in others. What is too challenging or too obvious to each reader is mostly dependent on the experience and understanding of that individual reader. It is my intent to never speed through this exposition at such a rate that you cannot appreciate the scenery, nor to wallow about repetitively in the obvious – though from time to time, depending on your background, some may feel that I occasionally do both.

Although the language of American English (the language in which these books were originally written) is decidedly poor in nonphysical conceptual descriptors, it does have the advantage of being unusually rich in communications and information technology descriptors. The latter, oddly enough, is what allows me to convey the former. As strange as that may seem, it is the pervasiveness of modern science and technology, especially communications and data processing technology, that provides the conceptual tools required to produce a model of the larger reality the Western mind – or more broadly, the Western attitude or more accurately, the Western belief system – can relate to, understand, and work with.

Science and technology have advanced to the point where their applications and understanding have begun to mirror some of the fundamental processes of existence. We of the twenty-first century have only recently acquired the necessary concepts to understand and appreciate the nature of the larger reality within the context of our contemporary Western point of view. Previously, knowledge and understanding of the Big Picture and our existence in it was comprehended and described by ancient sages in terms of metaphors that were pertinent to their cultures and specifically created for the benefit of their specific audiences. Today, we find these once practical descriptions to be largely symbolic and irrelevant to a modern scientific view of reality. Philosophy, theology, and science find themselves at odds over what is significant.

I am a scientist. This trilogy is the result of a long and careful scientific exploration focused upon the nature of reality and the individual. Preconceived notions will be more of a hindrance than a help. It is the task of this trilogy to clearly and completely construct your consciousness, your world, your science, and your existence in a general, logical, scientific way that comprehensively explains **all** the personal and professional data you have collected during a lifetime.

An overarching Big Picture theory that explains **everything** may seem highly unlikely, if not absolutely impossible, but it is not. Take heart: Good science and human ingenuity have consistently delivered the impossible for at least two hundred years. Be open – history repeatedly demonstrates that the appearance of impossibility is most often the result of limited vision.

Patience will be required. This adventure of mind, science, and spirit is complex and will take significant time to properly unfold. If it were immediately obvious, it would either be old news or you would be reading a short journal article instead of a trilogy. A keen mind that is skeptical and open is the only ticket you need to take this journey.

Based on the feedback from those who have preceded you, I expect that you will find this voyage into the depths of elemental consciousness and fundamental reality to be personally enriching. You will be pushed to think a few big thoughts and ponder a few big ideas, but the conclusions you eventually come away with will be entirely yours, not mine. These are not books that set out to convince you of anything, or persuade you towards a particular point of view. At every turn you are strongly dissuaded from becoming a believer. Data, facts, and measurable results are the exclusive currency upon which this trilogy trades.

Reading the *My Big TOE* trilogy is not likely to be a passive experience. If you decide to seize the opportunity to climb out of the box, you will likely end up doing some difficult work. You will always be encouraged to think for yourself and come to conclusions that are based on **your** personal experience. Despite all the serious cogitation, we are also going to play, laugh, and have some fun as we go.

Much of what you believe to be true about yourself, your existence, and the nature of reality will be challenged. If you are open to exploring a bigger picture, these books will make you think, and think again. Most readers will not consider this trilogy to be an easy read – merely following the logical processes and sequences as they swallow up old paradigms will require some focused effort. On the other hand, significant growth and learning is rarely easy – if easy, it is rarely significant.

Contrary to my best efforts, Sections 2, 3, 4 and 5 remain somewhat conceptually interdependent. Each section will be better understood and make much more sense after reading the other sections. That could not be helped. Reality is a unified whole thing with each of its parts inexorably intertwined with the others.

This trilogy's three books and six sections develop the conceptual content of *My Big TOE* more or less sequentially. Consequently, reading the books or sections out of numerical order provides a less than optimal experience. However, understanding *My Big TOE* is much more dependent upon reading the entire trilogy than it is upon reading it in any particular order.

The nature of reality and of the typical reader is such that we must sneak up on *My Big TOE* one concept at a time. We will examine the Big Picture from multiple perspectives to ensure the design and structure of the whole becomes clearly visible. If things seem to get a little far out every once in a while, hang tough until it all pulls together into a coherent complete picture.

For the reasons stated above, a slow and careful reading will optimize your investment – take your time, and meander through these books at a relaxed and unhurried pace. If you become bogged down, it is better to go on (and come back later if you want to) than to feel as if you must read every word in the order in which it appears. It would be unfortunate for you to miss seeing a part of the forest that may be important to you because you became lost, exhausted, or discouraged wandering unproductively among the trees in another part.

Throughout *My Big TOE*, I have used a seeding technique to sneak up on some of the more difficult ideas. I often plant conceptual seeds (that briefly bring up or introduce an idea) within the sections, chapters, pages, or paragraphs that precede a full and thorough discussion of that idea. I do this because many readers will find the concepts presented within *My Big TOE* to be totally unfamiliar. Comprehension and understanding of this trilogy is significantly improved if the reader is at least somewhat prepared for the major conceptual discussions.

Questions may occasionally leap into your mind as you read. Hold on to your questions, or better yet, write them down as you go. Most will be answered within a few paragraphs or pages. If you have unanswered questions after completing Section 5, these can be productively used as the initial focus of your own quest for Big Truth – a subject that is taken up with gusto in Section 6.

Be careful not to lose sight of the Big Picture as a result of being overly focused on the details. It is easy to get twisted around details that strike an emotive resonance with your beliefs. The winning strategy here is to get a glimpse of the entire forest, not to argue about the color of the moss growing on specific trees. Control your passionate interest in the coloration of moss or you may entirely miss what is important.

One final note before you begin. Those who know me well, along with a few of the initial readers, have suggested that I forewarn you about my sense of humor. If you read something in these books that could be interpreted as humor, sarcasm, condescension, arrogance, silliness, inanity, or all of the above, it is probably only humor, or occasionally, humor with a touch of sarcasm. If you find yourself unsure of how offended you should be, I suggest that you temporarily suspend your judgment of the author's mind-set. I am told that eventually (by the end of Section 4) you'll be familiar with my stealthy humor and informal chatty style. Consequently, a later judgment may be more accurate.

The structural anatomy of *My Big TOE* is laid open like a frog on the dissecting table in the paragraphs below. Most readers will find that this overview provides a helpful perspective on how the book you are now reading fits into the overall *My Big TOE* trilogy.

My Big TOE is designed as a three book trilogy. It is packaged as separate books for those who prefer smaller packages or are not sure of how big a first bite they wish to take, and as a more economical three books in one binding for those who are confident they want it all. Each of the separate books contains the same dedication, synopsis, table of contents, acknowledgements, preface, and foreword, as well as its own acronym list and two unique sections of content. Though the table of contents within each of the separate books displays the contents of all three books, the contents belonging to the other two are cast in light gray instead of black. When the three books are bound together into one large book, the page and chapter numbering, as well as the section numbering, run continuously from beginning to end.

Book 1: *Awakening* contains the first two sections. **Section 1** provides a partial biography of the author that is pertinent to the subject matter. Its function is to shed light on the origins of this unusual work by providing a look at the author's unique experiences and credentials that eventually led to the creation of *My Big TOE*. **Section 2** lays out the basic building blocks needed to develop this TOE's conceptual foundation. Many of the concepts initiated in Section 2 will be more fully explored in later sections.

Book 2: *Discovery* contains the middle two sections. **Section 3** takes the information gained in Section 2 and develops its implications in more detail and depth while relating it more directly to the reader's personal experience. **Section 4** pulls together the ideas of Sections 2 and 3, while developing the additional concepts required to bind it all together into one consistent whole. Sections 2, 3, and 4 are carefully designed to sequentially work together to produce the fundamental understanding that is necessary to comprehend Section 5.

Book 3: *Inner Workings* contains the last two sections. **Section 5** presents the formal reality model in detail. **Section 6** is the wrap-up that puts everything discussed into an easily understood perspective. Additionally, Section 6 points out *My Big TOE's* relationship with contemporary science and philosophy. By demonstrating a close conceptual relationship between this TOE and some of the establishment's biggest scientific and philosophic intellectual guns, Section 6 integrates *My Big TOE* into traditional Western science and philosophy.

There is a place in cyberspace [**http://www.My-Big-TOE.com** This URL is not case sensitive but the **hyphens may be required**] set aside for you to share your experience, exercise your intellect, voice your opinions, vent your angst, or simply hang out with your fellow travelers. You can send email to both the author and the publisher from the **my-big-toe.com** web site, as well as acquire all Big TOE books. There, you can keep up with the latest in Big TOE info, happenings, chitchat, reviews, research, and discussion groups.

— Tom Campbell
Dec. 9, 2002

Foreword: A Conceptual Orientation

Without the proper perspective, clear vision produces only data. The point here is to give the reader an initial high altitude peek at the forest before we begin our trek into its depths. In this foreword, I will describe where you will be going and what you should expect to accomplish. It is always helpful to know where you are headed even if you have no idea of how you are going to get there. This conceptual fly-over is designed to minimize the disorienting effect of totally unfamiliar territory.

Both the structure and the content of your perception of reality are culturally dependent. How a Tibetan Buddhist monk or an American physicist would describe reality is as vastly different as the words, expressions, and metaphors that each would employ to make such a description. What would make sense and be obvious to one would seem to be lost and out of touch to the other. If we can rise above our cultural bias, we have a tendency to ask, or at least wonder: Which description is right and which one is wrong? They seem clearly incompatible – certainly both could not be equally accurate and correct. If we are more sophisticated, we might ask which portions of each description are right or wrong and search for areas of possible agreement as well as define areas that appear to be mutually exclusive. That is a better approach, but it is still wrong-headed.

Neither of the above approaches, though the second is much more expansive than the first, will find truth. Which is right or wrong is the wrong question – it represents a narrow and exclusive perspective. Which works, which helps its owner to better attain his or her goals, which goals are more productive and lead to growth and progress of the individual – to happiness, satisfaction and usefulness to others? These are somewhat better questions because they focus on practical results and on the measurable effects that each worldview has when applied to individuals – as well as the secondary effects those individuals have on others. However, something important is missing.

How does one define, realize, and measure the satisfaction, personal growth, quality of life, and fulfillment of individual purpose that is derived from each worldview? What is the standard against which the achievement of these goals is assessed? Now we have a set of questions that have the potential to lead to personal discovery in pursuit of fundamental truth. Big Picture Significance and value have replaced little picture right and wrong as the primary measure of worth.

Fundamental truth (Big Picture Truth or simply Big Truth), though absolute and uniformly significant to everyone, must be discovered by each individual within the context of that individual's experience. No one approach to that discovery is the right one for everybody. The significance of "little truth," on the other hand, is circumstantial and relative to the observer.

Truth exists in all cultures. It is only understandable to an individual when it is expressed in the cultural language (symbols, metaphors, and concepts) of that individual. It is the intent of *My Big TOE* to capture scientific and metaphysical truths from multiple cultures and multiple disciplines and present them within one coherent, self-consistent model that the objective Western mind can easily comprehend. After all, a TOE (Theory Of Everything) must contain and explain **everything**. That is a tall order. A Big Picture Theory Of Everything or Big TOE must include metaphysics (ontology, epistemology, and cosmology) as well as physics and the other sciences within a single seamless integrated model of reality. That is what the *My Big TOE* trilogy is all about.

Truth is truth, but communicating a truth to another is a difficult undertaking fraught with misunderstandings of meaning and interpretation. Big Truth, like wisdom, is not something you can teach or learn from a book. It must be comprehended by individuals within the context of their experience. Each of us comes to an understanding of reality through our interpretation of our physical and mental experiences.

The experience of others can at best provide a useful model – a framework for understanding – a perspective that enables us to comprehend and interpret our experience data in a way that makes good practical sense. The best teachers can do no more than offer a consistent and coherent understanding of reality that helps their students find the larger perspective required to self-discover Big Truth. Such a model is only correct and comprehensive if it accurately describes all the data (physics and metaphysics) all the time under all circumstances for everyone who applies it. The usefulness of a model depends on how correctly it describes the data of experience. A good model should be predictive. It should explain what is known, produce useful new knowledge, and provide a more productive understanding of the whole.

If *My Big TOE* communicates something of significance to you by resonating with your unique knowing, then this particular expression of the nature of reality suits your being. If it leaves you untouched, perhaps some other view of reality will speak to you more effectively. The form your understanding takes is not significant – it is the results that count! If you are prodded to a more productive understanding, you are on the right track. The expression of reality that most effectively nudges your understanding in the direction of learning, growing, and evolving a higher quality of being, is the right one for you. *My Big TOE* is not the only useful expression that Big Truth can take. Nevertheless, it is a uniquely comprehensive model of reality that speaks the language of the Western analytical approach. This Big TOE trilogy fully integrates a subjective, personal, and holistic worldview with objective science. East and West merge, not simply as a compatible or mutually reinforcing mixture, but as a fully integrated single solution.

When some people hear the word "model," they imagine a scale model – a miniature version of the real thing. *My Big TOE* has nothing to do with scale models. A model is an intellectual device that theoreticians use to achieve a more concrete understanding of an abstract concept. Models are often developed to describe an unknown function, interaction, or process (something that lies beyond our current individual experience) in terms of something more comprehensible. The model itself may closely resemble the reality it describes or merely describe its inputs and outputs. In either case, **do not confuse the model of reality with reality itself.** Please repeat that twice before going on.

If you have enough direct experience and a deep understanding of what is being modeled, the model becomes superfluous. With no direct experience, the model enables an understanding that is otherwise impossible to attain. With limited direct experience, the model allows you to place your limited experience within the context of the consistent logical structure of the model. To those with enough experience to incite curiosity and formulate practical questions, the model brings a meaningful interpretation and explanation to data (experience, information, fragments of truth) that otherwise seem hopelessly random and unconnected.

The model of reality developed within this trilogy enables you to understand the properties and characteristics of reality, how you interact with reality, the point of reality, and the boundaries, processes, functions, and mechanics of reality. It describes the what, the why and the how (the nature, purpose, and rules) of the interplay and interaction among substance, energy, and consciousness. You will discover the distinction between the objective physical outside world and the subjective nonphysical inside world of mind and consciousness is wholly dependent upon, and relative to, the observer.

My Big TOE describes, as any Big TOE must, the basic oneness, continuity, and connectedness of All That Is. It systematically and logically derives the natural relationships between mind and matter, physics and metaphysics, love and fear, and demonstrates how time, space, and consciousness are interconnected – all with a bare minimum of assumptions. Additionally, it describes in detail the most important processes of our reality – how and why reality works. You will find the results of *My Big TOE* to be in consonance with current data – and that it solves a host of longstanding scientific, philosophical, and metaphysical problems.

The model of reality developed within *My Big TOE* is not the only valid metaphor or description of the nature of the larger reality. Nevertheless, this model is perhaps more understandable to those of us who are accustomed to understanding our local reality in terms of the processes and measurements of objective causality. A materialistic or scientific definition of reality is sometimes referred to as "Western" because the notion that reality is built upon an inviolate objective causality lies at the core of the Western cultural belief system.

My Big TOE is written to be especially accessible to this Western mindset or Western attitude. The West does not now have, nor has it ever had, a monopoly on a process oriented, materialistic, and objective approach to existence and reality. We in the West have perhaps pursued science and technology more religiously than others, and have no doubt added a unique cultural slant to our particular brand of consumer-based materialism, but the basics of what I am calling a Western attitude are thoroughly entrenched worldwide and expanding in every direction.

The stunning success of science and engineering in the twentieth century would seem to prove the usefulness as well as the correctness of this Western view. The result is that many people, whether from the East, West, North, or South of our planet, view reality from an objective and materialistic perspective that often coexists with some culturally based traditional form of religious and social dogma.

Thus, a balance, or standoff, between our inner and outer needs evolves into a practical worldview that encourages Western material productivity. A pragmatic materialism that depends on objective causality is used to generate the appearance of a manipulatable, rational stability on the outside, while a belief-system of some sort provides the necessary personal security on the inside. To eliminate the discomfort of conflicting worldviews, the two ends of this bipolar conceptual dichotomy are typically kept separate and do not mix or integrate to any significant depth. Each supports the other superficially as they together produce a materially focused, responsible, upwardly striving worker with a good work ethic, cooperative values, an inclination toward dependency, and a high tolerance of pain.

Because the Western mind-set is growing and spreading rapidly, and because the human spirit often withers on the vine before beginning to ripen in such an environment, it is particularly important to blaze a trail to the understanding of the larger reality in the terms, language, and metaphors of this mind-set. As a product of American culture myself, and as a scientist, I have endeavored to craft a model of the larger reality that not only appears rational to the objective Western attitude, but also provides a comprehensive, complete, and accurate model that Western science can build upon.

My Big TOE provides an understanding of reality that can profitably be used by both science and philosophy – one that provides an original perspective, and makes a significant contribution to physics and metaphysics as well as to several other traditional academic and practical disciplines. By the time you have finished Section 6, you will have been exposed not only to Big Picture physics and Big Picture metaphysics, but also to Big Picture psychology, biology, evolution, philosophy, computer science, artificial intelligence, and philosophy of science. There is even a TOE-bone to toss to the mathematicians – they will find new fractal concepts, and discover why geometric fractals successfully reproduce the likeness of natural objects. You will learn why Albert Einstein and others were unable to successfully develop Unified Field Theory, and why contemporary attempts to produce a successful TOE have been likewise frustrated.

The problem physicists are currently having describing a consistent reality is primarily because of the way they define space, time, objectivity, and consciousness. Their current ideas of these basic concepts contain limitations derived from erroneous cultural beliefs. It is this belief-induced blindness that creates scientific paradoxes (such as wave/particle duality and the instantaneous communication between an entangled pair). As Einstein pointed out more than half a century ago, space and time, as we interact with and experience them, are illusions. Many of the best scientists of the twentieth and twenty-first centuries realize this fact, but did not and do not know what to do about it or how to proceed. Their problem is one of perspective – their conceptualization of reality is too limited (only a little picture) to contain the answer.

Albert Einstein's space-time field (as described in his Unified Field Theory) asserted a nonphysical field as the basis for matter specifically and reality in general, thereby moving science closer to the truth, but he did not appreciate the discrete digital properties of space and time or the role of consciousness (instead of space-time) as the primary energy field. Einstein's student and colleague, the great quantum physicist David Bohm (along with a few of the best quantum mechanics theorists including Niels Bohr, Werner Heisenberg, and Eugene Wigner) made the consciousness connection but missed the digital connection and the Big Picture.

Contemporary physicist Edward Fredkin and his Digital Physics movement make the digital connection (quantized space and time) and are heading in the right direction, as were Einstein, Bohr, and Bohm, but they are missing a solid connection to consciousness. Digital physics has not yet discovered that consciousness **is** the computer. All are missing an appreciation of the natural limitations of our physical objective causality and a coherent vision of the Big Picture that ties everything together. You will be shown not only all of the pieces of this both ancient and contemporary reality puzzle, but will also see how they fit together – philosophy and science, mind and matter, normal and paranormal – into a single unified coherent Big Picture.

You will hear more from the above-mentioned gentlemen of science, as well as many of the top Western thinkers of all time, in Section 6 where I integrate the concepts of *My Big TOE* with the knowledge-base of traditional Western science and philosophy.

My Big TOE represents a scientific and logical tour of reality that goes considerably beyond the point where Einstein and other top scientists gave up in frustration. As limitations are removed from your thinking, you will see the source of their frustration clearly, how and why they got stuck, and the solution that they could not find or understand. That this is a non-technical exposition, devoid of the mathematical language of our little picture science, is actually not a weakness at all – even from a strict scientific perspective. How could that be? As you progress through *My Big TOE*, you will come to understand the **natural**, fundamental, and unavoidable limitations of little picture logic, science, and mathematics.

I will show you how physics is related to, and derived from, metaphysics. Additionally, you will find that mind, consciousness, and the paranormal are given a sound scientific explanation that stands upon a solid theoretical foundation. Not necessarily in the way hoped for and expected by traditional science – however, as you will discover, being nontraditional is a necessary strength, not an unavoidable weakness.

The evolution of knowledge demands that sooner or later, truth must succeed and falsity must self-destruct. Although the consensus of culturally empowered opinion may carry the day, measurable results will carry the day after that. The value and success of *My Big TOE* must be measured in terms of the personal and objective results that it produces. Only truth can produce significant consistent results. In contrast, falsity excels at producing assertive beliefs, arguments, and opinions. Open your mind, remain skeptical, pursue only significant measurable results, and let the chips fall where they may.

My Big TOE is in the form of a reality model at a level that is necessarily unusual, but easy to understand. It provides an exploration of the scientific and philosophical implications of consciousness evolution, a subject that holds critical significance for everyone.

Because this material must develop entirely new scientific and reality paradigms, it requires an extensive presentation to shed light upon the limitations of culturally habituated patterns of thought – a goal that cannot be both quickly and effectively reached. Such an in-depth multi-disciplined analysis is better suited to a trilogy than to the condensed formal structure of a traditional scientific paper.

The focus of this trilogy is directed toward the potential significance that *My Big TOE* holds for each individual reader. These books were written for you – you will find their tone to be more personal than general, more of a sharing of experience and concepts, than a lecture by an expert. It is your potential personal interaction with this material that has initiated, as well as driven, its development.

You will find an open, logical, and skeptical mind with a broad depth of experience is much more helpful than a technical background. The details of little picture reality are by nature highly technical, the exclusive territory of modern science and mathematics. On the other hand, Big Picture reality is available and accessible to **anyone** with an open mind and the will to apply it. There are no requirements for formal education or technical credentials in order to understand what is presented here.

There are three main challenges that must be met in order to deliver a shrink-wrapped Big TOE to the general public. First, with shirtsleeves rolled up and the lights turned on, I must turn some portion of metaphysics into physics because I intend to describe the whole of reality – mind and matter, normal and paranormal – not merely the matter and normal part. Consequently, metaphysics is where I must start – our contemporary physics will naturally flow from the metaphysics. The second challenge is to package this unavoidably far-out subject in a way that is interesting, easily readable, intellectually engaging, and non-threatening. To this end, I use the format of a one-on-one, peer-to-peer, informal discussion between the reader and me. The third challenge is to make and keep *My Big TOE* credible – to stay tightly logical while straightforwardly explaining the data of our collective and individual experience.

Culturally conditioned mental reflexes may need to be re-examined, generalized, and expanded. The fact that some of the content of this trilogy is likely to lie far beyond the comfortable familiarity of your personal experience creates a difficult communications problem for both of us. *My Big TOE* not only requires you to think out-of-the-box, but out-of-the-ballpark (if not out-of-the-universe), as well. You will be challenged to overcome deep seated knee-jerk cultural drag in order to climb high enough up the mountain to get a good view.

Modern science and technology are only now providing the combined knowledge by which metaphysics can be understood. It should not be too surprising that science, in its relentless explorations of the unknown, would one day arrive at the roots of existence itself. As it turns out, the nature of reality has both an objective and a subjective component. *My Big TOE* provides a thoroughly scientific description of an objective Theory Of Everything that covers all aspects of reality in an entirely general way. Additionally, it provides a remarkably practical, personally significant understanding of subjective consciousness, and explains how you individually are related to the larger reality. To appreciate and deeply understand the personal or subjective nature of consciousness, you must grow your own Big TOE. One of the major goals of *My Big TOE* is to provide the logical conceptual framework, materials, tools, and direction that you need to independently grow your Big TOE.

My Big TOE will provide the foundation and structure that you need to make sense of both your objective and subjective experience. Your personal Big Understanding of Big Truth must flow primarily from **your** direct experience – not from your intellect alone. This trilogy will bring your objective and subjective experience together under one coherent understanding of the whole you.

Please understand, I did **not** put the "My" in *My Big TOE* to flaunt pride of authorship. Nor does the "My" indicate any lack of generality or applicability to others. The "My" was added to be a constant reminder to you that this reality model cannot serve as your **personal** Big TOE until it is based upon your **personal** experience. On the other hand, personal or subjective experience is only one piece of the reality-puzzle. In the objective physical world of traditional science, *My Big TOE* delivers a comprehensive model of reality that subsumes modern science, describes our objective material reality, and is universally applicable. Contemporary physics is shown to be a special case of a more general set of basic principles. After reading the *My Big TOE* trilogy, you will better understand the universal (objective) and the personal (subjective) nature of perception, consciousness, reality, and Big TOEs. You will learn to appreciate the fact that the larger reality extends beyond objective causality, beyond the reach of intellectual effort, into the subjective mind of each individual. **My** *Big TOE* is the launch pad. **Your** Big TOE is the final destination.

A personal Big TOE is necessary because the larger reality, like your consciousness, has a subjective component as well as a collective objective component. The larger reality cannot be fully appreciated or understood merely by studying, or reading about it. You must experience it. Additionally, your understanding of the Big Picture must be sufficient to integrate your subjective experience with your shared objective knowledge or both will remain superficial. To the traditional scientist and other left-brained analytical types, what I have just said sounds suspiciously like a mixing of real science and hocus-pocus, touchy-feely, belief-baloney. It is not, but a properly skeptical mind may need to digest all three books before that becomes apparent.

Arriving at conclusions based upon the assumed infallibility and apparent truth of culturally, personally, and professionally embedded paradigms and dogmas will make it difficult to understand the larger reality. Change and new ways of thinking are often traumatic, difficult to integrate, and generally unwelcome. Resistance to change is automatic at the gut level; we cling to familiar ways for the security and comfort they provide. We do not easily see unfamiliar patterns. You must be willing to overcome fear and rise above self-imposed belief-blindness if you are to succeed in getting a good look at the Big Picture.

In the pages ahead, we are going to explore the reality-wilderness. This trilogy is about the how, what, and why of what is. It is about physics and metaphysics, your world and other worlds. It is about beginnings, endings, mind and matter, point and purpose – it is also about the quality of your personal consciousness.

Your intellectual understanding of the reality you exist within, and are a part of, is only the beginning – a place to start. The most important action, the real fun, begins **after** you have finished the trilogy and begin to apply what you have learned about reality and the Big Picture to the rest of your life – both professionally and personally.

Though you will soon learn there is more to reality than theory and facts, here is one fact that you should consider before you begin: Big Truth, once understood and assimilated, always modifies your intent, and invariably leads to personal change.

List of Acronyms, Symbols, and Foreign Words and Phrases

Foreign Words and Phrases:

gedanken experiment – thought experiment; a logical experiment performed only in the mind.

über alles – over all; over and above all else.

número uno – number one

No problema – No Problem

Que será, será – whatever will be will be

MY BIG TOE

BOOK 1:

A W A K E N I N G

Section 1
Delusion or Knowledge:
Is This Guy Nuts, or What?

Section 2
Mysticism Demystified
The Foundations of Reality

Section 1

■■■

Delusion or Knowledge:
Is This Guy Nuts, or What?

■■■

1

Introduction to Section 1

Beginning with a partial biography of the author may appear to be unnecessary and off the point. Ordinarily, an author's background is adequately covered by a few paragraphs on the inside jacket of the cover. However, understanding the origins of the author's experience is where this adventure must begin. If you do not know anything at all about me, the rest of this effort may wither on the conceptual vine before it is ripe. This trilogy sets forth a Big Picture Theory Of Everything (more concisely known as a Big TOE) that contains, as any Big TOE must, a comprehensive model of reality.

Because *My Big TOE* is so unusual, so far away from the thoughts you are probably used to thinking, it is important that you have some understanding of the seed, soil, and roots from which this unusual exposition sprang. Knowing its origins, the mental mettle from which it has been hammered out, may help provide the context required to assess the genuineness of *My Big TOE*.

This trilogy presents a working model of the larger reality based upon the data I have collected through a lifetime of careful scientific exploration. Section 1 describes how fate sent me down this highly unusual path and then delivered me up to the extraordinary experiences required to formulate the concepts presented in these pages.

The intent is to give you some insight into who I am, where I am coming from, and how I ended up with such an unusual Big TOE. This trilogy is about reality, not a biography about me; accordingly, please pardon my skipping here and there about my life and flying swiftly over things that you might wish I would explain in more detail. For now a quick trip through the formative years is enough. I will not attempt to explain the last twenty years of my becoming, but will instead focus on the ten years before that. That is when the mold was made and the direction of my life set.

Although I have learned much during the preceding thirty years, and have mellowed and gained in wisdom as one typically does with age, I have not changed my basic approach to learning, knowledge, or science since the events that I am about to share with you transpired.

2

■ ■ ■

Hey Mister, You Want to
Learn to Meditate?

■ ■ ■

How did a nice scientist like me end up in a strange place like this? My
family had no unusual interests – we were normal through and through.
Perhaps we were more problem-free than normal, but no strangeness at
the roots. I did well at college, graduating with a double major in physics
and math. Then on to graduate school where I picked up a master's
degree in physics and started working on my Ph.D. I passed the prelimi-
nary exams and qualifiers on the first try and settled into thesis research
in the specialized area of experimental nuclear physics. It was 1968 and I
was your typical twenty-three year old physicist – excessively cerebral, out
of touch with my feelings, analytical, precise, curious, and above all, intellectually motivated and driven.

Not a hint of strangeness anywhere. If you have known any brash young
physicists, you are probably aware that they tend to be unusual – somewhat skewed off the norm, one might say politely. If you are not that
polite and were not snowed or intimidated by their domineering intellectual style, then "arrogant" rather than "unusual" would probably be the
adjective that would most naturally come to mind.

On a warm day, late in the spring of 1971, while walking into the physics
building on the way to teaching an undergraduate class, I noticed a large
poster advertising a free lecture on Transcendental Meditation – or more
simply, TM as most people referred to it. As a graduate student, poverty
was a normal and given circumstance; consequently, anything free caught
my attention. "Control your mind," the poster shouted in large block let-
tering. "Learn how to relax deeply and lower your blood-pressure."

"OK," I thought, "that's nice, but who cares?" I continued reading. "Improve your concentration and decrease your need for sleep."

What! I read that line again. Now it was interesting as well as free.

A few days later, I attended the free lecture. TM was presented as science – technique and results, stimulus and response. There was no theory. One regularly performs this uncomplicated meditation process and physiological benefits automatically accrue – period. There was no dependency on eastern philosophy, mysticism, or any belief system. I wanted no associations with things that were non-scientific, mystical, or belief based – in other words, goofy. That sort of hocus-pocus was for gullible people with uncontrolled emotional needs.

The two TM presenters backed their claims up with some serious looking research papers and informal studies by reputable individuals at reputable institutions. It seemed straightforward enough – for the special student rate of only $20, I would be taught to meditate in four hours spread over one week. Better concentration, clearer thinking, improved memory, reduced stress, and I would need less sleep – all for $20 and four hours.

I was skeptical but if it worked half as well as the presenters claimed, it would be a good investment for any student. In those days $20 was a lot of money – serious money for a struggling grad student. Still, it was the 70s, I was a student, and a short walk on the wild side – doing something counter to the main culture – seemed almost obligatory. I signed up.

"Just follow these directions," the TM guy said with a friendly smile, handing me a packet of paper as he took my check.

3

■ ■ ■

The Sacrificial Banana

■ ■ ■

A week later, I was on my way to the TM Center to pick up my personal meditation sound called a mantra. It would take less than five minutes they promised – stop by, get your mantra, then attend the four training sessions – what could be easier?

As per the directions, on the appointed day I showed up at this seedy little house with a clean handkerchief and a banana. A clean handkerchief and a banana? What a crock! I began to doubt the wisdom of what I was doing and felt more than a little foolish walking up to this house with my banana and hankie. Anybody who needed fresh fruit and a handkerchief to teach me how to meditate was likely to be either goofy or fraudulent. That was the first sign of strangeness. I had seriously thought about showing up without the handkerchief and fruit – just to see what would happen – but realism interceded. I decided instead that humoring them would be a better strategy. Their inane request seemed harmless enough, and more importantly, they already had my $20.

"What's with the banana?" I said derisively, holding my banana up to the first person I saw. "What's fruit have to do with meditation?" I asked. "They never said anything to me about fruit and handkerchiefs at the lecture I attended – this seems a little goofy to me." The person I was addressing evidently wasn't a hard-science type. He was as clueless as I, but questioning the interrelationship among meditation, fruit, and handkerchiefs had evidently never occurred to him. I found that difficult to understand.

His reply had been an exaggerated shrug of his shoulders, followed by a big broad smile. It was now obvious that this guy was doing what he was told without thinking about it. He was probably wondering why I was making such a big deal over a banana that wasn't expensive or much trouble to get. He seemed a little embarrassed and uncomfortable with my direct-ness. I looked away, silently wondering how people could be that uncriti-cal and unthinking about what they were doing. There was something familiar about that smile – I chuckled silently with amusement, and won-dered if he was perhaps related to Alfred E. Neuman.

There were about ten people standing and sitting in the waiting area. I found a corner to stand in and waited in silence, hanging onto my fruit, like everybody else. Finally, it was my turn. "What's the banana for?" I asked at the first opportunity. The initiator explained that they (the TM organization) had a ceremony of sorts that he was required to do – the fresh fruit was to represent the traditional offering a student would bring to his teacher.

"What do you do with all the fruit?" I asked. In my mind, I imagined a huge pile of rotting symbolic fruit – a sad wasteful testament to the use-lessness of ceremonial gestures.

"We eat it," he said with a smile.

Now I understood – it was merely a way to get some free groceries. If they wanted to add on a surcharge of one banana to the $20 price so that they would have something to eat for breakfast – well, that was all right with me. Nevertheless, I would have preferred that they had been more straightforward.

The man standing next to me was about my age, well dressed, soft spo-ken, and seemed intelligent and serious – he didn't look like a fruit hus-tler. Maybe it was not them, I hypothesized, perhaps the people who set this business up knew that they would have a difficult time making ends meet. I let it go. Having solved the banana problem, I went on to the next issue – the impending ceremony.

With the intention of being helpful, I offered a constructive criticism to my initiator. I told him I thought any ceremony was completely irrele-vant and that it distracted from the rational image they had carefully pre-sented at the public meeting. He politely nodded, indicating to me that he recognized the discrepancy, but made no other reply.

"I have to do the ceremony," he said quietly.

It was obvious that he did not care what I thought about his ceremony, or probably anything else – this was just the way it must be done. He looked at me for permission to continue. "OK," I said, "no big deal, go ahead, do your ceremony if it makes you feel better. I will go along as long as I do not have to profess any beliefs or make any promises." He smiled at me with amusement and immediately agreed.

At his direction, we knelt, side by side, on a small rug in the middle of the kitchen floor. On the floor in front of us was my banana lying in the center of a clean white handkerchief draped over a small plate – a poor man's altar, no doubt. The instructor chanted a little Sanskrit (at least that is what he said it was) mumbo jumbo for about a minute, then was quiet for about thirty seconds, finally he told me my mantra. I repeated it with him a few times until I got it right. In fewer than three minutes I had been given and had properly memorized my very own personal (probably shared by millions) secret mantra.

I was put in a large room with several others and told to silently practice my mantra so that I would not forget it.

"Just repeat it over and over in your mind for twenty minutes," he told me. "That's all, after that you can leave – but be sure to come to the first training meeting."

"What a crock," I thought, "I paid $20 and a banana for this magic Sanskrit mantra? What could repeating some word-sound do for my mind?"

I expected nothing, but having come this far, I was determined to follow directions and give it a fair try. The twenty minutes of practice went by slowly at first, but then I drifted off into a pleasant nowhere. Suddenly, I became aware that some of the initiates who were far behind me in the queue were leaving. "They must think that it is a crock too, and are not going to stay around to practice," I mused. I checked my watch. "Whoa!" It was way past time to go – I had been practicing some forty-five minutes! I checked my watch again to make sure.

I tried to get up to leave, but I couldn't move. My body simply refused to respond to my will and effort – that had never happened before. My limbs were heavy as if I had been asleep for a long time, yet I was positive that my consciousness had been unbroken – I had **not** fallen asleep. "This is weird," I thought, as I forced my thick, viscous, semi-solid body to slowly stand up and walk. Hmmm, maybe there was something to it, and I was doing it! Or maybe I was merely more tired than I thought. That was not a good explanation – I had gotten more sleep than usual the night before. "This is interesting," I thought, "very interesting."

I had done something strange, or at least goofy, and experienced something weird (my first trip into inner space) all in the same hour. I remained skeptical, which was (and is) my nature, but I also remained open minded. There were four one-hour sessions remaining and my curiosity was growing. I had undeniably experienced something dramatically unusual. That experience, along with my Scottish ancestry and the fact that my $20 was irretrievably gone forever, provided the necessary commitment to follow this TM adventure through to the end.

4

■ ■ ■

Watch That First Step,
It's a Big One

■ ■ ■

Two days later I was attending my first training class. There were about fifteen of us sitting on uncomfortable gray metal folding chairs. I looked around to assess my peers. We were a somewhat motley but otherwise normal looking crew of casually dressed students. I was hopeful that I was about to learn something remarkable and useful, yet at the same time I was a little incredulous, almost embarrassed that I had spent twenty perfectly good dollars to be associated with this Indian meditation thing. The instructor looked like us, except that he had shorter hair, was clean-shaven, and better dressed – obviously not a student. He told us about TM, explained the techniques we should use to maintain the mantra in our minds, and answered our questions.

Everything was process oriented. No theory, no more ceremonies requiring sacrificial bananas, no metaphysical mumbo jumbo – just technique. I liked it that way. I knew there was some Indian yoga-guy with a long name, and beard to match, who was the leader of the TM movement, but no one ever mentioned him. Other than the fact that this so-called guru had sold me this meditation technique through his organization, neither he nor his organization was relevant to what I could do with it. I verified that fact in no uncertain terms as soon as the instructor opened the meeting to questions. The only condition was that I could not give away or sell the mantra to others. I instinctively didn't trust anything that had to be kept secret, but this request seemed reasonable enough. The TM folks were not hiding something from the scrutinizing light of open critical review; they were merely protecting their source of income and fresh fruit. I considered it as a copyright or patent – income producing intellectual property to be protected by secrecy. No problem.

Finally, the last question was answered and it was time to practice our meditation technique. I slumped in my steel chair trying in vain to get comfortable and began to occupy my mind with the sound of my mantra. In a few minutes, I was nothing but a single point of conscious awareness existing in a void of nothingness – floating free, doing nothing – existing as conscious mind without extent or form. No thoughts, no body, no chair, no room, no instructor, nothing. It was a remarkably pleasant experience until a single thought began to violate the expansive peace and quiet: "Uh oh... I am not doing this right; I am supposed to be thinking the sound of my mantra!" That one critical thought interrupted my otherwise thoughtless float in inner space. I needed to get back on task by thinking my mantra. Suddenly, awareness of the outside world rushed in and I was startled to realize that I not only had a body, in a room with an instructor, but it was about to fall out of its chair!

Ooga! Ooga! Emergency! Emergency! My internal alarms went off. Quickly I tried to straighten up. My right leg awkwardly shot forward jerking my body back into balance. "Whew! That was a close one – I almost hit the floor like a sack of potatoes – wouldn't that have been a scene," I thought to myself after it was clear that I had successfully regained my balance. In my imagination, I could see myself thudding to the floor creating a ruckus. My next thought was more practical. "Jeez, I might have banged my head on something – meditation could be dangerous." I made a mental note to be more careful in the future. Almost simultaneously the instructor began to speak.

"Come on back," he said, "practice time is over."

"Great," I thought sarcastically, "he noticed, and now he is going to cut everybody's practice time short because I almost fell out of my chair." I looked up. To my surprise, the instructor was engaged in a conversation with a student sitting on the opposite side of the room. He appeared to be totally unaware of me or my near catastrophe.

"Time's up," he said.

I looked at my watch. What?! That's impossible! I had just lost twenty minutes of my life. What had I been doing for the last twenty minutes, where had I been?

"Any questions?" he said to the group in general.

I raised my hand. "How do you keep from falling out of your chair?" I asked. Everybody laughed. I was serious; it did not immediately occur to me that my experience was unusual. I don't think he had ever gotten that question before because he looked surprised and didn't know what to say.

"What happened?" he asked.

"I started with the mantra, then everything went blank, it was nice, but when I realized I wasn't saying the mantra any longer I almost fell out of my chair and the twenty minutes were already up after only a few minutes." Everybody laughed again, including the instructor. I suddenly realized that what I had said sounded confused, disoriented, or stupid – take your choice – and that no one else had had an experience like mine.

"Two for two – this meditation could be strange stuff," I thought to myself. The question and answer session wrapped up quickly. We were told to practice meditating twice a day for twenty minutes each time. He set the time for our next meeting and we were dismissed. I had meditated (or at least attempted to do so) twice, and both times I had experienced something very strange. What will happen the next time? I wondered. This was exciting – something seemed to be happening. My intellect and curiosity were being tickled. I wanted to experiment; I wanted to know if anything real was involved, would it work as they claimed? If so, then what else could I do with it?

The rest of the training classes resembled the first one, except I sat in a chair next to the wall at an angle so that I could not fall out. "I need to wear a seat belt and helmet, or find a bigger chair," I thought with amusement. I immediately saw a mental picture of myself sitting in a folding metal chair, in front of a big fan. I was in the middle of an otherwise empty room, wearing a seatbelt, a motorcycle helmet, goggles, and with a scarf flapping in the breeze behind me (thus, the need for the fan). I was thoroughly amused and grinned broadly at the bizarre scene. I immediately thought of Ace-Snoopy sitting on the top of his doghouse engaging the Red Baron in battle with his Sopwith Camel. I often saw pictures like that. Some were funny like that one, some were helpful. They never took but a few seconds, and disappeared as quickly as they came.

With each training class, floating in the blank state or existing in the single point of conscious awareness state became easier and more familiar. Shortly I would learn, through trial and error, to combine the blank state with controlled visuals. The two may seem incompatible, but they are not.

Two months later I was meditating twice a day, being careful to put my body in a situation where it would not fall. The point awareness or point consciousness experience continued. It was always pleasant. I usually wanted to stay longer, but did not because I was extremely busy. I also found it invigorating – it seemed to boost my energy. The biggest surprise of all was that meditation could be professionally productive! By adding the visuals, it turned out to be a great place to work.

I could solve physics problems, design experiments, analyze research data, and write and de-bug computer code ten times more quickly, and with better results while in a point awareness state than I could in a normal state of consciousness. Extremely complex matters seemed to become much simpler and clearer in my meditation state. When you work a full week unsuccessfully trying to find a bug in your analysis software, and then go home and solve the problem – not just imagining that you have solved it mind you, but actually solved it – in just ten minutes of meditation time ...well then, you know that you have found something that has real effects – and real value.

This type of illogical yet extraordinarily productive experience was not merely circumstantial; it did not happen only once or twice, off and on – it became a dependable routine. Eventually, I did not wait until all normal methods and efforts were exhausted before turning to meditation. It worked as well when I tried it first, saving a great deal of time and effort by skipping the first three steps of the process. Hard work, long hours, frustration, and exasperation were no longer prerequisites for solving difficult and complex programming and physics problems.

I was surprised and delighted that meditation had a direct objective practical value. That was an unexpected revelation. I tested it and tested it, and then I began to depend on it. I had gotten smarter it seemed, and the other graduate students noticed. They commented on the change in my abilities. Now, more than ever before, they wanted to discuss their research and came to me for help in debugging code.

I told them about TM, but none made an effort to try it. That surprised me. For some strange reason, I had thought physicists would be more open. Meditation appeared to require too big a step out of the comfort zones of their personal and cultural belief systems. "Oh well, it's their loss," I thought. I did not know it then, but I was on my way to becoming strange. I was, at the very least it seemed, unusually open minded. My perspective and reality was expanding.

Up to this time, I believed that meaningful existence was confined to an operational reality. That is, if something can be measured it is real. To be measurable, a thing must interact with our senses or with some device that interacts with our senses. If it is not measurable (can not interact with us or our devices), then its reality or existence (or lack thereof) is irrelevant. It was that simple; things were either operationally real or irrelevant. Things that are not measurable, but can be inferred from other things that are measurable, fall into the gray area of conjecture. All things theoretical or hypothetical fall into this gray basket.

Gray things are acceptable as conceptual constructs or ideals but are not to be confused with real things and are not to be taken too seriously in and of themselves. The primary mission of the academic research scientist is to collect enough valid, repeatable, measurement data to transfer a gray theoretical construct into a real object or effect.

At that time, quantum mechanical wave packets, black holes, quarks, justice, and love all fit into that gray area. It is important not to confuse hypothetical things with real things or you can easily end up chasing your own imaginary tail, or hallucinating the attributes of solidity to smoke. My mind was not closed, I allowed for the possibility of new information. With enough real measured data one might eventually move a thing, such as black holes for example, from the realm of the hypothetical to the realm of the real – but only with sufficient good quality data. That particular attitude has never changed. I continue to feel that way, work that way, and employ that methodology to sort out what is real from what is not.

Now I had to change my philosophy of reality. There were those things that were non-measurable yet functionally operational (including my meditation state, which is properly defined as an altered state of consciousness) that fell into the category of **subjective experience with objective results**. One can use these non-measurable states of mind to operate on real things. (Here, I am using the terms "operate" and "operator" in the mathematical or scientific sense). I had shown that an altered consciousness directed by intent can consistently and directly affect and interact with real things such as my computer code. Complex logical problems could be solved without the intentional application of a rational process. Somehow, a non-measurable subjective experience could be turned into a reliable and effective scientific tool by some sort of consciousness operator. Verrrry interesting!

In contrast to conceptual constructs such as justice and love, altered states of consciousness and their objective results seemed to be measurable, consistent, and reasonably well defined. They were more like the things of science, things that were amenable to research and experiment. If one is in this particular altered state, one can always do these things with it – similar activity will produce similar results for all experimenters.

I was not special; I was the same as anyone else. I wondered if other altered states of consciousness existed and what one could do with them. I was curious – it is my nature to be curious.

My reality expanded. I added the statement: "If a thing is well defined and consistently functional (it can profitably and dependably be used by anyone within the known operational reality), then it must also be real." It seemed reasonable that only real things could be functional operators within and on an operational reality. How could something not real directly affect things that are real? By definition, in an operational reality, things that are not real have no measurable effect, cannot interact with, and have no relevance to, things that are real.

My meditation had a measurable objective effect – I knew it, and a dozen others had clearly noticed it as well, even if they did not understand why. To me, it was as plain as day and totally obvious; this was no subtle effect being misunderstood by some mushy headed non-scientist. This was no hallucination. I knew what I knew. How many others, if any, agreed with me was not relevant. I had confidence in my mind and my science.

I began to analyze other common altered states such as daydreaming. Were daydreams (self-directed imagination or purposeful visual imagery) functional? Of course! Why had I not noticed that before? People have been preparing themselves mentally for all sorts of things since the beginning of time. For example, one might repetitively practice giving a speech in one's imagination – making points and fielding imagined questions. The relevancy criteria are: Is it consistently functional, does it actually help one's performance in objective reality, and are the effects measurable? Our directed imagination example would rate a definite "yes" on all counts.

Ask any top-notch athlete if focused intentional mental preparation is important to his or her success. Mental effort within the context of a particular altered state must represent a real thing because it produces real effects that are universal as well as specific to that particular mental state. "Altered," was defined as different from normal. "Normal" meant wide-awake and focused in the physical world – as you are now while you are reading this book. Each altered state has its functionality. Daydreaming is one specific type of altered state. Of all possible altered states, those with no universal and consistent functionality are, by definition, useless and therefore irrelevant. It was only the useful ones that were welcomed into my newly expanded reality – those were the ones I wanted to know about.

Real things, significant things, must now be either objectively measurable, or consistently and predictably interactive with real things. That was a major expansion of my real world. The word "objective" means that these real things must exist universally and consistently for others as well as for me. They must be independent of me and exist whether I exist or not. Others (potentially everyone) must be able to make the same measurements and find the same measurable functionality. Otherwise, they would be only my private hallucinations, not a part of the larger reality we all share.

When I realized the scope of the reality picture was much bigger than I had previously thought, I wondered if there were other subjective experiences that had consistent objective measurable results. Where were the boundaries? How much more reality was out there that I had missed? What other real and functional processes of mind were lurking beyond my limited awareness? I had been blindly unaware of a significant part of my reality for twenty-some years! That thought was a mind bender that weighed heavily on me.

That I had inadvertently imposed a major limitation on my operative reality out of sheer ignorance was unacceptable, inexcusable, and more than a little humbling. What other significant parts of my life was I missing? I had to find that answer. Being content to accept whatever is given without pushing hard against the boundaries is absolutely foreign to my nature. My mind had been forced open by the indisputable facts of my experience, and as a result, I had become less of a philosophical know-it-all. I realized there was much about life and reality I did not know. Arrogance waned as openness and curiosity waxed.

I have not significantly changed my philosophical approach toward defining reality in the thirty-three years that have come and gone since then; today my definition of what constitutes a real experience remains essentially the same. Any credible conception of reality must include subjective experience that can consistently and universally lead to a useful objective (measurable by anyone) functionality.

Much later it became evident that expanding my reality beyond a certain beginning level would require personal growth. I had to increase the quality of my consciousness to understand the bigger picture. Conversely, understanding the bigger picture helped me grow up. They worked together.

Thus, my journey began innocently enough. Interestingly, though nothing much has changed as far as my overall philosophy goes, the continual flow of incredible learning experiences has steadily accelerated. My understanding of reality continues to actively expand.

5

■ ■ ■

Is This Guy Monroe Nuts, or What?

■ ■ ■

"Get a job!" intones a popular song of my youth. Everybody must get out of school sometime. I was now almost twenty-seven, and had been in school continuously since I was five. With research completed, I settled on my first real job applying classical physics and mathematics to electro-mechanical and electromagnetic systems simulation. A real job with a real paycheck – imagine that! I continued to meditate more or less regularly, but had found that I did not need the mantra any longer. A little research and experimentation indicated that any two-syllable nonsense word ending in "ing" (a resonant sound) worked as well as any other, including my given super-secret mantra. There was nothing mystical or magical here, only a method of controlling thoughts by filling the otherwise active mind with fluff, nothing but science and technique – no bananas or hankies were required.

Repeating the mantra eventually seemed to get in the way and slow me down; consequently I dropped it. The meditation state was now familiar enough that I could go there in an instant and return as quickly. This level of control was handy at work. I could meditate, find solutions, and return without anyone suspecting that I was doing something strange – to the world I seemed to be deep in thought. That I was disembodied point consciousness adrift in the void – gone completely from their world with no residual awareness of their reality – was my secret. Sometimes people would try to engage me while I was gone. To them, it was as if engaging a dead body. Needless to say, I gained a reputation for being eccentric – with unusual powers of concentration or an unusual ability to sleep while sitting up – nobody could tell which.

My boss, Bill Yost, was a super person. He was smart – no frills, no BS, no hidden agenda, no tact, – an engineer through and through. Bright, honest, and straightforward – that is the personality type that I related to most easily. One day Mr. Yost came by my desk and tossed a book at me. I caught it in mid air and read the title. *Journeys Out Of The Body*, by Robert A. Monroe. "What's this?" I asked, surprised by the strangeness of the subject matter.

"Read it," he said, "and tell me what you think."

When your boss says, "Read it," you just read it, you do not ask why.

I read *Journeys* during the next few days. The book was configured as a diary. It was a "This is what happened to me" type of story wherein Monroe claimed to have collected hard evidence in support of the reality of the out-of-body experience. The experiences, and the evidence of their realness, were laid out matter-of-factly with no theory or belief system attached. It was a wild concept – a more or less independent reality reachable only through the mind. Having prior experience with functional altered states of consciousness, I was probably more open minded than the average scientific type, but not gullible. I knew what was real to me, and my measurement data (experience) included none of that.

"It was very interesting," I told my boss, "but I don't know what to make of it. Is this guy (Monroe) nuts; is he trying to sell books to the gullible; or is he for real?" I asked rhetorically. I continued with barely a pause: "How can you tell where he is coming from? If his story is for real, if you take him at face value, a new aspect of reality opens up that I have never before considered. That would be a definite Wow! But for now, it just sounds wild and I have no way to judge the veracity of it."

Having spit all this out in rapid fire, barely pausing to breathe, I now took a deep breath and waited for the reaction. I watched my boss carefully to see if I had made a fool of myself by being too open.

As a student, I was used to having to get the right answer. Was I supposed to condemn it as foolish rubbish or believe it as a strong possibility? I had no idea where he was coming from. He had given me no hints. I was working as a civilian in a military organization, it was 1972, and the people here were conservative. I was a longhaired, wild-eyed kid-physicist recently out of graduate school. I had about decided that my openness had been a political mistake when Mr. Yost finally spoke.

"I agree with you," he said thoughtfully. "It is a wild concept isn't it?"

"Yes it is," I agreed, "very wild."

"But consider what it would mean if it were true," he continued with an enthusiasm that indicated that he had thought about it seriously. I did not reply. "Think about that," he said. "What does it logically imply **if** it is true – if the evidence is real and not made up?"

"Yeah," I said, "Pretty strange stuff – but how can you ever know if it is true or made up?" He nodded in agreement and changed the subject. That was the end of it. Not another word was mentioned about *Journeys*, at least not for a few weeks.

I had almost forgotten about Monroe's book, having put it out of my mind as something that could never be logically confirmed or denied – and therefore was irrelevant.

"Would you like to go with us to see Monroe?" my boss asked when we were alone.

"Huh?" I muttered, not making the connection.

"There is a group of us from work going to see Monroe – you know, the guy who wrote *Journeys* – this Friday after work. Do you want to go with us?" he asked.

"Where?" I asked in reply.

"Just outside town about forty-five minutes from here," he shot back with some excitement in his voice. "Sure," I said. "I would really like to meet this guy and see if he is crazy or sane, honest or a hustler, delusional or rational." "Me too," said Yost with a twinkle in his eye – "me too!"

6

■■■

Face to Face with
the Wizard of Whistlefield

■■■

Late Friday afternoon finally came. While our co-workers headed home to begin their weekends, twelve of us piled into three cars for the trip to Monroe's. I did not know these people; I was a relatively new employee within an organization that employed about five hundred people. We were a strange crew, male and female; young and old, very conservative – mostly professional technical types. We were not the sort of people one would expect to be eagerly converging on Mr. Out-of-Body. I was impressed there were this many open minded people where I worked. As always, anything unexpected demanded an immediate reassessment. I hypothesized that living almost exclusively among hard-core scientists for the past seven years had inadvertently skewed my judgment of people in general – clearly, there were bias errors in my analysis algorithms. That was a serious problem. Within a few seconds I had laid out a tentative plan for debugging my assumptions and made a mental note to observe these people more closely.

Most were skeptical, one of the ladies was a little frightened, everyone was enthusiastic, and nobody knew what to expect. They all jabbered nervously and endlessly – the scene seemed hyperactive and irrational to me. As usual, I said nothing. I was not a good mixer. I did not relate to unfocused bubbling emotion or to anxiety, and I did not understand these people. Their lives seemed to be driven, or at least animated, by random irrational feelings. They were strangely affected by uncertainty.

At the time, I had no idea that they were the ones who were actually normal. Years of graduate school and a lack of mainstream social interaction had nudged my vision of normalcy a tad off center. I thought that

Spock was normal while the rest of the Enterprise crew were hopeless, eternally lucky, mush-heads. I heaved a sigh, "This is going to be a weird night," I thought to myself, "with all these weird people, on this weird excursion to see Mr. Weird." As it turned out, except for the people I was traveling with, it was not actually that weird, but it changed my life forever.

Bob Monroe lived on an estate named Whistlefield – five-hundred acres of lakes, forests and fields. A large country manor house elegantly perched on top of a hill, half dozen horses, a barn, and two small lakes thrown in for good measure. It appeared to me that Mr. Monroe was a relatively wealthy Southern gentleman. We slowly motored down the half-mile long driveway that was bordered by a freshly painted white board fence. A few horses trotted along with us. "Whoa, this is classy," I thought. "This guy is no poor raving lunatic – that's for sure." My analysis continued, "Weird books don't pay that well – it doesn't appear that duping the gullible for their money is going to be a likely motivation." Nevertheless, I reserved final judgment on that issue until I could meet the man for myself.

The car finally stopped in front of the house. Several large dogs came bounding out to meet us – two Dalmatians and a large German shepherd were vigorously sounding the alarm that intruders were in the driveway. My riding companions thought it better to stay in the car until the friendliness of the dogs could be verified.

Nonsense! These people were so strange – like frightened children. I wondered what could have possibly happened to them to make them that way. Dogs bark at strangers because that is what they do – it means nothing. I quickly popped out of the car to say hello and rub some ears.

I was immediately mobbed by three wet tongues and wagging tails. It was love at first sight. They acted as if they had not been petted for weeks. It felt good to be out of that crowded car, and to be surrounded by rational beings that knew what they were about. I immediately felt more centered.

I should explain that popping out of that car had been neither brave nor foolish – those dogs were **obviously** friendly. What I did not understand was why that fact was not obvious to everyone else. I surmised that either I happened to be riding with a group of people who were not familiar with dogs, or that one dog-challenged person's fear had influenced the rest.

By now, everyone was getting out of their cars and looking around, wondering what to do next. "Perhaps I shouldn't have changed into these old jeans," one of the ladies said apprehensively – she was obviously intimidated by the classiness of Monroe's estate. All three women present were reflexively fumbling in their purses for fresh make-up. "Why do they always do that?" I wondered silently. "Haven't they figured out by now that it won't make any difference?" I was always amazed and amused when people were internally driven to blatantly irrational behavior.

I had, for many years, been curious about the root causes of "cultural insanity" – those absurdly illogical attitudes and actions that our culture considers normal. Some, including "makeup urgency" are entirely benign; others range from mildly dysfunctional to terribly destructive. I had come to the tentative conclusion that the key motivators of cultural insanity were fear-based and emotionally driven. I could not relate in the slightest to either. Nonetheless, I intuitively knew that this was not a good thing, and it did not speak highly of our society's overall level of rationality. I was curious about it, and took careful mental notes whenever I noticed such a display. I did not feel superior. I had no inclination to make comparisons. I was an impartial observer with an insatiable curiosity – that's all. I was merely different, not better or worse than others because of that difference. I had been born, it seemed, a perpetual outsider – and I liked it that way. Outsiders have a more objective and impartial view. As a scientist, nothing was more important than logical clarity and objectivity. Being different and having an outsider's view had generally been comfortable for me – I saw it as an advantage. It suited me well.

My reverie into the characteristics and causes of illogical social behavior was suddenly upstaged by something more important. The large white door of Whistlefield Manor began to swing open. Conversations were instantly terminated in mid sentence. All heads turned with silent expectation. The one, the only, the Amazing, Out-Of-Body-Man, was about to turn to flesh and blood before our eyes. We would all soon know if this guy was nuts, or what.

Out stepped Mr. Monroe into the doorway. For a second or two he seemed the slightest bit tentative – like a man who clearly knew he was about to be examined and evaluated like a captive alien or a strange animal at the zoo. He gazed out at the crowd of nameless heads staring silently back at him. After the briefest of pauses, he stepped fully out onto the elegant open stone porch with confidence and a solid presence. He was not wearing a white suit with matching hat and string tie like Col. Sanders (the only Southern gentleman I could bring to mind). Instead, he looked comfortable, informal, and friendly – more like the dogs than the house.

Robert Monroe was a heavyset man of medium height; he wore a big smile and had a twinkle in his eye. Just looking at him made you feel relaxed. He greeted us all individually as if he were an experienced politician – making quips and jokes as he went. "This guy could be Santa Claus," I thought with more than a little amusement, "a jolly old elf – passing the summer sipping mint juleps on the veranda of his country estate."

"What are **you** smiling at?" he demanded good-humoredly as his attention suddenly focused in my direction.

He was now looking directly at me with a knowing impish grin. For a moment, I had the feeling that he must have read my mind and was amused by my vision of Santa Monroe.

"Oh nothing," I replied, lamely brushing off the question. Before he could react, I instantly followed that dodge with a question of my own. "How and when did you first go out-of-body?" I asked. Up to this point, nobody had been that direct. I did not know how to be any other way. It suddenly grew quiet and more focused, everybody was now intently listening.

"It just happened," he said. "It just started happening about fifteen years ago for no apparent reason."

"How did you react to that experience?" I continued without pause.

"I thought I might be going nuts," he said. "It worried me initially, but I couldn't help experimenting with it – that's my nature."

He had consulted with psychologists, psychiatrists, and a parapsychologist. All found him rock-solid sane which made him feel better and gave him confidence.

Monroe seemed to have inadvertently stumbled into an altered state of consciousness that gave him access to a larger reality, producing some amazing evidential data under controlled circumstances with the parapsychologist. Because it had no deleterious affect upon his mental soundness and competency, he was encouraged to pursue, record, and eventually control his unusual experiences.

"What sort of evidential data do you have?" I shot back.

"Most of it was in the book," he said, "remote viewing sorts of things for the most part."

"What exactly is remote viewing?" I asked.

"Obtaining information paranormally by going somewhere in the out-of-body state to collect the target information – without taking your body along, as it were."

"Oh, I see," I said sheepishly, realizing that I had asked a dumb question that should have been obvious. My momentary pause gave others a chance to horn in on my private conversation. No one was shy any longer.

Toward the end of the evening Monroe took us to the facility that he hoped to soon turn into a lab dedicated to the study of altered states of consciousness. It was obvious to me that he desperately wanted to legitimize what had spontaneously happened to him. He wanted to remove the stigma of nutty and replace it with the approval of an accepting science. He was earnest, serious, and willing to put his money where his mouth was.

He was not posturing or hustling – he was genuinely interested in real science. He wanted legitimacy, not recognition, money, or fame. He was a successful local businessman. Additionally, he was the CEO of a growing cable company and appeared to me to be totally sane, intelligent, under control, and conservative. Best of all he was a rational type. He had an engineer's personality. He was more straightforward and intellectually precise – less emotionally driven – than most of the technical professionals who were now pelting him with questions. The quality of the questioning was erratic – clearly, he was a polite and patient man. If you did not know he wrote about out-of-the-body experiences, you would never have guessed it from his circumstances, appearance, or demeanor.

Then came an offer I could not refuse. Gazing at the bunch of us lounging on the back deck of the nascent lab, Monroe challenged us. "You folks are technical scientific types, aren't you?" He asked rhetorically. We all nodded our heads and mumbled our concurrence, wondering what was coming next. "I am looking for some hard-core science and engineering types," he continued. "Someone with good professional credentials who could help me do real and proper science that would be acceptable to other scientists." "So that's why we were invited," I thought with mounting expectation, "Great!"

My hand shot into the air – a reflexive act conditioned by over twenty consecutive years as a student – I couldn't help it. I could not, it seemed, respond to a question with my arms at my side. Monroe looked directly at me, amused by my waving hand. Feeling a little stupid, I pulled my hand down out of the air and said, "I am a physicist and I am very interested in your research into altered states. If you will teach me what you know about out-of-body and altered states of consciousness, I will help you do legitimate scientific research."

Almost immediately, from the other side of the deck another voice spoke up.

"I am an electrical engineer and I would like to work with you... if you would try to teach me what you know."

I strained to see who was speaking. It was a young guy, maybe a few years older than me, but not much. I did not know him; he wasn't riding in the car I had come in. Monroe looked at us intently – a long and pregnant pause ensued as he assessed the situation and weighed his options. Everyone was quiet, waiting to see what would happen next.

I think that Monroe would have preferred older, more established scientists to staff his lab. Someone more mature, with an established reputation – instant credibility. But we both knew that those types were not likely to be interested, or willing to work for an exchange of knowledge. If they had established professional reputations, they had reputations to protect and would never allow themselves to become associated with something this far out on the fringe – this far away from the safety of the crowd. Scientists, contrary to their own press reports, are mostly just sheep of a different sort. All credibility flows from peer review and the more notable peers of the scientific community would treat altered states of consciousness and out-of-body as intellectual leprosy. Monroe had run headlong into that brick wall of closed-mindedness previously – which is probably why he craved respectability and acceptability.

He quickly evaluated his chances to do better. Finally, he broke the long silence, "What kind of degrees do you have?" Dennis Mennerich, the other volunteer, had a master's degree in electrical engineering. "OK," he said confidently, "You've got a deal! Call me in a few days and we'll set up a meeting." He rattled off a telephone number. The subject changed. I stopped paying close attention to the conversation. I was as excited as someone like me can get – that is, I felt a mild surge of anticipation.

Where would this lead? What could he actually teach us? How would he attempt to teach us? What scientific protocols would be applicable? What kind of data would we be collecting? What would we be measuring? Question after question poured through my mind. What if it turned out to be bogus? If I found out he wasn't actually interested in real science, I would quickly and politely bow out – that would be easy enough.

The evening had turned out better than I had expected. I was going to be studying altered states – something I had wanted to do for a year or more, but did not know how to start. "This could be a great opportunity," I thought... "Well, maybe, I'll just wait and see what happens, if anything happens at all."

I did not know it then, but my life was about to take a sharp turn. Strange and stranger (all carefully scientific of course) was about to become as common as air.

After a few days passed, I called the number scrawled on the scrap of paper that I had hastily obtained that night on the deck at Monroe's. It had been dark, we were outside – at the time the number seemed clear. I dialed it again. There was no answer. I tried it later, no answer. The next day was the same, as was the day after that. No answer. No answer. No answer. I let it go the rest of the week. The next week I tried again. No answer. I decided that perhaps the three was actually an eight and tried that. No answer. I tried information – Monroe had an unlisted number. I let it rest a few more days. Two weeks had gone by since our visit. I tried it with the "three," then again with the "eight" ... hold on ... somebody began to talk ... jeez, it was just an answering machine! The machine mentioned no names. I left a message. Nothing happened, no one returned the call.

A few days later I decided to try one more time. I was so surprised when a woman's voice offered a polite "hello," that it took me a second to focus on what I was doing – a real flesh and blood person! Wow! I asked to speak to Mr. Monroe.

"What is this in reference to?" she asked politely.

I was on a roll. "Is this Robert Monroe's residence?" I asked.

"Yes," she said, "who is this?"

With immense relief at having made the connection, I quickly explained who I was and said that Mr. Monroe had asked me to call him in a few days and that was two and a half weeks ago.

"Just a minute," she said.

"Finally," I thought, "he'll probably be glad to hear from me." I imagined that he was concerned that I had perhaps changed my mind. He had seemed excited, even somewhat anxious, about getting his lab up and running.

"Hello," the voice said on the other side of the line with no sense of familiarity.

Maybe she didn't tell him who I was, I reasoned. "This is Tom Campbell," I said, "We talked on the deck at your lab a few weeks ago – I am the physicist – you asked me to call."

"What?" he said. "Physicist? What kind of physicist are you?"

What sort of question was that? From his tone it was obvious that he did not know who I was, and that he didn't remember our deal – or he was pretending that he didn't? He was obviously not sitting around worrying because I had not called sooner. I detailed the visit and the offer that we had agreed upon.

"Oh, that physicist," he said with dramatic inflection. "What is your name again?"

I told him my name a second time.

"There was another guy with you, wasn't there?"

"Yes, there was another guy – his name is Dennis," I replied tersely and waited.

"Why don't you two come to the lab next Thursday," he said after a short pause.

"That is good for me," I replied, "I'll check with Dennis and let you know."

"Just come on," he said, "no need to let me know – just come on up to the lab at seven – you and Dennis – is that OK?"

"Sure," I said a little puzzled.

"Do you know how to get here?"

"Yes," I answered, "I can get there. I'll see you this coming Thursday evening at 7 at the lab." I paused to make sure there was no misunderstanding.

He mumbled a gratuitous "OK," sounding mildly annoyed that he had to listen to me repeat the arrangements, and hung up.

"How does he know Dennis will be able and willing to come this Thursday?" I wondered. I sure didn't know that yet. How could he be that sure? I pondered the circumstances. Does he want me to come by myself if Dennis cannot make it? He didn't seem particularly eager to get started. Or... is his mind a little loose? "Now **that** conversation was definitely strange," I mused. In time, I would eventually get used to Monroe being distracted and knowing things paranormally. His mind was not loose, he was always a step ahead and usually right. Unlike me, he did not need to wait for the facts.

I told Dennis the next morning.

"No problem," he said.

I told him that Monroe had seemed to know he would be able to come Thursday evening. We looked at each other and shrugged our shoulders.

"Had you told me this yesterday or anytime last week, I would have told you that I couldn't have made it," Dennis added as an afterthought. "But just this morning, things changed, now I have no conflict."

This was going to be an interesting adventure – I just knew it.

"Who is going to drive?" Dennis asked.

"I planned to take my cycle. Do you want to ride with me?"

"OK," he said, "I have my own helmet – I used to have a bike a few years back."

"I have a big four cylinder Honda – two people won't be a problem."

"Great," he replied, "that should be fun."

"He is pretty courageous," I thought. "I hardly know this guy, he knows nothing about me, and he is willing to get on the back of a motorcycle with me? Maybe only once," I chuckled to myself. I got directions to his house and agreed to pick him up at 6:15 p.m.

7

■ ■ ■

The Adventure Begins

■ ■ ■

The trip to Whistlefield was a combination of interstate and country roads. Most of the mileage was on a brand new, and lightly traveled, interstate. I loved my motorcycle. I loved speed. I loved acceleration. I loved the feel of finely controlled and responsive raw power and I loved the presence, the sense of being alive, and the focus in the present moment that you get on a big motorcycle. You, the bike, the environment, and fate – one tightly integrated package – a shared destiny. That was fun. With Dennis on the back, I resolved to be conservative; nothing over eighty-five miles per hour on a regular basis was my plan. It would not be responsible, polite, or friendly to be reckless with somebody else's life.

At 120 mph, my bike was rock solid and smooth as silk. It was made for speed, and I was addicted to it. I had driven cycles ever since I was a teenager. With this particular bike, it was love at first sight – the biggest, the best, the fastest. Dennis was fearless, he never once complained or flinched – except once when the drive chain broke while we were humming along at eighty miles per hour and he almost lost a few fingers. A near miss, but when you are young enough to be immortal and invincible, any miss is as good as a mile – we never skipped a beat. Dennis was always ready and relaxed. Mounted on this trusty steel steed we cut the travel time to Whistlefield to less than half an hour.

Once we got through the initial getting-to-know-you data exchanges, schedules were quickly worked out and routines established. Dennis and I would go to the lab two or three times a week and sometimes on weekends. We would spend the first hour or so setting up equipment, soldering wires, designing and making measurement devices – in general, wiring and outfitting the place to be a lab. After a while, Monroe would join us at the lab and then the real fun began.

Under Bob Monroe's guidance, Dennis and I would begin a systematic exploration of altered states of consciousness. We were constantly working towards consistent repeatable, evidential experience. After a few hours of exploration, Bob would invite us back to the house for discussions, chitchat, planning, or perhaps to meet some other investigators that were working in related areas. His lovely wife Nancy, the ever proper, polite, and most congenial hostess, would often join our discussion. Dennis and I were so bright eyed and bushy tailed in our dogged pursuit of the outer edge of reality that our constant state of total amazement, night after night, amused her to no end.

It seemed as if Bob knew everyone in the country, who was investigating or experiencing anything unusual. They all came to Whistlefield eventually to meet Bob and share the results of their individual efforts. Bob was like a magnet in this disconnected community of leading edge researchers, experimenters, and freelance kooks, because of his no nonsense, straightforward manner and wonderfully open mind. There was no snake-oil being hawked at Whistlefield. Because of Bob's reputation, and the operation and reputation of the lab, there was a steady stream of tremendously interesting visitors. I was impressed there were so many intelligent and sober individuals, sometimes with impressive credentials, who took this area of endeavor seriously. These were not whacked out druggies doing their counter-culture thing. Bob had zero tolerance for that sort – he did not want to tarnish his legitimacy by being associated with drug users. The Timothy Leary types were out. Other than that, Bob was open to almost everything anybody took seriously. However, he was also always skeptical. Open minded **and** skeptical – he wanted to see hard evidence – claims were interesting, but never enough.

Most of the visitors were middle-aged, serious professionals looking for serious answers to serious questions. They, for the most part, were looking for validation and hard evidence. There were the occasional groupies trying to increase their credibility by associating with Bob and his research effort, and a few whose main object was to impress him with their unusual talents. Bob had little patience for either. He politely but firmly sent the pretenders and non-contributors on their way.

8

The Science of Altered States

The lab building contained, among other rooms, a control room and three isolation chambers. One chamber had been constructed with complete electromagnetic shielding so the earth's magnetic field and other stray radiation would not wash-out or overpower the effect that carefully controlled electromagnetic fields might have on altered states of consciousness. Each chamber was constructed to provide as much sensory deprivation as possible and was connected to the control room by audio and a host of measuring devices.

I borrowed some unused, sophisticated and expensive electrostatic sensing equipment and audio signal generators and Bob purchased a complete EEG (electroencephalograph) setup and a professional audio mixer. Bill Yost brought in an exceptionally sensitive high input-impedance voltmeter. Dennis and I designed and made a device for tracking Galvanic Skin Response (GSR). Before long we had a reasonably well-equipped lab to work in, even if it did have solder splatters all over the floor.

We measured Bob. He measured us. One key datum Bob had derived from his personal study of out-of-body experience (OOBE) was the perception of a 4 Hz oscillation within his body and consciousness just before exiting his body and sometimes just after returning to it. Experimentation showed that when an instrumented person was caught in that pulsation state in our lab, the EEG indicated the brainwaves collected by multiple pairs of electrodes were unusually coherent (in phase), collectively synchronized, and modulated at 4 Hz. The GSR reading would begin oscillating at 4 Hz as well. Bob intuitively knew that this pulsation state was a key artifact, and we set out to reproduce it – capture it and hold it steady, on demand, and under our control.

A literature search had turned up a few old scholarly studies of out-of-body experience, which was also known as "astral projection." A half dozen highly respected, well-credentialed, medical and technical professionals had been seriously studying out-of-body experiences for decades, mostly around the turn of the last century. A book by Dr. Hereward Carrington and Sylvan Muldoon entitled *The Projection of the Astral Body*, published in 1929 by Samuel Weiser of New York, suggested the pineal gland was somehow involved. *Astral Projections: Out of the Body Experiences*, by Oliver Fox and *The Study and Practice of Astral Projection* by Dr. Robert Crookall (University Press, 1960); agreed that the pineal gland was perhaps a key organ affecting the out-of-the-body experience.

We were in the cut and try mode of operation and would try anything at least once. Dennis and I always applied any unusual experimental devices to ourselves first. Only after we tested how it affected us, and became convinced of its worth and safety, would we try it on others to gain a wider sample – friends, visitors, passers by, anyone, we were not picky. We were looking for something that would work with anybody and everybody.

For example, one of the things we tried was to shake the pineal gland at 4 Hz. We built a huge capacitor with 2-ft^2 plates to generate a uniform and strong electric field. We were committed and dedicated to the pursuit of our quest – risk taking was not an issue. With something like a 250,000 volt, 4 Hz AC signal being fed to the plates, I stuck my head between them and tried to reach a working altered state. I stayed there about an hour or so experimenting with different voltages and frequencies against different altered states, hoping for a resonant effect to occur that would have a dramatic effect on the state of my consciousness. Suddenly I began to feel woozy. My head started to wobble dangerously between the exposed metal plates. The experiment was stopped. I had a terrible headache for about three weeks.

We worked with negative ion generators to provide controlled backgrounds, and used ultra-high impedance input voltmeters to study the changing electrical potentials generated by a body in an altered state vs. a normal one. We measured the dynamic buildup of static charge around our heads with borrowed equipment as we eased in and out of various brain-wave configurations measured by the EEG.

It may seem a little like mad scientists toiling in their hilltop laboratory at the midnight hour but we were as serious, sober, and straight as our counterparts working traditional problems in universities everywhere. We were careful about our science. Our methodology was good. Were we cautious and conservative? No, we were not cautious. If there was **any** reasonable chance of gaining knowledge, we took it. We were hard driven to find honest answers – real, verifiable, repeatable results. We wanted to know, and this was the chance of a lifetime to find out. We were fearless in our pursuit of truth because the risks were totally invisible to us – sometimes, the naïveté and brash enthusiasm of youth has its advantages.

Meanwhile, while Dennis and I were working in and on the lab, Bob was leading us and teaching us to experience and explore the nonphysical. He would first lead us into a deep relaxation state, then using visualization we would begin to focus our thoughts, center our attention – let go of our bodies and the environment. These exercises produced various states of consciousness that were similar to the meditation state I had learned to achieve with TM. Bob thought that perhaps there was opportunity on the boundary between being awake and asleep. We practiced hovering on that edge. Put the body asleep and keep the mind awake simultaneously. Eventually we got good at it and it did not take us long to get there. After developing a basic competency in defining and establishing willful repeatability of a half dozen altered states of consciousness, we began experimenting and exploring the functionality of each state. Each mind-state had its own unique functionality – things you could do, abilities you had, while in that state.

9

Breakthrough

One day while I was at work, Dennis dropped by and showed me an article he had found in the October 1973 *Scientific American*. It was a short article, by Gerald Oster, titled "Auditory Beats in the Brain" that described a phenomenon called "binaural beats." Simply put, if a pure tone of say 100 Hz was put in one of your ears, and a pure tone of 104 Hz in your other ear, you would perceive a 4 Hz beat frequency along with the 100 and 104 Hz tones. Dennis waited while I read it. "Let's try that at the lab," he said. "Sure, why not?" I replied. Dennis had been gathering information on the binaural beat phenomena for some months and had created a binaural beat audio-tape for us to experiment with. Our hope was that the beat frequency, occurring in the corpus callosum between the brain's hemispheres, would drive the brainwaves.

Dennis' intuition was correct. The binaural beats obviously affected our state of consciousness. During the next week we begged, borrowed, and bought the necessary equipment to expand our experiments. Bob had gone out of town for a week or two; subsequently, we experimented with the effect of binaural beats on altered state of consciousness on our own. After a week of trial and error experimenting, we were more excited about the possibilities. The effect was powerful. Using the binaural beat to entrain brainwaves as measured by the EEG was a fact. The effect on one's state of consciousness was dramatic. Bob came back and we started testing what this technology could do. The good news was that by trial and error we were able to significantly optimize the effect we were looking for. The better news was it seemed to work as well on everybody as it did on us. Now we had a technique for putting people with no training into specific altered states of consciousness, at will, on demand, with consistent results.

We focused on the 4 Hz beat and created a set of audiotapes that guided the listener into what Bob called "Nonphysical Matter Reality" (NPMR). "Physical Matter Reality" (PMR) contained my body, the lab, the house where I lived, and my daytime job. Once in NPMR, the fun began. Now we had the potential to collect evidence that would be based on a much larger sample of subjects. As we gained experience with more people, we continually improved the effectiveness of the audiotapes. In about eight months we were ready for the world to give us a try. Bob put out the word that we needed a limited number of guinea pigs to try out our binaural-beat brainwave entrainment techniques for facilitating the projection of one's consciousness into the nonphysical as an aware operational entity. The response was overwhelming.

Soon, Bob was booking every room at the nearby Tuckahoe Motel. The Tuckahoe management, having seen better times, agreed to let us string wires throughout their facility. Dennis and I had a lot to do before the big weekend when we would discover if our methodology was as effective as we thought it was. We expected about twenty totally naïve subjects – and we planned to keep them that way by telling them nothing. We did not want to lead their reactions and experiences by giving them any expectations.

Building mobile measuring equipment and audio equipment for large groups was a challenge. We barely made our deadlines by working evenings and weekends for three weeks, but with the help of Bill Yost and Bob's stepdaughter, Nancy Lea Honeycutt, we had the equipment installed, checked out, and ready to go late in the afternoon of the last day. What a panic! In a parallel activity, Nancy Lea, who had joined our research team after her graduation from college, orchestrated and administered all the necessary arrangements. Somehow, at the last moment, everything pulled together. It was worth it. During Friday night, all day Saturday, and half of Sunday, the attendees had the time of their life. There were so many paranormal happenings that weekend that we had a difficult time getting them all recorded. These naïve subjects were reading numbers in sealed envelopes, remote viewing, manifesting lights in the sky, visiting their relatives, reading next week's newspaper headlines, and much more. It was a circus! Fun, but exhausting. Dennis, Nancy Lea, and I ran the show, with visits from Bob off and on throughout the weekend.

We collected lots of solid evidential data – the results were more dramatic than we had expected. Things were never the same after that. When word got out about the effectiveness of Monroe's program, Bob was swamped with requests from people of all sorts wanting to participate. Bob began to see the makings of a business and Dennis and I, along with Nancy Lea, became trainers more than researchers.

10

■■■

But Is It Real?

■■■

Let's slip back in time and view the whole from a slightly different perspective. My association with Bob Monroe presented a fantastic opportunity. With those years of practice, Dennis and I could easily differentiate among the various altered states of consciousness and get to them, shift between them, and come back to a normal state at will. However, it was not that easy to begin with.

We worked hard and modified the rest of our lives to accommodate our work. I had decided that while I was working at the lab, I would take no mind-altering drugs of any sort. It was going to be confusing enough without that variable floating around in the equation. I had never used any illicit drugs as a student because it did not seem rational. I lived out of my mind, it was my ticket to success – I didn't want to mess anything up. But now I swore off even an occasional beer. Not a drop – socially or otherwise. I became a devout tea-totaler for the cause of clarity.

A few years later, food additives, preservatives, caffeine, and sugar were permanently banished from my diet. I reasoned that subtle natural effects might be washed out by the impact that these substances had on consciousness. I was right – the difference was dramatic. The success of our research hinged on the clear perception of subtle shifts in consciousness, anything that could potentially muddy those waters was dropped by the wayside.

We logged thousands of hours exploring and probing the limits of reality, produced a huge pile of measured data, and filled up boxes full of audiotape that recorded every word of our sessions. The mental space we practiced in was nonphysical – bodiless. Unlike my previous TM meditation, we were active, willful, autonomous agents within this larger nonphysical reality. We went places, did things, communicated with nonphysical beings.

It was fun, but neither of us could take it too seriously. Bob was careful to never lead the witness. He played the part of neutral observer – never hinting at what we might experience or how we might experience it. He didn't want his experiences to influence or bias us. As far as we could tell, he had no expectations of what we could, or would, accomplish.

Bob knew that if we were to experience the larger reality as he did, we would have to get there on our own. He could guide, but not lead – that would ruin the independent quality of our effort. He wasn't looking for an echo – he wanted to accomplish real science. Initially, Dennis and I had the same problem. "Is this stuff real?" we would ask each other. How could we tell if what we were experiencing was inside (we were imagining it), or outside (had its own existence independent of us)? That was the burning big question for both us – and for Bob as well.

Eventually we gained enough mental control and facility in working with altered states that Bob thought that we were ready to begin collecting some evidence to determine the operational significance of what we were experiencing in NPMR. Dennis and I were excited about the possibilities, and willing to accept the facts however they came out. We had been eager to objectively test the operational significance of our subjective experiences for some time. Bob had wanted us to wait until he thought we were ready. Neither of us was particularly optimistic or pessimistic – we wanted to know the truth. We were in the discovery mode and open to all possibilities. As long as our methodology was sound, we were confident that eventually enough results would accumulate to tell their own story.

One of our first experiments was for Dennis and me to take a trip (experience) in the nonphysical together. Our independent descriptions of what we were experiencing should correlate closely if the experience were real and independent of either of us. From the beginning of our training, we had learned to give real-time descriptions of whatever we experienced. A microphone was suspended from the ceiling above each of our heads. What we said was recorded on tape. Dennis and I could not hear each other because we were in separate soundproof chambers.

Dennis and I quickly achieved the appropriate altered state, left our bodies, and met in the nonphysical as planned. It was a long adventure. We went places, saw things, had conversations with each other and with several nonphysical beings we happened to run into along the way.

Bob had let us go a long time before he ended the session and called us back. We pulled off our EEG and GSR electrodes and stumbled out of the darkness into the hallway of the lab.

In the control room, Bob was waiting for us. After a quick exchange, we knew that this would be a good test because we both had experienced many specific interactions. But were they the same interactions? Bob looked at us deadpan. "So, you two think you were together?" he asked, trying to sound disappointed. We looked at each other and shrugged our shoulders.

"Maybe," Dennis said tentatively, "at least we perceived meeting each other."

"Listen to this!" Bob said emphatically. The tapes, rewound as we disconnected electrodes and climbed out of our chambers, began to roll forward. We sat down and listened. The correlation was astonishing. For almost two hours we sat there with our mouths open, hooting and exclaiming, filling in the details for each other. Bob was now grinning. "Now that tells you something, doesn't it?" he exclaimed beaming. He was every bit as excited as we were.

I was dumbfounded. There was only one good explanation: THIS STUFF WAS REAL! My mind searched for some other more rational explanation. "Perhaps only one of us imagined the trip and the other was reading his thoughts telepathically," I said trying to cover all the possibilities. That was almost as far out as the first explanation, but not quite.

The undeniable fact was: We had seen the same visuals, heard the same telepathic conversations, and experienced the same clarity. "This stuff might actually be real," I said aloud to no one in particular. Dennis and I sat there wide-eyed, incredulous, and at a loss to explain it any other way. I said those same six words: "This stuff might actually be real," over and over to myself fifty times during the next few days. I could not believe it, but I had to. I was there. This was my own experience. I was not reading this in a book about somebody else. In the vernacular of the times, I was blown away. You cannot understand the impact something such as this has until it happens to you. One more data point was in. My reality was about to get broader and stranger.

We repeated that experiment with similar results. It wasn't a phenomenon that depended on the two of us. Nancy Lea and I shared equally astonishing joint experiences. We tried other things as well. We read three and four digit numbers written on a blackboard next to the control room. Somebody would write a random number and we would read it while our bodies lay asleep. Then they would erase it and write another one, and so on and on. We went places – to people's homes – and saw what they were doing, then called them or talked to them the next day to check it out. We traveled into the future and into the past. We tried to heal people's illnesses with our minds and intent because that was a good technique for interacting evidentially with the energy of others.

We designed, generated, and tested intent-focusing tools for our use in the nonphysical. We diagnosed illnesses in people we never met, but that somebody else knew well. The evidence poured in. Now there were hundreds of data points; later evidentiary experiences tended to be more clear and often more dramatic, than the initial ones. We began to discern subtleties of the altered states where things worked well and where things did not work well. We refined our processes and improved our efficiency slowly during the next three years – it was a painstaking trial and error process.

Dennis and I were the same demanding and skeptical scientists that had started this adventure, but we had stopped asking if it was real. We now knew the answer. We also realized that one has to experience it oneself to get to that point. Nobody else can convince you. You simply must experience it yourself. All the data in the world, regardless of how carefully taken, become suspect if you are not there to participate and know the truth of the matter firsthand. Old beliefs must be shattered before you can begin to imagine a bigger picture. Until the inescapable logic of unambiguous firsthand experience hits you squarely between the eyes, the truth does not sink in deeply. That is the way I was, and so is most everybody else.

I suppose by now, Dennis and I were certifiably strange. We were strange because of what we knew to be true by our carefully evaluated experience. We could not deny what we had seen, heard, and measured – even if it was incredibly strange. We knew how careful, skeptical, and demanding we were. We knew how high our standards of evidence were. We also knew that nobody else could possibly understand unless they experienced these truths for themselves. Once you find true knowledge, ignorance is no longer an option – and if the knowledge you find is unusual, then strange becomes a way of life.

Our activity was not entirely internal. For example, Dennis and I were encouraged to volunteer for some remote viewing experiments at a well-known sleep & dream lab. The object was, under controlled conditions, to describe pictures being displayed in another room. As it turned out, being able to describe all the pictures correctly was not the most remarkable thing that happened.

When the EEG scrolls were returned from Duke University (where they had been sent for more detailed analysis) a higher level of strangeness was evidenced. We were told that Dennis's EEG results produced the highest levels of alpha-waves ever recorded at Duke. Mine exhibited strong simultaneous levels of alpha and theta unlike anything they had ever seen before. Both were singular events previously unseen by the Duke researchers because of the narrowness of the peaks. This was particularly meaningful because during the 60s and 70s, Duke University was recognized worldwide as the leader in parapsychological research.

Our brainwaves were, it seemed, tightly focused to specific, nearly single frequencies. We were not particularly surprised by the tight focusing, but duly noted with interest that out of thousands and thousands of EEG analysis results, ours stood out as blatantly unique ("Your data blew them away at Duke," we were told by the researcher). We had for some time felt that what we were learning and developing was uniquely effective at producing specific altered states, but now we had corroborating evidence – an independent lab at Duke had substantiated a physical manifestation of this uniqueness.

Once the mental door of indisputable fact is pried open, the light begins to flood through. The old questions returned with new meaning. Now my reality, my picture, was bigger than I could have previously imagined. Nevertheless, I continually wondered if there were other subjective experiences that could produce consistent objective measurable results. Where were the boundaries – how much more reality was out there that I had missed? Could there be other operational states of consciousness hiding in the darkness of my ignorance?

I was driven to understand how everything was related, how it all worked together. Surely, there was some sort of science at the root. We had lots of data, but no self-consistent model to explain the how's and why's of it all – to define the interactions. How did reality function? What were the processes, the limits, and the rules? Is this the way it is, or only the way it seems? What did the Big Picture look like – where all the data are consistent and makes sense? How could any self-respecting physicist not ask these questions?

Bob, Dennis, and I would discuss it down at the house after the work at the lab was done. We informally came up with some "the way things were" and "the way things seemed to be" statements, but they lacked deeper understanding. We surveyed the existing models – mostly a mishmash of emotion laden, belief focused, unscientific balderdash with little or no hard evidence that was reproducible. That was not what we were looking for. This was a scientific inquiry, not a new-age gathering of the faithful.

Finally, we ran across a candidate model – a place to start. Though imperfect, it was more or less rational, consistent, and coherent most of the time – that made it much better than the rest. Its explanations and descriptions were not complete, nor necessarily a place to end up, but it did provide a theoretical basis from which to tentatively and skeptically begin. This model came to us in the form of *Seth Speaks*, by Jane Roberts. That the material was channeled was not a problem for us. By then, we were all personally familiar with the nonphysical and its host of sentient beings. In fact, it was a plus. Would you ask a fish about mountain hiking trails? No, not if you expected an accurate or useful answer.

We began to spend much of our training time at the lab testing and interpreting Seth's concepts, and procuring information from our own nonphysical sources. We worked on these issues for several years, slowly gaining ground. It was sometimes confusing, sometimes clarifying, but always interesting and always evidence was required.

I worked harder at these particular models of reality issues than the others. I was the theoretician of the group (what you might expect from a physicist), Dennis was more into applications (what you might expect from an engineer). Bob was a practical man focused primarily on what-ever worked and upon gaining and maintaining objective credibility. Bill Yost contributed his engineering insights, management skills, encourage-ment, and support. Nancy Lea did much of the daily support work, and became a full partner in our explorations of nonphysical reality (as had her sister, Penny Honeycutt, a few years before). It was a good team.

We were all aided and abetted all along the way by our families (who for the most part participated in our research from time to time) and many unmentioned others. The research flowed in whichever direction seemed most productive at the time. Bob did not direct as much as he facilitated. Having perfected the wise and knowing smile of all good teach-ers who know how to let their students figure it out for themselves, he managed to float above the day to day effort and let our individual research take us wherever it would.

11

■ ■ ■

If This is Tuesday, I Must Be in Physical Matter Reality

■ ■ ■

Meanwhile, back at the lab, Dennis and I were putting in about fifteen to twenty hours a week. After I would get home from the lab, often at two or three in the morning, I would lie in bed practicing what I had learned or continuing that evening's experiments. After two or three hours of sleep, I would get up and go to work. The evenings I didn't go out to the lab, I would continue experimenting after everyone else fell asleep until a few hours before getting up and going to work. I was putting in forty-five hours a week studying altered states and the larger reality while simultaneously putting in fifty hours a week at my day job and raising a family.

My son Eric was about five years old at the time. Like most kids that age, he had frequent spontaneous out-of-body-experiences (OOBE). We would go out-of-body together – I would go by and join him – we would have a blast. One time we were exploring the oceans together when a huge whale approached us. As our bodies slipped easily through the whale, Eric's head for some reason bumped against each of its ribs, one after the next. It frightened him a little; typically we did not interact with our surroundings. We came back immediately.

Eric usually had total and clear recall of our nightly adventures. We would often discuss them in the morning – it was great fun for both of us. Exploring the larger reality turned out to be an excellent father and son activity, though perhaps somewhat unusual. Do not misunderstand me. I was not warping Eric's tender perspective, or jerking him out-of-body. At about five years of age, most children naturally and **spontaneously** have lots of OOBEs. I was merely joining him so that we could go together. It was comforting and reassuring to Eric to have me along – he was going with or without me. I was able to structure the experiences to be both fun and educational (such as exploring the oceans).

Instead of denying and discarding his experiences as foolish dreams (typical parental reaction), I was shaping and sharing them with him. He thought it was cool and looked forward to our outings. Eventually he was no longer a natural, and our forays into the wilds of nonphysical matter reality (NPMR) ended as easily and naturally as they had begun. He, by the way, now has an advanced degree in aeronautical engineering and to this day clearly remembers bumping his head on those whalebones.

I have always been a sleepy head – nine to ten hours a night is about right for me. Yet by spending so much time in altered states where my body was deeply relaxed, if not officially asleep, I got by on two or three hours of sleep per night – night after night after night – year after year.

At work, I was exceptionally productive, but becoming stranger. I was spending almost as much time in NPMR as I was in physical-matter reality (PMR), and it showed. I soon earned a reputation for being an absent minded professor. PMR and NPMR seemed to blend into a continuum and I found I could live in both realities simultaneously; it was no longer a matter of leaving one and going to the other. Now, it was merely a matter of shifting and splitting my focus – I lived and was continuously aware, sentient, and conscious (except when sleeping) in both reality systems simultaneously and permanently.

At first, I could only sequentially (albeit quickly) switch between them. Then I learned to engage mentally in NPMR on one thing while carrying on a conversation **and** driving a car (or motorcycle) at the same time. Most of the time there was no confusion between reality frames, but now and then, for a few seconds, until I forced myself to differentiate between them and get my bearings, I was occasionally not sure which reality I was in. Both were equally real, they were just different and had different functions. I began to marvel at the mind's capacity for parallel processing.

For one relatively short (about six months) period, I was spending more time in NPMR than in PMR. I was a space cadet and obviously needed a keeper. Luckily, being a physicist, and maintaining high professional productivity, I could get by with being eccentric. Nevertheless, I soon realized that I needed to regain a better balance. With a little experimenting, the optimum balance was obtained. I remained eccentric, but didn't need a full-time keeper to remind me of what was coming next in PMR.

With the two realities so completely inter-mixed, I began to notice connections between the two. One spring day while walking back to the office after lunch, I noticed that golden-white foam was draped over the trees in a nearby park. A quick reality check indicated I was solidly focused in PMR. "Wow," I exclaimed with mild surprise, "that is really pretty, but what is that stuff?" By now I was so used to being amazed by the larger reality that what was normally strange had become strangely normal. I studied the white foam; it had the texture of cotton candy. It connected all the trees into one large luminescent mass. It reminded me of a grove of cypress trees along the Gulf coast loaded with glowing Spanish moss.

I thought it was very interesting but had no idea what it was. I wondered if other people could see it. I made an effort to be obviously looking at something. A few passers-by turned their heads to see what I was looking at and then went on about their business without any noticeable reaction. I knew that they must not have seen what I saw because what I was looking at was not ho-hum in the least. It was massive and beautiful. If others could see it, there should have been a crowd forming.

I went back to work, and looked out of my third-story window to see if the light-foam was still there. It was. I closed the door to my office and began to study the phenomenon I was experiencing. I discovered that I could make it disappear and reappear by adjusting the state of my consciousness. Within a few days, I noticed that everything living had this fuzzy light around it, and that there were strands of this nonphysical cotton candy connecting everything to everything. What about inorganic matter, I wondered. I moved my attention to buildings, telephone poles and power-lines.

To my astonishment, there was a smaller more uniform close-cut off-white light around everything! The light around the power-lines was in motion and bushier than what was around the poles. I was incredulous and I looked repeatedly to make sure. I shook my head, then closed my eyes and opened them again. What I saw remained the same. I had hypothesized this odd light as some representation of life energy. Buildings, telephone poles, and wires with life energy? I knew I had to throw that idea out. The light around the wires danced. I immediately wondered what I would see around an electrical appliance. Would inside things have an aura too, or was it related to sunlight? I looked at the clock on my wall. It not only had light around it, but the light was highly structured and in steady motion. I looked at my programmable calculator and saw a finely structured complex pattern. I turned it on and set it to work – the patterns changed and scintillated as it worked. Now I was amazed all over again. What was I looking at?

Within a few days I noticed that people had auras around them that changed and scintillated as their owners talked to me about important things in their lives. A movie theater not only contained ordinary people, but also rows of swirling colored forms. I could turn all of it, or any of it, on or off by shifting the state of my consciousness. Years later, I would only need to shift my intent.

The connections linking living things became visually obvious. I could literally see that everything was connected. Even inanimate things such as clocks and computers had their complex moving nonphysical energy pattern. This same experience did not happen to Dennis. Perhaps he did not immerse himself in the exploration of NPMR and its theory to the extent that I did. I was extreme in my dedication to the effort. We often grew in different ways at different times and had usually, eventually, ended up with similar experiences. We were in this thing together and I had discussed my experiences – seeing energy forms – with Dennis as they happened.

One day he brought me a group photograph of five people and dropped it on my desk.

"These are all Soviets," he said, "one of them is supposed to be leading research in psychic activity in the Soviet Union. Which one is the psychic?"

I had never looked at pictures in this way before, but with focused intent, their auras blossomed up exactly as they did with flesh and blood people. "That was fascinating!" I thought. Conscious intent is everything – space and time are not fundamental. Wow!

"Which one is the psychic?" Dennis asked again.

I looked back at the picture, sure enough, one had a much more developed energy body – particularly around and above his head – than the others. "This one is different from the others," I said pointing to one of the men in the picture. "I am not sure what the difference means yet," I cautioned, "but this one is definitely different from the others." Clairvoyance was still a new experience and I did not know the significance of much of what I saw. At this point, I was more into formulating basic connections and had not thought about auras having unique meaning.

Dennis looked at me and grinned. "That's the one," he said with enthusiasm.

I was surprised. Dennis knew the answer – this was a test! I didn't mind; actually, I was pleased, another data point was in and I had learned something valuable and amazing about time and space being a subset of a larger reality. "I have so much to learn," I thought to myself, suddenly overwhelmed by the unfathomable depth and complexity of reality. Dennis went back to his office. I took a deep breath and wondered what would happen next, where was all this going, what else was out there waiting to be discovered? I felt small, humbled by the enormity of my ignorance. It was clear that I had barely begun to scratch the surface of something so immense and fundamental that I could barely imagine it.

At the same time, I was excited by the possibilities and determined to discover whatever I could about the nature of reality. I am a physicist and science and discovery are my passions – I was born wanting to know why and how. After twenty-two years of continuous education, I realized that I had studied only one small subset of the natural world. I was young, my learning seemed to be accelerating, and reality was far cooler, more complex, and more interesting than I could have ever imagined. To someone like me, it doesn't get any better than this – I was energized to discover any truth that would yield to my experimentation.

12

■■■

End of an Era

■■■

ack at the lab during the middle to late '70s, running the seminars dominated everything. We were overwhelmed with demand. People from all
over were clamoring to experience Bob Monroe's tapes – and all from
word of mouth. Bob saw an economic opportunity on the horizon. He
was a businessman, and this business (supporting the lab facility) had
been a constant financial drain. Perhaps, he thought, he could get two
birds with one stone. He eventually succeeded, but basic research was the
first casualty for a few years.

Eventually he was able to add the basic research back at a much greater
level than it had been before, as well as provide a life changing and enriching experience for thousands of people. But all that took time, and the
era of Bob, Dennis and Tom working until the wee hours of the morning,
trying to make science out of the strangeness they discovered, was gone.
Its time was rightfully over, fate had been extraordinarily kind, and we
ended on a long sweet high note. We were each ready to broaden the
scope of our efforts in our own way. It was time for us to soar, coast, or
crash on our own.

In the end, Bob was proved right, as usual. He captained his ship flaw-
lessly from the initial tentative launch, through the tricky undercurrents
of closed-minded rejection by the larger society, while at the same time
skillfully avoiding the shallows of easy, safe, generally acceptable answers.
With Bob at the helm, high standards of proof drove off pirate charlatans
who wanted to co-opt his success and commandeer his hard won credibility. Through dedication to honest science, personal integrity, and an
intuitive knowing that was steady and reliable, Bob optimized his gifts for
the greater good.

I do not wish to leave the impression that Bob, Dennis and I were the only explorers at Whistlefield Research Laboratories during the early seventies; there were others as well who made important contributions to Monroe's overall effort. A few became regulars making extended connections of various durations, while others were merely passing through trading knowledge like bees pollinating wild flowers. Nancy Lea had joined the research effort with Dennis and me after her graduation from college and soon became an integral part of the team, collecting evidence, testing concepts, participating in singular as well as joint explorations – even soldering wires on occasion. She began to carry more and more of the workload as Dennis and I reached and passed our limits of available time. Eventually Dennis and I needed to go home to our families. Nancy Lea took over the seminar operations and after a few years of successfully building and managing the business, she became the director of The Monroe Institute of Applied Sciences. The truer picture is that the overall effort at Whistlefield was a joint one. It was a busy place with a lot going on and many talented, interesting, and dedicated players.

The end of any era must necessarily share time's stage with the beginning of a new era. With the demands of the activities at Whistlefield winding down, I had more time to integrate and assimilate the continual whirlwind of extraordinary experiences that I had encountered. The nature of reality, a Big Picture that brought coherency to the wealth of collected data, began to take form in my mind. Any model or theory had to consistently account for, and accurately contain, the entirety of my experience – the roots of which ran deeper than I had previously imagined.

13

■ ■ ■

Once Upon a Time,
a Long, Long Time Ago

■ ■ ■

I have often told people who were inquiring about the possibility of learning what I have learned that if a bone-headed physicist like me could do it, anybody could. I would point out that I began from a cold start with no particular natural talent and learned everything from scratch the hard way. If I could, they could – and probably with less trouble. Dragging Spock from the deck of the Enterprise into the Twilight Zone, with logic fully intact and uncompromised, was a slow and tedious process. Most people could probably learn more quickly than I, even if they did not have the time or inclination to thoroughly immerse themselves as I did.

The point made above remains fundamentally true – anyone can learn what I have learned – but the picky fact is the previous statements contain one little white lie. My start wasn't as cold as I first thought. After I became familiar with the out-of-the-body experience and familiar with NPMR, I realized that I had done this sort of thing before. Old memories returned clear as crystal now that I had the knowledge and perspective to understand them.

When I was between five and eight years old, some friendly nonphysical beings helped me get out-of-body. It was not a random prank. They had a purpose. At first I played with it, slipping out through the wall of my second story bedroom and whooshing around the yard. They would get me out and I would play and soar. I well remember the first time I found myself outside floating a foot or two above the yard, gliding toward this monstrously thick hedge and realizing that I did not know how to steer or stop. I grabbed my head and curled up into a ball expecting a terrible and painful crash. To my utter amazement, I glided right through it and out the other side without interacting with it. Wow! Neat-o! That was fun!

> ▶ A short aside follows to help you find the proper perspective. Most children, particularly those younger than seven, have spontaneous, fully conscious out-of-body-experiences. Their parents tell them it is just a meaningless dream and they forget about it. These experiences are usually non-threatening and fun for the kids. You may recall some of your out-of-body-experiences if your memory is good and the experiences were dramatic.
>
> Most adults have spontaneous out-of-body voyages as well, but they are typically **not** fully conscious and therefore do not qualify as experiences. Out-of-body, as a phenomenon or happening, is much more mundane and common than it is strange. OOBE seems strange because the limitations we place on our concepts of consciousness and reality force us to reject many of the mental functions that are naturally available to our species and to misunderstand the purpose of the mental activities that take place while we are asleep. It is as if we are born with two good legs, but never learned to walk because the ability to crawl precedes the ability to walk, and because everybody is completely adapted to crawling around within a social structure that stigmatizes non-crawling non-conformists as wackos. ◀

Eventually these helpful beings and I became conversant. "How can I do that (get out-of-body) whenever I want to?" I wanted to know. They taught me several techniques. I practiced diligently and got results. Each time I would lose consciousness for a few seconds, regaining consciousness in the out-of-the-body state. "I want to be conscious the entire time," I complained, thinking that I would prefer to be more self-sufficient.

"You will not like it," they said, "We black you out for a short time during the transition to make the process more comfortable."

"I want to do it anyway," I protested.

They finally agreed. I immediately started applying one of the techniques they had taught me. As the here and now began to fade away into oblivion, a much fuller and richer awareness took root and began to blossom. Abruptly my body began to vibrate. The amplitude of the vibrations steadily grew larger and larger – my body had become plastic and was being shaken like a loose canvas awning flapping in a strong wind. Whoa! The violence of the oscillations had startled me. I immediately returned to a normal physical reality lying still in my bed. "OK," I said, "let's try that again." The same sequence repeated itself several times – I would pop back into physical reality after the oscillations became large, fast, and violent. "OK," I said, "you do it." I was almost instantly awake in the out-of-body state.

They won. I never bothered them anymore about doing it my way. Now I realize that I had been had. Those violent oscillations were not necessary. They simply did not want me to become too independent. I noticed that when I was out-of-body, I was an adult, not a little kid. That was neat. When I came back, I was a kid again. I hung out with my nonphysical mentors almost every night. They loved to teach and I loved to learn – we had a great time.

One night my adventures in inner-space took an unexpected turn that subsequently left nothing the same. Unknown to me, phase one was over and phase two was about to begin. Without forewarning, I was set up to begin a battery of exams that would determine the quality of my being. How evolved an entity was I? How much had I learned and grown with the help of my mentors? What was the limit of my understanding? The questions, or more accurately situations, presented me with multiple choices that became progressively more difficult. The first question was mostly verbal. "Would you rather have this treasure (I got a picture) or learn something new?" The answer was obvious; knowledge was far more valuable than goods. It was so blatantly obvious in fact that I decided to make a joke. "Just gimmee the loot," I said sarcastically in my best gangster voice, "I can always learn something new on my own." At the time, I did not realize that this was the first question on a long and especially serious test. My world exploded.

BZZZZZT!! Wrong! I was instantly transported to an entirely different place.

"He failed the first question!"

"He is not ready!" I heard someone exclaim with surprise and disappointment.

"Send him back!" someone else shouted, "he failed the test."

"Test?" I said, feeling a little like Alice at the Queen's castle. "I didn't know this was a test. I was joking, I knew the answer, I thought someone was playing around with me – making up goofy choices – so I was being goofy to get them back." The panel of judges that were to administer and evaluate my exam slipped back in time, inspected my mind for my true motivations – they evidently weren't expecting a smart aleck.

My advocates approached the bench vouching for my readiness. I was relieved that somebody was taking my side. I did not know who they were, but I was glad they were there. The tone had suddenly become heavy as if something terribly important was going on. Two judges said "Failed is failed – send him home," while the other three said "Continue the test, let's see what he can do." A higher-level judge was consulted. It was decided. I was to be given a second chance. It was obvious there were some serious hardball politics involved that I did not understand.

The tension was thick as tar; these judges did not like each other and there was a strong competitive hostility between the two groups. I knew this was very serious business, but I didn't know what I was doing in the middle of it. It was clear that whatever was going on was important to me and important to others for reasons I did not understand. I was thankful for the chance to show what I actually knew. My advocates who had evidently been working with me for a long time to get me ready for this first test were almost apoplectic. Their relief at my second chance was immense but they remained worried. It was as if their most important plans, careers, and reputations hung in the balance – and there I was, a somewhat unpredictable embodied bone-headed human.

Instantly, I was back in test-space. The first question was repeated – it was a precise replay of what I had previously experienced. I surmised that this was a fixed or standardized set of questions and that they had to restart the series from the beginning even though the first question and its answer had been revealed. Question after question, situation after situation, was put before me. I evidently did well because when you get one wrong, it's over. There were problems that tested ego and desire with sexual enticements – some of which were bizarre. There were choices between helping others and pursuing your personal path. They played to my emotions and ego, tried to instill fear, and probed how well I truly understood love.

Eventually, I was clueless as to how to approach the problem – I went with my best guess and the test abruptly ended. I was back in my body, turning into a young kid again though my mental space retained its adult nature while in the altered state. "Jeez," I thought, "what was that about?" As a kid, I often retained a clear memory of what happened, for a short while at least. But, I did not relate to it. At least five times since then I have been hauled in front of panels of judges because of my unpredictable quirky human nature. Thus far, I have prevailed. I must, as they say, have a rap sheet a mile long.

Some twenty-three years later, the event that had jogged this childhood memory occurred. I was at Bob's house one weekend afternoon when he began to comment on a training exercise I had been involved in the night before. Regular training sessions designed to develop my effectiveness within NPMR began again in earnest shortly after I resumed sentience within the larger reality. Bob had been in the audience watching my performance. He was telling me about what I had done. I was surprised, not that he could know, but because usually he was not involved in my OOBEs.

He knew every move I made and started kidding me about a show-off display I had performed at the successful end of a particularly difficult series – the way football players sometimes dance in the end zone. We were laughing about it, when he told me about an especially difficult test he himself had recently been given. He started to describe it. After he described the third test in this long series, I stopped him. He knew something was up. Now it was my opportunity to puzzle him.

My experiences of twenty-three years ago had come flooding back in a gushing torrent as he described the first three tests. I made him wait until I had collected my thoughts. I told him what the fourth test was. His jaw dropped. I had never seen Bob speechless before. I described the fifth test. He was dumbfounded. We went through the rest of the test, alternating who would give the description of the presented problems. Oddly enough, we had bombed out at the same place but with different answers. Evidently neither one was right, or that was the last question. Without a doubt, that **was** a standardized test. Since that time I have run into several others who have experienced some sequence of events while in NPMR that were **identical** in form, function, and content to events I experienced.

Back to the early 1950s. Immediately after that first big test, I was put in regular training classes. Every night for most of a year I was run through situations, given jobs to do, and further tested by my trainers. I was never told what I was being trained for. I worked hard. After that first major evaluation almost turned catastrophic, I was serious. It was more work than fun. I was learning to control my mind, to manipulate non-physical energy, to make the right choices for the right reasons, to think fast, and act fast. I was learning to follow directions and to break old PMR conceptual habits. I was becoming effective and efficient in NPMR – I was learning focus and control. Eventually, after much practice, I began to feel competent and strong, like an athlete ready to walk into the arena.

One night there was no more training. I would not resume the effort until some twenty years later, but I didn't know that then.

14

■ ■ ■

This Kid is Weird Enough

■ ■ ■

I did not know it, but my learning time in the nonphysical was about to come to an end for a long time. My mind would typically shift into adult mode as I drifted on the boundary between awake and asleep. It had been a week or more since my instructors had been by to take me to class and for some reason I had not gone on my own. It was time to get back to it. I began to slip out-of-body, unassisted by my instructors. Now it was easy. I wondered where my teachers were – why hadn't they come?

"Oh well, this will be fun," I thought with great anticipation. I paused for a moment, "Where should I go, what should I do?" My training had been highly structured; I wasn't used to this much freedom. "Perhaps they want to see what I will do on my own," I thought. I felt that they were probably watching.

Rather suddenly, someone stopped me cold in my tracks, forcefully shoving me back into my body.

"Hey," I said, "what are you doing – you can't do that to me!"

"Yes I can," the reply came back in a firm authoritative tone. "In fact," he continued, "I have been instructed to seal off this passage. You will no longer be allowed to exit your body."

"If this is a test, I don't get it," I thought to myself as I struggled to find the correct response. It was not a test. There were two of them but only one spoke – workers sent to close and seal the door on my experiences in NPMR. "Are you sure you have the right person?" I challenged, "This must be some mistake." I waited while he checked out that possibility.

"No mistake," he said looking at his work order, "you are the one."

"But why?" I pleaded.

"You are a young kid aren't you," he said.

"Yes, but just my body," I replied. "That has never been a problem before."

"You have completed what you were doing," he said matter-of-factly, "and now they don't want you to grow up too weird. You know what I mean – you need to grow up like a normal kid, develop a healthy personality. Too much of this other-world experience would make you strange and not fit comfortably into the world you will grow up in."

"I can handle it," I protested, "I haven't had any problems so far."

"You have no choice," he said emphatically. "Trust me. It may be easy now because you are only in one reality at a time, but as you get older, it would become a problem for you to juggle multiple realities simultaneously."

I did trust him, and I intuitively knew that he was right. I also had no choice; he was much more powerful and knowledgeable than I was. I knew I could not slip thoughts into his mind. I didn't try. "Will I ever be allowed to get back out again? Will you ever unseal this door?" I asked hopefully.

"I don't know," he said flatly as if it were none of his business.

"Can you find out for me?" I pleaded, "Please look and see." I don't think they were supposed to do this sort of thing, but I must have sounded desperate and sad. Neither worker said anything, but they both looked far away into the distance. It was a long and intense look. Suddenly, they both gave a loud gasp – as if they were simultaneously surprised and shocked. They were momentarily stunned, and remained quiet for a few seconds.

"What is it?" I asked. "What do you see?" I strained to see the picture that was in their minds, but could not. "Will I ever get through this door again? Tell me, please." Both workers now had the demeanor of someone who has inadvertently blundered into something they were not supposed to know about. Both had become tightly focused on what they were doing and were now non-communicative and in a hurry. They glanced at each other with concern. I could sense that they were more than a little worried. I could feel their feelings and could pick up a few surface thoughts, but the startling events they had witnessed were securely and purposely unavailable.

How could the information they had inadvertently found out about **my** future put **them** in a jam? I was dumbfounded. My concern and overall puzzlement was growing rapidly. They had a sense of impending personal jeopardy. After some quick discussion they briefed higher authority on the situation – a few others who were somehow connected were brought in. Security was tight. Decisions were made quickly – a plan B was set into motion and everyone but the two workers immediately disappeared.

The problem was not that the two workers had broken a rule by looking ahead for me – that was a relatively small issue that could be, and was, quickly forgiven. The problem was that now they knew something that they were not supposed to know. I was perplexed, and could not even make a wild guess at what was causing all the concern. "What did you see... what is the problem... what's going on?" I asked the workers in a quiet and serious tone trying to sound as if I were a team member who had, for some unavoidable reason, just happened to miss the preceding conversation. They took a few seconds to finish what they were doing and then slowly turned toward me, paused, and stared with quizzical amazement for a few moments. They looked at each other for a brief moment, then back at me. Saying nothing, they slammed the door hard. I was shut out.

As it turned out, I was not shut out entirely. There were no more out-of-body experiences, to be sure, but I still had friends. In my mind, I could easily communicate telepathically with my guides. I did not call them that but that was their function.

15

■■■

With a Little Bit of Help From My Friends: How's Your Love Life?

■■■

I always knew growing up that there were nonphysical entities available to help me. They looked after me and I knew it. They were nonphysical friends who were older and wiser and knew what was coming. I think that many people have this sense of having helpers. Whether you interpret it as a religious manifestation (God, or guardian angels perhaps), or simply let it exist unnamed within your imagination seems primarily to be a function of your cultural belief system and temperament.

When in need, I intuitively sought out my guides and depended on them. When I did not need them, I forgot about them. Whether I was aware of them or not, they were always with me, they were at times my intuition, my luck, my counselor, my pals – and they made a special effort to keep me alive. Risk and daring had always been two of my favorite playmates and most steady companions – I was not an easy case.

Several times in particular these guides provided detailed information about my future. Both times were in response to questions of relationship that contained strong emotional and intellectual content. Once, at fourteen while waiting for the school bus to take me home, I asked in frustration about the possibility of ever having a successful love life. Instead of being amused at my pathetic pubescent condition, they proceeded to lay out the relationship map of the rest of my life. Every significant relationship with pictures and detailed description followed in rapid succession. I repeated it over and over in my mind so that I would not easily forget it.

To this day, I remember most of the conversation clearly. Accurate predictions were made that penetrated twenty years into the future! How

108 | *My Big TOE* | With a Little Bit of Help from My Friends: How's Your Love Life?

could they do that? I do not know for sure – I must be an exceptionally predictable person. In particular they predicted that "the one," the final lasting connection (significant other) would be with a woman who was now only two years old!

"Two," I blurted out, "you have got to be kidding! I am paired with some two-year-old baby! Come on guys, I can't wait that long! This must be a joke! Right?" It was no joke. At fourteen, this was not what I wanted to hear. "That is so far away, how can you predict that?" I asked. They did not answer. "Do I know this baby?" I asked.

"No," they said, "she is not in this state."

"Great," I thought, "a monk forever!"

"There will be others before her," they said in consolation.

"Yeah, sure, but what's the point, they are not 'the one'," I shot back with obvious disappointment.

"No, they will not be 'the one', but they are important and necessary," my guides replied with great patience.

I heaved a sigh of resignation. They never kidded around and were, as far as I could tell, always right. Arguing with your guides was even more futile than arguing with your mother.

I paused and turned inward. I began to remember an experience I'd had a couple of years earlier, when I was twelve. I had sneakily taken a look at the foldout of the month in a girly magazine at a news stand while the owner was busy ringing up sales. Being exceedingly impressed with what I had seen, I wondered if I would ever have a woman like that to have and to hold.

To my surprise, a reply began to stream into my mind.

"Yes," I was told, "you will have one that looks very much like that one... she will meet all of your desires," my guides added with a hint of amusement. "She will really love you – the two of you will be tightly connected." I was given a sense of the quality of the relationship.

Wow! I thought. I was excited – the luckiest guy alive – because my guides always knew what they were talking about. They were always sure and confident and never guessed or speculated. "When?" I asked excitedly; "when do I find her?"

"When you are thirty-five years old," they said flatly.

I was totally devastated. "Awww man, that's really old – I mean **really, really** old! Do you have any idea how far it is to thirty-five when you are only twelve? It's forever!" I whined.

"That's the way it is," they said with no trace of emotion. I had the sense that they were ever so slightly amused at my reaction.

| http://www.My-Big-TOE.com |

As that memory poured into my consciousness, the connection was immediately obvious. That two-year-old baby and my foldout-princess were one and the same. "Interesting, but totally useless," I thought. The immediate romantic future that I was most interested in looked as bleak as ever.

Another instance occurred when, at twenty-one, I was about to get married after graduating from college. I was having second thoughts, when a similar experience took place. The future was outlined. Again "the one" was described and predicted. The problem was, it did not happen to be the person I was about to marry! Yes, I was supposed to get married. Yes, I would have a son. No, it was not going to last forever. "But why do it now if this is not 'the one'?" I argued, and then continued with, "It is not fair to her or the child."

The answer came back clear and strong: "It is what you are supposed to do; it is part of a larger plan. Marry her. It is the next step for you, and will be the best thing for everyone including her. Everything will work out." I did. It did.

Everything they said happened as they said it would, except for a few items at the far end of the given timeline that have either changed their probability of occurrence over the years (which is what I think) or they are still in the queue.

In the event you are curious, at the age of thirty-five (I am writing these words in the year 2000 at the age of fifty-five) I began a relationship with "the one" who is twelve years younger than I am (was two when I was fourteen). We have three children (exactly as foretold) who adore their big half-brother Eric. Life is good.

As you see, it is not exactly true that I had a cold start with TM at the age of twenty-four. Life had been good to me and I had completely forgotten most of what I just told you. To an exceedingly skeptical, young, bone-headed, hard-science-type who had made a concerted (and relatively successful) effort to purge his system of all beliefs, and who was naturally wary of anything he could not directly measure, it certainly seemed as if it were an ice-cold start at the time.

16

■ ■ ■

Here We Go Again!

■ ■ ■

By the early 70s, that door to NPMR that had been slammed shut 20 years earlier was thrown wide open again. I was back in NPMR with a mission to understand it all – from a scientist's perspective this time. By the end of the 70s, Bob and the lab were doing their thing, and I was doing my own thing independently. I had learned that the so-called phenomena or powers (paranormal events) were only an artifact of a path well taken, not the objective or destination. Becoming too enamored of paranormal phenomena can distract you from more important issues and retard or prevent your further development.

I also learned that the bigger picture was centered on the quality of your consciousness, the evolution of your being, and that my experience could be scientifically explained. Eventually, it became apparent that the rate of learning accelerates. The more you know, the faster you learn. Up to 1980, I was on the launch pad, getting ready, learning, training. Then things really took off. The pace has never slowed and my interest and dedication have never waned. More data points continue to pour in today.

From another point of view, nothing exceptionally strange happened. Many children around the age of five to seven have unusual experiences in their dreams. I was able to later remember mine and understand them within a larger context. Likewise, many people have nonphysical guides or advisors. In our culture, this sometimes occurs within a religious paradigm. I was given an unusual opportunity to work in a fascinating field with a good teacher and I took full advantage of it by working hard and being dedicated and conscientious in my effort. That was simply good fortune (or good planning) – being in the right place at the right time with the right credentials, the right interests, and the right attitude.

What I learned was more by virtue of hard work and a focused interest than by anything fantastic, amazing, or strange. There was no magic, no bump on the head, no aliens from outer space, no near death experience, and I did not find deep mystical secrets hidden in a golden urn buried under the rubble of an ancient monastery in Tibet. It was good science and dedicated effort applied to an opportunity – that is all. I essentially worked my way to where I am today and I have no regrets. Indeed, I feel extremely lucky to have found the path that I am on. The word "strange" is a relative word. Anything not mundane must, necessarily, **seem** strange to those who have not experienced it.

Today I work as a physicist in an engineering services company specializing in missile defense research and development, and if you saw me or worked with me you would not think that I was the slightest bit strange. Yet, in another environment, away from the workplace, I help guide the development of a small group of students and other people interested in increasing the quality of their consciousness. I have learned to work harmoniously with the environment I find myself in. Because most of my real work has been focused in the nonphysical, it has been easy to maintain a low profile in PMR.

I have been strongly encouraged to share some of the results and conclusions of my explorations of the reality we live in. That is what *My Big TOE* is about. I hope that by knowing how I came to have the somewhat unusual knowledge and experience that makes this book possible, you will be able to view the unusual concepts you are about to delve into with a broader perspective. Having been faced with the same dilemma of analysis (is this guy nuts, or what?) that faces you now, I fully understand your position.

There is no easy or satisfactory way to judge the quality of the information upon which this trilogy is based without knowing me personally. Nevertheless, I hope the above account at least helps a little. Those with your own experience will find familiarity in what I have to say. Good luck, I hope you make out as well as I did, or at least learn something useful. Remember, the evidence, as well as the key to understanding, lies within your own experience – and nowhere else.

Section 2

■■■

Mysticism Demystified:
The Foundations of Reality

■■■

17

■■■

Introduction to Section 2

■■■

What we call mystical is relative to the extent of our knowledge and understanding. If some process, phenomenon, or conceptualization **appears** to lie beyond our potential ability to explain it within the context of PMR (physical-matter reality), we describe it as mystical. Much of what was considered mystical a thousand years ago is considered science today and much of what is considered mystical today will be clearly understood by a future science.

As our current accumulated **objective** knowledge reaches its limits and begins to dissolve into the seemingly unknowable, what lies beyond our **presumed** theoretical reach is defined as mystical. Such presumptions well up from our beliefs about objective reality; thus, what appears to be mystical or unknowable from the view of Western culture simply reflects Western cultural beliefs and the limited understanding of contemporary science. Consequently, it is a double dose of ignorance (cultural beliefs and limited scientific knowledge) that defines what appears to be beyond our serious consideration. Think about that the next time you roll your eyes and snicker because you universally associate the concept of mystical with ignorance (other people's, not yours), foolishness, and unscientific blather.

For an individual, the process is more personal. What may be perceived by an individual to be mystical is relative to the individual's understanding, knowledge, and ignorance. It is our personal beliefs that determine what we consider to be mystical. Whether an individual knows little or much about either PMR or NPMR (nonphysical-matter reality), what lies beyond the reach of his or her personal understanding and knowledge may be: 1) interpreted by that individual as mystical, 2) construed by belief to be something that suits the needs of the individual, 3) regarded as a temporary ignorance of something theoretically knowable, or 4) regarded as a permanent ignorance of something theoretically unknowable. An individual's conclusions regarding what lies beyond his objective reach are necessarily belief-based unless, of course, the conclusion is that there can be no conclusions. Nevertheless, most of us embrace a multitude of both culturally given and personally derived belief-based conclusions with a degree of certainty that only a deep bone-level ignorance could sustain.

Belief is a conclusion based upon a mystical premise. Scientists might **believe** that what is unknown must be contained within the PMR data-set and follow ordinary objective causality, but that **belief** or article of faith simply expresses a more accepted form of mysticism. Mysticism that supports our cultural beliefs is accepted as obvious fact. By definition, such a **belief** necessarily appears to be the most reasonable assumption that a rational person (within that culture) can make. This is how ordinary mysticism expressed as cultural and personal belief is transmuted into an unquestioned philosophical foundation. We see that the objective causality of Western materialism must necessarily spring from a foundation of mystical assumption. Voilà! Faith becomes science, or at least an integral part of the scientific attitude.

The results of this illogical transformation continue in chain reaction. Next, science becomes truth – or at least the sole judge of truth – and is given the job of defining reality. Thus, from the Western perspective, the world of ideas and concepts bifurcates, with science and truth on the one hand and philosophy and conjecture on the other. Though science is important because it produces a marvelous array of useful physical products, philosophy is marginalized because it produces nothing but useless arguments. In the final act of this comedy of errors, we see that science, unable to overcome the barrier of its ancient, no longer useful, faith-based paradigms, becomes imprisoned by the limitations of its core beliefs. As science struggles with its self-induced myopia, philosophy tries in vain to mimic science's illusion of objectivity in order to appear relevant. Does this not remind you of some preposterously convoluted French farce?

Ahhh, but wait...the action isn't over – there is a surprise ending! However, you will have to take this journey with me and My Big TOE – from here to the end of Section 6 – if you want to discover how this drama plays out.

▶ No doubt about it, I am putting you off. Why? Because, if I told you the ending now, before thoroughly discussing some of your current beliefs and paradigms, you would not, could not, "get it." I know that you are experienced and brilliant, but that is not the point. Seeing the Big Picture requires more than intellectual capacity. It also requires transcending ingrained belief systems. That is something that many people are not willing or able to do – at least not quickly or easily, if at all. That is why a trilogy is required instead of a technical paper.

Academic papers provide a medium for communicating something of value only if one is working within the accepted cultural and scientific belief systems. Consequently, significant new knowledge is typically generated by meticulously taking countless tiny steps of ever expanding detail toward some specific goal. Digging out details requires a totally different mental process than discovering Big Picture paradigms. Today, almost all scientists remain focused on prying details from a reality exclusively circumscribed by traditional scientific belief. Fact is: You will never be able to see the Big Picture as long as you focus on individual politically correct pixels.

Because *My Big TOE* must go beyond traditional belief systems to introduce an entirely new understanding of reality, much of this trilogy must necessarily be spent broadening your perspective. Have patience and resist the urge to come to what appear to be obvious conclusions before you have consumed the entire trilogy. Transcending old paradigms and belief systems is as inherently difficult a process to facilitate as it is to undergo. ◀

Will *My Big TOE* come to the rescue and re-energize this languishing drama of stultified science and marginalized philosophy by dissolving the Gordian knot of limiting belief that condemns both to little more than picking the bones of the past (simply taking the next logical step or digging ever deeper into the detail of yesterday's discoveries)? Can *My Big TOE* successfully raise science to the next plateau of understanding while simultaneously returning philosophy to its rightful place leading the parade of human progress with meaningful and useful insight? Stay tuned!

What I hope to accomplish in this section is to develop the basic concepts necessary to support a rational theory of existence – a conceptualization that provides the foundation for a reasonable Big Picture Theory Of Everything (Big TOE). In the process, I hope to push your mystical edge (where knowledge meets ignorance) back to a point where the residual unknown has no potential significance and is of no practical interest.

A TOE that does not reach beyond PMR is only a little TOE (Little picture Theory Of Everything) confined to PMR and limited to a local causality. Can you imagine the elements and limitations of a Tiny TOE from the perspective of bacteria living in your intestines? Would a seemingly complete and comprehensive (from the view of the bacteria) Tiny TOE be concerned with the nuclear fusion taking place within our sun or the density and composition of our atmosphere? No, of course not; these things have no **direct** significance to the bacteria in your intestines. Although all things that live upon the earth depend on the sun's energy and the composition and density of earth's atmosphere, the bacteria in your intestine cannot directly experience either – their Tiny TOE needs only to describe everything that is potentially knowable by the bacteria.

To a given awareness, the **practical** definition of the word "everything" means: everything knowable, important, meaningful, or significant that can directly interact with that awareness. Thus, what constitutes a comprehensive and complete Big TOE is relative to your perspective, knowledge, and limitations. That Tiny TOE would seem complete from the view of the bacteria, even if it neglected to include the money in your bank account, the light bulb in your refrigerator, or the car in your garage. However, these items may **indirectly** have a profound effect on the current intestinal environment.

Money, refrigerators, light bulbs, and automobiles are too far removed (far beyond the practical, functional, or theoretical scope of a bacterium's Tiny TOE) to be comprehended by the bacteria, or to be of any **direct** importance to them. To the bacteria in the intestine, the source of digested food descending from the stomach would seem mystical. The economic, social, and physical circumstances and processes that indirectly result in a particular food being deposited in the stomach would be beyond mystical. The causal mechanisms that drive and order these apparently mystical events and processes are necessarily invisible to even the most brilliant intestinal bacteria. The forces and relationships that govern the growing of wheat as well as the making and marketing of bread falls beyond a bacterium's theoretical ability to imagine, and therefore forever lies beyond the largest reality it can possibly comprehend. Do not be too surprised to find Homo sapiens in a similar situation.

This is a difficult pill for many, especially scientists, to swallow. The concept that there may be a natural practical limit to the extent of our knowledge – a limit beyond which our perception cannot penetrate – is based upon the notion that we are only a very small part of a much greater reality. This humbling thought runs counter to the significance and self-importance we humans place upon ourselves. If our experience is limited to a small part of a larger reality, it is only reasonable to assume that beyond the limit of our possible knowing there may well exist a host of phenomena, interactions, relationships, and ordered happenings upon which our reality and existence profoundly depends, but of which we cannot **directly** perceive. Allowing the outside theoretical possibility that our beloved PMR may be a local reality (a subset of something larger) is the first step toward comprehending a bigger picture.

This possibility breaks the conventional paradigm of PMR being all there is and replaces it with a more expansive paradigm that forms a logical superset – the limited little picture concept is fully contained within a more general bigger picture concept. If the larger and more general paradigm provides a better and more concise understanding of the available data as well as produces valuable new knowledge, approaches, and processes, then the more general conceptualization is also a more accurate, productive, and truer representation of the whole. An improved reality paradigm is one that broadens and deepens the available solution space relative to existing data and problems in a way that is practical and useful.

It is as easy to understand the limitations of a bacterium's perspective as it is difficult to understand the limitations of our own perspective. That is natural enough. We cannot be aware of what is beyond our awareness. However, we can be open to learning new things, and in the process, expand the scope of our awareness – and therefore our reality – to its outer limits.

At the top level, where any Big Picture must first be clearly drawn, *My Big TOE* will attempt to encompass everything known and knowable. What lies beyond *My Big TOE* will remain practically unknowable (forever mystical) because of the inherent limits of our vision. Nevertheless, both the mystical and the beyond mystical will be explored as we have some fun with logical extrapolations that reach well beyond our capacity to comprehend.

> ▶ "Encompass everything knowable? Derive physics from metaphysics? Explain the paranormal scientifically? No way! Even at the most general theoretical level that's completely impossible – what's the catch? Is this guy nuts or what?"
>
> I know the idea of encompassing or bounding "everything known and knowable" sounds unlikely at this point – as does the rest – but it is not as far-fetched as you may think. *My Big TOE* is not presented at such a high level of generality, nor is it so off the wall as to hold little direct scientific or practical value to the real world. No catch, no megalomania, no hypothetical wackiness, no goofy beliefs – just straightforward science that better describes the measured data and provides a wealth of practical results and new understanding that can be applied personally and professionally by scientists and nonscientists alike. By the end of this trilogy you will be able to assess the accuracy of these seemingly impossible, unsubstantiated statements. Until then, open minded skepticism is the only approach that retains the possibility of success for either of us. ◀

Let us begin to explore the outer limits of our reality and see how far we can push back the boundary. We are generally somewhat smarter than bacteria and are theoretically limited only by the capacity of our minds. Unfortunately, we are limited by much more – beliefs, pseudo-knowledge, preconceived notions, attitudes, fears, desires, needs, and cultural biases. For this reason, we must talk about some of these first.

The biggest picture must cover everything – everything objective, everything subjective, everything normal, and everything paranormal. Mind and matter, consciousness and concrete, all the true data and the facts of existence (the personal as well as the scientific) must be accounted for, compatible with, and contained within this single Big TOE – **if** it is a comprehensive and correct Big TOE. If it is only a little TOE, or incorrect Big TOE, it will not support or explain **all** the data. It is your job to assess the extent to which this Big TOE explains **your** experience and knowledge. However, before reaching conclusions, it is important to understand the difference between knowledge and belief. In the chapter after next we will thoroughly explore that subject. To set the stage for that challenging epistemological adventure, we must first take a look at our collective beginning. It is always a good idea to start at the beginning.

Let's get started on this strange journey. Oh yes, it will need to be strange or it could not possibly be a correct Big TOE. Trust me, unless you have invested many years of experience in this area, what you are about to read will greatly challenge the elasticity of your mind. To make things worse, my insistence that this Big TOE is primarily derived from and based upon carefully evaluated scientific data, rather than theoretical conjecture, will require your sense of my credibility to stretch even further than your mind. Unfortunately, a reader with a simultaneously stretched mind and stretched sense of credibility puts me on thin ice from the beginning, but that is the way it has to be. Such is the nature of this topic, the facts of my experience, and the results of my research. Something less strange or more widely credible would be easier to convey, find a wider audience, and be more acceptable to almost everyone, but it would not be correct.

People generally **believe** that they know **almost** everything that is knowable, that the final few things to be figured out will constitute small steps compared to the distance already come. For example, scientists toward the end of the nineteenth century often lamented the **obvious** fact that everything important (in science and technology) had been discovered. Little more than a century later, that claim is laughable. By definition, it is clear that you cannot be aware of what you do not know. Yet, we almost always let our egos trick us into believing that we are much less ignorant than we actually are. It is often said there is nothing as outrageous or strange as the truth. The truth of that statement, clearly demonstrated by modern physics, demands an open minded approach.

Hopefully *My Big TOE* will stimulate you to consider some important things in an entirely new and beneficial way. Take from it whatever you can. The picture and perspective may be very big and initially the credibility (for those who do not know me) may be very thin, but if you **feel** your way through it with your **intuition** as well as think your way through it with your intellect (all the while collecting and applying the data of your experience), it might just make some sense to you.

Appreciating and understanding the Big Picture is always the first step in focusing and directing the effective investment of the resources you have at your command. The point and purpose of this work is to offer an expanded view of reality and of your relationship to that reality that is **useful** and helpful to you in a direct and practical way. In *My Big TOE*, you will find a reality model that provides a unique perspective which can be profitably applied to your professional and personal life.

One final caution. Section 2 lays down the conceptual foundation for much of the rest of the trilogy without explaining how these concepts link up later to complete a meaningful pattern. It may be difficult to get a firm grip on some of this material because it is predominately abstract and may be relatively distant from your common everyday experience. Understanding and validation will accrue slowly as you digest the next four sections and bring your own explorations of Big Truth to bear on the conclusions drawn.

Though there are many abstract tunnels to explore in Section 2, there is precious little light to be seen at the end of any of them initially. Nevertheless, a logical approach requires many of the conceptual building blocks to be defined and set out before Big TOE construction can be initiated. Delay final judgments until these concepts are more fully developed and applied in subsequent sections. If you can hang tough, the method to the madness should become fully apparent by the time you finish Book 3.

If *My Big TOE* entices those who are open minded and intrepid explorers to experience and investigate the far reaches of their reality on their own, I will be delighted. To be optimally effective, my discovery must lead to your discovery. I must not only help you to objectively understand the nature of your reality, but must also help you discover it in personal terms.

In the end, if these books have no effect other than to cause you to reassess your beliefs, concepts, knowledge, and attitudes, regardless of what the outcome of that assessment is, my effort will have been worthwhile. Have at it. Good luck on your journey – may you find value of lasting significance.

18

■ ■ ■

In the Beginning...
Causality and Mysticism

■ ■ ■

Beginnings are always difficult. Wherever one starts, there is always the question: "What was before that?" This question comes from our sense of objective causality – that everything must be preceded by its cause. Must everything have a cause? If "no," then one leaps immediately to invoking mystical beginnings. If "yes," then the beginning is a logical impossibility. There can, by definition, be no beginning if everything must have a cause. By the logic of causality, beginnings are illogical. The logic of causality requires (because we **do** exist) the **initial** existence from which we are derived to erupt spontaneously from nothing. Clearly, the notion of objective causality must violate its own logic in order to get started.

The other alternative – there is no beginning, existence is **somehow** infinite and perpetual, is itself a mystical assertion that comes from nowhere and goes nowhere. Such an unbounded mysticism offers its supporters no possibility of either answers or clues. Beginning with a premise that our ignorance of beginnings is total and perpetual is not a particularly clever way to begin an analysis of beginnings. Easy perhaps, but not useful. This logical alternative provides a trivial solution that leaves no foundation upon which to build a reality.

Thus, the logical result of invoking an objective causality is a mystical beginning. Likewise, the logical result of denying an objective causality (our beginning began without prior cause) is also a mystical beginning. Although the logic of our objective causality would seem to indicate that our beginning **must** be mystical, that is not necessarily so. It depends on the reality in which such a beginning is taking place, and the reality from which we are viewing it. Causality is system specific – the logic of causality (the logic of PMR physics for example) holds within a given causal system. The logic of causality only requires that a given system's beginning appears to be mystical from a point of view that lies within that system. The logic of causality can say nothing about the beginnings of its own system because those beginnings lie outside that system – beyond the reach of its own causal logic. Beginnings belong to the next higher level of causality and are beyond the purview or scope of a subsystem's own causal logic. Imagine a hierarchy of causal systems, each being a subset of the next. Thus, mysticism may be removed from our own beginning if we can obtain the perspective of the superset to which we belong.

I am not saying that our objective causality should be tossed out. The logic of our objective causality has been, and remains, the philosophical foundation of our science. It has motivated us to ask: "How does this work?" or "What caused that?" It has led to the technology and understanding that is now begging us to take the next step beyond the purely material. I am not putting down the logic of our objective causality. I am a scientist – I live and work by it. I am simply putting it into the proper perspective. I am pointing out its logical limitations, the boundary of its meaningful application, the fact that it requires its beginning to violate its own logic.

Thus **our** beginning, from the point of view of **our** objective causality, must be indefinable, or equivalently, mystical. If you do not logically equate "indefinable" with "mystical," that is fine. Given that the subject is the creation of our reality (our beginnings), the terms "necessarily undefined" and "unknowable" quickly morph into "mystical" in the minds of many – thus I use the word "mystical" to generally describe the unknowable. By the time you reach the end of this trilogy, the veil of mysticism will be logically removed from our beginnings and you will clearly understand the roots of our existence and how and why those roots came into being.

Once we realize the causal logic that gives us science also limits our understanding of the larger reality (and its beginning), we are free to begin exploring the larger truth. Without this realization, our perspective and capacity to understand is trapped in a conceptual prison (a belief trap) of our own making.

The erroneous belief in a universal causality (as opposed to a local causality) is used repetitively to make those who would dare rationally tackle the questions of beginnings appear to be ignorant and incorrect. The repeatedly and iteratively asked question "What was before that?" inevitably must end with a confession of complete ignorance existing at the foundation of an otherwise rational discourse. The position is taken that logical arguments built upon a foundation of ignorance are highly suspect and can be dispensed with immediately as foolish or unsubstantiated conjecture.

Our physical space-time causality **is** local and simply does not apply to "what was before that" – otherwise we would either be stuck with no beginning, or we would have spontaneously popped out of nothing. Either of those alternatives lead to mystical beliefs that are not scientifically or logically productive – neither makes good sense, nor provides a rational foundation from which to build a scientific Big Picture Theory Of Everything. A major paradigm shift described within this section provides another alternative (necessarily mystical from the PMR point of view) that is **not** belief-based, that **does** make good sense, and that **is** scientifically and logically productive.

Our beginnings appear mystical to us because of the limitations of our logic and because of the limitations that our belief-based perspectives impose upon our minds. If you raise science, as well as your vision and understanding, to the next higher level of causality – to the supersystem that contains PMR as a subsystem – the ever-present mysticism will recede to the outer edges of your newly acquired knowledge.

If your picture (worldview or understanding of reality) is significantly bigger than your neighbor's picture, your neighbor may see you as a mystic. Indeed, you will always appear to be a mystic from a viewpoint that is greatly limited in its understanding regardless of how rational, complete, or scientific your understanding is. A mystic appears to be animated by unknowable interactions that lie beyond rational understanding. Perhaps your dog thinks that you are a powerful irrational mystic. If your neighbor also finds you to be a particularly good, loving, wise, productive, successful, and capable person, he should try to understand what you seem to understand. If, on the other hand, you appear arrogant, condescending, manipulating, or begin proselytizing and asking for donations, he would do well to keep his door locked and avoid you.

The quality of your being expresses the correctness of your understanding. Think about that a moment. What does the quality of **your** being say about the correctness of **your** understanding?

19

■■■

Beware of the Belief Trap

■■■

Most of us are awash with beliefs of all sorts. We are steeped in the common sense and prevailing wisdom of our culture, traditions, communities, profession, family, and friends. Because belief is very personal for each of us, I will approach this discussion of belief and knowledge from many different directions. Hopefully, at least one of these approaches will connect with your unique experience, intellect, and inner-self.

When encountering something complex and unfamiliar, repetition is usually required before we feel comfortable with it. Likewise, reiteration is often needed to punch through deeply held and ingrained ways of thinking and being. Most of us have deeply held and ingrained ways of thinking and being whether we are intellectually aware of them or not. What is deeply ingrained in us is nearly impossible for us to notice – it becomes part of the invisible inner core of our being. It is a fact that subtle belief systems circumscribe our personal reality. It is also a fact that most of our beliefs lie beyond the easy reach of our intellects. Outside our awareness, they literally define, and thus also limit, what we allow ourselves to perceive and interpret as reality.

I am sure that you can process Big Picture information intellectually without difficulty, however, because your core beliefs are profoundly ingrained, it is far more difficult for you to successfully integrate that information into deeper levels of understanding. Because you find certain material to be **intellectually** easy to understand or conceptually obvious and repetitious, does not necessarily indicate that the **significance** of that material has actually sunk in to deeper levels of knowing.

What passes for intellectual understanding is often shallow and incomplete because we have no means to accurately assess the extent of our ignorance. The logical result of an awareness of our ignorance relative to some very important issue is an uncomfortable anxiety – the anxiety of not knowing what you desperately need to know. We generally feel compelled to produce an **apparently** solid assessment of the problem regardless of how much we know or do not know. In order to ensure that our assessment appears solid enough to significantly reduce the fear-based anxiety brought on by our ignorance, we make assumptions about the degree, quality, and completeness of our knowledge that invariably lead us to interpret "shallow and incomplete" as "sufficient and conclusive." The resultant intellectual judgment, regardless how ill conceived, will always produce a conclusion that appears (to its creator) to be reasonably certain as well as obviously correct. Presto-change-o! The discomforting anxiety disappears as pseudo-knowledge is manufactured by the ego to deny ignorance its due.

Do you see how the fear of not knowing thus assuages itself by creating a believable self-satisfying story that provides an alternative to acknowledging and accepting ignorance? That the story may be false is invisible to its author because it is based upon assumptions and beliefs designed to meet the author's pressing needs for reassurance and security. Have you ever wondered how **other** people can come to the strangest conclusions about all manner of things? From religious fanatics to your sometimes exasperating manager or significant other, it works the same way. Be careful that **your** intellect does not trick you into believing comfortable and seductive conclusions that are primarily designed to reduce your anxiety, reassure your ego, and maintain your current self-satisfying worldview.

Whenever you feel reasonably certain that you are obviously correct even though you have no real data to back it up, you should at least consider the possibility that you may be stuck in a belief trap of your own creation. Only open minded skepticism will allow you to assess that possibility.

Big Truth must be understood deeply to be effectively applied. Wisdom resides more in the heart and soul than in the intellect. Your intellect can only take you so far in your exploration of Big Truth; it can direct your search but cannot cause you to learn anything of deeper significance. On the other hand, your intellect can cause you to squander every opportunity to know Big Truth.

To focus our discussion of belief traps, let me give you a more precise and clear understanding of what I mean when I use the words "belief" and "knowledge." Beliefs may be cultural, religious, scientific, or personal. Belief is generated and necessitated by ignorance. If you know for sure, belief is not required. In that case, you have real knowledge. Knowledge is derived from knowing what is true. If your apparent knowledge is false, you only **believe** that you know. In this situation, belief is masquerading as knowledge. Belief posing as knowledge is pseudo-knowledge, not real knowledge.

▶ Because many are now wondering how you can tell pseudo-knowledge from knowledge, I think a short aside is in order. Throughout this trilogy, you will find discussions of how to discriminate between the wise and the foolish, between the real and the apparent, between falsehood and truth, between knowledge and pseudo-knowledge (belief). We will approach this issue from several directions over the course of Sections 2 through 6.

This is a pudding thing and a science thing. "The proof of the pudding is in the eating" implies that truth and knowledge can be evaluated by the objective results of their application. Science is the primary tool and process that enables you to avoid belief traps while assessing objective results.

Typically, it boils down to the fact that you need personal experience (knowledge must be **applied**) and measurable results (you must taste the pudding) to become a discriminating connoisseur of reality. If you are not careful, you can be deluded about the results, as well as the experience. That is why you must be a good scientist in your explorations. Being a good scientist requires only that you have the right approach and attitude – no degree or formal scientific training is required. You must wait until you have collected enough high quality evidential data before converting potential possibilities into actualities or knowledge. These potential possibilities, with their associated probability of being true, must always be reassessed and recalculated as new data come in. Apparent knowledge remains potential and tentative – truth is absolute.

Your list of potential possibilities will, for a very long time, if not always, be a much, much, longer list than your list of absolute truths. If you are careful to remain simultaneously open minded and skeptical, you will be unlikely to inadvertently make a major investment in false knowledge. On the other hand, you might pursue a hypothesis or potential possibility to a dead end – or to the conclusion that your hypothesis is wrong. That is how good science works. There is no way to guarantee that your hypotheses will be proved correct. Proving a hypothesis wrong also produces useful information.

Always remain skeptical and open minded so that you won't wander too far down too many blind alleys. It is those who abandon the open minded skepticism of the scientist, in pursuit of easy and quick answers, who end up leaping into belief traps. They lose their way and unwisely invest their time and energy moving in non-optimal, unproductive directions based on pseudo-knowledge. The truth is not delicate; it will stand up to vigorous testing.

However, you must be careful that your tests are valid. This is not as easy as it first appears. The inherent difficulty scientists have in validating concepts and results reflects a standard problem of science. By definition, it is always difficult for you to design tests to evaluate something that you do not understand. Exploring NPMR is in many ways the same as exploring PMR. Research in NPMR requires, more often than not, a long, slow, sometimes frustrating and tedious process to find tentative new knowledge. There, as with serious research anywhere, dogged perseverance, careful analysis and steady effort yields results better and more surely than any other method.

If you repeat your questions about how to separate knowledge from pseudo-knowledge while at the same time pretending that you are talking about the knowledge of PMR instead of the knowledge of NPMR, most of the same obvious answers will apply. Science is science, in both PMR and NPMR. Scientific methodology has the same difficulties and attributes in both "places." The primary difference is that in NPMR (as viewed from PMR), science is a personal or **subjective** activity with **objective** results. This contrasts with PMR (as viewed from PMR), where science is a directly shareable objective activity that likewise produces objective results. For the record, NPMR science, as viewed from NPMR, is also a directly shareable objective activity that produces objective results.

This discussion will be continued in much greater detail, and from several different perspectives, at other places within the *My Big TOE* trilogy. In this aside, I simply want to point out there is no magic formula or shortcut for finding and assessing truth. **Believing** what I say, or what anyone else says about NPMR or the nature of reality is tantamount to jumping to conclusions without doing the science or investigative work oneself.

Belief is not a shortcut that will actually take you to a significant destination. Believing what others say is a risky business. You must discover truth and knowledge for yourself or it will not be your truth or your knowledge. Your truth and knowledge lives deeply and vibrantly within your being while someone else's truth and knowledge can penetrate no deeper than your intellect.

Listening to others may greatly improve the efficiency of your journey, or send you off to wander aimlessly. In either case, you must make the trip and experience, assess, and validate the reality you find.

I do not wish to imply that you must accomplish everything by yourself. Including others on your journey to a greater understanding is generally a good idea. Interacting with others, if wisely chosen, can help you develop a broader perspective and provide much needed encouragement as well as guidance and direction. But others cannot learn and grow for you — you must do that for yourself. Wisdom, maturity, and the capacity to love are all personal attributes that dwell within the core of your being.

Some may be wondering how you go about collecting, evaluating, validating, and applying data in thought-space or how you can experience the reality of NPMR. These are good questions. I am talking about the subjective experiential data you gather while you carefully explore the realm of interactive consciousness within inner space. I am also talking about clear, objective, evidential results. At the boundary, where NPMR activities influence and modify PMR activities and vice versa, evidential data can be collected easily.

At this boundary, being aware in NPMR is somewhat like participating in a narrowly focused mental activity (somewhat like a totally absorbing interactive daydream) that has a strong causal (evidential) connection with PMR and a dynamic existence that is independent of you and your conscious or unconscious mental processes. Your mind must be calm, clear, and steady without unruly chatter or noise or you cannot differentiate between what your mind creates and what exists independently of your mind. That is why finding (and learning to exist and operate within) the calm unperturbed center of your consciousness is always the first step.

To a clear, low-noise, operative consciousness, collecting data in NPMR will be like moving through an unusually clear daydream. An exploring consciousness should initially busy itself doing experiments and looking for evidence of NPMR activity creating effects in PMR (and vice versa). One day, after enough hard evidence is gathered through personal experience, it will become clear that NPMR is both independent of you and causally connected to PMR. Realizing that fact will be your first, biggest, and most amazing discovery!

I still clearly recall my feelings of incredulous amazement when Dennis and I first began sharing identical simultaneous experiences while exploring NPMR (see Chapter 10 of this book). After consistently verifying the accuracy of remote viewing experiments and experiencing firsthand the efficacy of dramatically affecting the health of a physical body with a focused mind, my reality was forced to broaden. As the paranormal becomes normal, one has no rational or logical option other than to seek a bigger picture that contains the whole of one's carefully evaluated evidential experience.

You can always assume that **other** people are lying or confused but when your own experience, consistently and on demand, carries you to a logically and scientifically inescapable conclusion, the truth of that experience will demand a larger and deeper understanding of the reality within which you exist. Simply labeling the paranormal as something experienced by the delusional, the diabolical, the weird, and the wacky will no longer provide an easy way out of dealing with the existence of a reality that flies in the face of your deepest beliefs and assumptions.

When the experience is yours, and the processes used to gain and evaluate that experience have been careful and scientific, you must deal with the **objective** facts of what you have discovered – denying them is to cling to ignorance and limitation out of habit, insecurity, and fear. To deny such experience is to give up difficult knowledge for the comfort of a mindless dogma.

Allowing ignorance and fear to define the possibilities eliminates opportunities for growth and squanders potential. That this is true is obvious when applied to others whose ignorance and disadvantage we understand, and nearly impossible to see in ourselves.

After the collection of evidential data (NPMR-PMR psi effects) becomes routine, one generally moves on to determine the culture, laws, and physics of NPMR by careful experimentation and observation of cause and effect. Imagine that you are an alien scientist from another dimension teleported to earth to learn what the earth and its life-forms are like: exploring NPMR is like that, but without the limitation of a needy physical body. ◀

▶▶ Let's take a short break to define the word "psi." Psi is a familiar term among parapsychologists and others who are engaged in studying and exploring the capabilities of mind. Psi events generally refer to unusual artifacts of consciousness, specifically to paranormal events associated with mental abilities or altered states of consciousness. Psi is often used as a synonym for Parapsychological – thus the term "psi phenomena" refers to measurable physical phenomena that are produced by some characteristic or ability of mind that is presently beyond traditional scientific explanation. For example, telepathic communication, precognition, and remote viewing are a few commonly experienced and researched psi events. The terms "psi energy" and "psi forces" are often used to imply some unknown theoretical causal mechanism that is assumed to lie behind psi phenomena. ◀◀

▶ Because I exhort you at every turn to collect your own experience data, I suppose that I should tell you how to go about accomplishing that task. To this end, in Chapter 23 of this book, we will discuss techniques that you can use to reduce the mental noise, gain mastery and control of your mental energy, and begin exploring NPMR and the inner-space of consciousness.

However, before we get to that, it is imperative that you first discover how your beliefs limit your reality and learn to appreciate the difference between your knowledge and pseudo-knowledge. Otherwise your attempt to explore NPMR could end up exploring nothing more than your own ego.

Exploring your own ego might be a productive first step if it leads to a better understanding, and thus dissolution, of the fear and belief systems that limit your natural potential, but it is not at all the same as exploring NPMR – a reality that exists independently of your belief systems and personal mind.

There is a natural and necessary order to any developmental growth. Skipping steps almost always leads to frustration and blocks, not to rapid advancement. You must first deal successfully with your fear and limiting belief systems before you can productively explore NPMR. Conquering all fear is not required – far from it – but some minimum threshold of competence in managing your consciousness is a prerequisite.

The required basic competence is usually developed through meditation, courage, and an energetic dedication to the discovery of truth. The underlying principles and mechanics

of meditation are discussed in Chapter 23 of this book. Though there are man
meditate successfully, fear can only be overcome by courage and determination.

I do not have a special weird science that can help you cut corners and find truth in
giant leaps of intuition. Discovery and science are all about making steady progress by
taking one small tentative step after another, all the while "tasting the pudding," check-
ing the evidence, and producing **objective** measurable results. There are, as far as I
know, no "worm holes" that let one tunnel through to enlightenment, or sign posts
clearly defining the best route for each individual to take. The choice of path and the
effort applied must be the result of your own volition.

Think deeply about how the material in this chapter applies to you – though it may
get intellectually repetitive, this discussion of belief, fear, and knowledge represents a
crucial step that must be assimilated at a level much deeper than the intellect. It is as
difficult to over-emphasize the subject of fear, knowledge, and belief, as it is easy to
speed through it intellectually without digging in too deeply. The concepts themselves
are deceivingly simple and thus, repetition quickly becomes tedious. However, their
implications for each and every individual are extremely difficult to grasp fully.

What fears lurk deeply hidden that push me this way, and pull me that way, like a small
boat on an angry ocean? What beliefs limit my reality? What can I do about them? These
are difficult and complex personal questions that many egos are all too eager to sidestep.

Typically, those who most need to answer these questions are also most likely to zip
by without taking much notice. As these folks skim over the surface, they may find this
chapter to be as tedious and annoying as a parental lecture. On the other hand, those
who have little difficulty in this area will eagerly review the issues once again because
they understand the importance, are not discomforted by the process, and do not fear
the results. They too may notice the intended repetition, but will use it to improve their
mastery rather than suffer through it. ◀

Everyone understands the terms cultural belief and religious belief, but
some may be wondering what is meant by scientific belief and personal
belief. Scientific belief is the belief that the larger reality and **all truth**
must be solely defined by, and limited to, objective, repeatable-on-
demand, consistent, PMR hard-science measurements. This is the narrow
view of the scientific method cast as an exclusive PMR-only dogma.
Though this belief holds true and is wonderfully productive for a certain
subset of reality, it does not hold in a bigger picture that contains consciousness – much as classical mechanics fails in a bigger picture that contains very high velocities (which requires relativistic mechanics) and very
small sizes (which requires quantum mechanics).

Personal belief encompasses all those things you believe about yourself
and other people, places, or things – your apparent reality extrapolated

beyond your certain knowledge. Personal beliefs, if not correct, are often distorted by hope, fear, guilt, need, desire, misinformation, and misunderstanding. Personal beliefs contain many personalized versions of your cultural beliefs as well. Many are derived from the beliefs of your parents, peers, and associates.

Cultural beliefs represent those beliefs that you assume true because everybody around you assumes they are true. Racism, for example, is an expression of a cultural belief. A belief in a universal objective physical causality is a cultural belief as well as a scientific belief. Beliefs that telepathy, mental or faith healing, psychokinesis, or precognition are all totally impossible are also culturally derived.

Religious belief, on the other hand, is a belief in the creed, dogma, and articles of faith of some organized religion. Scientific belief is like religious belief. It is a belief in the creed (PMR causality rules all), dogma (only PMR exists), and articles of faith (the subjective contains no fact, only opinion) of "objective" Western science.

This discussion about belief traps is relevant to all belief – religious, scientific, personal, cultural, political, economic, and any other category you might conjure up. These categories of belief do not have distinct boundaries and, in many instances, overlap greatly. Some people associate belief primarily with religion and are unaware of the pervasive and significant role that belief plays outside the typical religious context. Neither I nor *My Big TOE* is particularly picking on religious belief. It is the characteristics and properties of belief itself, rather than any particular type of belief, that are being scrutinized in this chapter.

Sorting knowledge from belief is the function of science – both objective and subjective science. Knowing the difference between knowledge and belief relevant to any particular piece of **subjective** information is called wisdom. Knowing the difference between knowledge and belief relevant to any particular piece of **objective** information is to know the facts.

Many feel compelled to either believe or disbelieve all information they come in contact with and quickly pass a judgment on everything accordingly. Such a process leaves little room and little time for actual knowledge and shows no particular interest in truth. For these individuals, pseudo-knowledge is good enough, especially if it also happens to reduce anxiety and be widely accepted. This approach to information is unfortunate and produces a tendency to jump to conclusions based upon erroneous feel-good assumptions.

The result of understanding, appreciating, and accepting the limits of your knowledge is that you neither believe nor disbelieve much of the information that **initially** lies **beyond** your knowledge. Judgment should be suspended until **sufficient** data are collected. That method of approaching information is called open minded. The quality (rigor) of the conditions and processes that define "sufficient" is dependent upon how scientific your exploration is. Good science produces actual knowledge whereas bad science produces only pseudo-knowledge.

Knowledge, ignorance, truth, falsehood, good science, bad science, wisdom, foolishness, fact, fiction, open mindedness, and closed-mindedness almost always exist simultaneously in differing proportions as they pertain to developing (as in creating this trilogy) or evaluating (what you think of this trilogy) any piece or set of information. Rarely are knowledge and science perfect and pure. It is more a matter of degree and proportion. Perhaps all public thoughts, ideas and published papers need to be clearly marked by the Federal Knowledge and Belief Administration: "This concept contains 80% knowledge, 20% belief." An amusing thought with frightening overtones. Obviously, the only valid assessment is yours and you must make it as correctly as possible – the quality of your mind and being hangs in the balance.

You should **not** depend on experts, professionals, or anyone else to distinguish knowledge from belief for you – even if you trust them more than you trust yourself and are **willing to believe** what they say. Conversely, you can only discriminate between belief and knowledge for yourself, not for others. Think about it: How do you discriminate actual truth tellers from those who only believe they are telling the truth? Hint: Comparing their beliefs to your own is not the answer.

You should take responsibility for separating belief from knowledge for yourself (and only for yourself) because you will reap the rewards of being correct or suffer the consequences of being wrong. Herd instincts – going along with others who are themselves simply going along with others – are counterproductive. There is no safety in numbers with regard to discovering Big Truth. Failing with the majority provides no consolation because all successes or failures are personal. No one can drag you along to success by thinking or experiencing for you. On the other hand, you may allow others to retard your progress by not thinking for yourself.

It is true that to trust and assume the truth is often necessary at a mundane level and can be a useful shortcut in a world of ideas where we are time and experience limited. Nevertheless, you must be careful not to inadvertently absorb limitations on your mind's ability to expand or modify what you initially trust to be the truth. Be forever watchful for, and open to, new data. Do not block out or creatively reinterpret information that conflicts with your beliefs or what you desire or need the truth to be. Good science starts with honesty, and honesty is most easily applied in an ego-free and fear-free environment.

Belief is created when one who lacks scientifically evaluated knowledge puts faith in the premise that things actually are as he or she supposes them to be. Dogma is a fixed set of beliefs that must be accepted on faith in order to join the ranks of the believers who share that particular dogma. Dogma can be cultural, religious, scientific, or personal; it is an integral part of any category of belief. Knowledge that **appears** to be scientifically or objectively evaluated (to a given individual) may actually be incorrect. This is because we are not omniscient (our knowledge and data are limited), and because we each create our personal reality (objective and subjective) by **interpreting** our experience.

Belief and knowledge can be either false (incorrect) or true (correct). Both (in either state of correctness) can strongly motivate action. If you are presented with new information, new ideas, or new concepts that you think **may** possibly have merit, it is far better to maintain open minded skepticism while collecting your data on the subject (even if it takes a lifetime) than to jump to conclusions based on some previously held belief or by adding a new belief. Hold on to **all** the possibilities, old and new, until you have produced the knowledge that **correctly** evaluates the issues by means of direct experience. The important thing is: You need to get out there and collect the data. Laziness or fear of incompetence on this issue produces high-risk results and dramatically reduces the possibility of significant gain or progress.

The proof of correctness of any piece of knowledge lies only in the results its **application** produces. That is true of any knowledge (objective or subjective) offered up from any source about anything – including any astute cerebral gems that you may find in this Big TOE trilogy. If knowledge cannot be applied, or its application produces no practical results, that knowledge is, by definition, useless and irrelevant.

Before drawing your sword of truth and hacking away at pseudo-knowledge, let me remind you of something. If you cannot productively apply a particular piece of knowledge or a new concept, then that knowledge or concept **may** be pseudo-knowledge **or** you may be ignorant and basing your evaluation of that knowledge or concept upon belief or pseudo-knowledge. For advice on how to deal with this logical dilemma, re-read the aside at the beginning of this chapter.

Results can be objective, subjective, complex, obvious, abstract, or concrete but they must be real actual results – they must eventually produce

objectively measurable effects or changes by interacting with something that is real. **Knowledge is only as significant as its effects**. The proof of the pudding is in the eating – you will hear more about how to apply this results-oriented truth-testing concept in Sections 3 and 5. For now, it is enough to understand that "tasting the pudding" refers to testing the value of your experience, truth, or knowledge by evaluating the objective measurable results it produces. If what you consider to be truth cannot honestly produce objective measurable results, then remove it from the truth bin and put it back into the interesting possibilities bin. Continue collecting pertinent data and always maintain high scientific standards when evaluating results.

If the results are not clear and obvious to yourself and others, you are either shooting blanks or playing with a toy gun. You should always keep in mind that results need to be measurable and meaningful. Here, "meaningful" includes advancing your personal development, increasing the quality of your evolving consciousness, and improving the correctness and depth of your understanding.

Most of us are thoroughly dominated by beliefs, most of which lie outside of our intellectual awareness. How should we go about reexamining our beliefs? The proof of correctness of any belief lies in first removing the ignorance that necessitated the belief in the first place and replacing it with knowledge or open minded skepticism. Whenever sufficient knowledge has been accumulated to support logical scientific conclusions, apply that knowledge and observe the result. If the ignorance that defines the belief and upon which the necessity for the belief is based cannot be replaced by testable (or, using our pudding metaphor, tasteable) knowledge, then the belief simply remains a belief and its falseness and correctness remain unproven. No intelligent comment can be made either way and you should remain skeptical as well as open minded until enough data are collected to provide testable knowledge.

Beyond the edge of your knowledge and the outer boundary of your 3D PMR understanding lies your personal unknown. Some of the potential knowledge that remains unknown to you may appear to be beyond the theoretical reach of your knowing (mystical), and some may appear to be only a lack of information. In either case, you can leave the unknown alone, ignore it and accept it as the forever unknown, or you can probe it and explore it with the intent of eventually converting at least some of it into knowledge. Most people do a little bit of both, typically choosing a small part of the comfortable unknown to explore using objectivity and belief as tools and ignoring the rest. They pick the low hanging fruit and convince themselves that anything that isn't relatively easy to reach isn't worth the effort or the social risk.

Using subjective experience tightly coupled with objective results is not often explored because it is more difficult and because it is an individual rather than group experience. Those who cannot take a step without the reassurance of others are frightened away. These individuals erroneously believe that what remains naturally mystical to them is forever beyond their reach. The companion belief is that what remains naturally mystical **to them** is forever beyond **anyone's** reach. This soothing belief is manufactured to absolve themselves of the responsibility of mustering the courage and making the commitment to do the hard work required to turn the unknown into knowledge. Because of its absolving nature of this belief, it is held most passionately and is a great closer of minds.

Another serious error is to intellectually and emotionally deal with that naturally existing (what lies beyond your knowing) mysticism by assuming a dogmatic set of beliefs. Dogma and belief are straitjackets on the mind, blindfolds to the awareness, limitations on the thoughts you are capable of thinking and the understanding you can obtain. Dogma creates a small, usually incomplete and distorted perspective that cannot be expanded beyond the confines of the belief.

Let us look at disbelief for a moment. Disbelief typically represents a negative reaction to a competitive belief. Whether you **believe** something is true, or **believe** it is not true, you are using belief to smooth the discomfort of ignorance. Whether believing or disbelieving, the ignorance is shared. It is only the jumped to conclusions that are different. Most vocal non-believers and closed-minded skeptics are as wrapped up in their beliefs as those they ridicule or disagree with. The main difference is that they are more apt to deny that their beliefs are beliefs.

The more you are **committed** to your belief, the more that belief appears to be knowledge and represents absolute truth. Pseudo-knowledge can be passionately held if it meets a powerful need or stems from a powerful fear. The relationship linking need, the discomfort of ignorance and fear, and the faux salve of ego is established in Section 3.

The statement, "I know that what you believe is untrue" most often expresses a conflict of differing beliefs and pseudo-knowledge, not a statement of truth. How can you tell which is which? Here is how most people tell. If **someone else** is saying those words ("I know that what you believe is untrue") to us, **they** obviously are suffering from delusionary pseudo-knowledge. On the other hand, if **we** are saying those words to someone else, it is again **they** who are obviously confusing pseudo-knowledge with knowledge. This is an easy rule to remember: If others agree with you, they possess real knowledge but if they disagree with you, they are afflicted with delusion and pseudo-knowledge.

If you want to stay in the mainstream and play it safe in the center of the well-beaten path, applying the above rule is the standard technique for discriminating real knowledge from pseudo-knowledge. What rule could be more simple or satisfying to apply? Other people are always the idiots!

The fact is, disagreement on Big Picture concepts is most often the result of a conflict of beliefs regardless of who is talking, pretending to listen, agreeing, or disagreeing. If you are to avoid jumping to conclusions in the absence of knowledge, you must maintain the state of open minded skepticism – there is no other reasonable or logical alternative. Anything else is a trap.

By now, you may be wondering if there is such a thing as good belief. I can best answer that question with another question. Is there such a thing as good ignorance – is there any situation where ignorance is better, more valuable, than knowledge? If there is, then wherever and whenever ignorance is best, that is where you will find a good belief. In the short-term and in the little picture you might find some advantages to ignorance in a few special cases. Ignorance is perhaps not so bad if the problem is of little significance and of minimal importance, or one you can do nothing about. If you are trying to trick, use, or manipulate others to your advantage, **their** ignorance is always very helpful.

In the long run and in the Big Picture, if you are not trying to manipulate others and your ego is small, ignorance has little to no value. If the issues are significant, the stakes high, or the outcome important to you, then ignorance and belief will leave you vulnerable and looking like an ostrich with its head in the sand. In substantive matters of long-term significance, there is no good belief.

The main use and function of belief or pseudo-knowledge is to deny the existence of ignorance, sugarcoat fear, and to manipulate others. Knowledge, on the other hand, provides you with the opportunity to optimize your given potential in any situation. A head in the sand may make you feel better in the near term, but it prevents you from going anywhere actually helpful or productive, and it lets your you-know-what stick out unprotected.

If what you happen to believe is Big Truth, you will be saved by the good luck of being born into the **correct culture**. A correct culture would necessarily, by definition, be composed almost entirely of impeccably wise individuals of stellar quality. Does that description resemble the culture in which you are immersed?

Because the quality of your being expresses the correctness of your understanding, it is easy to determine if you and the members of your culture or sub-culture (including those who share your religion, profession, association, gang, or neighborhood) are enlightened. Simply taste the pudding – look at the people around you. Look at the **average** people in **your** culture and look at yourself. If you primarily see goodness, wisdom, wholeness, and love everywhere, then your belief system needs no further adjustments and you are spared growing the quality of your consciousness in order to **appear** grown.

If that is, by some unfortunate circumstance, not your situation, or if you are more interested in actually being grown than in appearing to be grown, then temporarily suspend any limiting beliefs (that is to say **all** beliefs) at least long enough to ponder a few big thoughts. If you succeed, you will have greatly raised the probability that you will figure out how to improve the quality of your being.

Don't worry; these unusual concepts cannot stretch your mind beyond its elastic limit. Your mind has an almost unlimited capacity to take in, as well as shut out, new information and new relationships between pieces of information.

Some individuals believe that their belief systems are perfect – that their only problem is an imperfect implementation of those beliefs. No way! You are who you are. You absolutely reflect your **actual** beliefs completely and accurately. The quality of your being necessarily reflects the quality, the correctness, of your beliefs and understanding. Perhaps you don't know what your actual beliefs are (the real ones, not the intellectual ones you talk about). That is normal enough. Cultural, religious, scientific and personal beliefs can be extremely subtle and are often invisible to the individuals and to the members of the group that share them.

Religious, personal, and cultural truths are typically so ingrained and so obvious that they appear to define reality itself and thus are never called into question. Therein lies a major limitation of belief – when you believe that you have the right answers, there is no need to continue to seek truth or ask questions.

▶ People who continually question the obvious truth are annoying to those of us who know the answers. If we could only find an effective technique for reeducating the problem people who don't understand the **real** truth as we do, the planet would be a much better and safer place for everyone and everything. Gentle and kind terminations of the blatantly uneducable would clearly be justified and would go a long way toward making our world a better place for our children and future generations. We could ensure a continuing bright future for all by finally and effectively neutralizing the most undesirable and negative elements that are the root cause of all the trouble. God is counting on us to manifest his will. We will be the heroes of all future generations! Are you with me comrade?

If you found this book laying on top of the john in a public restroom or abandoned in an empty subway seat and opened it to this page, the previous paragraph was meant to be sarcastic. I have attempted to use a little generic religio-politico-historical humor to make a deadly serious point about the siren song sung to one's ego and fears (the bait), and the uncompromising iron jaws (peer pressure) of the belief trap – and the debilitating effect it can have on the common sense of **someone else.**

If **ego** (I must be right, my needs, opinions and beliefs define the truth), **fear** (of the unknown, being wrong, disapproval, imperfection, failure, God, or of the unholy enemy), and **peer pressure** (this is the way everybody else thinks, therefore it must be right, or at least safe) have influence or veto power over what thoughts you can honestly and seriously entertain, then you are caught in one or more belief traps – even if you don't want to terminate those degenerates who are screwing it up for the rest of us.

On the other hand, if one actually thinks it is a good idea to terminate the unredeemable degenerates among us, then such an individual is not only caught in a belief trap, but is potentially dangerous as well. Violent or forceful interdiction as a solution to a problem almost always produces the opposite of the effect intended; it usually makes the original problem much worse while greatly reducing the credibility of the forceful individual's viewpoint. ◀

Unfortunately, the wisdom and intended meaning of the ancient sages necessarily seem obscure from the viewpoint of those who share neither their culture nor their experience. Additionally, such wisdom and meaning are easily lost and twisted by the belief systems that others quickly establish around these individuals in order to express their ideas at the lowest and widest level of understanding. Furthermore, the self-serving concept of "holier than thou" often dilutes the significance of such knowledge further as a movement or ideology forms to codify and extend what is essentially an individual quality of understanding and being to a more marketable group-certification.

No group, regardless of how small or large, can possibly create and bestow experience-based understanding, integrity, and personal growth (the basis of wisdom) upon an individual. The individual must accomplish that. However, there are some things that groups and organizations **can** create and bestow – power, influence, wealth, and prestige come immediately to mind. These attributes, delegated primarily to the group's leadership, are created by recruiting and maintaining large numbers of members or supporters. The group's members find mutual support, approval, status, political power, and security.

The power of numbers is so compelling that groups spring up and are organized around every conceivable interest or idea that can support a viable membership. Large groups, movements, and organizations – from science to religion to professional societies to politics – often end up being about ego, power, money, prestige, and influence, regardless of what their original intentions were. Guilt, fear, intimidation, tradition, security, acceptance, identity, shared values, socialization, and acculturation become the tools of choice to grow, maintain, and strengthen the organization and its power.

In contrast, it is the **personal** science, philosophy, and quality of the **individual** that must supply the fire at the creative core of human existence. Only the individual can bring content, direction, quality, and value to the power of numbers. Though *My Big TOE* is about science, philosophy, and the general organization and mechanics of reality, it is simultaneously about you – the individual. You are a vitally important element of the Big Picture because your individual consciousness plays a key role at the core of reality.

20

■■■

Causality in Every Dimension can Potentially Transform a Mystic into a Scientist from Inner Space

■■■

Scientists often believe that everything must have an objective cause. This, as it turns out, is not a fair or reasonable expectation. It stretches the concept of our PMR causality beyond the bounded intellectual or logical region to which it applies. A more limited statement is: Everything that we can objectively understand from our PMR base of knowledge must have a cause. We can clearly agree with that one, fully understanding the limitations implied by "objectively understand from our PMR base of knowledge" which is based upon measurements made exclusively within PMR. From this viewpoint, it is only those things that we can assess from our limited PMR perspective that logically must have accessible causes. What is beyond PMR may seem mystical to us and can, within the causality of its own dimension, logically violate our 3D objective causality resulting in **measurable** PMR effects that we often label as paranormal. Paranormal essentially means acausal – beyond the normal cause and effect relationships defined by our limited PMR physical science. Once the limitations of PMR science are surmounted, what was once defined as paranormal becomes a normal part of a larger scientific understanding that answers to a higher (more general) level of causality.

▶ Does the concept of "beyond PMR" seem strange, unscientific, and reek of non-provable goofiness? If it does, you are probably in the majority. The assumption that nothing exists beyond PMR is a normal, self-fulfilling, self-perpetuating, illogical belief. I intend to examine this belief thoroughly over the next four sections of *My Big TOE* and provide a rational alternative that more fully, accurately, and consistently explains the available measured data. The unfolding of something as unusual and complex as this TOE must necessarily be slow and methodical – for this reason it may be a while yet before you can begin to see the Big Picture come into focus. If you can maintain an attitude of open minded skepticism until the end of Section 6, you will be in an excellent position to apply your own personal data and specific knowledge to verify the value of this model and develop accurate conclusions. Unfortunately, the paradigm busting and rebuilding process must necessarily introduce concepts that seem dubious and are initially incredibly difficult to fathom – it can appear no other way. ◀

Normal events and interactions within NPMR must take place within the constraints of a uniform causality. There is well-defined action and reaction – similar processes must consistently produce similar results for all experimenters. The major difference between the causality that is local to (and defines science in) NPMR and the causality that is local to (and defines science in) PMR is that within NPMR the range of possible causes is far less restricted. PMR and its causality is a subset of NPMR and its causality. The rules that govern NPMR physics and the interactions between NPMR beings are of a higher order (more general, less restrictive). Thus, NPMR can interact with PMR in ways that violate PMR's causality (such an interaction may produce paranormal activity from the viewpoint of PMR), yet maintain NPMR's own causality. Stepping up a level, beyond-NPMR also has its own unique causality and answers to a yet higher order of less restrictive rules. Similarly, beyond-NPMR can interact with NPMR in ways that violate NPMR's causality, but maintain beyond-NPMR's causality. And so on and so forth as each larger dimension of existence supports, and is a super-set of, the next one down.

▶ Eventually we will come to understand that whether a reality appears to be physical or nonphysical is relative to the observer. The property of being physical or nonphysical is simply the result of one's perspective and has no real significance of its own. For the time being, the concepts of PMR and NPMR provide a useful conceptualization of the larger reality from the perspective of a PMR resident who has experienced no other reality save the physical one in which he or she is now reading this book. ◀

From our viewpoint, PMR appears to be the final downhill stop for this inter-causal reality train (unless one counts the fictional *Flatland* as the next dimensional stop below us). The book *Flatland*, by E. A. Abbot, provides a wonderful understanding of the scientific, philosophic, and social difficulties involved in perceiving higher dimensions. Anyone can easily understand the limitations of the dimensions that exist **below** their normal perspective; at the same time, looking upstream reveals nothing but mystical confusion. Though *Flatland* deals only with geometric or spatial dimensions, the difficulties encountered in perceiving and understanding a dimensionality that is different from one's native perceptual construct are much the same.

▶ The second revised edition of the book *Flatland* was published in 1884 by E. A. Abbott and is currently available from Princeton University Press. The book describes, in a light-hearted and humorous manner, the fundamental technical, epistemological, social, and political difficulty in expanding your awareness of reality beyond the dimensionality of your physical perceptions. If you have not yet read this book, I strongly urge you to do so. It will help you understand how the apparent logic of your reality and the analytic quality of your thinking process is limited by the dimensionality you **believe** you live in – and, it is a hoot. *Flatland,* in its entirety, can be accessed on line at: **http://www.geom.umn.edu/~banchoff/ Flatland.** ◀

Perhaps beyond-NPMR is the outermost layer, or perhaps beyond-beyond-NPMR is outermost. I will describe and discuss both in great detail later, as well as explain what dimensionality actually is and how it is generated. Hold on to these thoughts. We will pick this discussion back up and continue to peel the reality onion after we have more thoroughly developed the conceptual foundation required to support the construction of a Big TOE.

Though I have not yet explained the origins and nature of dimensionality, it is not too early to discuss a few of its properties relative to causal hierarchies or reality subsystems. We see that beginnings belong to, and are governed by, the rules of causality of the next higher dimension. Each dimension of existence births and nurtures the child dimensions it spawns. A child can (but is not required to) become a parent. One parent can birth many children. Each child exists within its own dimension. Dimensionality is like your family tree, it has the property of breadth as well as depth. However, in this discussion we are only looking at depth – the creational hierarchy. From the perspective of the child, its birth (beginning) must appear mystical. To the parent, the process and circumstances of the child's birth are well understood and not the slightest bit mystical.

From the viewpoint of the child's own local objective causal system, the child's reality logically requires a mystical beginning. In other words, any system of objective causality is insulated from other causal systems by the

local logic through which it defines itself. Reality subsystems, each with their own local causality, can be likened to the software components and subroutines of a large complex simulation – all run interdependently within the same computer as long as they have self-consistent rule-sets to define their internal and external interactions. There may be relationships and interactions between causal systems, but comprehension and understanding normally flows in only one direction – from the superset to the subset. The subset does not have what it takes to understand the superset. To understand the superset, one must first become a member of it.

If you have read *Flatland*, it will be clear that the ordinary residents of a given reality can only observe and understand interactions within their own reality and the interactions of residents of realities that are more highly constrained than their own. Residents of a more constrained reality cannot comprehend a less constrained reality because it lies beyond the limits of their normal perception.

Each dimension of reality has its own rules that define its objective science. Additionally, each dimension of reality experiences the next higher (less limited) dimension as subjective and mystical. Consequently, your mysticism may be another's science: It depends on how big a picture you live and work in, and the degree to which constraints limit your perception. The perspective from the next higher dimension provides a bigger picture with a more complete understanding. This more comprehensive, complete, and less restrictive knowledge is only accessible to lower dimensional beings (those with a more constrained awareness) through the experience of their individual locally-subjective mind.

Consequently, a mystic could be a scientist from a higher dimension, or a delusional fool hopelessly caught in a distorted web of belief. How do you know which is which? A good question! We will go through the differentiating process in great detail in Section 3 (especially Chapter 48, Book 2). First, read *Flatland* to help you appreciate the problem of understanding higher dimensions. Second, carefully and scientifically gather your experience as you progress, step by step, along your path toward increasing the quality and capability of your mind, consciousness, or being. Then simply taste the pudding to separate the wise from the foolish. If you can't tell a high quality consciousness that is wise and loving from one that is not (you have uneducated taste buds and cannot correctly interpret your experience), repeat step two as often as necessary. To some extent, it takes one to know one, and you may need to develop (evolve) your consciousness before you get good at discrimination.

The notion of local realities within separate dimensions and of a hierarchy of dimensional existences is probably a difficult concept to grasp. Have patience – the seed has been planted and later we will learn where these dimensions come from, what they mean, how they are created, and what love, wisdom, and physics could possibly have to do with any of it.

21

■ ■ ■

Cultural Bias

■ ■ ■

Objective causality is the fundamental philosophical underpinning of PMR science. It has been extremely useful to us in understanding and manipulating the material realm. Unfortunately, we PMR beings of limited comprehension have become so committed to our belief in a physical objective causality that we force everything into the PMR causality straightjacket. Why do I love you? Why do I enjoy music? Why do fractal images look like natural landscapes? Why am I obsessed with frogs? There **must** be some good reason that falls within the PMR causality model. Even if there is no cause within PMR (my feelings and behavior simply erupted spontaneously, mystically, or from some interaction or association with a larger reality that constitutes a superset of PMR), reasons will be hypothesized and rationalized in order to make our PMR causal model appear inviolate (no doubt some unknown neurological or psychological function or dysfunction explains all but the fractals). Invoking the unknown to serve as a logical explanation for some difficult to understand event is not logical or even particularly rational in most circumstances.

▶ We routinely adjust our interpretation of events and our scientific theories to satisfy the dogmatic requirements of our beliefs. Theories that violate our cultural and scientific beliefs are preposterous by definition and are not taken seriously by the majority of scientists. Our beliefs set the boundaries and define the limits of our science – they always have and any reasonably accurate history of science will verify that fact. Most scientists, from pre-history to the present day, feel that though belief obviously blinded their forebears, it does not seriously inhibit their own clear vision. As time passes, the belief-blindness of those who came before appears more and more ridiculous yet current belief-blindness remains as invisible as ever. If you think that we of the modern world – we who have come so far in our understanding and knowledge – are no longer seriously and dramatically limited by our beliefs, you are mistaken.

Major conceptual breakthroughs in science and philosophy must always lie outside the solution space defined by what is generally accepted. If you wish to leap ahead, be prepared to transcend your present notions of reality and possibility and to rip old limiting beliefs and paradigms up by the roots.

Thinking that you can effectively live and work in the middle ground between bold leaps and dogmatic limitations is no more than a comforting delusion. **To get out of the box, you first must step over its edge** – an act too frightening and intimidating for most box dwellers who will always find plenty of good reasons why it is actually better to stay safely in the box. It is a mistake to let the fear of going from the frying pan into the fire prevent you from ever getting out of the frying pan. Open minded skepticism, careful science, and a willingness to work and learn can enable you to get out of the box (or frying pan) without getting hurt, burnt, or deluded. ◀

We **believe** there are always objective causes for every effect and every event whether we know what they are or not. Determining what those causes are and discovering their rules is what we call science. Given an effect, if we do not perceive an objective cause we believe that our science is simply incomplete. Our belief in the supremacy of our local causality will not allow us to consider there might not be an objective **local** cause – that the effect may have at least one component that lies beyond PMR objectivity. Such an effect would be called paranormal and would appear mystical when viewed from a PMR perspective. This possibility is immediately rejected because it conflicts with our cultural and scientific **beliefs** about reality.

The Western commitment to the universality of our **local** objective causality is a dogmatic (nonnegotiable belief) attitude which is culturally ingrained at a deep level. As we have seen (Chapter 18 of this book), this belief **requires** our reality system to have a mystical beginning. At the same time, the Western cultural and scientific belief in universal objective causality condemns every effort to investigate that mystical beginning as irrational, illogical, and superstitious – an objective Catch 22. Science simultaneously logically demands and rationally denies a mystical beginning.

The problem is our belief that objective causality is universal (applies to all reality) instead of just local to PMR. When one sees the bigger picture and realizes that PMR is a subset of a larger reality, the logical and operational difficulty of our beginnings appearing to be mystical immediately disappears. Now our beginning is simply the result of a more general causality working within the rules of its own science – better yet, it is amenable to our analysis and open to our understanding **if** we can gain the perspective of that more general causality. Ahhh ha! A solution and a plan to effect that solution begins to emerge from the logical possibilities.

In other less technically focused cultures, what appears from the PMR perspective as mystical is not necessarily associated with, or defined as irrational, illogical, or unscientific. However, to most Western ears the phrase, "Assume the existence of an apparently infinite absolute unbounded oneness" sounds less credible than the phrase "assume the existence of a spherical chicken."

We have shown it to be logical that if there is such a thing as higher, more correct and complete knowledge that reflects the science of the "place" of our beginning or beyond, then it must necessarily appear mystical to us. We have also shown that such knowledge is only available to us through the expansion of our perspective into the next higher dimension of existence, where our origins are ordinary, mundane and well understood. Nevertheless, a material-based Western culture steadfastly labels mystical (from the PMR viewpoint) thought and experience as unsubstantiated useless blather that is beneath serious consideration because it cannot be understood within the purview of our limited (applies to PMR only) scientific method.

It is a goal of this Big TOE to take what appears to be mystical and beyond knowing, as seen from the PMR-only viewpoint, and, through the use of impeccable logic applied to two reasonable assumptions, turn it into hard science in broad daylight under your watchful and properly skeptical gaze.

A more general science is believed not to exist because it cannot be derived from a portion of science that limits itself exclusively to local, physical, objective phenomena. Do you see the logical inconsistency of this cultural belief? Is it clear that the self-referential circular argument that is primarily responsible for closing twentieth century minds to the possibility of a bigger picture is simply the result of being caught in a belief trap?

Such belief-based, circular, non-logic posing as obvious truth within Western cultures severely blinds and restricts the growth options of those who are caught in that particular trap. The only remaining logical possibility is that although there must have been a mystical beginning, now for some unknown reason, the substance and intent (force) behind that mystical event has disappeared leaving nothing else to exist beyond the local objective measurable reality. Do you find this a plausible, **objective** explanation, or does it seem more like limited thinking desperately trying to justify its limits? A logical possibility perhaps, but it leads to an irrational conclusion.

The intent and the implementing power behind our seemingly mystical origins must represent a source more capable and powerful and more fundamental in its existence in order to give birth to our local reality. Our parent reality must necessarily be operational at a higher (more general) level of existence or dimension. It must necessarily represent a superset to which our local physical reality belongs. Assuming that this creative source has somehow disappeared is like the ice-cubes in my automatic ice-maker bucket believing that the compressor responsible for their freezing must have stopped working years ago. Those delusional ice-cubes obviously do not understand the bigger picture.

Does it seem likely that this higher-level creative force, for unknown reasons, just dried up and blew away, leaving us to exist alone like deserted orphans? That premise assumes we could exist independently from our initiating source. As it turns out, our source is both initiating and sustaining – we cannot exist independently from it any more than our internal organs can exist independently from our bodies. Does this assumption (the source of our mystical beginning no longer exists) appear to be the result of scientific analysis – or does it seem more like one of those mystical beliefs that are considered credible because they support the accepted individual, cultural, and scientific dogma? Popular pseudo-wisdom says if we (our individual selves, our culture, and our science), with our impressive understanding and knowledge do not understand it, cannot clearly grasp it much less measure it, then it must not and cannot exist. Does this appear to you to be a scientific conclusion, or the expression of a little picture belief?

Coming to the logical and rational conclusion that a larger reality within a bigger picture could **possibly** exist beyond the confines of our present physical reality defines what I have called open minded skepticism. Simply allowing this possibility (regardless of how remote you might **believe** it to be) and having the gumption and commitment to explore that possibility honestly and scientifically is all that is necessary to grow your Big TOE – a **personal** Big TOE that has the ability to accelerate the evolution of your consciousness.

I have mentioned the terms personal growth, consciousness evolution, and improving the quality of consciousness several times without defining what they mean. These presently vague terms will be precisely defined after we have more fully developed the conceptual basis required to support their meaning.

Should you expect our collective mystical (only from the view of PMR) origin, which is necessarily initiated and sustained from a higher level of organization and a higher dimensional existence, to be obvious, easy to understand, and just like us? Do our machines, computers, pets, designer viruses, intestinal bacteria, and internal organs have a difficult time understanding human experience and motivation within the context of their existence? They cannot begin to comprehend anything but a shallow one-dimensional sense of us. The knowledge, understanding, and intents that animate our actions and fuel our seemingly awesome power are unfathomable to them.

If the understanding of the larger reality that contains our beginning is so difficult and beyond the tools of our objective science, is it any wonder that many people, having seen a fleeting glimpse of Big Truth (derived from their own experience or, more commonly, delivered to them by others), have anthropomorphized all manner of beliefs and gods to fill the void created by ignorance and fear?

> ▶ Deep ignorance and deep fear produce a long and varied list. Sun gods, Tree gods, Moon gods, Fire gods, River gods, War gods, Fertility gods, Tribal gods, Ocean gods, Storm gods, Animal gods, even Booze, Sex, and Party gods (Bacchus) – to mention just a few of the probably thousands of gods people have conjured up for their own needs, in their own image, or in the image of their fear. What else would account for the myriad of false gods that other people believe in?
>
> Do you realize that people from every religion of the world will agree with the preceding sentence? Wow! Question: Does unanimous agreement from such a contentious group constitute a miracle? ◀

Left with a total unknowing of something so fundamental and important as the circumstances of their beginning, the nature of their reality and purpose, we can forgive other people for anthropomorphically projecting what they did know into a plausible (to them, at the time) answer. That is a typically human, if not rational, response. Unfortunately, it also sets the stage for much mischief, agony, guilt, intolerance, fear, confusion, and violence.

Undeniably, a little knowledge can be a dangerous thing. That is particularly true of knowledge gained from others when the recipients do not have the personal quality to have derived that knowledge on their own, thus guaranteeing that misunderstandings will occur and that it will be impossible for the recipients to make distinctions between knowledge and pseudo-knowledge. In matters of Big Truth, you cannot teach, much less force, someone to get it.

As a teacher, it is better to wait until your students are ready than to push Big Truth into an apparent position of mystical misconception within their minds. As a student, it is better to wait until you are ready (have grown up enough) to understand Big Truth at a profound level than to leap headlong into a belief trap, thinking that you have taken a short-cut to knowledge and wisdom. You cannot access understanding and wisdom that is beyond what the quality of your consciousness can naturally support. Every individuated unit of consciousness must develop in its own unique way, powered by the free will that drives its intent.

Given an **important** question in any dear-to-the-heart subject, it seems that humankind (this is true for individuals as well as groups and cultures) vastly prefers any plausible (at the time) answer, even if it is likely to be wrong, to no answer at all. When knowing seems to be important, the only thing worse than a wrong answer is no answer. It is far easier and more rewarding in the short-term to calm anxiety with pseudo-knowledge than to face ignorance with open minded skepticism. Unfortunately, growth over the long-term, which is what is important, is severely stunted by an almost universal preference for the short-term feel-good solution.

If faced with no answer to an important question about almost anything, we humans tend to make up an answer that suits our emotional and intellectual needs and then believe in it with a force of conviction that is equal to the power of the original need. That is how we humans are – fearful of what we do not know or understand – ill at ease with not knowing, uncomfortable with uncertainty. This is why open minded skepticism, as an approach to learning and growth, is rarely implemented. Though open minded skepticism is obviously and logically the most correct, beneficial, and productive approach to evaluating new ideas and experiences, it does not provide the immediate closure and false confidence of a believed in conclusion – and it requires further work. Jumping to conclusions, particularly if they are widely held and therefore a socially safe short jump, is much easier and immediately more satisfying than doing the long difficult work of honest scientific research.

It is these tightly held beliefs, fantasies, and delusions of convenience that drive the day-to-day behavior (dysfunctional and functional) of most of us. There is almost nothing more important to us than our fantasies or beliefs. Beliefs appear to make life easier, less work, and happier, at least in the short run. Without them we must face our ignorance, our uncertainty, our inadequacy, and our fear – anything is better than that. The unfortunate fact is that in the long run, from the perspective of the bigger picture, beliefs and fantasies almost always have an effect that is opposite to what is intended. The process of denying a fear generally causes what is feared to manifest in your reality.

The richness, importance, and meaningfulness of our **subjective** existence directly conflicts with the notion that if you cannot measure or **physically** experience it, it is either non-existent or irrelevant. Likewise, the abundance of reputable scientifically collected data documenting paranormal happenings also flies in the face of our limited reality. Given the accepted scientific facts of wave-particle duality, paradoxical entangled particle pairs that instantaneously communicate, and statistically based material existence, modern physics itself is pushing the notion of our cherished objective reality into the subjective mind-space of the experimenter.

If you are thinking that "subjective" and "rational" are mutually exclusive concepts and wondering what reputable data or modern physics I could possibly be referring to, it might indicate you need to assess your beliefs (spot the traps), open up, look around, and get out (of the box) more often. Credible information speaking to these issues is out there by the basketful and not difficult to find.

▶ Examples of reputable data are available in *Mind-Reach – Scientists Look at Psychic Ability* by Russell Targ and Harold Puthoff, Delacorte Press, 1977 and *The Conscious Universe: The Scientific Truth of Psychic Phenomena,* by Dean I. Radin; Harper Collins, 1997.

As far as modern physics goes, an excellent **non-mathematical** description of the theories of relativity and quantum mechanics written especially for the non-scientist is *The Evolution of Physics* by Albert Einstein and Leopold Infeld, published by Simon and Schuster in 1961.

These are only a few of a large selection of books that you might use as a starting point in broadening your knowledge of the boundary between physics and metaphysics. From these books, you will learn that the material reality that you think you live in is actually much stranger than you ever imagined. You will also gain an appreciation of how little science actually knows about the fundamental characteristics and properties of reality. The one thing that most modern physicists agree on these days is that what we generally take for our local 3D time ordered causal reality is merely a perceptual illusion. Some sixty years after quantum physics destroyed the widely accepted material foundation of physical reality, what lies behind this persistent perceptual illusion remains as mysterious as ever to a traditional science trapped in the little picture by limiting beliefs. ◀

The reality paradigm is shifting under our cultural feet. East and West, North and South, are increasingly exchanging information and inextricably intermingling their cultural and philosophic values as information and communication technologies continue to link and integrate the mind-space of our planet. This is an especially propitious time to ponder these issues and to figure out what is real, productive, and non-delusional.

It will be important to maintain solidity and balance as the cultural ground shifts and shakes beneath you. It is also important to filter out the truth from the inevitable cacophony of conflicting concepts to which everyone is about to be exposed in the coming cultural implosion (a spin-off of the information-computer-networking revolution). To achieve the most efficient personal evolution and growth, you must find an optimal synthesis of the available concepts and then add to or customize this information to suit yourself. Having a correct and comprehensive Big TOE has never been more timely or important to your future growth than it is right now at the dawn of the Information Age.

That some individuals refuse to make any effort to explore subjective truth says something about those individuals and the limiting power of their belief systems. Those who seriously take up the challenge of exploring reality and growing their awareness rarely come home empty handed. They inevitably find a greater reality beyond objective PMR, and it is almost always worth far more to them than the considerable effort required to access it.

Again, the proof of the usefulness and quality of any pudding you cook up is in the tasting, evaluating, experiencing, and sharing the results of that particular batch of pudding. You can start anywhere. Be skeptical, be open minded, and demand measurable objective results after a reasonable effort. If there are no obvious measurable results, try a different recipe. Assess the results, adjust the recipe, and go make some more, slightly better, pudding. Repeat the cycle continually. Before long, you will be winning prizes at the county fair for the quality of your consciousness.

It is important to be aware of how your cultural biases and beliefs can severely limit the scope (breadth, depth, and quality) of the thoughts you are able to think, as well as the size of the picture you are able to comprehend. The struggle to reach beyond a belief-limited perspective is usually immensely difficult and only determined and serious explorers doggedly pursuing the truth down whatever path it takes are likely to be successful. Unfortunately, the belief-limited blindfold we all wear feels so natural and is so obviously correct, deeply ingrained, and widely held that we are not aware of it, and may vehemently deny it exists.

A **belief** that PMR is "all there is" is extremely limiting and makes some very important and interesting questions absolutely impossible to answer without invoking additional limiting belief systems. The argument between science and belief – a more general version of the argument between science and religion – is a self-energizing, endless loop of non-logic bouncing uselessly in endless debate back and forth from one limiting belief system to another. The illogical excesses of each create the rational necessity for the other. These arguments violate the Rule of Rationality by forming a perpetual wasted motion machine within a logical black hole! Campbell's Third Law of No Motion (otherwise known as the law of inaction-reaction) accurately describes these arguments: For every irrational rationalization there exists an equal but opposite irrational rationalization.

Certainly, most religions do not **believe** that PMR is all there is, and religion, as well as science, is a significant part of our cultural heritage. Nonetheless, the organization and codification of mysticism into various religious doctrines and dogmas is of little value. That we as a culture permit a limited and narrowly focused mysticism (various religious dogmas) to coexist and blend with our scientific dogma serves only to confuse, distort, and restrict our ability to deal with the real issue of consciousness quality.

Science and religion, each in their own way, preach the gospel of hope and promise deliverance to the Promised Land of good and plenty. However, as a general rule, neither provides a significant boost to the inner quality of an individual's life. The quality of your consciousness must grow as an independently evolving entity in the shadow of both. Consciousness quality is a personal achievement that can only be developed by an individual – it is not a group endeavor. It has absolutely nothing to do with creed, dogma, or belief. An individual's quality cannot be increased one single iota by any belief, or by accumulating information about anything, or by doing good deeds that are not properly motivated, or by talking to others or reading books.

Again, I seek your indulgence – as horses must remain in front of carts, logical progressions must take the time to develop their logic one step at a time. The concept of consciousness quality and its relationship to spiritual quality will firm up later when it is more precisely described and given a technical definition. In the meantime, let me say this: Spiritual growth, personal growth, improving the quality of your consciousness, evolving your being, increasing your capacity to love, and decreasing the entropy of your consciousness are all essentially synonymous and equivalent. Many readers have a good idea (or at least think they do) about what these terms mean, but there are some who are not at all sure, and some of those are now becoming a little worried. This is as it should be – properly skeptical minds need logical clarity. Establishing logical scientific connections that interrelate physics, spirituality, consciousness, and love is not as goofy or impossible as it appears – in fact it is something that a comprehensive Big TOE must necessarily accomplish. Hang in there with me – these ideas are more logical and rational than you might guess.

Do not misunderstand me. I am not denigrating the potential spiritual quality that can be found within religion by **individual** seekers of truth. When I use the word "religion" here, I am speaking only of institutionalized dogma or organized religion, which represents how the great majority of people are connected to religion. **There are individuals and organizations that flourish outside of this generality** and it is highly likely that you count yourself and your associations to be among them.

Is it not a simple fact that other people are usually the ones who don't get it, and that the phrase "great majority of people" usually does not include you? Do you not find it logically intriguing that the great majority of people feel rather strongly that they cannot be grouped with the great majority of people? We humans are generally as aware of our individuality as we are blind to our conformity – that is our nature. The unquestionable truth of the bold sentence above should give other people a logical loophole big enough to squeeze through in order to get back into their personal comfort zone.

There are those few who after subtracting organization, socialization, status, tradition, habit, dogma, creed, ritual, and belief from their religion still have something left over that is very significant. For these people, religion is a personal spiritual experience that enables them to evolve the quality of their consciousness as effectively as any other spiritual path. That they choose to integrate this honest spiritual experience within some traditional religious setting merely represents the individual path they have chosen – there is no **intrinsic** benefit or penalty in doing so. All paths have benefits and challenges.

Contrary to popular belief, I do **not** condemn belief and dogma as useless and harmful merely because they are illogical and unnecessary. Much of what we do every day – particularly our habitual activities – is illogical and unnecessary. Inefficiency is not a crime; if it were, we would all be in jail. Condemnation typically flows from arrogance and is not likely to be part of a helpful process nor is it likely to be a good technique for fostering understanding or improving communication. This trilogy is about being helpful, improving understanding, and reducing ego and arrogance.

Be careful not to jump to conclusions. I welcome you to walk your dogma in my neighborhood as long as you clean up after it and keep it under control. Don't let it bite, harass, or intimidate anyone. Make sure it does not dig in our gardens, kill our flowers, bushes, or children, or leave piles of poop in our yards. Finally, do not allow it to terrorize or bully the many vulnerable critters and beings that peacefully live and play in the surrounding environment. If you are a responsible owner of a friendly dogma, you and your dogma are welcome in my neighborhood anytime.

It is possible, though exceedingly unusual, for an individual to lose ego and gain consciousness quality in pursuit of a favorite dogma. For most of us, dogma erects barriers on our path to personal growth, distracts us from what is truly important, confuses our sense of what is right and wrong, arbitrarily limits our reality by snagging us in belief traps, and tends to make us more egocentric, arrogant, and self-righteous. We readily embrace dogma because it soothes our fearful ignorance with a comforting salve of easy to obtain pseudo-knowledge, and because its downside always falls outside our awareness. However, there are a few who outwardly appear to be in pursuit of dogma because of their habits of ritual and association, yet inwardly they have grown beyond its limitations. For these individuals, the dogma (along with any associated ritual) becomes a familiar pattern of doing that is similar in function to a meditation mantra.

▶ Offering either science or religion to one in dire need (as is everyone) of internal substance is like giving a starving person a rubber chicken. It looks good and he immediately feels better. Now filled with hope and confidence, he chews and chews and chews but continues to get thinner and thinner all the same. If his preoccupation with, and his belief in, the nutritional value of the rubber chicken prevents him from procuring real food, the situation becomes worse.

That some real food may be securely hidden within the rubber chicken only makes the situation more pathetic (he can smell it, but doesn't know how to get access to it). The hungrier the man gets, the more he becomes fixated on the rubber chicken, and the less capable he is of eventually figuring out the puzzle. He eventually dies of starvation, forever grateful to the Great Benevolence that provided him the precious gift of hope in the form of a wonderful rubber chicken. ◀

We are, it seems, like the citizens of Flatland (two-dimensional imaginary beings) who can not understand the connection between solid geometry and their stomachs. They had to go inside (through) themselves to begin, what was for them, a mystical and metaphysical journey toward an understanding of the third dimension. We need to make a similar journey to understand the larger reality. They had to transcend their beliefs which were born out of their objective experience within their local reality and limited causal system. We need to do the same. They found the reality of the third dimension through subjective experience combined with careful scientific reasoning – not by dedication to dogma (old or new, religious, cultural, scientific, or personal). We need to do the same.

Some things cannot be comprehended from, or conversely, translated to, the perspective with which we beings, seemingly trapped in this Physical Matter Reality, have to work. If this journey to understand reality appears mystical and metaphysical to us, that is an artifact of our perceptual limitations and small space-time perspective, not a condemnation of the realness of our vision. Solid geometry and the third dimension are real even though they exist beyond the comprehension of belief-limited Flatlanders stuck in their local objective causality.

Our ignorance does not impose limits on the larger reality – only upon our understanding of it.

22

■ ■ ■

The Right Attitude

■ ■ ■

Significant personal benefit can be gained simply by developing an explorer's attitude. You must be courageous and open minded enough to contemplate the unknown and then step into it (experience it) to find out for yourself. There is no other way. Your experience, your time, your effort, and your mind compose your only doorway to an understanding that is not belief-based. A belief-based understanding is only slightly better than no understanding at all if, and only if, what you believe turns out serendipitously to be true. Do you think the correct belief system can get you to the finish line without running the race? Unfortunately, it cannot – even if your belief turns out to be true.

Some people believe that they do not have the time, energy, or ability, to gain consciousness quality on their own. They think that if they find the right religion, organization, book, teacher, guru, or advisor, they can minimize the effort required to develop their personal experience of Big Truth because the teacher will explain what is true and they will simply believe it. Do you think this strategy will work? No, it cannot! You cannot believe your way into consciousness quality any more than you can believe your way into being a master violinist, sumo wrestler, or the president of your country. There is yet another problem. You obviously must choose whom to believe very wisely. How can you do that without great wisdom of your own? Though knowledge can be passed from person to person, wisdom is derived only through your personal experience and is non-transferable from others.

You need to **be wise** to choose the belief system that can make you **appear to be wise** so that you do not have to earn wisdom through experience and actually **become** wise. There are no shortcuts. You must develop the quality of your being through your personal experience. Attaining wisdom, choosing paths to spiritual growth, improving the quality of consciousness, discerning pseudo-knowledge from actual knowledge and discriminating good teachers from bad teachers will all be discussed in more depth in Chapters 47 and 48 of Book 2.

Your beliefs (cultural, religious, personal, or scientific) are for the most part not relevant to the quality of your consciousness, except that they may retard it by limiting what your mind can think. Evolution is not a matter of passing an exam. It is a matter of how you are, the quality of your being, not what you believe. Beliefs don't often, if ever, translate into quality of being; they are only about using pseudo-knowledge to fill in for unavailable (or difficult to obtain) real knowledge. You can talk **about it**, know **about it**, and believe anything you want to **about it** for free, but you can't **be it** without **paying** the price (extensive diligent experience generated from rigorous dedicated effort). That is true of a pro football player, a brain surgeon, a nuclear physicist, a master carpenter, a sumo wrestler, or a concert violinist – as well as a spiritually evolved being. No pain, no gain.

There is no free lunch. You either pay the price or forgo the benefits. Because the benefit is the growth and evolution of your consciousness, forgoing it would not be a wise choice. Why? Who cares? What is the cost of an opportunity lost? We will discuss that later. Let me say for now that the costs are severe and once incurred cannot be sidestepped – but its effects **are** reversible. The costs should not be construed as a punishment; they are merely the logical result of not evolving.

For those who are wondering what evolution, belief traps, and consciousness quality could possibly have to do with deriving a more comprehensive physics, be patient and the connections will eventually become clear. First we must develop the conceptual landscape more broadly and in more detail before PMR physics can logically be birthed from a more general level of causality.

I do not want to leave anyone with the impression that spiritual growth (improving the quality and thus decreasing the entropy of your consciousness) is like working in a salt mine. Besides being useful, spiritual growth is also thrilling, interesting, rewarding, fun, and joyful. Once begun, it is so exciting an adventure that you will gladly want to put more of your time and energy toward it. It is also practical: Increasing the quality of your consciousness immediately increases the quality of your life.

23

■ ■ ■

Who Ya Gonna Call?

■ ■ ■

True enough, in matters of evolution there is no free lunch. Nevertheless, contemplating and evaluating the ideas of others can be an immensely helpful **aid** to your progress, and to your effort to grow the quality of your consciousness. You do not need to figure everything out for yourself. The advice of others can be like having a map to guide your explorations. An incorrect map can send you off on a wild goose chase. You must evaluate the correctness of the map **as you go** – because, **before** you go, you can only guess and assume your way through a shallow evaluation of any map. A useful map must necessarily be somewhat general, whereas each journey must be individual and personal.

Before going on to the wholly new concepts of the next chapter, let's first pull together what we have learned about the origins and consequences of belief and the requirements of personal growth so that those who are so inclined can get started on developing the experience base you will need to construct your personal Big TOE or, at least, evaluate this one.

In the preceding chapter we determined that you must do your own exploring and grow your own wisdom. You cannot progress by letting others do the work. To believe what someone else (including me) tells you (to become a believer) is lazy, risky, and amounts to accepting **someone else's** belief or knowledge in place of **your** knowledge. Copying the behavior or beliefs of others, or reciting or memorizing their knowledge, cannot produce **significant** spiritual or personal growth for **you**. Although some guidance by a fellow explorer may help you better understand your challenges and choices, discovering Big Truth and increasing the quality of your consciousness is fundamentally an independent individual effort. Talking about it all day and all night with the greatest of gurus won't produce one iota of real progress. Your lasting progress must be the result of your personal effort.

Personal growth is most efficiently and effectively the product of good science. This is subjective science or personal science, not to be confused with either organized or personal religion or objective science. Subjective science is the mother of objective science. Real personal science requires real, verifiable, measurable, **objective results**. Here, the word "results," at the most basic level, refers to significant, continuing verifiable progress toward the improvement of the quality of your conscious being, the evolution of mind, the growing-up and maturing of spirit. Why? Because that is the nature of the reality we live in. You will see that the physical nature as well as the spiritual nature of our reality is straightforwardly derived from the natural process of consciousness evolution. By the end of the next two sections, science will have logically derived the origins, nature, purpose, and mechanics of both spirituality (increasing consciousness quality through evolution) and your physical world.

You will eventually discover that our reality is fundamentally nonphysical (from a PMR perspective) and is animated and driven by profitability toward states of lower entropy. If your efforts do not produce measurable, significant growth, your personal science is only illusory. The knowledge gained through personal science continually and dramatically modifies itself as it **grows and changes**. On the other hand, cultural, personal, religious, or scientific belief systems require only a sincere **belief** in the assumed truth of their associated dogma, doctrine, and creed.

A belief system requires faith in the correctness of its beliefs. Because correctness is simply assumed, actual results are not required (correctness cannot be objectively demonstrated – that is the nature of belief). Mature and stable belief systems, including those generated by cultural, scientific and religious belief, once in place, do not tend to change. There is a logical disincentive to modify significantly what is, **by definition**, assumed to be complete and perfect. In contrast, the knowledge gained from mature personal scientific experience is always in continual flux. Open minded skepticism and continual scientific exploration for new data make sure of that. The search for truth is, by its nature, in a constant state of discovery, refinement, assessment, and reassessment because new data continue to pour in as long as the individual is aware and interested in growth.

Honest truth seekers never become know-it-alls – there is always room to improve yourself as well as your knowledge. When you know it all, when you believe that you have all the answers, you have, in fact, lost it all – nothing remains but a hollow shell.

You do not need any particular belief, disbelief, or faith to motivate you to start on this journey. You need only to grasp the **possibility** of a greater reality of some sort. After that, the desire to discover the truth should be motivation enough. Additionally, if this just-perhaps-possible larger reality is also **potentially** very important and significant to your life and being, nothing should hold you back from expending the necessary energy to explore the truth of the matter for yourself.

You can and should learn from others to the greatest extent possible, but you must grow yourself. Learning from those who have gone before can speed your progress; however, choosing those that you think you can best learn from is an iterative process that must constantly be reevaluated in light of your experience and your results. Those who can be most helpful, at any given time of your life, will change as you and your situation changes.

Do not get stuck in patterns, habits, or rituals. Do not look to groups or organizations to tell you what to do. Do not fall into belief traps. Have confidence in yourself. You not only **can** do it yourself, but you **must** do it yourself eventually, quickly or slowly, easily or with great struggle. We are all constantly evolving our consciousness. Evolution **forces** choice and change. Remaining the same by choosing the no action option is not possible. Change cannot be avoided. Change can take place as either positive growth or negative deterioration; the individual choices you make ultimately determine the direction (positive or negative) of your growth.

A good teacher provides encouragement, makes the learning experience more intense and more concentrated, and gives the student an **opportunity** to learn more quickly. Unfortunately, the more you need a good teacher, the less likely you are to be able to tell a good one from a bad one.

A good teacher focuses your effort to speed up your progress; a bad one misdirects your efforts and inhibits progress. Always stay skeptical, open minded and belief free, and most of all, taste that pudding – continue to require and evaluate actual measurable results. If six months go by with no **obvious measurable results**, this indicates that you need to buckle down and get serious, or change your approach.

Results, results, results, results. Actual, clear, un-subtle, measurable results – that is how you must evaluate the efficacy of your process. Intellectual knowledge and intellectual results are **not** the results I am referring to. These are no substitute for the real results of a growing, changing being. **Knowing about it** can be interesting and helpful, but it should never be confused with **being it**.

A change in the quality of your being, growth in the quality of your consciousness, evolution of your spirit: these are the results I am talking about – results of the being, not results of the intellect. It is about who you are, not what you know. It is about **why** you do what you do, not what you say, or what you do. When you start intending, doing, and being differently, you will produce measurable results. The tests you must pass are not written ones. Great factual knowledge cannot help you pass a test of the quality of your being. You are who and what you are – **and it shows** – no matter how good you might be at controlling your behavior with your intellect.

Truth is absolute, but how to discover it, and express it within your being, must be personal. Develop your tentative road map by applying open minded skepticism to the experience and conjecture (theory) of others. Then **modify** that map as **you** collect **your** data. This makes good sense, and offers you the **possibility** of leveraging the accumulated knowledge and wisdom of others as you define your unique growth path. Adopting a set of beliefs is a comparatively unproductive and risky approach to the evolution of your being and the quality of your consciousness.

How do you go about increasing the quality of your consciousness? How do you purposely pursue the evolution of your spirit? If dogma, ritual, and intellectual or emotional group-gropes are out, how do you get from here to there on your own?

For the scientists who are wondering what consciousness and all this blather about spirituality has to do with physics, let me assure you that I have not lost my focus and that this discussion is directly on the path to a scientifically legitimate, more general theory of physics. However, we are now, and will be for some time, developing the necessary basic concepts required to construct this Big TOE. Because this is a Big TOE and not a little TOE, a larger perspective supported by several wholly new paradigms must be developed. This process may appear, from time to time, to wander through irrelevant, ridiculous, or far-out ideological territory but if you can maintain open minded skepticism through the end of Section 6, you will eventually understand these unusual connections and their significance to science.

Because this is science and not theology, let me digress in the following aside on the process of getting from here (wherever you happen to be) to there (an increased quality of consciousness). The journey to higher quality consciousness is more simple and straightforward than you might think. I cannot promise easy and quick, but I can promise easy-to-do techniques and exercises that are simple and effective. For some it will be as easy as learning to swim, for others progress may come slowly; nevertheless, all dedicated and courageous explorers can succeed superbly if their desire to do so is sufficient.

▶ Because improving the quality of your consciousness (spiritual growth) is not, and cannot be, an intellectual achievement, it makes little difference how you intellectually approach the initiation of such improvements. **How** you start or what you **do** to improve the quality of your consciousness is insignificant compared to the act of starting. Additionally, an improvement in the quality of your being does not **automatically** flow from any external activity or practice. All you need is the will and the insuppressible drive (energy) to grow your being and the path, the process, to do so will appear before you. You are surrounded by opportunity to grow; your optimal path starts from wherever you are. I am talking about changing your being, intent, motivation, and attitude; modifying the quality of your interactions with others – changes in behavior and action (what you do) are secondary (results, not causes) and will follow on their own. Primary changes, when significant, are clear, obvious, and measurable to you and to others.

The evolution of consciousness is an extremely difficult concept for the Western mind to grasp because we are exclusively focused on the materially productive fact that right results are the products of right action. Westerners want to know what action they should take to get the results they want. Because they deal almost exclusively with external actions designed to produce external results they do not appreciate that internal results follow a different logic. What you are presently doing, how you live your life from day to day, is probably good enough as it is – what you need to change or improve is **why** you are doing it. When the "why" – the motivation and intent of what you do – is right, the "what" will take care of itself. Improving the "why" can start anywhere any time because it requires modifying internal variables, not external variables; nothing must change but you.

You can hope and pray for someone else to provide you with enlightenment (trust me, that won't happen), or you can take the steps to develop and grow it. Do not expect to find shortcuts through the flypaper realms of religious, scientific, or personal dogma, or along the midway of a New Age carnival. You must keep your mind free to change and grow. The right question is: How has the fundamental quality of your being changed. The answer to that question defines the metric of your progress. Self-proclaimed success means nothing; progress must be demonstrated by clear and obvious results.

The answer to how the fundamental quality of your being has changed is either totally obvious to everyone (including yourself), or not much progress has been accumulated. Genuine results are not subtle. You and most other people, given enough time with an individual, have the capability to tell the difference between a wise and loving being and one that is only trying to appear that way. This is not rocket science; it is not difficult to determine if you are making real progress. A significant change in your capacity to love is as subtle as the healing of a badly broken leg – nobody, including you, could miss noticing the change.

Are you like a deer caught in the headlights of an oncoming car – frozen, unable to take the first step? Because of our cultural belief that we must **do** something in PMR in order to affect change (even if that change is within our consciousness), most people are effectively paralyzed and cannot take that first step. "What should I **do**? Where should I start?" they ask, looking for a prescription or set of clearly directed "how to" steps. Improving the quality of your consciousness, energizing spiritual growth, and gaining a Big Picture perspective are not accomplished by changing what you **do**, but by changing what you are. Reread the previous sentence at least twice and think about your need for a physical process to develop your consciousness. You are a product of your culture – you cannot help that.

Spiritual growth, improving the quality of your consciousness, is about changing your attitude, expanding your awareness, outgrowing your fears, reducing your ego, and improving your capacity to love. To succeed, you must change your intent, and modify your motivation. The problem (and the solution) is one of being, not one of doing. You can **do** everything by the book, meditate regularly, be conscientious, try very hard, go through the proper prescribed motions and still make little progress. Going through the motions does nothing if the mind is not open to, and in pursuit of, fundamental **internal** change.

The prospect of fundamental internal change can be very frightening. When change begins to occur, many people run away because they are terrified of the unknown. They are afraid of where the changes may lead (which is often directly into the face of their fear) and of not being able to intellectually control the process. They find that shaking the foundation of their being at its deepest level is too unsettling an activity. What if the entire I-structure comes tumbling down into ruins? The ego begins to fear its own dissolution and death.

Fear and belief cause many well-meaning people to reject fundamental internal change, particularly if their beliefs are incompatible with the required changes. Instead of embracing change and facing fears, many would-be spiritual seekers focus on the external rituals associated with some type of mental or spiritual discipline: They go to church or learn to meditate. Many meditators and a few churchgoers hope to produce measurable **external** changes and to have cool internal metaphysical experiences.

In the West, meditation is acultural and an individual, rather than a social, activity. Most church goers continue their attendance out of social convenience, habit, duty, or cultural expectation – whereas most meditators eventually decide that meditation does not **do** anything for them, or at least not enough to be worth the effort and time required for a long-term commitment. A few of each group pretend their effort has made them superior. The more honest and objective of the failed meditators give up in frustration or due to a simple lack of interest and soon forget about it. "I tried spiritual exercises, and they didn't work for me."

Practicing some form of meditation to effect external change, gain paranormal abilities, placate the guilt of doing nothing, or simply because you think you should, is analogous to a carpenter trying to build cabinets while holding the screwdriver and hammer by the wrong ends. All the pieces are there, but the execution is flawed. Make the required internal changes and the measurable external changes will occur on their own. You have to grab the screwdriver and hammer by the wooden end or you will come to the erroneous conclusion that they are useless tools that only someone else can effectively use. Or, more arrogantly, that nobody could use such stupid tools, that cabinets are a logical impossibility, and that all carpenters are delusional frauds and fools.

You must realize that you cannot modify **being** merely by taking physical action within the local physical reality. Westerners have a particularly difficult time understanding this fact and often feel helpless without a way to compel results from the outside. The opportunity to bolt to personal success and freedom by employing a more complete knowledge is lost in a culturally conditioned false commitment to the little picture. Belief traps are bigger problems than most of us think they are.

I know, after all that, you still want to know what you should **do**, how you can best modify your being, and what the most effective techniques are. Let me guess, you feel that you could use a hint – a little help, a little direction to get started. All right, all right, I give up! To help you get started here are some things you can **do** that **may** lead to opportunities to grow your being; however, it is entirely up to **you** to recognize, seize, and develop the opportunities that come your way. You already have plenty of opportunities, but let's pretend that by doing what I am going to tell you, more obvious and easier opportunities will appear before you. That will get us started with a hopeful, positive attitude.

What is more likely to happen is that by conscientiously working at the following exercises, your perspective will change, enabling you to see opportunities that are now as invisible to you as water is to a fish that lives two miles down in the middle of a four mile deep ocean. With no light and only a dim awareness, the fish knows nothing of water. Water just is, has always been, and is taken for granted. The fish does not ponder the nature of water, it swims in it. We swim in an ocean of consciousness. We are not aware of the ocean, but only of our local interactions with it.

The first and most necessary ingredient is a sincere desire to grow the quality of your consciousness – to evolve your being – to permanently change yourself at a deeply personal level. The second most necessary ingredient is to have the courage to change – the courage to face your fears – to face death and personal destruction, for that is the story your ego will tell (and try to get you to believe) when it comes whining to you with its tail between its legs hoping to dissuade you from increasing the quality of your consciousness.

Why would your sweet little ego do a mean thing like that? Because the ego's main job is to keep you feeling good by managing various systems of belief that are designed to keep your fears beyond the reach of your intellectual awareness. Increasing the quality of your consciousness requires you to face your fears, overcome them, and dissolve your ego. You should expect the ego to struggle mightily. ◀

▶▶ Ego does not necessarily imply arrogant self-centeredness. Ego comes in an infinite array of expressions – arrogance is only one. Being timid, unsure, or a worrier are also manifestations of ego. Insecurity and anxiety about that insecurity are common. How each personality expresses that insecurity and anxiety reflects individual quality and style. The strategies that are used to deal with fear, though common at the top level, are uniquely applied to each individual. Great ego reflects great fear; it does not necessarily reflect great arrogance or great pride, though it may reflect both. Self-centered, self-focused, and self-absorbed are three of the many possible aspects of ego – each of these three can be directed either inwardly (producing timidity) or outwardly (producing arrogance) to create personality traits that appear to be opposite.

Courage and determination will grow sufficiently to overcome fear if the intent to succeed is sufficiently strong, steady, and clear.

I will more carefully define ego and explain its functions (how it works and achieves its goals) in an aside in Chapter 42, Book 2. Go there now if you are seriously confused.◀◀

▶ The most obvious pathway to the exploration of consciousness is through the exploration of your personal consciousness – a scientific investigation of your subjective experience. Studying consciousness from the outside (objectively) is like studying biology by looking at pictures of zoo animals. Consciousness is fundamentally individual and personal. Our objective sense of consciousness is derived from the reflection of our personal consciousness from the uniquely curvy surface of a mirror that we call "another."

Our objective experience of other consciousness is the result of an interaction of our personal consciousness (representing one set of possible choices or ways of being) with another, which suggests to us new configurations, interactions, and possibilities for **our** being. We project our awareness of consciousness into "other," define the nature of "other" in terms of ourselves, and thus see only a reflection of ourselves in the mirror of interaction with "other."

To preserve the symmetry of interaction, we also serve as a uniquely shaped mirror in which others can see themselves reflected in challenging new ways. Within this funhouse hall of interactive mirrors, your consciousness is a singular actor. Opportunities for change arise, choices are made, reality is actualized, and progress or regression in terms of personal growth is achieved. Your conscious awareness defines your personal reality. There are as many different shades and levels of personal reality as there are of personal awareness. "Other" provides opportunity for the improvement of the quality of your consciousness by accurately reflecting the truth of you.

If improved consciousness quality along with personal effectiveness, growth, and power are your goals, approaching consciousness from the inside, from the scientific exploration of inner space, is the only logical approach that delivers results. An approach from the outside will limit you to collecting the facts about the shadow that consciousness projects upon the wall of PMR.

We project our personal consciousness onto the field of action of a multi-player interactive reality game whose point is our individual growth and learning. The experience of consciousness, as well as the evolution of consciousness through choice, is entirely personal. However, an awareness of a larger (source) consciousness and an understanding of its properties are accessible through scientifically probing and objectively assessing the value and operational characteristics of the subjective experience of personal consciousness.

One method of accomplishing an assessment of subjective inner space is through meditation. Learning to meditate is like learning to play a musical instrument: It takes a serious steady effort before you should expect to make music instead of screeching noises. It takes dedication over a much longer time before you can master the basics of the instrument and play it well. Unfortunately, most people who pick up an instrument and give it a try give up before they ever learn to play it well. So it is with meditation.

As mentioned previously, going through the motions, or in this analogy, **pretending** to play an instrument, regardless of how perfect or impressive the visual (external) display, produces no significant results.

There are many effective paths to personal growth – meditation is only one. Within the wide range of practices that circumscribe what we have loosely defined as meditation, there are many different types, approaches, and methods. Because it is the easiest, most effective, and universally applicable, a simple mental-awareness meditation is the path of choice for most teachers and students who have no dogma to propagate. Within this subset of meditation, there are many differing techniques. The technique you choose is not as important as the application of steady effort – so choose a technique that suits you. Within this genre of meditation, you do not actually have to learn how to meditate; you need only to learn how to stop blocking the meditation state from occurring naturally.

Though we are pursuing the dubious subject of what you can **do** in order to undo what you have inadvertently done, I will help you out here because I know your cultural beliefs force you to begin with a physical process. It will be helpful to your doing and undoing if you understand meditation – its purpose and how it works. With this understanding, you can custom design your own personal spiritual growth **doing** thing – a physical and mental process that may lead you toward a higher quality of being. The doing process cannot get you there by itself, but it **can** serve as the on-ramp.

The meditation state that I encourage you to achieve represents a condition of inner attentiveness wherein you become aware of your personal consciousness. This, in time, leads to the awareness that you are a unit of consciousness among many such units. Eventually, you will regain your fundamental identity as a spiritual (nonphysical) entity – as well as understand your relationship, your oneness, with all consciousness. Personal growth is a natural result of meditation.

Becoming aware of your consciousness is analogous to that fish becoming aware of water. The fish is aware only of its interaction with water. It experiences water through doing, through action, through its objective causal interactions with water. Yet water has existence and significance in its own right beyond the interactions of that and other fish. To become aware of water, one must differentiate between water and a **subset** of the properties of water. The fish is aware only of the latter.

The fish experiences water only in terms of its limited interactions (experience). It experiences variations in current, temperature, salinity, viscosity, and dynamic limitations, but does it actually experience water in a fundamental or broad sense? Is the fish right? Is water nothing more than the sensed **variations** in its properties? Does water with no variations in its properties cease to exist as water, or does it simply become an invisible background to the fish because the fish can no longer perceive it? To appreciate your and the fish's limitations, imagine the perfect sensory deprivation tank where your local environment disappears because of zero input to your senses. Granted, this is not a perfect analogy, but you get the idea. When you are totally immersed in something, such as cultural belief systems for example, that something often becomes invisible because you cannot differentiate it from the background of your local reality – there is no contrast to bring it to the attention of your senses. Consciousness is like that.

Like the fish, we define our consciousness in terms of our doing – in terms of the physical actions it allows us to take. The major attribute of consciousness can be summed up as awareness, yet we and our fish brethren are aware only of what we can physically do with it, how we interact with a **subset** of its properties. Moreover, we can only interact with that subset of properties that is contrasted enough against the invisible background of primal consciousness for us to notice. We create a foreground of contrasts, relationships, and variations in the fabric of absolute consciousness that we define as representing ourselves. "See that cute little wad of wrinkles in the fabric of consciousness? That's me!" But you are more than the wrinkles; you are also consciousness, a piece of the whole. Meditation lets us experience the invisible background of consciousness. It lets us notice the water itself, not just variations and contrasts in its local properties relative to an invariant constant.

The point of meditation is to enable you to become aware of your consciousness and thereby introduce you to your larger self. Becoming aware of your consciousness at a fundamental level will eventually lead you to see the real you, the complete you, the whole you, the sacred and the soiled – fears and all. Without the ego to hide the scary parts by inventing an attention-getting "I vs. other" delusional contrast, it is not always a pretty sight.

How does meditation lead you to experience your consciousness? By turning down the contrast, noise, and other activity that make up the busy foreground – by turning down, and eventually turning off, the cacophony of mental interactions, judgments, and operational processes. To become aware of your consciousness (as opposed to being aware of the thoughts that inhabit your consciousness) you must eliminate the obsessive preoccupations most of us have with ego based self-definition – the contrasts that you use to define yourself against the relatively unchanging, invisible background of your individual consciousness. Meditation is thus an act of **not doing**. It is an exercise in removing enough of the contrasting clutter of your mind to get a glimpse of the real you.

Individual consciousness is a subset of absolute consciousness. You are not **only** the clutter, the wrinkles, the ego, the thoughts – even if that is how you unwittingly define yourself. You are much more than that. Meditation allows you to discover that fact in a uniquely personal way. That is its purpose – self-discovery – a glimpse of the fundamental reality of which you are an integral part.

This discovery is possible for humans because, at least theoretically, our memory capacity and processing capability are somewhat greater (and contain less entropy) than that of the average fish. The fish will never directly experience or contemplate unvarying water (the fish equivalent to total sensory deprivation), but you can experience the fundamental nature of your consciousness if you truly want to. If your desire to know yourself and to know the truth at the deepest level of your existence is not strong enough to provide the necessary focus, energy, and persistence required to succeed, you are not yet ready to begin that journey. There is no rush and no penalty for not being ready. It is much better to wait until you are ready than to push yourself into a state of self-limiting frustration.

Do you see why meditation is almost universally prescribed as the first step – the doorway to understanding and exploring consciousness, as well as to the attainment of spiritual growth? It makes sense that a program to develop your consciousness should naturally start with finding and becoming acquainted with that consciousness. There are other methods, but they apply less universally, are more difficult to learn, and are much more difficult to teach. Meditation will work wonderfully when you are ready. You may first need to work on getting ready by developing an honest desire to grow spiritually and the courage to pursue Big Truth to its conclusion. You may need to first overcome some of the fear and cultural beliefs to which you have become attached.

How does meditation clear out the clutter and reduce the noise level of a mind caught in a self-referential endless loop of obfuscating circular logic? The technique is simple and straightforward – the trappings of ritual, dogma, belief, and physical process are mostly irrelevant. You simply stop the incessant operational, self-referential, contrast producing chatter of the mind by filling the mind up with something less distracting, less self-focused and less obsessively driven. While the mind is preoccupied with non-operative busy-work, you can experience the still center of your being. Eventually, after much practice, you can let go of the mental busy-work and explore the larger reality of consciousness from an imperturbable, still, and quiet place that will slowly develop and grow larger at the center of your being.

Some traditions call this mental busy-work assignment a "mantra." Traditionally this is a sound of some sort, but in this Big TOE we are bound only by science, not tradition. We quickly move to toss belief, dogma, and ritual out of the window and focus, by experimental result, only on the active ingredients of mantra. Science allows the concept of mantra to be generalized to accommodate the various ways we take in and process information through our five senses. Typically, people tend to take in most of their experiential input data through their ears (auditory), eyes (visual) or sense of touch (kinesthetic). Many people absorb information more effectively through one of these avenues of data input than they do through the others. Over the previous decade or so, the popular literature is full of assessments of personality type and characteristics by data input preference. It makes no sense to force everyone down the traditional auditory path – some people simply do not get it that way.

If you are not now successfully meditating, and have no idea where or how to find a suitable technique to do so, I will provide to you, free of charge – for this one time only – a mentally calming busy work mantra custom made for your personal mind that is based upon each of the dominant perception types. Simply use the one or combination that seems to work the best for you. For those more heavily into smell and taste than the average humanoid form, I am sure that you can follow the three examples given to custom fit a smelly or yummy mantra to suit your individual preferences.

After explaining each mantra, I will, against my better judgment, tell you what you can **do** with them. Oh, no, not like that – I wouldn't be that rude! I understand that your Western mind-set needs to begin everything with physical process whether it makes sense or not.

Those who seriously want to get started on their spiritual journey, but find themselves caught in the headlights of physical action-reaction causality, will now have something to **do**. It may or may not help you improve the quality of your consciousness – that depends on you – but it will give the committed doers a place to start. Often that is what is needed – a place to start – a **do**able approach to the problem of how to modify the quality of your being. This could be the step you need to break free from the mesmerizing glare of those cultural beliefs that reduce, rather than extend, your vision. Try it: You may surprise yourself with some dramatic results.

For the audio types, we need a sound that means nothing, is two syllables, and ends in a soothing or vibratory sound. Here are a few examples of proven quality – take your pick or make up one of your own: "sehr-ring", "da-room", "ra-zing", "ca-ouhn", "sah-roon", and "sher-loom." For a simple multi-syllable repetitive string (chant), try: "ah-lum-bar-dee-dum – ah-lum-baa-dee-dum." When the "bar" and "baa" regularly inter-change themselves effortlessly, you will be well on your way. These are sounds, not words – it is important that they carry no intellectual meaning. The point of this exer-cise is to quiet your operative intellect so that you can experience consciousness directly by reducing the variations, comparisons, and contrasts that your ego-intellect imposes upon consciousness.

Feel free to mix and match – put any of the first syllables in front of any of the sec-ond syllables to produce no fewer than thirty-six unique mantras. For most people, it won't make much difference which sound is used, but if one sound feels more natural than the others, use it. Obsessive-compulsive types should take care not to get wrapped around the axle trying to find the best one – any will do.

Lighten up; do not be intense and serious. **Have no expectations.** Sit in a com-fortable quiet place where you will not be disturbed, close your eyes, and fill your mind with the sound of your chosen mantra – no need to make an actual sound. Focus your attention on the sound. Let the sound fill your mind – think of nothing else. Use what-ever devices you need to stay focused on the sound – merely **listen** to it repeat itself. The repetition may be simple and straightforward or occur in interesting ways – per-haps with complex variations.

Eventually, let the sound of the mantra slow to a rhythmic, bland repetition and then slow and smear further into a continuous background sound. If thoughts creep in, gen-tly put them aside and refill your mind with the sound. If intruding thoughts constantly stream into your awareness, give the mantra a more active form. As thoughts disap-pear, leaving your mind empty, simplify and soften the sound of the mantra. Continue the meditation process uninterrupted for at least twenty minutes, twice a day for three months before evaluating the results. If the sound slips away, but no extraneous thoughts appear, let it go and drift in the quiet blankness of your consciousness – you will love it.

Visual types need a non-personal visualization that begins with complexity (but not detail) and ends with simplicity. You may start with a black and white soccer ball – then let the colors change to red and blue, let the ball begin to rotate slowly, let the colors change. Your image should be as clear as a watercolor painting, not as precise as a high-resolution photograph. Switch to a series of simple geometric shapes such as spheres, cubes, circles, triangles, cylinders, rectangles, and lines. Let them rotate slowly. Slowly change their size and colors. Choose one shape and let it change very slowly. Watch your images intently – think of nothing else.

Gradually progress your images toward greater simplicity and slower motion. Do not force the images; let them do what they want to as long as they do not disturb your tran-quility. Look at your images uncritically and dispassionately, as if you were watching a plotless movie. If thoughts creep in, gently put them aside and refill your mind with more active images. Continue the meditation process uninterrupted for at least twenty minutes, twice a day for three months before evaluating the results. If the images slip away, but no extraneous thoughts appear, let them go and drift in the still oneness of your consciousness.

If you enjoy natural places, you might start with a scene – perhaps a generic beach. Hear the waves, feel the sand, smell the salt spray, listen to the sea gulls. Be there with all of your senses. Slowly simplify your image and focus on a few items at a time. Eventually you may narrow your focus to a single grain of sand. Go in close to inspect the tiny crystal from every angle. Choose the viewing angle you prefer and see how the light plays off the surface of the crystal. Back away until you can barely see its surface features. Hold that view as only you and the grain of sand quietly coexist within the void.

Choose images that particularly suit you. Be careful not to try too hard, and do not struggle with high resolution, image quality, or anything else. Images may be felt as well as seen. Struggling to make your meditation be how you think it should be is always counterproductive. No expectations. No struggle. No demands. The point is not to force your will on the process, but to let the process unfold naturally as it captivates your attention.

Remember that what you are trying to do without trying is to not do. Read that sentence again – don't you just love it? If it makes sense to you, you are on your way. If it sounds like idiotic gibberish you should go back to the beginning of this aside and start over – but don't get stuck in an endless loop – twice is enough.

For the kinesthetic types, we need textures that are non-personal, interesting and pleasant. For example, feel a rich velvet or fur coat as you mentally rub your hands slowly over it. Dig into it with your fingers, feel it rub across your arms and face. Explore the buttons or zipper, the seams, sleeves, and collar. Become tiny (or create a giant coat) and roll around on it, crawl into a pocket. Slowly let your sensing of the coat become simple and rhythmic. You might do the same thing with walking barefoot in squishy mud, or walking in the rain, or swimming in a pool filled with grape jelly. Start with complexity and progress to more and more simple rhythmic sensory stimulations. If thoughts creep in, gently put them aside and refill your mind with the sensations. Continue the meditation process uninterrupted for at least twenty minutes, twice a day for three months before evaluating the results. If the sensations slip away, but no extraneous thoughts appear, let the sensations go and drift aimlessly in the boundless depths of your consciousness.

Smell and taste mantras would work similarly to the kinesthetic mantra above. Use your imagination. Do not be afraid to mix and match the senses; combine them in ways that work for you. Have no intellectual or emotional connection to your mantra. Maintain only enough complexity to keep extraneous thoughts away – nothing should be in your mind except the sound, sight, feel, taste, or smell of the mantra. As intruding thoughts become less of a problem, simplify your mantra. When you no longer need it to maintain a state of blank thoughtless existence, let it go. ◀

▶▶ Speaking of experiencing a state of thoughtless existence through meditation – let's do another ego-tweak – a real blatant one this time – and have some fun.

Hey ladies, why do you think most spiritual gurus are men? Think about it. Do you give up? Because, men are born thoughtless and remain that way the rest of their lives! Why else?

Oooh…what a low blow!

Uh oh, easy fellows, I was only kidding – just a little double-entendre word play. Come on now…what's the matter guys? Remember the rules: peace and light – no violence until after I leave.

Oh well, at least the ladies thought it was funny – they get it. ◀◀

▶ The point here is to learn to control your thoughts and your operative mind so that you can experience your consciousness. This is a first and necessary step. Later you can learn how to direct that consciousness once you have freed it from a noisy, frantic, ego serving, perpetual tail chase. Do not try to direct it too soon – that will only delay your progress – get in touch with, and follow, the source of your intuition. Do not pursue or chase after specific or general results. All results must come to you. If you go after them, it will only delay your progress.

Continue to experiment and to taste the pudding periodically. Natural, easy, patient, and gentle are the hallmarks of a successful process. Result driven, ego driven, success driven, frustrated, forced, fearful, and having preconceived notions and expectations are the hallmarks of a wrong-headed flawed execution of the meditation process.

Experiment to find what works best and what feels most natural to you. After you find it, stick with it for a while. If thoughts intrude, as soon as you realize that your mind is no longer exclusively working with the mantra, put them gently aside. If thoughts continue to come, increase the complexity of your mantra a little. As thoughts disappear and do not return, decrease the complexity. Never try too hard. If you ever become frustrated, you are trying too hard. This is most important: Have absolutely no expectations and no specific goals.

This is also important: Do not **begin** to judge how well or poorly your meditation is working until you have found and implemented a productive meditation process twice daily for at least three to six months – **then** taste the pudding.

Do not analyze or compare, just experience – this is not an intellectual exercise and your analyzing justifying intellect will only get in the way. Never force the mantra – go with it, flow with it, and let whatever happens happen – this is a gentle activity with no preconceived notions of what the outcome should be or feel like. There will be plenty of time for evaluation and pudding tasting after you gain some basic competence. There must be a time to be critical, but not now – you do not know enough to be productively critical yet.

Let every meditation be an entirely new and unique experience. Do not force every meditation experience to be like a previous experience that was judged to be a good one. Continue tasting the pudding at three month intervals. Look for the existence of measurable results in the form of objective changes in your being. After six months, ask people who are close to you if they notice a change in your approach to life. Be aware of your mental state, and how that state changes as the meditation progresses. Customize your meditation to suit yourself. Your meditation should become easier, more effective, and more efficient over time. Be patient, do not rush the process – trying to speed-up or push the process will only delay your progress.

Pay careful attention to the choices you make throughout your day. Examine your motivations and intent relative to those choices. By an act of your will, modify your intents to be more giving, caring, loving, and to be less self-serving. Shift the focus from you, from what you want, need and desire, to what you can give to, and do for, others. In the same manner, change where and how you invest the energy that follows your intent in your relationships and interactions with other people.

Examine your motivations and intent as described above immediately before and after, but not during, each meditation. You must be consistent – that is most important. Once you get used to the exercise, thirty minutes twice a day is enough to accomplish both the meditation and the examination of your choices – take more if you wish, but much more is not necessary.

If you constantly end up in a state of frustration instead of a state of expanded awareness, let go, back off, and take a break until you can find a different perspective, a different attitude, or a different intent. Try a different mantra. Perhaps you are trying too hard. Perhaps you are limited by your belief and fear, or lack the necessary courage and drive. Perhaps you need to read and follow the instructions more carefully. Perhaps you are using a meditation technique that does not suit you. Perhaps you are not ready at this time. Don't worry: Everything works out in its own time. There is no blame, no reason to feel badly, and no failure on your part. Continue to apply the meditation process gently and consistently and one day, when you relax, success will take you by surprise.

Everyone grows in their own way and in their own time. No one faults children for not being adults, though most children wish they were adults. There is no practical technique that allows you to skip steps. You are who and what you are – accept that gracefully. Work on getting ready by continuing to practice the given exercises gently and with no expectations. There is no faster process or better way to get ready than that.

Thus, we see that getting prepared and ready to grow, as well as actively progressing along a growth path, as well as optimizing the growth path that you are on, all follow the same prescription. It matters not what your initial conditions are or where your starting point is, the same set of meditation exercises is optimal and appropriate for all. That is why virtually everyone who wishes to follow the Path of Knowledge toward spiritual growth, toward improving the quality of their consciousness, is instructed to begin with daily meditation. Each individual will naturally extract from their meditation what they need for their next step. This personalization of the growth process takes place naturally because each individual is essentially engaged in a "bootstrapping" (pulling themselves up by their bootstraps) operation with his or her own consciousness. The meditation experience is as individual and personal as is your consciousness.

This is a life's work; it takes significant time to take root, blossom and bear fruit. Results will accrue in proportion to the energy that is invested productively. For example, with moderate effort, significant results should become obvious within six months to a year. Continue to apply the meditation process with gentle resolve; there is no rush, no test, and no diploma. You have all the time you need to get it right. Some will get it right away; others may take a long time. Gracefully accept however it comes to you – you have no choice. A teacher can only encourage and facilitate the evolution of your consciousness by helping you find opportunities to exploit on your own – spiritual growth, as any growth, is an internal process and cannot be forced from the outside.

Hopefully, these meditation exercises have addressed your need for a physical process to facilitate positive consciousness development. However, in doing so, I may have created a new problem for you – how to deal with the frustration that is often created by the inadequacy of **doing** to produce dramatic spiritual progress quickly. The Western attention span is notoriously short. To make matters worse, dramatic results are often required to overcome strongly opposing cultural belief systems. The fact is that progress in meditation, like progress in playing a musical instrument, usually accrues slowly and only becomes dramatic after significant time and effort has been invested. Progress accrues by the accumulation of many unnoticeable tiny successes. Take the long view and have patience.

Westerners caught in the glaring headlights of their cultural beliefs desperately need something to **do** before they become spiritual road kill – run-over by mindless conformity and a blind obedience to the cultural norm. Thump! Splat! Oh jeez, what a mess! All the king's horses and all the king's men will have a difficult time getting that one back on the road to spiritual progress again.

Actually, it is unfair of me to pick on Westerners as being particularly limited by needing to **do** something in order to **be** something. Most Easterners are in the same doing-fixated boat. Their do-boat may **appear** to be bigger – not as confining perhaps – but just as limiting. **Doing** within a spiritual-cultural tradition is as problematical and unproductive as **doing** within a material-cultural tradition.

If you are so inclined, you now have something productive to **do** as well as an expanded perspective on the limitations and personal nature of that doing. ◀

You now know what to **do** and how to **do** it, and if you settle in for the long haul with a serious commitment to finding Big Truth, you will succeed beyond your wildest dreams.

24

■ ■ ■

Two Concepts

■ ■ ■

The previous chapters should have destroyed the illusion of a quick and easy fix, leaving you with your feet firmly planted in the personal as well as the shared nature of reality. With your attitude, perspective, and focus now properly adjusted, let us get back to the job of developing a credible Big Picture Theory Of Everything (Big TOE). The best process that can be employed when exploring new theoretical ground is to: (1) minimize the number of assumptions required, and (2) simplify all remaining assumptions to the most basic level possible. Albert Einstein put it succinctly when he said: *"The grand aim of all science is to cover the greatest number of empirical facts by logical deduction from the smallest number of hypotheses or axioms."*

I think that Dr. Einstein would be proud of us – we are going to derive a comprehensive Big TOE that explains the empirical facts – mind as well as matter, philosophy as well as physics, the normal as well as the paranormal, consciousness as well as concrete – by logical deduction from just two simple assumptions. And only one of those two assumptions is extraordinary.

Theorists can build a complex speculative structure, but the foundation should be as simple and straightforward as possible. The first assumption, referred to as the Fundamental Process of evolution, is readily understandable and is well within our experience. The second, often called the "Absolute" or the "One Source" by philosophers and theologians, may require our spatially limited worldview and our cultural beliefs to stretch beyond their normal patterns. As it turns out, this "One Source" is simply consciousness – primordial consciousness – the fundamental energy that is the medium of reality.

If the concept of primordial consciousness cast into the form of a finite, but practically infinite, monolithic undifferentiated form of potential energy makes you feel uneasy, that is as it should be. You should remain skeptical of any premise and doubly skeptical of any mystical premise – even if logic demands that at least one mystical (beyond objective knowing from the PMR point of view) premise must be at the base of any successful Big TOE.

Things beyond **objective** knowing, from the PMR point of view, are by definition things that we cannot hold with a firm intellectual grip. If this concept fell within your objective experience, if you could be intellectually and rationally comfortable with it, it would no longer qualify as the mystical or metaphysical foundation upon which we could build a Big TOE capable of logically accounting for our beginnings, for mind as well as matter, for the paranormal as well as the normal.

The Fundamental Process of evolution along with primordial consciousness as a fundamental source of structurable energy are the two basic assumptions on which *My Big TOE* is based. Everything that follows will be logically derived and explained from these two fundamental assumptions.

I am **not** asking you to **believe** in the truth of these assumptions, nor to have **faith** in their correctness. I am not trying to start a religion here. These are the logical underpinnings of this Big TOE. They are the assumptions, assertions, logical premises upon which the structure of *My Big TOE* is built. *My Big TOE* is a model. **Its usefulness**, and the value of these assumptions, **is based exclusively on its ability to explain the data – period!** Please reread the previous two sentences enough times to commit them to memory. By the end of Section 6 you will have a much better idea about this reality model's usefulness, and the potential correctness of these assumptions.

The data, or "empirical facts" in Einstein's words, are any and all of the carefully evaluated experience or truth that you and others have accumulated about life and the nature of reality. These data come in two types – **personal data** that are unique to your subjective experience and **shared data** that represent the objective physical experience of PMR (the realm of PMR science). If your personal data have been objectively evaluated and are un-warped by ego, attachment, belief and fear, they can be used to either corroborate or invalidate this model. For these types of data you are the sole judge, the only possible judge. Because of the personal nature of subjective experience, you can only be the judge for yourself.

Shared data are also important. Physical reality and PMR science must be an integral part of any Big TOE. We, after all, are at this moment interacting in what appears quite convincingly as a physical reality. No one can reasonably deny the importance, or the necessity, of physical experience, even if consciousness is at the core. Physical reality and its science must be a necessary, rational, and derivable part of the Big Picture. Furthermore, it is the interactions that take place in PMR (as defined by our physical experience) that enable one to test and measure the **objective** results of an increased quality (decreased entropy) of consciousness.

▶ This is where the metaphysical rubber first meets the road. What I am offering in this chapter is **not** what I **believe** to be true. This is not about what I believe about reality. This is about a **model of reality** based upon my experience and research. If the difference between the two is not crystal clear, I have not communicated as clearly as I need to.

Some may **believe** there can **never** be a significant difference between the two – that belief and subjective knowledge are the same. Not so. Results, objectively measurable results **can** differentiate between the two. **If there are no objective measurable results, there can be no solid or scientific conclusions.** My Big TOE represents a serious attempt to describe and model the larger reality in which you exist as a digital consciousness. If what you have read thus far seems to be far removed from hard science, be patient. The merger of physics and metaphysics, which is a logical requirement of any correct **Big** TOE, requires substantial background development.

From the little picture perspective of our local objective causality, all one needs to do to make a strong scientific argument is write down the appropriate equations. Math is merely symbolic logic. The limited local logic of the little picture is clearly expressed by little picture mathematics. Within the PMR causality system, if the assumptions are correct and the math is correct, the results will be correct – that is how little picture physics works. The Big Picture, responding to a more general causality, cannot be adequately described in terms of little picture logic. I know that this is a difficult paradigm for many (particularly scientists) to transcend, but it is logical that the Big Picture with its more general causality cannot be fully described from a little picture perspective. In other words, the larger reality (NPMR – consciousness) cannot be fully specified by a limited local logic and knowledge that belongs to a small subset (PMR) of that larger reality. If I could describe My Big TOE in terms of little picture (PMR) mathematics (logic), it could not possibly be more than a little TOE. That is one of the invisible walls that Einstein ran into with his failed Unified Field Theory – he tried to describe the Big Picture exclusively in terms of little picture logic. At least he knew there was a Big Picture even if he did not know it constituted a superset of the little picture.

The traditional belief-blinded little picture perspective will maintain old paradigms and preserve cultural and scientific dogma by simply denying the existence of a larger reality. To be consistent, it must also deny, or concoct excuses for, any scientifically validated data and experience that directly conflict with its beliefs. That is the old tried and true head in the sand, butt in the air trick. I can assure you from firsthand experience that attempting to engage these elevated butts in intelligent conversation is often not productive – low IQ, bad breath, no vision. Fortunately, minds can sometimes be pried open and perspectives can sometimes change with additional experience, knowledge, and understanding.

If the above ideas are fuzzy, hold on – understanding will improve and become clear as you continue. This model, as any model, is constructed mostly of knowledge and experience with a little conjecture or theorizing to bind (integrate) the various discrete data together into a coherent and consistent whole. I approach this model with open minded skepticism as you should. If I believed in it, it would limit my ability to grow the model in new directions as new data accumulates. If it makes you feel better, I don't **believe** any of this stuff either.

I either know it as fact (knowledge), or regard it as the most likely possibility or best hypothesis thus far (based on the scientific data available to me as of this writing). In other words, this trilogy represents the **tentative** results and conclusions of thirty years of my personal, serious, careful, scientific exploration of the physical and the nonphysical. I have been strongly encouraged to share the results and conclusions of my experience with you. These books are it. What you do with it, get out of it, or take from it is entirely up to you.

Models should always remain tentative to preclude shutting themselves off from the possibility of further evolution. For the sake of argument and to give proper credit to those who **believe** that belief and knowledge can never be entirely separated, let's take the opposite view. Even if this trilogy represents my belief, and is an effort to convince you to believe what I believe, the potential it holds for you is undiminished as long as you approach it with open minded skepticism and resist the compulsion to either believe it or disbelieve it. There is no reasonable, rational, or logical alternative to open minded skepticism.

My Big TOE, among other things represents a personal map of the reality I have explored and the explanations I have created to make coherent sense out of my individual experiences. Because of the personal nature of experience, your experiences will be somewhat different from mine. However, If your experience is interpreted and gathered scientifically and not created out of belief or fear, then the underlying truth from which both of our experiences spring will be perceived as the same. The underlying truth is absolute (the same for everyone) and is not a function of our experience or our existence (or anyone else's). That is why my map, though it reflects my individuality, can serve as a valuable **guide** for you. Though the destination (Big Truth) is the same for everyone, you must take your own journey and find your own way.

A word about truth. Many people think of truth in terms of local and universal truth. Local truth may be relative. It may be in the eye of the beholder. It is sometimes dependent upon the perspective and beliefs of the individual. This is **not** the truth I am referring to. Universal or absolute truth is the same for everyone – it is timeless and unchangeable. The paths to absolute truth (Big Truth) and the individuals who walk those paths can be so different that the description of the same absolute truth may appear to be very different – particularly to those of less understanding. Individual interpretations are, as they should be, a reflection of that Big Truth within the mind and experience of that individual.

A group of similar (by depth of understanding) but different (by path, culture and personality) truly wise and knowledgeable individuals would view the differences in each other's descriptions of the same absolute truth as insignificant and trivial. Unfortunately, in their absence, their less knowledgeable followers may actually kill each other over the **apparent** differences as they vie for relative power and superior correctness within their own and competing organizations.

The actual differences between descriptions of the same absolute truth are – after differing language, cultural, and religious modes of expression are removed – much smaller than you would likely imagine. This is because all spiritual paths converge on the same absolute truths by means of reducing ego and fear, which are the primary generators of confusion and divisiveness.

Models can be a practical representation of an underlying truth or process. The shell model of the atom is not based on a theory of why atoms, like mollusks, should have shells. The shells are simply assumed; the model is useful because it explains some of the data better than any other viewpoint. **Believing** in the shells is intellectually limiting silliness. What are atomic shells made of and where do they come from are not reasonable questions – atomic shells are only conceptual tools – they are a metaphorical structure describing (modeling) an underlying reality.

My Big TOE is also conceptual. We know that from our limited 3D perspective and local objective causality, a Big TOE that reaches beyond PMR absolutely **must** have at least one mystical leg to stand on. That mystical leg must be beyond the reach of our objective PMR based logic. As you recall, that same logical requirement was placed upon the understanding our local reality's beginnings (see Chapter 18 of this book). Causality and rationality require that we enter the metaphysical or mystical realm (from the PMR point of view) whenever we go beyond PMR. The existence of an apparently infinite absolute something seems both mystical and vague from our PMR perspective, nevertheless, that is not a weakness of this model. Quite the contrary, it is a requirement.

A **Big** TOE that contains no mystical (from PMR view) assumptions must necessarily be incorrect or incomplete and can logically never amount to more than a little TOE focused solely upon PMR. A little TOE restricted to PMR objective causality can never, even theoretically, grow up to be a Big TOE because it has, by definition, shut itself off from the solution space required to span the data. As we have shown, little TOEs can have nothing logical to say about our beginnings or our relationship to the circumstances of those beginnings.

We shall see that we are inextricably bound up with, and an extension of, our beginnings. A higher-level objective causality (of which PMR's objective causality and physics is a constrained subset) demands that we be solidly connected to what came before us. As a very young child is connected to, dependent upon, and extrapolated from its parents, we are connected to, dependent upon, and extrapolated from our beginnings. The inclusion of an assumption that is beyond the objective causal logic of PMR is an essential and necessary ingredient of a successful Big TOE.

With believers of all sorts dismissing every datum that conflicts with their dogma, self-limiting, self-inflicted belief-based blindness becomes a common social disease. So common, in fact, it defines the normal and therefore the rational view within any given culture. The point is: One should not necessarily expect the normal or traditionally rational view to shed light on new paradigms. When it comes to evaluating new scientific paradigms, one must look forward for answers and only glance backward long enough to make sure the new properly contains the old as a special case.

The only issues are whether or not this apparently infinite absolute something that I have chosen as my one mystical (from the PMR point of view only) assumption produces the desired results, and whether or not it is the simplest, most basic and fundamental place to begin our beginning. We will more precisely develop its characteristics and properties as we proceed.

The most important question is: can these two assumptions (the existence of consciousness and the process of evolution) deliver the goods? Can they provide a logical foundation broad enough and solid enough to support a comprehensive model of reality? Can a comprehensive, honest, and straightforward Big TOE, reflecting the elegant simplicity of our reality, be built upon them? Is the model based upon these assumptions both general and accurate? Does it make sense and fit **all** the data? Is it useful, practical and predictive? Does it produce objective measurable results? These are the proper criteria for judging the correctness and usefulness of the two given assumptions. As these two concepts are more fully developed throughout the next three sections, you will understand them and their implications more clearly and precisely. After you have completed Section 6, you will be in a much better position to judge the efficacy and value of the *My Big Toe* reality model.

This model, is results oriented, as are all models. An important question is how well does it fit the present knowledge-base including physics (all science). An equally important question is how well does it fit **your** personal data and the personal data of others? Consciousness has both public and personal components; consequently, both questions must be answered fully. However, though you may not see it this way now, you will probably come to the conclusion that it is the personal side that contains the most value and significance to you as an individual – even if you are a scientist. Make out of it (interpret it) and get from it (use it) whatever you can. The *My Big TOE* reality model will help you understand your life, your purpose, all of the reality you experience, how that reality works, and how you might interact most profitably with it.

Each reader brings to the table his or her personal data – their carefully evaluated experience. Most people will approach this model from one of four initial conditions. (1) They have no data to either support or contradict the model. (2) Their data supports the model. (3) Their data contradicts the model. And (4) their data supports and contradicts the model.

Let me address each initial condition separately: (1) If you do not have scientific or trustworthy personal or public data with which to assess the model, you should nevertheless find its exploration a thought provoking journey through a hypothetical landscape of very unusual yet coherent and reasonable conjecture. With no trustworthy data of your own, everything necessarily **appears** to be conjecture; however, this Big TOE offers a set of **potentially** useful maps that may help guide you toward the further evolution of your being and a better understanding of the world around you. Perhaps an intellectual or logical appreciation of the model will help you effectively evaluate and understand future data that you will no doubt discover now that your mind is open to the possibilities.

(2) If this model fits your data, it will provide a rational structure for your experience; a context wherein the whole of your experience makes good sense. You can either enhance your pre-existing Big TOE, or grow a new Big TOE. Either choice may potentially redirect and refocus your path of growth, development, and evolution.

(3) If your data clearly and directly conflict with this model, you should continue on in order to understand your own conceptual model more clearly so you will be better able to protect yourself from the confusion of some of the delusional fools you must share this planet with.

(4) If you find that some of your data support the model, yet some of your data contradict the model, you are most likely going to be confused. In this case, you need to pay close attention to understanding the limitations of your beliefs, reevaluate the data in conflict, make plans to collect additional data, and above all, stick to your guns until you have developed clear evidence and experience to the contrary. No one else's data or experience can be profitably substituted for yours.

The following advice is to all readers, irrespective of what their initial conditions are. (1) Always keep your mind wide open and remain skeptical – two traits common to all great scientists and explorers. (2) Do not leap to conclusions that are familiar, convenient, or easy because of belief or faith; instead, rely only on your personal experience data. (3) While contemplating **your** Big TOE, develop some ideas about what additional data you might need to collect and how you might go about collecting them. (4) Begin a program to collect the data you need to develop the knowledge and understanding you want to have.

You can always change your mind no matter which of the above four initial conditions you start with. Whether the *My Big TOE* trilogy **appears** to you as belief, knowledge, conjecture, or fact, or however you perceive the relative proportions of each, is entirely and only a function of **your** perspective – your belief, your knowledge, your experience. If you are "normal," you strongly believe (consciously or subconsciously) that your perspective and judgment is nearly flawless. Take care not to limit yourself. Maintaining an attitude of open minded skepticism will maximize the probability that you will not inadvertently or purposefully cut yourself off from the solution-set that contains the truth. ◀

Now that you have been properly initialized and calibrated, let us get back to work. Let's start with the most difficult concept first, the one that appears mystical from the PMR point of view. This concept assumes the existence of an apparently (not necessarily actually) infinite absolute something: A Oneness that is uniform, plain, and without differentiation. This **apparently** infinite source of our larger reality has no **discernable** boundaries, edges, or limits (as a perfectly calm ocean would appear to you if you were floating in the middle of it). This undifferentiated something would seem to be, and has often been called "All That Is," "The One," "The Void," "The Primordial Big Dude," and other similar philosophical descriptions. It represents the fundamental core of existence. It is a familiar concept to various religious and spiritual conceptualizations. An absolute no-thing, pervasive, yet existing beyond our space and time. It is simultaneously everything (in potential) and nothing (no-thing, no differentiation or boundaries).

Unfortunately, this is how we 3D creatures must express this concept. It is not an especially satisfying concept in light of our need to seem objective. It sounds mystical like curved space-time sounds mystical to those uninitiated in the mathematical basis of the general theory of relativity. From the larger perspective of a wider (beyond PMR) experience, it is as real as moon rocks. "Mystical" in this context means beyond our personal and collective ability to make an explanation within the context of PMR. Thus, to explain something (such as reality) that is a **superset** of PMR, one must not be afraid to explore the metaphysical, nor to offer serious scientific consideration to the seemingly mystical. That is the only doorway to a larger perspective, to peeking over the cultural-scientific-causality blindfold.

If exploring the conceptual mountains and swamps of metaphysical and mystical thought is at this point still discomfortingly unfamiliar or **believed** to be unproductive or just plain dumb, continue reading. Later in Section 3 (Chapters 47 and 48 of Book 2) and Section 5 (Chapter 74, Book 3) we will discuss a few issues of science and philosophy that may enlarge your perspective to the point that you feel more intellectually comfortable with being part of a bigger picture. Unfortunately, there are many who reflexively shake their heads and roll their eyes upon hearing the words mysticism and metaphysics: those individuals should reconsider the limitations of their basic assumptions and beliefs.

Each of us needs to continually make an effort to maintain an open mind, suspend our limiting beliefs, and raise our intellectual courage to the point that we can honestly explore new and uncertain territory: Vigilance and unending effort are the nonnegotiable price of intellectual freedom. A mind that is not free is simply a self-referential belief machine that continuously spins off useless and unprofitable thought energy. Belief and fear are the only ties that can bind a mind, while unconditional love and open minded skepticism set it free. A body may be enslaved by others, while a mind can only be enslaved by itself.

Open yourself to the possibilities and remain skeptical. Even if this journey takes you right back to the point at which you started, your time will have been well spent. Big Picture thinking will have taken place, concepts defined, possibilities raised, new perspectives viewed, old conclusions re-evaluated, and goofy ideas debunked.

Like Flatlanders, you must step through your mind, with a careful scientific evaluation of your subjective experience to understand what is beyond PMR. This simultaneously ancient and modern metaphysical concept of the infinite absolute unbounded oneness is a very useful idea, even if it is disconcertingly fuzzy. It is naturally impossible to get a firm intellectual grip on this idea because our PMR perspective is necessarily limited. It is a big concept, one that appears to fly in the face of two of our most sacred cultural dogmas, causality and objectivity. There is no point in struggling. Even if you were brilliant, (actually, I am sure you are – it is all the others I am referring to) it would still be a fuzzy concept. Do not despair; it is our nature as 3D beings in space-time to be removed from an **objective** sense of infinite absolute oneness.

I did not say that this absolute oneness exists everywhere. Space in this concept doesn't exist yet. Location, even relative location is not yet defined

– everywhere, in this context, has no meaning. I also did not say that this primordial consciousness was all knowing, omniscient, self-aware, or even vaguely intelligent – it is not. At this point, this rudimentary form of consciousness is not capable of forming or holding a single coherent thought. Imagine an immense unstructured, but structurable form of digital potential (the potential to be more profitably organized) energy.

You will soon discover that this seemingly infinite Absolute Unbounded Oneness (AUO) is not infinite. Nor is it absolute, unbounded, or a oneness – but only appears to be an infinite absolute unbounded oneness from a limited point of view within PMR. Although the name Absolute Unbounded Oneness is a descriptive misnomer, I use it because many readers will find the concept to be comfortably familiar. It would appear to be a better communications strategy to begin with a familiar concept and expand the basis upon which it rests, rather than the alternative (begin with an unfamiliar concept and then attempt to explain it in more common terms).

> ▶ The third logical possibility, to begin with an unfamiliar concept and explain it in terms of other even more unfamiliar concepts, is reserved for sophisticated university professors, physicians, and other specialists who are trying to set themselves apart, emphasize their superior knowledge, and mightily impress lesser beings such as you and me.
>
> This trilogy is not sophisticated – you have probably noticed that by now – the only big word I know is "bifurcate" (to split in two) and I have used that one already in Chapter 17 of this book. So, I'm done – that's as impressive as I get. Sorry Mom.
>
> Because we are using humor to delineate a few common communication styles, I would be remiss if I did not also mention lawyers and politicians who explain simple and familiar ideas in the most complex and convoluted terms possible in order to create the illusion that they have special knowledge that is critically important.
>
> The point is: There are many communication strategies, each designed to meet specific personal needs and purposes, and these strategies determine not only what you say and how you say it, but also what you hear and understand.
>
> Your purpose and needs dictate your communication strategy whether you are transmitting or receiving. Think about that. Does your personal communication strategy open you up, or close you down with regard to receiving and understanding a concept like AUO?
>
> It always takes at least two to communicate – each through selective send and receive filters of their own making. Receivers (listeners, watchers, or readers) appear to be in a solely passive role, but that is not so. They, knowingly or not, actively put a spin on every datum that they receive in order to make the interpretation of that experience-datum fit

satisfactorily within their current worldview. No doubt about it, the internal spin doctor is there to make you feel confident and secure in your knowledge and beliefs about how the world and the people in it work.

This subject will be taken up again from an entirely different perspective in a discussion of ego in Chapter 42, Book 2. I encourage you to make an effort to become aware of the particular implementation of interpretive listening that necessarily colors and filters **your** end of any attempted communication.

I will eventually paint a detailed high resolution Big Picture for your consideration, but if your beliefs distort it or filter it out, your awareness will never get the opportunity to evaluate its general significance or assess its personal value.

You will have to rely on open minded skepticism to deal with transmission spin that you suspect might be placed on the signal you receive. ◀

The word "infinite," used as a **practical** descriptor of this absolute oneness, does not refer to size in space. It refers to something we 3D humans don't experience directly – perhaps the concept of an apparently infinite potential merged with the concept of a totally blank mind will get you as close as you are going to get. Neither space nor time has been invented yet. I will get to those clever and necessary inventions later, but first this seemingly boundless something must evolve.

The concept of an **apparently** infinite Absolute Unbounded Oneness (AUO) is only half the ingredients we need to grow a new Big TOE. The remaining ingredient is what I call the "Fundamental Process." The Fundamental Process is the basic process of evolution. It is the repetitive pattern of trial and error, of spontaneous change and choice, of random events and encouraged behavior that enables evolution to result in progress. Progress is defined as change within an entity or group of entities that is in some way immediately beneficial or profitable to that entity. Profitable change leaves the entity or its group, or both, in a comparatively better, stronger, more functional, capable, and successful position relative to its internal and external environments.

An entity is a well defined, self-contained (bounded) interactive system. It can be an atom, molecule, rock, technology, computer, worm, monkey, human, organization, city, nation, planet, or an aware individual non-physical consciousness. The interaction of an entity with its internal and external environments is constrained by what those environments will support, encourage, or discourage. Thus, constraints that reflect the demands of the internal and external environments define the criteria for profitability and are the source of evolutionary pressure that pushes every evolving entity forward. When I describe evolution as an imperative, as a force that moves an entity along its evolutionary path, as a pressure, or a driver of change, remember that I am talking about a self-initiating natural process that represents how a self-modifying system or entity interacts with its internal and external environments. We will go over these concepts again in more detail.

The Fundamental Process, as it is applied to a complex entity, moves that entity toward those internal and external states of being that are the most immediately profitable. This motion is the result of external and internal pressures. Changes (evolutionary motion) that result from external pressures can be viewed as the cumulative results of the dynamic interactions between an individual entity and everything else, which in aggregate constitutes its external (outside) environment. Evolutionary motion that results from internal pressures can be viewed as the cumulative results of the choices that an individual entity makes relative to all available internal configurations and potential states of being.

People genetically engineering better people would be one example of evolutionary motion produced by internal pressures. Fear, ego, love, purpose, stress, pleasure, growth, contentment, ambition, self-improvement, satisfaction, confidence, self-esteem, and social interaction are a few of the internal constraints creating evolutionary pressure. Internal physical environments (internal to your body) may also pressure subsystems (internal organs) and specific tissues to modify themselves in order to gain individual or system efficiencies.

Consciousness primarily evolves by responding to internal pressures; simple biological systems primarily evolve by responding to external pressures. Internal and external evolutionary pressures may interact with, and influence, each other.

As individuals strive for maximum profitability, groups of interdependent entities may evolve together interactively to form a larger system or ecosystem. Interrelated systems interact with each other to form larger systems (such as earth's biological ecosystem and the solar system). Nothing stands alone – all are interrelated from the perspective of the largest level of organization. This applies to all evolving systems and systems of systems – animal, vegetable, mineral, technological, or organizational – and to subsets of systems (including your brain, stomach, or a networked computer), and to subsets of subsets of systems (including your blood and bone cells, or a microprocessor), and so on and so forth (cell parts and transistors followed by their supporting molecular and atomic layers).

Everything within a system evolves together, layer upon layer, within one big web of interaction. All systems exist, survive, and thrive by virtue of a local ecology that defines how their various interdependent internal groupings interrelate. Consciousness, the earth and everything that exists on it, governments, and the Internet all evolve their own supporting ecologies. To understand the dynamics of these systems, one must first understand each system's **local** ecology. To understand their significance, one must first understand each system's **larger** ecology.

In Section 5, where we discuss evolution as a process fractal, you will find the same basic evolutionary process turns up again and again at every level of reality – physical and nonphysical. You will see it is the Fundamental Process that drives the smallest sub-atomic particle to populate the states available to it and it is the same Fundamental Process that leads to Our System (OS) spinning off multiple probable realities containing multiple universes within nonphysical-matter reality (NPMR).

▶ Another short aside is now necessary to define two important terms that we are going to use repeatedly.

The term "Our System" (OS) is used often in this section as a high-level descriptor of the larger "place" we live in (OS is discussed in more detail in Chapters 79 through 81, Book 3). OS is defined to be PMR (our physical universe) **plus** the subset of NPMR$_N$ that is interactive with PMR. Here, the notation NPMR$_N$ simply represents a specific subset of NPMR.

If this terminology seems confusing, don't be concerned; it will become clear later. In the meantime, all you need to know is that OS is our home-base reality – the reality system in which we primarily interact. OS is composed of interactive elements that are both physical (such as sentient beings, information, energy, physical matter, organizations, PMR physics, all physical forms, and assorted critters) and nonphysical (such as sentient beings, information, energy, nonphysical matter, organizations, NPMR physics, all thought forms, and assorted critters).

The second concept to be discussed is that of entropy. It is my guess that most readers are not particularly brushed up on the meaning and significance of entropy – a word I use throughout this trilogy. Actually, I slipped in the first use of the word "entropy" toward the end of Chapter 21 – I hoped you wouldn't mind a few isolated instances of physics-speak. However, from now on, you are going to hear more about entropy; accordingly, it is time for me to be more helpful. This aside will theoretically save you the trouble of looking up the word "entropy" and give you a good idea what entropy is all about without becoming too precise or technical. ◀

▶▶ I say "theoretically" because almost nobody actually takes the time and trouble to look something up these days. We are always in a hurry despite how much time we have. We have a strong need to press on with or without the necessary information. Fuzzy understanding seems to always be preferred over delay or loss of continuity – detours are **fundamentally** annoying beyond the extra trouble they cause.

Before I get back to the task of defining entropy, let's wander a little farther down this particular rabbit hole and see what we can learn about how we personally relate to information, and how we are affected by a changing Western culture that is in the process of redefining itself through a revolution in electronics and information technology. The cultural metamorphosis accompanying our transition from the Industrial Age to the Information Age is as much about who and what we are as it is about what we do and how we do it. As our tools and relationships change, we and our reality change with them.

From radio to TV to video games to computers, the evolution of consumer electronics has modified how humans think and interact. Most of us, like kids watching *Sesame Street,* need constant stimulation and constant input or our attention wanders and boredom quickly sets in. To the electronic generation, thoughtful pauses become superfluous interruptions, and serious reflection becomes nearly impossible. Under a "use it or lose it" evolutionary mandate, depth and quality of thought are quickly becoming unnecessary and obsolete.

In pursuit of short-term goals in our professional lives and continual entertainment and stimulation in our personal lives, our attention spans grow shorter as superficiality dominates our mental processes.

I understand – we are all in the same boat – it is a Western cultural thing. This is how electronic entertainment and communications technology have affected our minds and thus our perception of reality. Our limited reality is slowly getting broader and shallower. The ecology of our minds is changing from a local lake to a global swamp.

To come full circle along the chain of cause and effect, you should note that the quality and significance of what the commercial media and book vendors produce for us to read is generally so unchallenging and superficial that missing a few words here and there does not make a significant difference. Guess and keep going is a time saving strategy that will always appear to work reasonably well because whatever we miss, by definition, is unknown and not worth the effort.

Our experience tells us that what we miss can be easily surmised from the context of what we get. Finding the most productive and efficient ratio of get-to-miss allows us to process the huge amount of largely inconsequential information that continually bombards us. In the Information Age, skimming instead of reading is a survival tactic. The calculated skipping and missing of random, difficult, or prickly pieces of content eventually becomes a well justified sampling technique that ultimately grows to exclude much more than irrelevant details. ◀◀

▶▶▶ In a practical sense, anything that we define as irrelevant quickly becomes that way. You do not miss what you are unaware of; consequently, all that lies beyond your vision appears to be irrelevant to you. That is why the uneducated care little for education and are unaware of what they are missing. The uneducated do not know or appreciate the deeper significance of being uneducated. A fact of life: The implications and consequences of your self-imposed limitations can only be seen and appreciated by someone else who does not share them.

If you wish to expand the paradigms that make up your worldview, the most salient question becomes: Who ya gonna call? (See Chapter 23 of this book.) Predicating the validation of a broader viewpoint upon the consensus of the people you respect and associate with is a particularly risky business because you like people who are like you – people who are likely to share your basic beliefs and limitations.

Among these people you feel solid, confident and self-assured. They confirm the soundness of your understanding and the correctness of your vision. You do the same for them. They define your community or social set at all levels of social organization.

Let me take a wild guess. As I look into my crystal ball, I see that you and your best friends are aware of almost everything that is **truly important** for you to know. Furthermore, it appears that you and your closest associates have an excellent understanding (probably better than most) of the truth that lies beneath the surface of most events and actions. Did I guess correctly? You are not alone. Almost everyone on every side of every issue (including wackos and fanatics of all sorts) feel exactly the same way.

"Hey, I'm doing well the way I am – what's the big deal over a little minor ignorance? Serious ignorance? Lots of it? Much important truth lies beyond my vision? Bullpucky! That's impossible! You must be talking about someone else. The people I hold in low esteem are like that, but not me, I know what's **really** going on." ◀◀◀

▶▶ Looking up unfamiliar words, concepts, or events is accurately assessed as an impractical exercise by busy people who are focused outwardly on entertainment and practical information. We focus on what we value. Don't blame the media guys; they produce whatever we are willing to pay for. We value what is quick, convenient, entertaining, easy, practical, and feels good or pays well. What else is there? For most of us, that is a difficult question, the honest answer to which is either the null set or an exceedingly short list.

Time management, a concept born in the Industrial Age, goes critical in the Information Age. If we are not sleeping, we are very, very busy and in hot pursuit of whatever is happening next. Most of us only value the present moment as a bridge to the future. Because we live on the run, we have a tendency to confuse life, meaning, and purpose with motion. The fact is, the choices we make (that ultimately define us) all occur in the present.

Talk about detours! This aside containing an aside that contains yet another aside is forcing you to think in multiple layers. I need to quit and get back on track before we both get lost and decide to go watch TV instead. ◀◀

▶ Entropy is a physics term that you learn about when studying thermodynamics – which is why it is probably not on your list of everyday words to use. Nevertheless, because of its dramatic implications, it is reasonably well known in many technical and philosophical circles.

PMR systems naturally dissipate their energy. If no new energy is put into a system, the energy that is available to do work within that system will eventually waste away. Everything, such as a new battery, has a shelf-life. Constrained energy (such as a flashlight battery or full tank of gasoline in your car), which is energy in a form that can do work, is either used up doing that work or eventually dissipates away (as its shelf life is exceeded). In either case, all the original highly organized and structured molecules and atoms making up the gasoline and automobile and the new battery and flashlight have been converted (after doing work or dissipating) to a new, less organized and less structured collection of atoms and molecules (including fumes, vapor, light, heat, worn off brake and tire particles, and metal flakes in the oil) that no longer can do as much work as the original configuration. We say that energy has been used up when actually it has only changed into a less structured, less usable form such as heat that is diffused throughout the system.

Any closed physical system must eventually run out of **useable** gas or batteries and become no longer able to do work because gas evaporates and both chemically degenerate over time even if they are not purposely consumed to produce light or provide transportation. We say that the entropy of the closed system has increased. "Closed" means that nothing is added or taken away – all the same particles and same total energy remain within the system before and after. Though the **total** energy must remain constant, it continuously, spontaneously, and slowly changes form over time (naturally degenerates as it grows older) as well as occasionally changes form (turn on the flashlight or ignite the gasoline) rather quickly. Eventually, one way or another, the available-for-use full potential of a new battery or fresh tank of gasoline becomes a comparatively useless collection of uniformly distributed vapor, radiation, and heat.

The change in entropy is a measure of how much energy is now (as compared to earlier) no longer available to do work. Equivalently, entropy is a measure of the disorder within our hypothetical system (less organization and less structure among the atoms and molecules in our closed system). More entropy means more disorder and less energy that is available to do work. Conversely, less entropy means less disorder (more order and structure within the system) and that more of the system's energy is available to do work.

The second law of thermodynamics states that the **average** entropy of the PMR universe (initially **assumed** to be a closed system) must increase with time. This means that all matter and energy in the universe will eventually move toward a state of inert uniformity – absolutely nothing left of our entire universe but hydrogen ions and elementary particles.

Don't sweat it, you still have time to finish *My Big TOE* and grow one of your own before you decompose. A universe that is as massively chunky as ours takes a long time to disintegrate. Look on the bright side: 1) the physical universe may not be a closed system and 2) the sun will expand and vaporize our entire planet long before the universe decays into a homogeneous soup of simple particles. ◀

You will get more detailed information about these concepts later. For the purposes of this section, you need only to understand that **all** things and systems of things are in an active process of change – animal, vegetable, mineral, technological, and organizational – and that the Fundamental Process is the basis for that change. All things evolve toward greater profitability. Profitability is defined by the degree of immediate success an entity has in dealing with the evolutionary pressures created by the constraints within their internal and external environments. This is true if the entity is an individual or a complex system of interrelated dissimilar individuals.

Non-living and non-growing things (such as uranium atoms, organic molecules, the house you live in, and the rocks in your back yard) generally evolve toward minimum energy states. When non-growing entities make a natural, spontaneous, or evolutionary change of state, the lowest **available** energy state is defined as the most profitable next-state to occupy. Most of the physical matter, on and in our planet, falls into this category. Thus, physical matter gradually follows a path to ever lower energy states, increasing its entropy as it evolves. To inanimate matter, natural spontaneous change due to interaction with its internal and external environments (evolution), disintegration, and increasing entropy are all related and moving together in the same general direction.

Typically, growing things (things that are in a state of becoming) can naturally and spontaneously – for at least some limited period of time – decrease their entropy. In contrast, the entropy of non-growing things always naturally increases. In PMR, the growing things ultimately depend on the non-growing things, thus fulfilling the grim prediction of the second law of thermodynamics. We will discover that in mind-space where AUO exists as dim consciousness, there are no physical things and **everything** has the potential to grow. In mind-space, seeking the lowest energy state is never a goal. Quite the contrary, for a consciousness system the most profitable next-state is one that enables that system to minimize its entropy. In other words, evolutionary change within a consciousness system is profitable for that system if it causes the entropy of the system to decrease.

There is a natural tendency for an entity to make those choices, or to exploit or succumb to those changes that move that entity toward its evolutionary goals. That is, toward a more profitable configuration or mutually beneficial arrangement – toward a lower energy state for physical matter (higher entropy) or a lower entropy state (higher usable energy) for consciousness. The Fundamental Process explores all the opportunities and possibilities for change, then inextricably and statistically moves each entity, each collection of similar entities (such as a species or a field full of rocks), or each collection of collections of diverse entities (community or ecosystem) toward its immediate goals by continuing to evolve the winners. The Fundamental Process of evolution is a recursive process that builds layer upon layer of interdependent organization and structure: a process that repeats and folds back upon itself at many different scales and levels of interaction.

Again, let me caution you to not be confused by the use of verbs in a sentence that has the Fundamental Process as the subject. That particular language construction is simply convenient and more succinctly expressive in describing the dynamics of evolutionary change. The Fundamental Process is a simple natural process – not an active sentient growth manager that can execute action verbs toward the achievement of some preconceived goal.

Each evolving entity moves into all the possibilities open to it that it can afford to occupy. It continues to invest in the winners by succeeding within those possibilities that provide the most value or profitability – not by deliberate **intellectual** choice-making – but by trying everything available and allowing whatever works (holds immediate value for the entity) to progress forward while letting whatever does not work (the losers) flounder. This is a natural selection activity from the inside executed by the evolving entity, not an intellectual activity or a directed response from the outside executed by the Fundamental Process. The only **external** forcing function is the external environment in which the entity exists.

From now on, when you read that the Fundamental Process does this or that, or makes this or that happen, or I use any action verb to describe an evolutionary effect, you will know what I mean, right?

This same Fundamental Process, when it is applied to earth-based biological systems, produces the evolution you learned about in school. The Fundamental Process applied to consciousness produces different results, because consciousness and its environments are different, but it is the exact same idea, the exact same process raised to a more general understanding.

For that matter, the Fundamental Process also drives change in organizations and technologies. All sorts of technologies have evolved – from the complex nest building of many insects, birds, and mammals to the transportation, medical, agricultural, communication, networking, and computer technology developed by humans during the preceding century.

The evolution of technology follows a similar Fundamental Process. It expands into the available states (needs, uses, applications) progressing the winners and dropping the losers. The evolutionary pressure, defined by the criteria for profitability, flows from the constraints of usefulness, feasibility, marketing, economics, and business. Here, profitability can be literal as well as figurative. Technological evolution (the evolution of a technical entity) is like the evolution of any entity. As technology evolves, it produces an increased richness and complexity of potential future states that creates more evolutionary opportunities for new or improved technologies. As evolution progresses, the pace of change accelerates.

Technological evolutionary pressure comes from the external economic environments of raw materials, marketing, supply, distribution, demand, and price, as well as the internal environments of feasibility, design, engineering, parts availability, manufacturing, quality, pricing, and finance. Mutations and new technical entities may seem to spring into being quickly and discontinuously, yet all are built upon the knowledge-base or genetic pool of earlier technical entities. The Fundamental Process works basically the same way with all types of entities – physical, nonphysical, human, insects, bacteria, molecules, rivers, mountains, rocks, organizations, nations, consciousness, automobiles, or computers. The differences in evolutionary patterns among animals, organizations, consciousness, and technology are not due to differences in the evolutionary process, but rather to the variety of entities and to the variety of environments and constraints that define the criteria for their profitability.

It is the Fundamental Process that encourages life and other types of evolvable systems or entities (including technological, organizational, social, and so on) to flow into any niche or configuration that can profitably sustain them.

For consciousness, living things, organizations, and technology, dynamic repetitive implementation of the Fundamental Process of evolution represents the natural statistical inclination of an entity, organism, species, or a complex interconnected population of entities to grow into and become whatever is most profitable from each organism's point of view. Simpler entities sometimes band together cooperatively to form more complex entities and systems that are more powerful and thus more profitable for all participants.

Groups of entities may develop mutually beneficial relationships and dependencies because they share an interactive synergetic environment. Eventually, after much integration, the view of the whole becomes more significant than the view of each part – they become one complex thing – a system that evolves as its parts evolve. Next comes systems of systems, and so on as complexity, specialization, and higher orders of organization (decreasing entropy) become profitable. Consciousness, alligators, governments, planet earth, and the internet all evolve the same way. Exactly what constitutes living and conscious and how such systems are uniquely defined will be described later in this section.

In short, the Fundamental Process explores every available opportunity and every possibility on all levels, and invests more heavily in whatever actually returns the best overall profits. In carbon-based biology, the assessment of profit has traditionally been based upon the constraints of survival and reproduction. The earth's biology represents one special case of the application of a much more general evolutionary process. For instance, for non-living and non-growing entities, what is most profitable is usually the lowest available energy state – the easiest and cheapest state to achieve. The **sequence** of these lowest available energy state "choices" made by an inanimate entity defines the path of least resistance, which that entity **naturally** follows if it is not impacted by external energy forcing it to do otherwise. That is why atoms decay, your house continually needs maintenance, and rocks roll downhill.

Even the most basic forms of PMR activity follow the evolutionary pattern of choice and profitability. Thus a radioactive atom decays and the shelf-life of a battery deteriorates according to the Fundamental Process. On the other hand, an atom is pumped up to a higher energy state, sand is turned into sandstone, and mountains are thrust up along fault lines because of the input and absorption of external energy. Even in these circumstances, the entity **plus** the absorbed and retained energy will arrange themselves into the most profitable minimum energy configuration.

The Big Picture Fundamental Process of evolution (or Fundamental Process, for short) is as follows. An entity starts from any point (level) of existence or being, spreads out its potentiality into (explores) all the available possibilities open to its existence, **eventually** populating only the states that are immediately profitable while letting the others go. It then iteratively progresses the successful states to their logical conclusions by probing new opportunities and limitations. As long as there are profitable opportunities, the process continues to iterate and the entity continues to evolve. Additionally, intermediate states can generate (branch to, or be the potential seeds for) new states. States that no longer hold potential for (do not support) profitable growth may persist or be recombined with others with which they are redundant. Potential is generally maintained, in the event that new initial conditions appear.

We will now apply the Fundamental Process to our assumption of an Absolute Unbounded Oneness (AUO). The first step in that process is to expand an entity's potential for existence or being into all the significant possibilities (available states). An apparently (but not actually) infinite, absolute, unbounded, undifferentiated, thing can most obviously find potential new states of existence or possibilities of being by changing its undifferentiated oneness. That is, it can most directly and easily interact with itself. Additionally, because at this point in our story AUO is the only thing we have posited to exist, it is logically necessary that it interact only with itself.

Most biological things are aware of their environment (external awareness of things such as hot/cold, acid/base, and wet/dry). Additionally, they are aware of themselves (internal or self-awareness of things such as hungry/full, hydrated/dehydrated, threatened/safe, peaceful/agitated, hurt/not hurt, pain/pleasure, joy/sadness, and fear/love). All life-forms (biological, consciousness, or otherwise) change themselves. Sometimes the change is in reaction to their environment – they adapt. Sometimes the change is internally generated – they mutate. Evolution typically moves toward more complex, capable, and highly organized lower entropy forms – building upon or enhancing the earlier simpler forms. More advanced products of evolution can sometimes successfully change their environment – like prairie dogs building cool moist underground homes in hot arid places.

▶ Beep. Beep. Beep. Beep. Beep… Uh, oh… that is the number three audio alarm on my new problemometer. Hang on a minute while I find out what's the matter…. All right. …I see…. *No Problema,* the solution is a simple one.

Relax, you are not brain damaged, retarded, or becoming mentally infirm with old age. If you have not been graced with a career that brushes up against government organizations awash in acronym stew, a poor memory for acronyms is normal. If you are not used to using acronyms, they seem, for most people, to have been given a special non-stick coating that causes them to constantly slip from memory. It is a natural brain wiring and symbol association problem that occurs when the name-symbol represents a string of associated words instead of a single common name. The solution? Forget the words; let the acronym be a single name.

For those not used to acronyms, here is how it works. The first time or two you encounter an acronym, think about what the words mean, get a sense, an understanding, of the concept being talked about – then forget about the individual words making up the acronym and use the acronym as a name associated with the meaning and concept the acronym represents. Subsequently "AUO" becomes a name like "Fred" – a name that you associate with a finite, but seemingly infinite, evolving primordial consciousness. It is that simple – AUO becomes the proper name of a particular conceptual entity.

If you do not treat the acronym as a name, you may feel obligated to look up what the constituting words are every time you see it. Dutifully looking up acronyms wastes time and effort and generates little value. The individual words are not important; the concept, the definition, the meaning is important – attach those to the acronym itself and go on.

If you forget the **concept** associated with AUO, look it up in the acronym list at the beginning of each book and follow the page reference to where it was first defined. However, if you have a firm grasp of the meaning and significance of AUO, then the individual words that AUO represents are not important – let them go. Do not struggle with how habitual language use has wired your brain. Rewiring is difficult and provides little value for the effort spent.

Acronyms are handy for condensing long **descriptive** titles into short names. I use them extensively throughout the text because an acronym has the nifty attribute that it carries its own definition within its name. Treat them as words and symbols, the same way you would treat any name of any noun and your memory will not seem so sieve-like. Problem solved. ◀

We now have our two fundamental concepts in place, and as required by our local PMR causal logic, at least one, AUO, appears to be mystical from the view of PMR. These two assumptions provide the necessary and sufficient foundation (called a "footer" in the construction trades) from which to build a new Big TOE. Neither is a new idea.

Darwin ushered in the theory of evolution by natural selection in the late 1850s. The concept of AUO is, and has always been (at least since about 600 B.C.), a fundamental concept in various philosophic, spiritual, metaphysical, and religious traditions. For our purposes, we utilize only the bare basics of this concept, no frills, no dogma, and no extraneous conditions or attributes – just a plain, simple, minimalistic Absolute Unbounded Oneness – a form of potential energy that appears to be non-physical from the limited PMR point of view. AUO is a seemingly infinite primordial consciousness whose beginning depends upon a causal requirement that is so far beyond our local reality that it is outside our rational consideration – as refrigerator light bulbs, sunshine, or stock market dynamics are outside the rational consideration of our intestinal bacteria – nevertheless, the bacteria are **indirectly** and powerfully affected by these things.

The Fundamental Process of evolution is what enables and necessitates growth, learning, and change. Without the Fundamental Process, nothing could become more than it is now, all change would be random, direction-less, and purposeless. We know from our personal experience and the diversity of biological systems that this is not the case; that there is a fun-damental imperative for all life-forms to grow and evolve is obvious. Furthermore, it is also obvious that exploring the possibilities and investing in what works is the engine that drives the process. The evolution of bio-logical forms on earth, though a special application of a more general Fundamental Process, demonstrates these facts adequately.

From the foundation of these two assumptions alone, we will construct piece-by-piece, a rational Big TOE based primarily on direct experimen-tal evidence with some reasonable conjecture added to bind it all together. It will be shaped by whatever common data points (accepted facts and truths) we can find. If successful, there will be no known facts or truths that conflict with this Big TOE. The degree of your success in corroborating this reality model with your experience depends on the correctness of My Big TOE and on the correctness of the truths and facts with which you test it.

The sub-foundation has been laid. Next we need to develop the basic logical results and consequences of our two basic assumptions. In the process we will improve and refine our understanding of each assump-tion. If things are not crystal clear at this point, they will become clearer as we proceed. By the end of Section 4, you will be able to pull these con-cepts together into a complete foundation for the reality model of Section 5. Developing basic concepts may be tedious, but this is where the Big Picture must necessarily begin. Getting the basics is usually the least fun part, but also the most crucial. Let's get started.

The adventure begins! Ladies and gentlemen...start your engines!

25

■ ■ ■

The Evolution of AUO:
Awareness

■ ■ ■

The awareness of a single cell, an amoeba, a worm, a bumblebee, a chimpanzee, and a human being are all fundamentally based upon the same process. Awareness differs mainly by degree. Somewhere along the path from simple single process action-reaction (or stimulus-response) to a more complex multi-process interaction-reaction, the capacity to remember (store information) evolved. It is this ability to store or retain information that enables learning. Even the simplest of biological life-forms have the capacity to learn by retaining or storing information in their physical structures (DNA for example). This learned information can be transferred to future generations through reproduction. Learning through the evolutionary process of exploring the possibilities and retaining the results of lessons learned (in your DNA or in the Library of Congress) so that they can be passed on to future generations is a natural self-reinforcing process.

Deoxyribo Nucleic Acid (DNA – a nucleic acid that carries the genetic information in the cell and is capable of self-replication and synthesis of ribonucleic acid), and other forms of bio-memory, are not as difficult to influence or modify as most scientists think. It is commonly thought that it takes thousands of years for evolution to modify biology substantially – that many millennia must pass before complex biological systems (like people or polar bears) evolve new ways of being – new instincts, new firmware. Quite the contrary, biological systems do not necessarily go through a long drawn out and difficult random mutation and selection process to modify bio-memory. In a relatively short time, significant change can be established or disestablished that will be passed from generation to generation if those latter generations reinforce the change because they also find it profitable. Memory, capable of inputting and retrieving data quickly enough to support cognition, is a primary enabler of both species learning and dim awareness.

Shorter data input and data modification times, as well as quick read and write access, are required for **individual** learning and to support a brighter awareness. It should be obvious to you that a brighter, more functional awareness is naturally associated with larger memory, shorter memory input and access times, and better information processing capabilities. What does that remind you of? A biological brain connected to a sensory apparatus – or maybe a computer? It appears as if biological brain consciousness or digital consciousness systems, operating with high bandwidth input-output (I/O) and very short processor cycle times and memory access times, might **theoretically** have an exceptional capacity for individual learning and brightness (evolving their own firmware and software). That theoretical potential for the exceptional brightness of a brain or a digital computer could perhaps become actual brightness if the data input-output systems connecting it to its internal and external data-source environments, along with its purpose-goal-intent-action-result feedback systems, generate a rich enough set of uniquely profitable possibilities to provide adequate evolutionary potential to support the development of a high level of mental function. In other words, if the entity **interacts** (at a minimum receives data and makes choices based on those data) within sufficiently complex environments and develops sufficiently high bandwidth connections to those environments, it will eventually evolve a higher, brighter level of awareness to better utilize that data. If the range of possible choices and the methods of selection between choices are also sufficiently rich and complex, the potential for brightness greatly expands. Hold on to these concepts: They will become more important later on as brains and computers are related to digital consciousness.

Thus, biological systems – living things – evolve, learn, grow, and change purposely. They make progress. Their existence is directed by finding more profitable ways of being and relating. Profitable strategies such as storing food for the winter, and technologies such as nest building, are developed, internalized, and "remembered" (retained by biological modification) and put to good use by later, more successful generations. Over time, a biological or genetic group memory may develop – a sort of hardwired cultural belief system.

Heredity is not the only transfer mechanism. Simple flatworms have been trained to respond to external stimuli. Untrained flatworms that are fed the trained worms seem to gain the knowledge of their consumed brethren. (Kids, don't try this at home. Unless you are a flatworm, eating your parents or teachers will not make you any smarter and could get you into serious trouble).

Most of the more advanced products of evolution eschew chewing on their intelligentsia and find it more efficient to exchange information by energizing each other's sensory inputs and outputs – sight, sound, smell, touch, and taste for example. This communicated information is processed, evaluated, acted on, stored, and retrieved. When a large dog faces you squarely, growls and shows his teeth while the hair stands up on the back of his neck, do you get the intended message? Is there a successful audio-visual information transfer?

Simple life-forms know when things feel better. For instance, they know what, when, and how to eat and reproduce **selectively**. The extent to which learning takes place depends on the extent and efficiency with which experience interacts with memory. The degree of complexity of the life-form allows for variations in the specialization of the organs of awareness. Such specializations define the breadth and depth of the sensory input and the speed and efficiency of storing, retrieving and using information to describe, evaluate and select possible action-reaction sequences. All life-forms react to both their internal and external environments. Every life-form is in the problem solving mode trying to improve its situation and increase its immediate profitability. Problem solving is the natural response to environmental pressure. Plants turn toward the sun, night-crawlers scurry down their holes as they assess minute seismic danger signals, and humans have invented whiskey and nuclear bombs.

Awareness is a characteristic of consciousness. To some extent, all the earth's life-forms are aware and conscious. All store and retrieve information more or less continually and therefore have the capacity to learn, to improve themselves, and to become more than what they presently are – to evolve. Some awareness is more or less efficient, brighter or dimmer, than others. It is a continuum. Jellyfish, raccoons, and people are all aware of some things and unaware of other things. It is more a matter of degree and focus than a fundamental difference in the nature of awareness. Limitations on the richness and complexity of the possible interactions and responses that an entity can have with its external and internal environments are what define each entity's capacity for awareness. The process of developing awareness is the same for all entities, whereas the limitations and the environments of each entity are unique.

By now, it should be clear that developing and evolving awareness is a standard, normal and natural happening among biological entities. Similarly to a one celled biological entity, AUO (Absolute Unbounded Oneness) develops awareness of itself in relation to itself by mutating or modifying its internal uniformity or structure. I am using the term awareness here relative to AUO in the same way that it is used above. Let's call it dim awareness. Dim awareness is fundamentally the same as any awareness, only very limited in its scope and complexity. At its dimmest, dim awareness is simply the potential for awareness.

Consider the first simple multi-celled creatures floating in our oceans eons ago. This dim cellular awareness figured out (in the evolutionary sense) that if certain cells would specialize in specific functions (eating, digestion, reproduction, locomotion, coordination, control), then more complex and more successful creatures could evolve (improved survivability and procreation). What a discovery!

This and many other pivotal evolutionary discoveries provide strong testimonial for the usefulness and efficacy of dim awareness as motivated by the Fundamental Process. One thing leads to another and, long before time runs out (our sun expands to incinerate the earth), evolution progresses earth biology from simple groups of dimly aware cells to a diversity of relatively clever fish and reptiles (complex biology that is more aware of, and interactive with, its environment). Eventually, this brain train moves on to a quirky species of naked apes capable of genetically engineering themselves. Awareness grows and evolves similarly to how physical bodies have grown and evolved. That is an important concept to understand.

How does this Absolute Unbounded Oneness indescribable something acquire dim awareness? The same way one celled things did, and simple multi-celled things did, and so on up the ladder of evolution: by interacting with whatever constitutes the external and internal environments. AUO is real and therefore not actually infinite and thus may have an external environment as well as an internal environment. Nevertheless, because this external environment and its description are beyond our comprehension, we will not use it to develop our TOE.

An external environment may or may not have contributed to AUO's initial dim awareness, but we have no logical need to include it. It is adequate for our purposes to assume that AUO interacts only with itself (mutation) to gain dim awareness. Developing awareness is the natural result of a self-modifying entity applying the Fundamental Process to challenging environments (internal or external) that contain large numbers of multiple possibilities.

Awareness at the cellular (dim) level evolves naturally from the existence of a rich and varied array of available potential states. These states are generated by direct choice, random events, and dumb luck – being at the right place at the right time with the right stuff. The entity explores all the potential states it can successfully populate. The more successful or profitable configurations continue to evolve while the less profitable ones either remain in their niche or die off. Over time, hardware (body), firmware (instincts) and software (mind – processing capability) are slowly improved.

Applying the Fundamental Process is natural and intuitively obvious to even extremely dim awareness. Try everything that is accessible and affordable, and go with what works. Is AUO alive? Certainly not by our 3D earth related (biological) standards and perspective, but then our standards are only relevant to PMR. I will show that AUO meets the requirements for developing awareness and eventually intelligence.

All that is required for AUO to evolve some type of limited awareness is that it exist interactively within the context of a challenging environment (external or self-generated) that contains a large number of multiple possibilities, and that there is available to it a rich and varied array of potential choices or states of being with variable profitability it can explore. Remember, we are conceptualizing and postulating something that by definition lies beyond our 3D comprehension. In our effort to understand AUO and NPMR, we are like Flatlanders struggling to visualize a tetrahedron slowly tumbling within a rotating hollow cube that is spiraling down the inner surface of an elliptical funnel. Go explain that to your two-dimensional buddies over flat beer.

Because the existence of AUO is the one mystical or metaphysical (from a PMR perspective) assumption that enables this Big TOE to discuss our beginnings logically, we need to understand this AUO-thing. In some ways, AUO can be thought of as roughly analogous to the first biological cells from which many of our planet's living things evolved. AUO was dimly aware of itself in the same way that primordial biological cells were dimly aware of (could interact with) themselves and their environment. Where did AUO come from? Our relentless and illogical sense of a universal causality forces this question. There is no good answer. The best bad answer: AUO came into existence the same way those first biological cells did. Good luck eventually ran into the right ingredients at the right time with a supporting environment and perhaps a spark of some sort of mental energy thrown in for good measure – I truly do not know. AUO's origins lie beyond the logical reach of our understanding. Our PMR causality can offer no rational input on its own beginnings, much less AUO's beginnings. The only logical failure is in the asking of the question.

Where did the ingredients, the environment, and the spark come from? Leave me alone – go ask your mother! It is not an appropriate question. Demanding answers to questions such as these is analogous to demanding that a Linelander (a one-dimensional being) describe a sphere to his two pointy-headed friends. He cannot do it. He does not have the capacity to comprehend spheres because of the fundamental limitations of his existence and his reality. The question is simply not relevant to his reality or to his existence. Likewise, intestinal bacteria cannot understand the stock market or the properties and value of sunshine – these are beyond rational consideration for intestinal bacteria. We humans are in the same position relative to the origins of AUO. There are some things, which may affect us indirectly, that we do not have the practical or theoretical ability to perceive or understand. Denial of that fact (or at least that possibility) may change our perception of reality and support our needs, but it cannot change the nature of reality.

You should continually work on developing scientific data about what you can understand while avoiding making assumptions and creating beliefs about what you cannot understand. Appreciating your limitations is the first step toward both knowledge and wisdom.

From some limited perspectives (like the perspective of our space-time PMR universe), the answers to some questions cannot be comprehended or conceived. You have to live with that fact. Metaphysics, mind, and a careful exploration of the subjective offer a less limited perspective that goes a step or two beyond our current objective science, but go too far beyond that, and truth dissolves into speculation while speculation degenerates into fantasy. To be sure, conjecture can quickly wander into fantasyland if it does not remain connected to a rational foundation of objective, measurable truth-data. How well the model performs is what lends credibility to the assumptions made at the beginning. Save your judgments until later.

You will discover by the end of Section 5 that everything we do in this reality seems to **follow** (is in the image of, one might say) AUO's pattern of being. It will become clear that our existence, being, and evolution are an expression of, and the result of, the existence, being, and evolvement of AUO. Our individual consciousness reflects the pattern of AUO as a piece of a hologram, or each individual element of a large fractal design, contains the pattern that defines the whole. If the evolution of our consciousness and our physical reality is a reflection of AUO's evolution, it would seem reasonable (as one can be addressing this issue) that AUO's evolution should exhibit similar processes to our own. The actual reasonableness of that statement depends on how closely we followed AUO's original pattern. AUO is primary consciousness, we are derivative consciousness. It is not unusual that a child might resemble its parent.

26

■■■

The Evolution of AUO:
Who Is That Masked Consciousness
Thing, and What Can It Do?

■■■

At this point in our exposition, AUO is apparently the only thing that exists (no space, no time), and its dim awareness interacts primarily with itself, in relation to itself. To be more accurate, AUO is apparently the only thing that exists that is interactive with, or relevant to, our reality or us. What may or may not exist beyond AUO is also beyond relevance to us. Thus, for now, we can ignore any hypothetical external supporting environment, or conceptualize it as a non-interactive void – whether it is or not.

What else do we know that can exist outside the bounds of space and time? If you think about it for a moment, you will immediately become dimly aware that thought, mind, consciousness or spirit are the only things known to us that are not bounded by our time and space. AUO is primordial consciousness. Could our individual consciousness as well as the physical universe be related to, or derived from, AUO? Absolutely!

> ▶ A quick glance forward may provide a more solid basis for you to properly structure the concepts that are now popping up in front of you. Sometimes it is easier to make sense out of what you are doing if you can see where it is leading. As this Big TOE slowly comes into focus over the next three sections, you will eventually come to realize that consciousness is the basic medium from which reality is formed – the fundamental non-material material of existence. It will also become apparent that a more structured form of this fundamental primordial consciousness energy can reasonably evolve into an ultra-low-entropy consciousness that can generate the operative digital mind required to animate and define the content of our larger reality.

Consciousness provides the basic organizational energy from which all individuated existence is derived. It provides the self-modifiable form, function, and potential energy the Fundamental Process of evolution optimizes. Eventually, something analogous to cell specialization evolves within AUO to create internal dimensions that support various realities populated by individuated units of consciousness. The Fundamental Process drives change at all levels including the growth opportunities that eventually define individuated consciousness content such as yourself. By the end of Section 5, you should understand how our physical reality, including you, is created, the process by which it operates, and what purpose it serves. You will also better understand the nonphysical dynamics underlying your physical existence as well as the nature of both personal and public reality. I do not want to get too far ahead of my story lest it tempt you to jump to conclusions that are wrong-headed and in error even though they might seem obvious at this point in the unraveling. ◀

Let's be careful not to fall into an anthropomorphic trap. I am not implying that AUO is (or ever becomes) conscious or sentient in the same way that a human being is conscious or sentient. I am implying that AUO is (or represents) primordial consciousness – an energy form that, like earth's early carbon-based bio-forms, has the potential to evolve awareness. AUO represents the primordial substance of consciousness itself, not the attribute of being conscious.

Consciousness is the result of evolution offering profitability to a self-organizing complex system of relational digital content. Another way of conceptualizing AUO is as an apparently infinite potential energy. The potential energy of digital consciousness is the potential to self-organize, which is derived from the potential of a digital system to reduce its average entropy through evolution. A system able to lower its entropy through self-organization will eventually host some level of consciousness. Over the next several chapters we will see how a digital model of highly aware consciousness can be derived from a single monolithic consciousness system that is driven to ever lower values of average entropy (higher values of usable energy) by the requirements of evolutionary profitability.

What can AUO do with itself to create the stimulating environment that drives evolution and how did it figure out how to do this? It figured out how in the same way those biological cells did. It may take an extremely long time (dim awareness can be very slow), but under the steady pressure of a relentless evolution constantly nudging all systems toward greater profitability, progress is eventually made as new possibilities are explored.

What could AUO do that was interesting and challenging and stimulating?

It can first create a duality within itself. This vs. not this. It can alter its uniformity to create a local disturbance or distortion. Then there exists, distorted relative to non-distorted. A duality has been created. It is as if that calm, still ocean you were floating in a few chapters ago to gain an intuitive sense of apparently infinite absolute unbounded oneness now has evolved to the point of making a tiny ripple. With lots of practice and some good luck, ripples eventually become waves and whirlpools. Distortion or any other non-uniformity exists only in relation to uniformity. How many different kinds or types of disturbance can AUO create relative to its uniformity? Who knows? For our purposes, we need only one.

A single localized modification of the uniform state is all that is required to derive ourselves, our physics, and all the rest of our reality (physical and nonphysical). This way (uniform) relative to that way (locally non-uniform) is sufficient. Given that AUO somehow discovers a disturbance or non-uniformity relative to itself, it can return to uniform, then intentionally reestablish the locally disturbed state, then return back to uniform and so on. We now have the possibility of a regularly recurring event that will eventually evolve into a precisely recurring event. Figures 2-1, 2-2, and 2-3 show three examples of a regularly occurring continuous change between non-uniform and uniform states. A precisely occurring event might eventually be used to invent time (the first tick-tock, tick-tock, tick-tock as it were). But before we could reasonably expect something as advanced as uniform vibration or oscillation, AUO has a lot more evolving yet to accomplish.

Considering biological evolution on earth, one would say that one celled critters discovered multi-celled configurations were more effective, and later discovered that cell specialization provided a more efficient multi-celled organism. Here, I am using the word "discovered" in the evolutionary sense. Such discoveries are not necessarily well defined events. They are more typically diffused happenings that occur over long periods of time and after huge numbers of more or less random internal experiments have been evaluated for profitability (improved survivability for bio-organisms or lower entropy for consciousness organisms).

The continuing evolution of AUO requires more and more complexity in the ways that AUO might express differentiation of one part of itself relative to another. What are its opportunities (the possibilities) for dualizing? There are many ways AUO could change state and then change back. Here are a few of the simplest ways. AUO could, for example, discover dualization in a bipolar as well as unipolar mode. The unipolar mode (uniform, positive-disturbance, uniform, positive-disturbance, uniform, and so on), could be in the form of $|\sin(\omega t)|$ (shown in Fig. 2-1), or

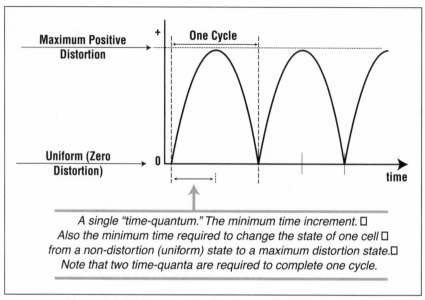

Figure 2.1: Unipolar oscillation in the form of |sin(ωt)|

perhaps [sin(ωt) +1]/2, or sin²(ωt). It could also be in the form of a linear sawtooth (Fig. 2-2), or any number of other useful forms.

AUO could also oscillate between positive and negative disturbance states

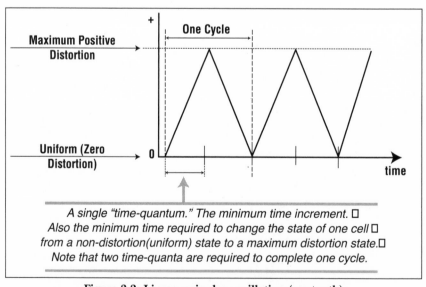

Figure 2.2: Linear unipolar oscillation (sawtooth)

(uniform, positive-disturbance, uniform, negative-disturbance, uniform, positive-disturbance, uniform, negative-disturbance, uniform, and so on). This type of oscillation is called "bipolar" and could, for example, be in the form of sin(ωt) (as shown in Fig. 2-3). Here, the words "positive" and "negative" represent opposite disturbances such as good/bad, hot/cold, more/less, sharp/dull, bright/dim, acid/base, hungry/full, thought/no-thought, or whatever suits AUO's nature and newly found sensibilities. Notice that I avoided using spatial opposites (such as here/there, up/down and in/out) because space is not a native construct of AUO and has not yet

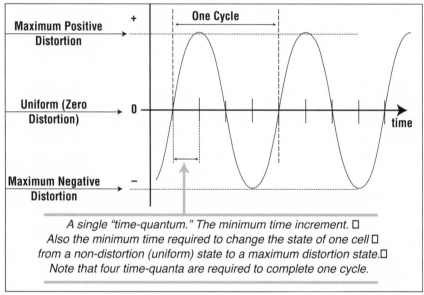

Figure 2.3: Bipolar oscillation in the form of sin(ωt)

been discovered by AUO.

If math (trigonometry) functions such as sin(ωt) or |sin(ωt)| are not a part of your everyday reality, do not be concerned. They are not particularly important to understanding the binary nature of disturbed vs. not-disturbed. I use them here only to demonstrate the wide range of simple as well as complex possibilities that AUO could employ to define this versus that and because the logical side of Mother Nature seems to be partial to sinusoids. That AUO, similarly to the primordial biological cells of our planet, somehow evolves to become dimly aware that it can affect changes in its state is the main message here. The details of how it changes state are not important to the conceptual Big Picture being painted.

AUO is now at a level that is analogous to the one celled or few-celled organisms floating in the primordial soup that represents our planet's biological beginnings. In fact, in honor of this analogy, let's call this singular relative disturbance in AUO a reality cell.

As the Fundamental Process dictates, awareness, however dim, always creates ever more possibilities for itself to explore. Likewise, complexity, however simple, always generates greater possibilities for the evolving entity to investigate. Thus, as evolution would have it, AUO eventually discovers it can simultaneously create two disturbances relative to its seemingly infinite uniformity, then three, then four, then hundreds, then billions, then.... After all, an approximately infinite, practically unbounded oneness should easily manage and support quantity far beyond our comprehension. We ourselves are composed of billions upon billions of cells. Continuing our analogy with biological systems, AUO is now at the stage of a huge multi-celled but still very simple behemoth. Here a cell, there a cell, everywhere a cell cell ... (reality cell, that is), but no structure, no order, no specialization of function – yet.

Growth in the type and number of reality cells should be sufficiently challenging for a long time because each increment of growth creates new possibilities for new configurations of the larger system. Likewise, each new state of the system represents new possibilities and opportunities for growth. Awareness, complexity, and useful control grow with the number of potential states or possibilities AUO can profitably explore (reach through evolution). What happens next? AUO is about to get organized.

27

■■■

The Evolution of AUO:
Patterns, Symbols, Information,
and Memory; Motivation and
Evolutionary Constraints

■■■

If you were AUO, what could you eventually do with a trillion billion gazillion reality cells? For starters, you could begin arranging your local deformations (cells) into patterns. Then patterns of patterns, and groups of patterns of patterns – whatever your mind could hold. Would a good memory be helpful here? Isn't it neat how the imperative to implement the Fundamental Process creates evolutionary pressures that deliver whatever is most profitable to the evolving organism (mega-memory is on the way)? Where many choices can be profitable, the outcome is great diversity.

To get a glimmer of the potential of AUO organizing its cells into multiple levels of patterns, look at what William Shakespeare accomplished by arranging and rearranging only twenty-six letters of the English alphabet. Consider the contents of the Library of Congress, and this book you are reading: together they represent a tremendous output from simply rearranging twenty-six letters of one language into groups of patterns of patterns. If every word **ever** written or spoken in **any** language were stored utilizing binary reality cells as memory (using distorted vs. not distorted, in the same way computers use 0s and 1s to represent letters and numbers), AUO would have used up only an infinitesimal percentage of the available reality cells. Such is the nature of seemingly infinite unbounded oneness. It seems limitless; its capacity, relative to ours, **appears** infinite.

I do not say that AUO appears to go on endlessly and forever because applying our space and time constrained words and concepts to ideas far beyond their reach creates confusion and paradoxes. AUO cannot be conceived of in terms of space and time. Yet we 3D creatures of limited knowledge must use the tools and modes of conception and communication that we have – and stretch them to reach out to the unknown that lies beyond our horizon. It is important not to let our words and their implicit PMR conceptual limitations trap us into believing that reality is, by definition, fully contained within three spatial dimensions and restricted to what we can physically perceive.

By now, AUO is (or parts of AUO are) considerably less dim. Continuing with our biological analog, it is now time to diversify and specialize. Within biological forms and functions, diversification and specialization are primarily driven by two basic pressures: survival and propagation. Applying the Fundamental Process of evolution to biological systems creates these pressures as well as creates the various individual strategies that are designed to respond to these pressures. The survival (continuing existence) of a particular individual member of a species is not only significant to that individual, but also contributes to the profitability and continuance of the species, evolutionary experiment, and ecosystem. Each individual, besides being an autonomous entity, is an integral part of a larger system. Nothing stands alone, disconnected from the rest. To misquote John Donne: *"No man or bacterium is an island."*

▶ During the eighteenth, nineteenth, and first half of the twentieth centuries, humans, in a typical display of little picture self-centered arrogance, believed they stood alone above all else – particularly in the rapidly industrializing West. Many, whose economic dynamics are still at the front end of the industrial revolution, continue to feel and act that way. That is easy to understand. Industrialization is a process that in the short-run turns natural resources into wealth, power, and a higher standard of living.

Unfortunately, short-term profitability often drives non-ecological utilization of resources. Rapid industrialization is a very difficult package to turn down for the sake of ecological responsibility – especially if you are one of the last looters to get to the scene of the riot. To update and urbanize an old maxim, "the early looter gets the best TV." Furthermore, the early looter is also the least likely to get caught and face the consequences.

It is an obvious fact that the easiest way to get some quick cash to support an immediate higher standard of living is to mug Mother Nature as she walks through her park. Fortunately, she is an exceptionally hardy and charitable sort and gracefully tolerates our abuse up to a point. However, to pass that point is to trade an endless supply of golden eggs for a single goose dinner. Even if you are extraordinarily hungry, that is a stupid trade.

As the bigger picture of a global ecosystem has come into view, some have realized that we have responsibility as well as rights of plunder by virtue of superior force. A viable and stable ecosystem must achieve and maintain balance. That we humans have the capability to seriously disturb that balance is the source of our responsibility.

Our natural environment does not constitute the only global ecosystem. Eventually, among the human population, we will discover (and learn the rules and dynamics of) global technological, political, social, and economic ecosystems. In global systems, individual players always have bit parts, yet all are vitally important, even from a system perspective. In general, the larger the system, the smaller the part any one participant plays. Each individual entity (including you) plays out its unique interactive part and absolutely affects the whole whether or not they are aware the whole exists. Some individuals and groups affect the larger system more than others and thus must assume a greater responsibility commensurate with their greater impact and influence.

A complex ecosystem can maintain balance only if a great ability to use, exploit, and destroy is tempered by an equally great responsibility to conserve, replenish, and protect. Ability must be balanced by responsibility, force with caring, and fear with love. With great responsibility comes great challenges and great opportunity. The bigger your perspective and understanding, the more effective and productive you can be within your system.

Your ability to take intelligent action depends upon the depth of your vision and the quality of your understanding. The smaller your perspective and understanding, the more likely you are to inadvertently shoot yourself in the foot. Low quality consciousness does not see the Big Picture and makes short sighted decisions for near term gain. To a being of low quality, feel-good appears more important than do-good which appears more important than be-good.

There is no rule that requires evolution to optimize profitability through only random trial and error activity. **We are allowed to use our brains.** No kidding, we actually are. I know there is little hard evidence to support that assertion but a bigger perspective and understanding of the larger system and our optimal interaction with it and responsibility to it is well within our intellectual capability.

Theoretically, we humans are intelligent and educable. However, let me remind you that intelligence and education are attributes of individuals, not groups – you cannot pass the buck or abdicate your **personal** responsibility (rely on somebody else to solve a shared problem) without becoming part of the problem.

Large ecosystems come in layers. To capture the point of this discussion, you must generalize the above thoughts to a much bigger picture. We are participants in an evolving consciousness ecosystem. The consciousness system in which we participate is the mother of all others. It is the largest, most complex ecosystem we can interact with. It supports all other ecosystems. Our physical universe is a small virtual habitat within this larger consciousness system. The same ecological issues, concepts, conclusions, and lessons-learned that we discussed earlier in this aside apply directly to you and your relationship to the consciousness ecosystem. Science is science, the principles are the same. Reread this short aside with each observation about us and our physical ecosystem being applied and reinterpreted to describe our interaction with the larger consciousness ecosystem – you will gain a new perspective. ◄

Any species, or any evolutionary experiment, continues to fulfill or actualize its future potential until there are no more significant potential states to investigate and the Fundamental Process grinds to a halt. The experiment is over, at least temporarily. This does not necessarily imply extinction, just that no future progress, growth, evolvement, or experimentation takes place until new external or internal conditions create new significant possibilities. In biological systems, the imperative or urge to implement the Fundamental Process is naturally energized and guided by the profitability of survival and propagation. For evolution to remain viable within a system, the system must maintain a **continuing** set of significant, useful, or profitable choices and possibilities. Specifically, for biological systems, evolution remains a viable process only if the biological organisms continue to survive and propagate. Nevertheless, survival and propagation are not the only test of evolutionary significance or profitability of biological organisms.

Both extremes of biological evolution are not explainable by positing survival and propagation (as we generally think of it) as the **main** motivators of choice. The first small groups of biological cells had no enemies (even the environment became friendlier over time) and they weren't interested in eating each other – not yet. Life was simple and good.

Toward the other end of the evolutionary spectrum, many humans are likewise no longer focused on survival issues on a daily basis. Technology intervenes in our behalf. For example, almost a fifth of us in the Western world have a genetic vision impairment that is correctable by technology (wearing glasses or contacts). Ten thousand years ago such a defect would have made us much less survivable. Are our modern bespectacled brethren still dropping like flies because of their inferior visual equipment? Is evolution weeding them out as genetically inferior? Does having to wear glasses or contacts make it more difficult to survive, find a mate, and propagate your genetic material? A little perhaps, but not as much as it used to.

Some futurists picture us as evolving into people sporting large brainy heads with spindly arms and legs – our physical weakness both created and overcome by technology. If survival and propagation were the only measures of evolutionary significance, why would we need such big brains? Have we not dominated our competition with the brain we have? We have only one natural enemy now – ourselves.

Perhaps we need bigger brains to avoid self-destruction. That proposition doesn't make us seem very smart. It is our brains that make us so destructive. A bigger brain, as the solution to the excesses of a bigger brain, is hardly a logical solution. I have read somewhere that we use only a fraction of our brain's capacity as it is. Bigger is hard pressed to justify greater evolutionary profitability. It is not a lack of intellectual capacity that is at the root of humankind's self-inflicted difficulties and dysfunctional behavior. The fundamental problem is the lack of quality in consciousness. Quality is the issue, not quantity or capacity. We obviously have more processing capacity within our consciousness than we have the quality, maturity, or wisdom to put that capacity to good use. As a species, we are out of balance – and that is not good.

Mother Nature, who is talented at evolving interdependent complex systems, always produces self-balancing systems. Natural systems are self-regulating because the evolutionary process of maximizing the profitability of the **overall** system must be a self-optimizing and self-balancing process in order to achieve and maintain stability. If a natural complex system is knocked out of balance beyond recovery, it self-destructs or regresses to a state where the unbalancing factor is brought under control. Growth toward an out-of-balance condition within a complex natural system will eventually be eliminated one way or another (future growth moves the system toward balance, the system regresses to a previous stable state, or it self-destructs).

Within biological systems, the most obvious example of an unbalanced system leading to self-destruction is a quickly growing cancer. In contrast, within consciousness systems, the most obvious example of an unbalanced system leading to self-destruction is ignorance, fear, desire, need, and ego – a lack of quality.

What is creating evolutionary pressure if we are no longer preoccupied with the traditional issues of survival and species propagation? We have become aware enough and capable enough that **self-improvement** is now our main goal. Most importantly, that means spiritual evolution, or equivalently, the improvement of the quality of our consciousness. Self-improvement could also mean the development of science and application of technology that might eventually result in spindly arms and legs, or merely developing a more civilized civilization. It could also mean something as quick and direct as human genetic engineering.

Self-improvement is the result of an internal pressure created by the Fundamental Process. Evolution expressed through a high level of sentience becomes the urge to grow (reduce entropy), the drive to increase profitability, and an inherent dissatisfaction with stagnation and decay (increasing entropy). Stagnation is at best an astable state for any life-form. Typically, if growth is halted and does not soon resume, decay sets in.

The internal environment produces the evolutionary pressure of self-improvement for the sake of greater individual and system profitability while the external environment specifies system constraints. Clearly, self-improvement represents the major evolutionary pressure driving change in contemporary humankind as well as the original clumps of cells in the primordial sea. Take note that inside environments can be a significant player in the evolutionary process and that survival and propagation are not the only important variables in the profit calculation supporting PMR evolutionary dynamics.

Profitability is a matter of dynamic evolutionary circumstance that is wholly dependent upon the nature of the inside and outside environments. Whether or not a driving evolutionary force leads to a more profitable or less profitable state for any particular self-modifying and evolving system depends on the collective choices made (path chosen) by that specific system.

▶ There is a bigger picture in which our overall profitability equation is shown to have additional terms beyond PMR related survival, propagation, and self-improvement that you will understand much better after you have completed Sections 3, 4, and 5 of *My Big TOE.* For now, keep your skepticism healthy, your mind open (free of beliefs), and realize that it is much too early to jump to conclusions – there will be plenty of time to do that later. ◀

Our level of awareness has delivered us to the point where we have within our grasp the potential to dramatically affect the direction and speed of our evolution. However, to use this potential to improve the profitability of our species within the larger system will require a dramatic increase in our ability to integrate technology with wisdom at the point of individual application. Without a doubt, improving the quality of our individual and collective consciousness is the most important and critical aspect of self-improvement (evolution) facing the human race as we move into the twenty-first century.

Balance is always important. If our technical know-how gets too far out in front of our ability to apply its results wisely, we will lose our balance and fall flat on our face – perhaps irretrievably. Ours is a particularly critical time. The choices we make during the next half-century will dramatically affect the outcome of the next half millennium.

Improving the quality of our consciousness is, and always has been, fundamental to our evolution but today it is also critical to our survival and to the continued success of the Homo sapiens experiment. If we accomplish a significant improvement in the quality of our consciousness first, if we can lead with our quality, the rest of our options will be guided by our wisdom and we will leap boldly ahead. If, on the other hand, the quality of our consciousness lags, and we do not accomplish a sizable measure of spiritual growth first, we will have the cart in front of the horse, so to speak, and it is going to be a wild and dangerous ride.

Isn't the evolution of consciousness a fascinating process? I suggest that we invest some serious effort in improving the quality of our consciousness. I also suggest that we fasten our seat belts.

We have been discussing our biological evolution because it is best to start with something basic that we at least think we understand. In doing so, we have defined the concepts of evolutionary pressure and specific criteria for evolutionary profitability. Having done that, it is obvious that survival and propagation are non-issues to an Absolute Unbounded Oneness. Instead, it is self-improvement, the reduction of system entropy, that provides the evolutionary pressure and profitability criteria that guide AUO's evolvement.

In order to understand the natural pressures that drive AUO's evolution we must first understand how consciousness improves itself by achieving higher levels of internal organization (entropy reduction). Let's review the evolutionary process. After the first step of exploration into **all** possible states, the second step is to assess which states are profitable or significant and to continue to explore those valuable states while letting the losers go.

Do not imagine that evolution, or an evolving entity, makes rational, cognitive choices using specific criteria that define profitability and significance (such as lower entropy, survivability, and propagation). Though I may use that language construction as a convenience, be aware that evolution must **begin** as a natural, not a cognitive or intellectual, process. Only after a certain level of awareness, competency, and tool use has evolved can evolution be influenced by individual intent.

Potential or possible states are continually explored. Those that happen to lead to reduced profitability, self-destruction, or dead ends are the losers. These losers either remain as they are or degenerate if there are no more unique possibilities or viable existences to explore that are also good investments. Every evolving complex system will eventually either keep growing (entropy decreases) or run out of evolutionary gas (entropy increases or stays the same). This is true of all large complex systems (consciousness systems, earth's ecosystem, the solar system, technological and organizational systems, computer systems, the internet, and biological systems including physical human beings).

However, keep in mind that evolutionary potential is dynamic. Investment opportunities come and go as evolution continually modifies the system upon which it operates. Change one thing and everything is affected, including the external and internal environments of the individual entities that make up the system. Change is constant ... and often chaotic. As long as there are large numbers of complexly interacting self-modifying entities, there will be evolving systems of those entities.

Survival and propagation represent the external pressures or constraints of a biological environment. Thus, if a particular complex system is biologically based and has constraining issues such as survival and propagation, choices relative to these constraints can put certain avenues of exploration out of business. On the other hand, where there are few external constraints and no competition is defined, individual systems, and systems of systems, simply continue (evolving) as long as they have someplace to go. Thus, with fewer constraints, evolution delivers more diversity because there is always someplace to go. For example, there were few constraints placed on offensive or defensive survival strategies or on modes of locomotion for biological life-forms. Consequently, we observe many, many types of survival strategies and modes of locomotion among the creatures born to Mother Earth, as well as many variations within each type. AUO naturally has few constraints, and therefore is granted great freedom of form and function in solving evolution's profitability equation.

Winning evolutionary strategies are not based upon some **predetermined** specific criteria or profitability calculations, they just happen and persist. Losing strategies just happen and then fade away or stop progressing. No one performs an evaluation. I know this is difficult to fathom, but accountants and lawyers are not required to implement the Fundamental Process – it just happens. No doubt some lobbying organization is trying to figure out how to legislate mandatory legal and accounting services into the evolutionary process and then charge high fees, but so far, all life-forms continue to evolve, change, and grow without professional help.

▶ In fact, Mother Nature so abhors bureaucracy that even if she gets raped, she will not call the authorities. However, there is little comfort in that fact for any of us because if the abuse she suffers goes too far, she will eventually get even – count on it. And when she does, all within her reach will be equally punished regardless of culpability.

On the other hand, as long as we are respectful, she is willing to let us have our way with her whenever desire and indulgence pique our insatiable appetites. She is an extraordinarily robust lady – at ease with poisonous snakes, killer bees, vampire bats, erupting volcanos, tornados, and earthquakes – but shortsighted lobbyists and self-serving politicians scare her to death because together, like a malignant tumor, their destructive potential may overpower her capacity to regenerate. Greed kills. ◀

Optimal profitability, as it is constrained and prodded by internal and external environments, defines evolutionary success. Profitable being, growing, and evolving constitute the criteria for evaluating the success of sentient entities. The Fundamental Process iterates individuals and systems toward success by improving upon previous results wherever growth and investment potential exists. Because no management and no intellectual efforts are required to control or manipulate the evolutionary process, it remains efficient. System within system within system – the whole must be self-balancing, self-sustaining, and self-correcting, or it becomes self-eliminating.

Because survival and propagation do not constrain our timeless, immortal, space-less, apparently infinite AUO, it would seem to be free to pursue every possibility open to it. If you happen to be apparently infinite, spatially unrestrained, timeless, immortal, multi-dimensional, and aware, you have tremendous potentiality. With so few constraints, there is almost no end to what you could do to keep yourself busy and avoid becoming dead-ended. (It should now seem more obvious to you why the less constrained NPMRs should evolve a more varied set of life-forms than the more constrained PMRs, but more about that later.)

A potential constraint on the capacity of a finite consciousness system might be the energy or focus that would be required to keep track of all the "*gedanken* experiments" in progress. To maintain and profitably organize all the information required to support a complex interactive virtual reality would require a good memory – or perhaps a big computer. The existence of limiting factors (because of AUO's actual finiteness) would inevitably lead a self-modifying computational system (digital organism) to set priorities and develop rules that optimize resource utilization.

Because there must be limiting factors associated with any form of actual or real finite existence, then "infinite" and "unbounded" do not accurately describe AUO. However, from the relatively tiny perspective of our local reality system (OS), AUO appears infinite and unbounded – there are no edge effects that we experience **directly**. Imagine how unbounded and infinite your brain would appear to a carbon atom stuck in the middle of it. We shall soon see that a **relatively** infinite Absolute Unbounded Oneness is sufficient for the purposes of generating all the reality that we can possibly be aware of – and much more.

It makes sense that evolutionary pressure operating upon a complex system of consciousness would encourage that system to invest in improving the extent and quality of its own awareness (the ability to profitably interact and organize). In other words, the Fundamental Process causes AUO to organize and configure itself to decrease (minimize) its entropy. Equivalently, we could say the Fundamental Process causes AUO to improve and develop the potential of its being, the quality of its consciousness, the level, breadth, and depth of its potential.

That appears to be similar to **our** mission, **our** reason for being, doesn't it? Does AUO's motivation and evolutionary purpose flow down to us? Of course; we are manifestations of it, so how could it be otherwise? We will hear more about that later. For now it is enough to understand that through increased complexity in relation to itself, AUO escapes atrophy and puts its limited energy (ability to modify itself, to change its internal environment) to the most profitable use – continually redefining and improving its overall capacity, quality, and awareness. AUO develops an awareness of its **internal** states and learns how to modify them to decrease its overall entropy.

Other types of interaction and awareness related to AUO's **external** environments may also be evolving but they are not **directly** relevant to us; consequently, we will continue to ignore them because we cannot experience or understand their processes or their significance. AUO's outside environment is to us like our outside environment is to our intestinal bacteria. Those little entities that live in our gut are so immersed in their own limited reality that they simply do not appreciate what we go through to earn money to purchase food – much less the roles that farmers, sunshine, rain, transportation, and economic conditions play in the larger food cycle. Jeez, some of them don't know there is a larger food cycle – can you believe that? They think food is manna from heaven. Bullpucky! They should deal with your boss and the line at the checkout counter!

Let's wrap it up with a short summary. Devising and implementing ways to optimize the process of making self-improvements, as judged by AUO's immediate sense of profitability, leads to improvements in the quality of

AUO's consciousness. That is in line with what those tiny globs of cells were doing in the primordial soup a few billion years ago on our planet. They were engaging in simple self-improvement – exploring the possibilities – trying to find a better, more profitable existence. The energy form and constraints are different, but the evolutionary process is exactly the same. AUO is constrained by its internal environment and the Fundamental Process to move toward a more profitable existence by decreasing its average entropy, or equivalently, by increasing its useful energy.

Self-improvement has become humanity's primary evolutionary motivator as well. The external environment for humans is essentially subdued; consequently, we have become our only major threat. The greatest challenge to our species today is to survive the self-destructiveness of our own low quality of consciousness. We must now learn to master the internal environment. Our success at gathering knowledge and making tools has placed a great capacity to destroy in hands animated by low quality, underdeveloped, immature consciousness. The need for rapid self-improvement has become critical. Will the individual conscious beings that collectively define humanity grow up (lower the entropy of their consciousness or equivalently raise their spiritual quality) enough to make the choices that will allow their species to prosper?

Might this maturing of the spirit or quality of mankind happen sometime soon or do we need to experience more pain before a significant number of eyes and minds begin to open? Remember, growing up is not a group or political activity, it is a personal activity. Groups raise their average quality level only as the individuals within those groups make a personal effort to increase their individual quality. Self-improvement of the species in general is up to you as an individual. Nobody has more potential to contribute than you do. Only individuals acting as individuals can make a difference.

Tough guys with big muscles have less real power and are less in control of their environments than brainy bespectacled nerds. Survival and procreation are giving way to self-improvement as the defining evolutionary constraint for human-kind. Genetic engineering and psychotropic drugs are a quickly growing reality. The control we can exercise over our physical bodies is rapidly increasing. If the maturation and quality of our consciousness were as precociously developed, it would provide grace, stability, and balance to our great leaps into the unknown. We are slowly taking the reins of physical evolution into our own hands. Climbing out of the Petri dish and into the lab, we are becoming a co-designer – a partner in the evolutionary process. We are undoubtedly following in AUO's footsteps because the Fundamental Process is the same for everyone and everything, and because we, like AUO, are primarily manifestations of aware consciousness.

The fundamental process (which develops the potential of an individual entity or system of interacting entities by continually exploring the available states of profitable existence) applies equally well to humans, other beings and critters, consciousness, ecosystems, technology, governments, and all other sufficiently complex interactive systems. Systems are simply an organization of individuals whose synergistic interaction produces collective results. Systems of systems are developed in the same way.

All growing systems, including consciousness systems, evolve toward minimizing average entropy by generating more profitable levels of organization. Inanimate things such as rivers, mountains, rocks, atoms, and molecules change through an evolutionary process that moves toward increasing average entropy by seeking the lowest available energy state. Non-living physical matter, objects, and systems do not have the capacity for self-improvement or entropy reduction that sentient consciousness has because their choices, interactions, feedback, and the ability to make self-modifications as a result of their "experience" are either nonexistent or too limited.

It seems that we have derived a new and more general method for differentiating between living and non-living objects or systems (virtually all objects are systems). Living systems are sentient and exhibit consciousness – no matter how dim or unfocused. They have choices. Their choices are guided by intent and profitability – they evolve. Any system that can intentionally (act, react, interact) improve its present situation or configuration, or purposely (by making the appropriate choices) decrease (or maintain) its entropy in the face of the constant tug of the second law of thermodynamics, is a growing, evolving, living system. The attributes of a consciousness system are given more precisely at the beginning of Chapter 41, Book 2. For now it is enough to know that the capacity and capability to effectively self-organize toward some purpose (evolutionary profitability) is critical to forming and maintaining aware consciousness and all life as we know it. Does that make your government, the earth's ecosystem, a national or global economy, digital computers, and the internet **potential** life-forms? Stay tuned.

It is the fact that non-living and non-growing systems tend to move naturally (evolve) toward minimum energy states, along the path of least resistance, that is ultimately responsible for the truth of the second law of thermodynamics. We will see later that this fact of PMR existence is one of the rules in the space-time rule-set that defines the laws of physics in PMR.

Later it will become clear that we are an integral part of AUO's evolutionary process – that we are evolving consciousness.

Where is this AUO thing going, how do we fit into this picture, and what does AUO have to do with deriving physics, the meaning of life, our physical reality, or the Big Picture of the larger reality? Are we lost wandering aimlessly in the metaphysical desert? We are not lost and the answers to these questions and many more will be forthcoming. Be patient, there are many basic concepts that first must be introduced before the Big Picture can begin to take shape. It is better to let these questions slow cook a while before we get back to them. Before results can be discussed, we need to follow AUO's evolution to its logical conclusion.

That is it for the summary. I expect that questions are popping up in your mind like toadstools after a summer rain. To what end is AUO motivated to expend resources? What defines AUO's top evolutionary priorities? Mental capacity? Growth? Power? Curiosity? What turns AUO on – art, science, knowledge, benevolence, love, entertainment, investment strategy, leisure, pleasure, or just having fun? Interesting questions to be sure – but also hopelessly anthropomorphic ones. Avoid the habit of thinking of AUO in terms of human attributes; there is a connection, but attribution naturally flows from the parent to the child – try thinking of humans in terms of AUO's attributes.

28

■ ■ ■

The Evolution of AUO:
The Birth of the Big Computer (TBC)

■ ■ ■

Meet AUO – a gazillion reality cells more or less randomly changing state, oscillating, appearing and disappearing. These eventually evolve into patterns and groups of patterns of patterns as awareness and complexity increase. Some patterns are more interesting and profitable than others. Recall that higher levels of organization leading to lower average entropy define self-improvement within a consciousness system. The Fundamental Process in interaction with consciousness produces an evolutionary pressure that pushes all consciousness entities toward self-improvement.

The concept of an entity organizing itself into a more profitable ensemble of patterns and patterns of patterns returns us to an earlier discussion of how and why life-forms diversify and specialize. Bacteria, guppies, alligators, kangaroos, people, arms, eyes, legs, breasts, brains, electric automobiles, the internet, and representative government are just a few examples of the advanced biological world organizing itself into more profitable ensembles of patterns and patterns of patterns. If somewhere within the far reaches of your intuition the phrase "patterns of patterns" suggests some sort of undefined fractal process as the ultimate creator of complex structure, you are on the right track. Hang on, we'll get there eventually.

Some patterns might indicate a subsequent or a preceding pattern or define pattern interaction. Rules are themselves none other than knowledge-based or experienced-based patterns for the reliable definition, repetition, and interaction of other patterns. Rules defining the formation of patterns, as well as pattern interaction and relationship, would evolve naturally from the evolutionary pressure of self-improvement. With the concept of rules come the concepts of control and hierarchy.

For instance, binary patterns of 1s and 0s, on and off, or for AUO, uniform and non-uniform can be used to do arithmetic, store information, and move data around between groups of cells. Simple binary patterns, and the rules (instruction-set) defining relatively few operations, have **evolved** into today's computer technology.

Unquestionably, it is the patterns, the patterns of patterns (and the rule-sets that evolve to define, regulate, and order them) that constitute the basic ingredients of almost everything we directly experience. For just one example, imagine a human brain's pattern of cellular organization, and its patterns of neuron and electromagnetic activity. For other examples, think of the city you live in, the economy in which you work, and the political and cultural patterns that order your life. Civilization is about ordered patterns defined by profitability and constraints – so is everything else, including consciousness.

It may prove to be a useful concept to describe consciousness as made up of discrete fundamental units (quanta). Some minimum group of complex interactive reality cells would constitute a quantum of consciousness. Given that groups of reality cells are at the heart and core of consciousness and given that the simplest form of a reality cell is a binary unit, then logic dictates that computers have the **theoretical** potential to become conscious devices. After all, they are made up of basic binary units that are like the reality cells that led to the development of consciousness, memory, and pattern processing within AUO. Clearly, having a collection of binary cells is not enough. That these cells interrelate, use and share information (patterns of data), and modify themselves around some purpose or intent (regardless of how dimly perceived) is critical to the formation of consciousness quanta.

AUO could specialize a small part of itself into gazillions of tetra-tetra-tetra bytes of memory because that would be an interesting and useful thing to do. Indeed, very useful! In the last chapter we concluded that evolving consciousness may be constrained by accessible memory. Evolving a memory and processing section could lead AUO to invent its **own** form of mathematics – which is nothing other than patterns, patterns of patterns, rules, operations, and relationships – a self-consistent system of logical process.

This computational and memory function of AUO represents the primordial Big Computer. The Big Computer (TBC) is an important metaphor used throughout this model of reality. TBC's function and operation are described and discussed in more detail throughout the rest of *My Big TOE* but especially in Section 4 (all), and Section 5 (specifically

Chapters 78 and 83 of Book 3). The Big Computer is just that – memory, processing, rules, operations, and content. It is not necessarily The Big Brain as we think of it in biological (discrete physical organ) terms. At this point, it is more of a memory-intensive computationally based process that provides improved organization to a relatively dim cellular mind. Think of TBC as a digital computational functionality (based upon discrete reality cells) that naturally evolved to improve organization, and thus reduce entropy, within the energy form we call AUO.

AUO itself now represents a form of digital consciousness based upon reality cells. AUO is able to reduce its entropy and brighten its awareness because it evolved gazillions of self-differentiated reality cells that eventually became organized into patterns of patterns, rules, and ordered processes – all maintained, tracked, and controlled by evolving groups of interactive binary memory cells whose actions and interactions are coordinated to facilitate a more highly organized and profitable system. Digital logic and memory are needed to apply the rules of interaction and to coordinate the whole toward greater profitability. Content sharing and intentioned manipulation evolves on the heels of digital logic, thus adding direction and purpose to the process. Purpose differentiates success from failure and provides the necessary rationale and direction to achieve lower entropy configurations that appear as system self-improvements.

Think of TBC as a special purpose digital processor and memory subset of a much larger digital consciousness. AUO is evolving specialized groups of cells in the same way and for the same reasons that biology-based physical critters did: The same Fundamental Process is universally applied to all interactive or intra-active systems complex enough to be capable of profitable self-induced change (growth). Nevertheless, the results of applying that simple process differ widely according to the capabilities and constraints that define each system. A system's capability is intrinsic to the nature of the system while its constraints are defined by its internal and external environments.

Just as evolving non-living, non-growing entities naturally move toward higher entropy, consciousness naturally moves toward lower entropy. Dimness gradually gives way to brightness as entropy is lowered. The quality of consciousness also increases as entropy is decreased – we call this growth of quality, spiritual growth. Just as **external** energy can **sometimes** drive non-living, non-growing entities to decrease their individual entropy for a time (they grow – like sand into sandstone, minerals into crystals, decomposing bio-mass into fossil fuels, or the fusing of hydrogen into helium), **internal** energy can **sometimes** drive consciousness to greater entropy and lower quality. More about this in the next section as we pit our favorite team, The Rats, against the Anti-Rats in the Reality Bowl.

▶ We now know almost enough to pull together a better understanding of the word "dimension." We have said the larger reality is a multidimensional reality and that PMR and NPMR represent two of those dimensions – but what does that mean?

Let's begin with what we know. I have used the term "three dimensional" or its abbreviation "3D" to characterize our PMR local reality. Some may be confused because they have been told that modern science uses a 4D space-time to model PMR. Indeed, space-time, as the term is used within general relativity, is referred to as a 4D continuum – three space coordinates (that are also functions of time), and one time coordinate (that is also a function of the space coordinates). There is mathematical justification for calling space-time four dimensional, but in the sense that I use the word "dimension," mixing time with the position coordinates does not constitute a new dimension.

Although time – the technology and process AUO will eventually construct for ordering events – is fundamental to defining and creating the larger reality, it is not particularly helpful to think of time as an independent dimension. It is more descriptively accurate to say that we humans directly experience a 3D time-ordered reality that appears to be the perception limited product of a space-time universe requiring four coordinates to specify individual events relative to one's frame of reference, wherein each possible reference frame is as fundamentally proper as any other.

Searching for a fourth and higher dimensions in terms of geometry is also not particularly helpful to the understanding of the larger reality. Reality is not fundamentally geometric. We think it must be geometric because our little picture view is centered in 3D geometric reality and we naturally tend to expand upon what we know. PMR is geometrically constrained by the space-time rule-set; however, the space-time rule-set is only a local rule-set and does not apply to the larger reality. This will become clear in Section 4.

What is this dimension thing – If it is not time and not geometric, what is left? Forget about all the sci-fi and fantasy movies you have seen, the dimensionality of reality does not function as the screenwriters for *Twilight Zone* or the *X-Files* would have you believe. However, it is true that most realities simultaneously exist within their own dimension and as a subset of a larger reality, and that travel between dimensions is possible – indeed, it is relatively easy once you know how.

We will discuss the concept of dimension in more depth later; for now, I want to peek ahead just enough to give you some sense of the nature of dimension without confusing you too much in the process.

Because the larger reality and the subset of the larger reality that serves as our local reality are constructs of consciousness, we could say metaphorically that they exist in mind-space or thought-space. Connect this with the idea that fundamental consciousness, in the form of AUO, is composed of reality cells and that reality cells may be employed as binary cells – cells that are either in this state or that state. Binary cells, like transistors in a microprocessor or the 1s and 0s on your favorite compact disc, are handy for creating memory, storing information, and supporting complex processing. Thus mind-space, in the form of digital consciousness, is beginning to look like a logical, rule-based, computational system – a generalized computer of some sort.

You will discover in chapter 30 of this book that the fundamental potential energy of consciousness (AUO) eventually evolves into brilliant digital consciousness as its dim awareness slowly brightens. Within digital consciousness, the creation of various dimensions is simply the creation of separate memory and processing subsets. For years, we have implemented partitioning in our commercial digital computers as well as figured out how to multitask and multiprocess. We can easily, for example, run multiple simulations and multiple instances of the same simulation (with differing initial conditions perhaps) simultaneously within our PMR mainframes. Think of each simultaneously running simulation as representing an independent reality within its own dimension and you will have a glimmer of the concept of dimension within digital consciousness, and within the larger reality. "Dimension" refers to a well defined processing subset within The Big Computer, a constrained thought-space or region of related content processing within the larger consciousness. A particular dimension or reality may be thought of as a digital simulation of a virtual world running in its own memory-space within TBC. The concept of dimension will become clear as these ideas are developed in later chapters. ◀

Now we must begin to contemplate things happening in sequence. There is an immense benefit to the organizational potential of a system if operations and content can be arranged in a specific order and sequence. AUO, in its pursuit of profitability, needs to "invent" time, thereby inventing ordered process. Like sea creatures needing to "invent" lungs and legs in order to crawl out of the oceans to exploit the resources available on dry land, AUO needs to invent time in order to exploit the increased order and organization (lower entropy) that comes with ordered process.

Time provides the indexing and sequencing scheme to support the next higher level of organization, awareness, and complexity within consciousness. With the possibility of indexing and sequencing interactive content, dynamics (time ordered causality) is born. The application of time to digital content causes the number of explorable possibilities available for the fundamental process to explode exponentially into a creative interaction of cause and effect.

Time is a digital technology that enables consciousness to organize its content (thoughts) more effectively. Improved organization implies reduced entropy and greater profitability for the system. The evolution of mind is about to shift into high gear as the potential of aware consciousness takes another great leap into the unknown.

29

■ ■ ■

The Evolution of AUO:
The Birth of Time

■ ■ ■

AUO is about to get natural rhythm. Time is as easy for AUO to invent (evolve) as it is for us to keep time by tapping our foot. To create time, our state flipping friend needs only to oscillate (repetitively change the state of) some individual or group of reality cells more or less regularly (uniform, non-uniform, uniform, non-uniform, and so on). These regularly oscillating reality cells become AUO's clock – like a metronome, they keep time for everything else. If the frequency of this group of clock cells is constant, it will be a more useful clock. The best clock would oscillate as efficiently, regularly, and quickly as possible (its natural frequency). Because "works better" is what drives evolution, we should expect that AUO would eventually evolve a process that produces a highly regulated constant frequency.

Time is a technology, a construct of a self-modifying evolving consciousness, an artifact of a system of energy improving its internal organization. When the potential energy of primordial consciousness (the potential to self-organize more profitably) evolves the ability to decrease its own entropy one infinitesimal smidgen, time is the byproduct of that internal change. Time separates the "before" state from the "after" state. Change creates the notion of time. Awareness of change necessitates the idea of a personal time. The concept of time is defined and created within the dim awareness of AUO when the Fundamental Process enables AUO to change something, to somehow modify its absolute oneness – even if that change is entirely random.

The initial modification may have been a quirky unplanned fortuitous event, or directly related to AUO's unknowable external environment – no one can know. We cannot with certainty specify the origins of the first bio-logical cells within PMR, much less the first reality cells within AUO; that was a long time ago and we were not there when it happened. As I said in Chapter 25 of this book: "...go ask your mother! It is not an appropriate question." Scrupulously avoid making assumptions and creating beliefs (pseudo-knowledge) to fill in for what you cannot understand.

Pseudo-knowledge is useless except as a pacifier for a needy ego. You and your intestinal bacteria must realize there are some things that you will never fully appreciate because they are, and will always be, beyond your limited reach. Appreciating your limitations is the first step toward obtaining wisdom. At the same time, creating apparent limitations where none actually exist by getting stuck in belief traps is a great waste of poten-tial. Read the previous two sentences again. Do you see the importance of discovering Big Truth, and why you should spare no effort to clearly understand the difference between actual and apparent limitations? Make a note: The ability to accurately assess one's fundamental limitations dif-ferentiates the wise from the foolish.

The incredible evolutionary profitability of time is immediately obvi-ous. Ordered events allow sequences to carry and propagate content. Complex interaction and causal logic chains evolve as the beat goes on. Entropy is reduced. Time acts as a catalytic agent. It dramatically enhances AUO's ability to self-organize, thus speeding up the interaction between the Fundamental Process and consciousness.

The evolutionary pressure of self-improvement moves consciousness toward higher quality and lower entropy states – which is equivalent to moving consciousness to brighter, more aware, and more highly organ-ized internal configurations. Lower entropy produces higher quality, which means that the consciousness system has more energy available to do work (to more profitably organize).

One divided by the frequency of oscillation gives the period of one cycle. The period of a cycle is the time required to change state and then change it back again. One period is often used as a handy measure or unit of time. For example, the period of the rotation of the earth on its axis is one day and grandfather clocks use the period of a pendulum to count seconds.

More possibilities and evolutionary potentiality can be generated if AUO's parts and patterns can communicate, if they can interact and be coordinated with each other (like the arms, legs, brain, tail, and eyes of a monkey trying to snag a banana). Coordinated activity can both propagate and regulate patterns of content. Signals and messages can be passed from cell to cell as fast as one cell can change state in response to an "adjacent" cell changing state. If the patterns encode meaning as our patterns of neurons, letters and words, gestures, or sounds do, soon AUO has parts of itself communicating and sharing data with other parts.Imagine something roughly analogous to a biological nervous system communicating between body parts using sequenced patterns of neurons.

▶ Do you find it an interesting concept that our central nervous system (CNS), which includes brain, nerves, neurons, synapses and electrical charges, mirrors consciousness in its information transfer processes? Contemplate the close connection between our central nervous system and our consciousness. The CNS is a highly constrained physical analog of consciousness functionality. The CNS hosts our consciousness as a computer hosts an operating system and applications. It serves as a transducer, a data port and bi-directional translator between the virtual experience of the physical body and the individuated nonphysical consciousness that defines your existence and motivates your intent within the larger reality. Those who wish to explore an expanded discussion of the relationship between consciousness and brains should look at the "Physics" / "Consciousness and Physics" topic in the discussion group at the **www.My-Big-TOE.com** website.

It is reasonable that the mechanisms of data-processing and information-transfer that take place within the cell-based mind we call AUO are similar to information processing and communications used by the CNS. Imagine a sophisticated top-end mainframe computer being, among other things, connected through a simple interface to a bank of custom made sensors. Because the CNS evolved to be the interface between the perception of physical experience and the consciousness of the perceiver that it hosts, it is reasonable to expect the communications technology on both sides (physical and nonphysical) of the interface to be highly compatible – which usually implies some functional similarity. Our view from the physical side of the interface is no doubt severely limited, but after we more fully understand how the brain and CNS work, we will perhaps have a small, physical, highly tinted window through which we can peek at the mechanics of the processes within nonphysical consciousness – the same processes through which the larger reality is eventually expressed. ◀

The technology of AUO's clocks could get complicated. We know for sure only that AUO can make one terrifically good clock in order to produce, regulate, and integrate patterns, and patterns of patterns of patterns. Evolution, through the Fundamental Process, creates the optimal clock solution for AUO – whatever that is. The details are not that important. Do not be intimidated by sinusoids or by the mathematical expressions that were given as examples of oscillating functions in Chapter 26 of this book; all you need to know is that a cell-based digital AUO can readily evolve an adequate clock (regularly oscillating group of reality cells) to meet its evolutionary needs.

The upper limit on the speed with which a given message can be propagated is determined by how fast a single reality cell can change state. This maximum state-change speed is directly related to the minimum clock time of one half cycle or one quantum of AUO-time. (A full cycle would require a reality cell to change state and then change back to the original state.) In other words, the maximum speed with which a cell can normally change state provides the definition of the smallest unit of time our AUO clock can directly measure. This minimum time unit is defined as the fundamental quantum of time.

I can see that many of the technoids in the reading audience have their hands in the air. If the state-change is smooth and continuous (in the form of $\sin(\omega t)$ for example), the smallest practical unit of time can be reduced to the smallest discrete **portion** of a state change ($\Delta\omega t$) that AUO can consistently and accurately measure (be aware of). However, that is a technical detail that is of little importance to the Big Picture.

Above, we have been using words such as "adjacent" and "speed" which denote distance or space, but space is not yet defined. These words are only metaphors – do not take them literally or you will be thinking anthropomorphically, back in habitual 3D concepts, believing that AUO takes up space and exists "out there" somewhere, somehow, in a different place apart from us. Consider the speed of your thoughts, the space between consecutive or adjacent ideas, or the patterns of thought in your skeptically open mind. Think about multiple, simultaneous, or regularly occurring thoughts and about having thoughts simultaneously exist in the foreground and background of your mind. Great big thoughts and little tiny thoughts are not spatially large or small. We must guard against imposing our extremely limited 3D conceptual patterns (as well as our sense of time and causality) on AUO.

The constant evolutionary pressure to maximize profitability by reducing entropy, made AUO progressively more complex and self-aware; biological organisms went through a similar process. The more awareness and complexity an individual entity or system creates through evolution, the more potential states and possibilities there are to explore. Learning, growing, and evolving are often accelerating processes. The actualization of a given potential gives birth to a greater potential that gives birth to an even greater potential that ....

Because the birth-survival-propagation-death process, which absorbs an enormous amount of our time and energy within PMR, is not part of the consciousness evolution process, the acceleration of the evolutionary process is much greater for a consciousness system than a biological one. Without a body to drag around and care for, AUO travels light and moves fast within a much larger set of possibilities.

For example, education and learning is a bodiless process that obviously accelerates. Given a constant level of intelligence, the more someone knows and understands, the more quickly and readily he or she can know and understand additional related material. The rapid increase in the complexity, value, power, and pervasiveness of technology provides another clear example of a bodiless evolutionary process (the expansion of knowledge and technical expertise) that clearly demonstrates evolutionary acceleration.

▶ The rate of technological evolution has been increasing for hundreds if not thousands of years. As always, each generation believes the pace of technological change must be about to slow down – since almost everything they can imagine has been invented. Hah! Those closer to the edge of knowledge know better. Forward vision is always clouded by limited imagination. Our science and technology have **barely** scratched the surface of the possibilities – we have only now begun to pull back on the throttle and get our training wheels off the deck – "Uhh oh …hey…how do you steer this thing!? Oh jeez, it's picking up speed! Hey! There is no way to get off, nothing to steer with…and no brakes! Criminy! Now what do I do?"

What we need to do is obvious: We must learn very quickly how to guide our rapidly accelerating technical know-how with some hastily gained and applied wisdom before the opportunity to do so is lost. Trying to halt or suppress innovation, discovery, and the development and application of new knowledge is like trying to permanently stop the flow of a major river – it is not a realistic option. Social and economic forces (religions, governments, unions, corporations, and various social movements) have never been successful in slowing technological progress by very much or for very long.

The bottom line is that high speed far ranging change guided by immature and inexperienced social dynamics over totally unfamiliar territory depends, for the most part, upon good luck to avoid disaster. Depending upon good luck to avoid a multiplicity of disasters that each have the potential to be catastrophic is not a clever plan. The only way out of this predicament is to quickly develop clear long-term vision and execute good planning based upon high quality judgment (wisdom) **before** the speed increases too much (system goes irretrievably unstable), or luck runs out – Splat!

Troglodytes, Luddites, religions, homicidal wackos, and governments have all tried from time to time to dramatically slow the pace of innovation and technology – all have failed, are failing, and will continue to fail because that is the wrong approach. Combating unwise applications of knowledge and technology by extolling the virtues of selective self-imposed ignorance is a losing strategy that will never work. Typically, this strategy is not even effective as a delaying tactic. From the smashing of labor saving machinery in the seventeenth century to the outlawing of stem-cell and cloning research in the twenty-first century – all attempts to control the application of knowledge will necessarily fail.

Attacking a specific result (particularly research and products that cannot be practically controlled) will never defeat the Fundamental Process: Cut off one head and two more will pop up in its place. Focusing energy on a failed strategy simply makes the situation worse – attention is directed away from the real problem while the available time for implementing an effective solution is squandered. A head in the sand always leaves an unprotected butt waving in the air.

Great potential is a double-edged sword. We are all passengers in this local reality rocket where the technological manipulation of our internal and external environments may soon produce more dramatically accelerating change. We must depend upon the **quality** of our political, ethical, economic, technical, and philosophic institutions to grasp the opportunities and deliver the advantages, while avoiding the pitfalls.

Before you begin lamenting the incompetence of your public institutions you should understand that quality, understanding, and wisdom are individual attributes. The quality of the individuals within a society defines the quality of that society's institutions. There is no one to whom you can pass the buck – collectively, the citizens of planet earth more or less get what we deserve. The average social, economic, educational, or government institution reflects the quality of the average individual that produces and populates it. You individually are either an integral and active part of the solution or you are a part of the problem – there are no innocent bystanders. The onus is on everyone to substantially raise the quality of their personal consciousness.

Compelling a solution by the application of external force is usually counter productive and never a good long-term solution.

The point is that the evolution of consciousness, as all big system evolution, is an accelerating natural process. To make that concept and some of its attendant issues more intuitively understandable, I have simply pointed to the evolution of technology as a well-known example. ◀

Nonphysical mind can change (grow, evolve) more quickly than the physical body because it contains more degrees of freedom and fewer constraints. There is, for a long time, a steady acceleration in the growth and development of consciousness. This means the rates of growth continually increase. This acceleration allows each newly evolving expression of consciousness to be more efficient and productive more and more quickly.

Learning, like any other cumulative function, cannot experience large positive acceleration forever, but you would be surprised at how far consciousness can progress before the Fundamental Process begins to channel the greater part of its energy toward only the better opportunities.

The phenomenon of accelerating learning and accelerating evolutionary process applies to the evolution of both biological and consciousness entities. In the biological realm, with the coming of cloning and genetic engineering (a result of our computer and other technologies), expect the biological evolution of certain sentient entities to make discontinuous hops through the available possibilities to accelerate the potential (positive or negative) of their species relatively quickly. In the consciousness realm, the acceleration effect represents a steeper and smoother function of self-improvement vs. time as brightness increases its capacity to learn as well as the quality and depth of its understanding.

> ▶ I have painted a picture of how sentient beings with aware consciousness evolve. However, I do not want you to let this understanding limit your vision as to how a particular subset of beings might have ended up populating a given dimension of reality like OS or PMR. The facts of consciousness evolution as they apply to communities within $NPMR_N$ do not logically preclude an implant of sentient beings or an implant of additional sentient capability within existing beings. Modifying existing OS entities through evolution, adding new beings to OS, or directly modifying selected OS entities represent three ways to change the OS consciousness sub-system. Because these are not mutually exclusive operations, existing collections of sentient beings may be the result of all three processes working together to optimize the whole. ◀

As awareness masters and refines its evolving mental capabilities, a new, more complex motivation arises as a result of, and in conjunction with, the imperative to implement the Fundamental Process. The complexity and interactive properties of the mental processes being evolved eventually produces the functions of intention, feedback, interaction, synthesis, and integration. As these important second-order functions begin influencing and driving the evolutionary process, the potential number of profitable states dramatically expands.

Earthbound mammals and many insects can provide countless examples of second order attributes. The slogan, "Homo sapiens do it better" (found etched on the rump of a prehistoric horse) makes the point that because of our ability to think and process information, we declare ourselves to be the current mental synthesis, integration, and interactive feedback champions of all earth-bound biological evolution. Even if

Flipper and Shamu, with their significantly larger brains, disagree with the preceding statement, humanity has evolved an extraordinary potential that is primarily fed by an impressive array of second order consciousness functionality. The AUO consciousness-system-thing (now also a digital-thinking-being-thing) has the right stuff to one day be many orders of magnitude better at these second order analytic functions than we are. Perhaps there are fourth, fifth and higher order functions we can't imagine.

The choices and complexities available to AUO are now staggering. The sounds that people can make and hear likewise represent a staggering selection of possibilities, though infinitesimal in depth and breadth compared to the possibilities available to AUO. What do we people do with the available sounds? Language and music are two applications that immediately come to mind. The first holds great practical value in direct response to the external demands of the Fundamental Process; the second provides for our internal well-being, our pleasure and enjoyment. "Just for fun" does **not** imply useless, or that no evolutionary profit is produced. Music evokes emotion, reduces stress, motivates and bonds people, communicates feeling, and increases milk production on progressive dairy farms.

Look at what we have been able to accomplish with only three primary colors, twenty-six letters and a limited array of sounds. All of our communications, art, literature, collective memory, science, and technology rest upon specific combinations, sequences (in both time and space), arrangements, and patterns of this relatively small set of fundamental variables – and we have only scratched the surface of what is possible. Can you imagine what AUO might be able to cook up without our severe constraints? No, of course you can't, but we can imagine that AUO would have a gazillion times more variables to arrange within a huge manifold of multidimensionality. Simple fact: We cannot even vaguely imagine how far beyond our comprehension AUO's possibilities are. Imagine your intestinal bacteria speculating about the larger organism within which they live.

It is these second order processes within consciousness that are responsible for the eventual development of values. When awareness and the complexity of choice reach a sufficient level, the concepts of enjoyment, aesthetics, ideals, and quality begin to modify our expression of the Fundamental Process and our motivations in partnership with it. These comprise a set of third order functions that subsequently influence the outcome of the Fundamental Process of evolution. Given enough awareness and complexity, we have fun. We make music. We produce art. We are the creator and the consumer. We have preferences, likes, and dislikes.

Good and bad become defined. We develop values and make moral choices. Birds sing and soar; dogs fetch tennis balls and chase Frisbees. Play is widespread among highly evolved creatures of all sorts. If having fun wasn't profitable to Big Picture consciousness evolution, it wouldn't come so naturally or be so popular. Sentient critters have inside environments as well as outside environments to pay attention to and interact with. In fact, the more sentient we become, the greater our quality, the more important our internal environment becomes to our evolution.

It is only reasonable and logical that AUO would have the potential to evolve values, compassion, purposeful choice, play, communications, music, mathematics, computer science, creativity, humor, and so on within a system that has billions of trillions of gazillions times more native richness, selection, complexity, memory, and awareness than we humans do.

Wow! Don't you wish you could have bought stock in this binary state-flipping baby before it took over the reality market? This Creative-Supercomputer-Being-Consciousness-System-Thing (with emphasis on the "Thing") is evolving values. Additionally, it has some extraordinary potential that goes far beyond our comprehension. If people evolved second and third order functions that influenced their subsequent evolution, why would AUO not be able to do likewise? After all, the Fundamental Process and the nature of consciousness are the same in all of their various manifestations. Values, choice, play, communications, music, mathematics, compassion, computer science, and creativity are a few of the natural results of the Fundamental Process being applied to consciousness.

Simply having these attributes of consciousness is not enough. In order to effect self-improvement, AUO needs a process and feedback mechanism that can accurately deduce what is profitable from what is not – and another to move the entire system toward a lower entropy existence. We will discuss these processes and feedback mechanisms (of which we are a part) in Sections 3 and 4.

There is the outrageously limiting anthropomorphic temptation to make this AUO being-thing into a bigger and better super-cool spooky humanoid. [Ahhh... just like us, how comforting. Beings that are truly great, significant, and powerful must be sort-of like us... inside, I mean... even if they look a little weird on the outside... right? I feel better now.] Resist that temptation. It is not at all like us, although we may be a tiny little bit like it.

You will soon be able to logically support the astonishing idea that AUO's consciousness and existence is fundamental while our consciousness and existence is derivative.

30

■ ■ ■

The Evolution of AUO:
The Birth of AUM
(AUO Evolves into AUM)

■ ■ ■

By now AUO (Absolute Unbounded Oneness) is ready to make space-time, physical matter, you and me, as well as begin several experiments in the evolution of consciousness. Each experiment is developed and evolves within a specialized part (memory and computation-space) of AUO. We perceive each of these special parts (unique mental spaces within AUO) as separate dimensions, separate realities. There may be dimension within dimension, reality within reality, like a set of Russian Matryoshka dolls or perhaps the concentric layers of an onion. For example, our beloved PMR universe is one of a very large group of both physical and nonphysical reality subsets – each within its own mind-space dimension of the larger digital consciousness. These realities are contained within, as well as derived from and dependent upon, the much larger, more general, less constrained (more degrees of freedom) reality of $NPMR_N$. Likewise $NPMR_N$ is one of a dozen or so subsets of NPMR.

▶ If the use of subscripts is confusing, simply think of the $NPMR_n$ as being members, subsets, or separate portions of NPMR (like neighborhoods, states or countries represent social, political, and geographical subsets on earth). The following is a list of some of the subsets of NPMR: $\{NPMR_1, NPMR_2, NPMR_3, \dots NPMR_N, NPMR_{N+1}, NPMR_{N+2}, \dots\} = \{NPMR_n\}$. Using subscripts such as $NPMR_N$ is a convenient way to refer to some portion, dimension, or neighborhood of NPMR which contains and supports OS, which in turn contains and supports PMR. Using a similar notation, we will sometimes refer to multiple PMRs as the PMR_k (where k = 1, 2, 3 ...). If the use of a subscript notation bothers you, just forget about it – it is not that critical to your overall understanding. A general idea of what is intended is all that is necessary. ◀

Using our digital simulation analogy (see the aside on dimension at the end of Chapter 28 of this book), we can easily envision subroutines within subroutines, partitioned memory and calculation space that is further sub-divided into smaller pieces where certain subsets (such as PMR) of the overall simulation are processed. We can imagine more general realities containing more constrained sub-realities, each in their own calculation space, mind-space, or dimension. Do not think of a solitary onion or a single set of Matryoshka dolls, there are many. For you couch potatoes with digital TV tuners, reality and dimension are somewhat analogous to picture within a picture within a picture – displaying or running multiple TV programs in their own sub-space of the larger TV screen simultaneously.

However the consciousness pie is divided up, however many realities may be running as simulations or as sub-sets of larger simulations, all are connected through (and are a part of) the One Source: AUO consciousness. AUO is the foundation of everything, because everything exists in the relation of AUO to itself. Consciousness is The One, while the many, the great diversity of realities and the entities that populate those realities, are specialized subsets of consciousness within their own thought-space or dimension. The process and purpose of subdividing consciousness into both dimension and individuated units of awareness is discussed in Sections 3 and 4.

When AUO eventually grows up and evolves into an extremely complex and highly ordered set of specialized reality cells performing specialized functions, it becomes an entirely different type of entity. Consequently we are going to give it a new name: The Absolute Unbounded Manifold (AUM).

AUM appears to us to be absolute and unbounded. For that matter, AUM is still an Absolute Oneness as well, but now also a manifold (one into many) of sequences, patterns, realities, dimensions, and existence. AUM is a more complex, lower entropy manifestation of AUO; no new substance or assumptions are added. At AUM's core is the same basic AUO consciousness energy sporting a more evolved level of awareness, lower entropy, and greater functionality.

There is no clear dividing line between AUO and AUM. The distinction is entirely arbitrary. The extent to which AUO grows up and develops structure, organization and complex communications, as well as content, meaning, self-awareness, memory, value, and purpose is what allows AUO to emerge renamed AUM. It is more a matter of degree than it is a matter of developing new fundamental capabilities – evolution, not revolution. Because the awareness, function, ability, and purpose of AUM is so different – more resembling the consciousness we are used to, than the consciousness we call dim awareness – I decided to give it a new name, reminiscent of how we call people "people" instead of upright naked monkeys with short arms.

Think of AUM emerging from AUO as the natural consequence of digital consciousness purposely evolving a highly-parallel, multitasking operating system and interactive application software. That concept should not be too difficult to understand: We humans may likewise soon begin to modify our software or firmware (relative to our physical systems) as our ability to apply genetic engineering to ourselves matures and we become better at manipulating the biochemistry of our mechanisms of conscious perception. Human genetic manipulations and psychotropic medications are at their infancy. We haven't seen anything yet – we cannot imagine what is coming – these fields will impact humanity with every bit as much force as the digital revolution in computer and networked communications technology.

We live at the dawn of a potentially explosive evolutionary and cultural transition. The pace of change is dramatically accelerating. Ahead lies a unique and powerful potential for an incredibly accelerated progression, regression, or self-destruction. After 250 years of the Industrial Age comes the genetics engineering, cloning, psychotropic drug, computer-networked communications, digital Information Age – all at once! Oh Jeez! Get ready! Put on your helmet, batten down the hatches and fasten your seatbelts – this baby is lifting off the launch pad.

The point is: Once a complex system is capable of directly programming itself (developing or modifying its original source code – which includes genetic engineering), the pace of evolution dramatically accelerates. The intellect, aided by the products of its increasing awareness (digital and other technology), replaces random mutation and natural selection as the primary driver; consequently, evolution accelerates from a very slow paced process to an ultra-fast one.

Exactly how much AUO-AUM's evolution accelerated as it gained the tools to become self-directing is unknown. Looking at the evolution of biologic systems we can foresee processes that would normally take hundreds of thousands of years being compressed into a handful of decades or only a few years through genetic manipulation. Faster evolution may seem like a good idea, but speed is speed whether it is in the positive (lowers entropy) or negative (raises entropy) direction.

If the direction turns out to be negative the question is: Has the evolutionary process changed too much too quickly to be turned around before the overall system self-destructs? Wisdom would say to go slowly until one knows for sure what the pudding tastes like. Is there enough wisdom with enough influence to have a significant impact on the quality and far-sightedness of our collective decisions? If not, we need to start the process of generating some because the day of decision is not far off!

The nature, quality, stability and balance of any self-programmable system determines if evolutionary acceleration will eventually cause that system to take giant leaps forward or backward. Great opportunity and great risk travel together down the road of self-design and chemical manipulation. Of course, digital consciousness does have some major advantages over biological systems. If the resulting cumulative profitability assessment goes negative, it can always preserve (save) previous states and then punch the undo or reset button (while retaining lessons learned) – one of those cool digital tricks that make carbon-based systems envious.

I know what you are thinking: You are coveting the degrees of freedom contained within digital systems. You desperately want your spouse, mother-in-law, and the neighbor's dog to have undo and reset buttons. Be careful of what you wish for, you might get it before you understand all the ramifications. You will learn in subsequent chapters that you, your spouse and his or her mother as well as the neighbor's dog are, at the most fundamental level, individuated sub-sets of digital consciousness. More amazing yet, you will clearly understand the how, why, purpose, and inter-workings of this Big Picture digital reality model. No, no, no, that is not true – you are being pessimistic. Such an understanding will not be a delusion, or serve as proof that I have driven you insane – it is not as far-out or as difficult a concept as it first appears. Some very well respected hard-science types are in my corner on this one and I will introduce a few of them to you in Section 6 (Chapter 91, Book 3).

▶ There is more to an operational consciousness than an on-off switch. You also have reset, pause, record, rewind, fast-forward, playback, slow-motion, instant replay, edit, picture in a picture, and repeat buttons – so pay attention to this AUM Dude and discover how to operate your consciousness fully. It would be a shame for your consciousness to sit there idly flashing zeros in its display for a lifetime because you never bothered to learn how to use the controls.

"Here comes AUM ladies and gentlemen! The One, the Only – the Source of All That Is, the generator of reality! All right! Let's give AUM a big round of digital applause! Get those digits moving folks!

Ladies and gentlemen and distinguished members of the press, this is the moment we have all been waiting for. As the highlight of our program tonight, AUM – the omniscient digital consciousness dude from the beginning – has agreed to answer one question from the human gallery. This is an unprecedented moment in the history of existence ladies and gentlemen – before us lies the answer to any of the great questions of our time – past, present, or future.

"Not all at once please! Quiet! Quiet please! Yes, Ms. Gumwrapper from the Seattle Sunshine, what is your question? Shhhhh! Quiet everyone! I cannot hear the question. Huh? What is AUM's favorite color? Come now Ms. Gumwrapper, don't you think that is a tad shallow – quickly, let's take another question! OK, Ms. Anchordesk from the Evening Views, what is your question? Does AUM wear briefs or boxers? Oh, good grief! OK, let's get a question from somewhere other than the press box – Uh oh … AUM's getting up… he's walking away shaking his head…. come back…. sir… we are not all from the press…we can do better….it's, it's,… just that, well, sir, to be honest, people want to know… sir… Sir? What's the answer sir? Briefs or boxers? You promised sir, you promised! Don't go! Please sir …briefs or boxers? Give us a sign!!

"He is gone folks. Well, you saw it. The Big Dude was here and left without saying a word. Not a single word. What do you think, Johnny?"

"Well Dan, when AUM walked off I thought I saw his shoulders bobbing up and down. Do we have that instant replay yet? See that, see that little jiggle there! Can we get a close-up of that? Look at that Dan – definitely bobbing up and down!"

"All right folks, there you have it – now pick up those phones and call in your opinion – That's 1-900-$$$-0000. Was AUM laughing or crying when he walked away shaking his head? Let's run that playback again Johnny…. Yep, that's a definite jiggle there! What does our studio audience think? Laughing or crying …that is the big-dollar question for tonight!" ◄

All right, enough silliness, let's settle down and get back to work. You get the picture: The digital consciousness AUO is renamed AUM when the degree of complexity and opportunity (possibilities) evolves sufficient awareness and mental function capable of creating, evolving, storing and manipulating self-modifying content, process, value, and purpose. In the next chapter, we will see how time generates space and how they both together generate space-time – all so we can experience. Why? Because as everyone knows, experience is the best teacher.

31

■ ■ ■

The Birth of Space-Time
How Space is Created by Enforcing
a Constant Speed Limit in PMR

■ ■ ■

As I will demonstrate, space and time are closely interrelated; accordingly, we will refer to them as a single entity called space-time. Space-time, in this context, is a construct of consciousness. It is not a physical substance or a thing – it is not a physical construct – it is created by imposing a set of constraints upon a subset of the larger reality. PMR scientists describe space-time as a continuum in which events or physical objects with associated times appear to be located or moving relative to some observer. It is also described as a manifold which is sometimes time-like and sometimes space-like, depending on how one interacts with it

Space exists as a conceptual 3D matrix of imagined (in the mind or consciousness of AUM) chunks of PMR volume (a 3D pixel or quantum of space) that define our reality's resolution and form our reality's underlying structure. This structure, along with successive increments of time, provides the conceptual infrastructure that supports the rule-set that defines our perception and thus our physical experience. Consequently, space-time, as viewed from within PMR, appears to have a structural component (space-like) and a dynamic component (time-like). Space-time, from the point of view of My Big TOE, represents a particular set of rules and constraints applied to interactive energy transfers between objects. It is the construct within which our fleshy bodies appear to live and interact. The space-time rule-set serves as a consciousness interface filter that defines the perception (experience) of Physical Matter Reality (PMR) to participating sentient conscious entities. The experiences of PMR space, PMR time, as well as PMR mass, energy, and gravity are all derived from the space-time constraints placed on the experience of interacting individuated units of consciousness participating in the virtual reality we call our physical universe. (These strange sounding statements are explained in detail in Section 4. By then they will seem much more reasonable and credible than they do now.)

When I say "AUM invented space-time," or "AUM... (followed by any action verb), I mean in the evolutionary sense, the way fish invented lungs and legs with which they turned themselves into amphibians.

Don't get anthropomorphic on me here and postulate AUM as a little old man with a long white beard making people out of space-time clay on day seven. That is a perfectly nice metaphor, but it is not where this discussion logically leads. This is an effort to improve science and model reality, not to expose and follow threads of Big Truth that turn up in various religious metaphors, though it may accomplish both simultaneously.

AUM utilizes a particular subset of its consciousness' digital capacity, a portion of its organizational potential, a chunk of thought-space, in which a uniform set of constraints is imposed on all energy transfers between individuated subsets of consciousness that inhabit (are voluntarily participating in) that particular chunk (virtual reality dimension). If the constraints are represented by a rule-set that defines the properties of space-time, the virtual reality dimension created is one of the PMR_k – perhaps our PMR. Thus, space-time is a special perceptual construct defined within a dimension or subset of AUM's consciousness. The space-time rule-set delivers a consistent experience to interacting individuated units of consciousness. Huh? Is this English? I know this description of space-time is somewhere between difficult and impossible to understand right now, but if you hang with me through this chapter and Section 4, what now appears to be off the wall and confused will eventually become both reasonable and crystal clear.

I decided to jump right in and tell you what space-time is before I developed the background necessary for you to understand it because I think that having some idea of where we are going will help you synthesize the concepts required to get there. The experience of space-time is produced by a consciousness-construct that constrains interactions between individuated chunks or subsets of consciousness. The point or purpose of space-time is to produce a specialized consistent virtual experience to individual constrained subsets of consciousness called conscious entities or beings. Don't worry; these concepts should be difficult to comprehend at this time. As always, open minded skepticism defines the correct approach. Be patient: Believe it or not, there is a coherent and logical process that leads to these descriptions but it may be a while before we get to it and through it.

Let's go back to building a logical model from the bottom up. AUM, under pressure to improve itself, must evolve profitable internal environments and processes that methodically reduce the entropy of consciousness, that is, the entropy of its own system. Space-time is **one** such environment. Within the space-time structure, the PMR rule-set defines the processes that make us and our space-time uniquely profitable to AUM.

Digital simulations, like those run on our mainframes or within TBC, generally model what appear to be analog events at the macro level. Likewise, we will develop a macro model of how AUM might conceptualize digital space-time in terms of space-time cells and the time required to pass information between those cells. This should be more satisfying and more helpful than simply stating that space-time and the PMR rule-set constitute an optimized design solution for a virtual reality simulator that serves as an interactive multiplayer consciousness-quality development trainer for units of individuated consciousness.

Let's start by conceiving space-time as a construct of space-time cells. Remember, these space-time cells are conceptual – the result of a *gedanken* experiment in the mind of AUM. We will show that space-time cells must be uniform in form, function, and makeup in order to produce the underlying structure for a simple isotropic experience-space (reality) that can be used to improve AUM's evolutionary profitability.

To help make this metaphor more concrete, let's say that these space-time cells reside in the space-time part of the mind of AUM. Next, let's drop down one more level of detail and separate space from time. The **structural** basis of this model is space-like, and provides the spatial concept or space-part of AUM's *gedanken* experiment, while time provides a dynamic basis for this conceptualization that leads to the possibility of ordered change and therefore growth. Space-time subsequently becomes the fundamental medium for a uniquely dimensioned sub-reality or virtual reality within the larger consciousness system. It would appear that space-time is more profitable, useful, or advantageous to AUM if the space-part functions as a simple 3D isotropic experience-space for interacting consciousness because that is how it has successfully evolved. It should be relatively easy to picture a three-dimensional matrix of uniform isotropic space-time reality cells in your mind – which is what I imagine AUM started with, but that was only the beginning.

No need to grab your dictionary for the scientifically challenged. "Isotropic" is one of those $5 cool-sounding techno-speak words with a ten cent meaning that we technoids love to use to impress the masses of business majors who make more money than we do. It simply means that conceptual space-time is not directionally unique; that within the consciousness of AUM, it is (reacts, interacts, propagates, behaves) the same irrespective of how AUM "looks" at it, applies it, thinks about it, or interacts with it.

Keep in mind that we are talking about the concept or idea of space-time within the digital mind that is AUM. Within a spaceless consciousness, "direction" has no meaning. As the concept of space solidifies around its defining constraints, "isotropic" provides a simpler more straightforward conceptualization of a virtual space. That the **implementation** of the space-time rule-set (governing PMR time and energy as well as 3D space) has non-Euclidean consequences presents no logical difficulty.

The relativity buffs are probably choking on this description of an isotropic, Euclidean conceptualization of space. However, the preceding assertions create no conflict with the relativity of our perception in the PMR subset where all coordinate systems may be moving arbitrarily relative to each other, each being as fundamental as any other. Nor does it conflict with the chunky non-homogeneous distribution of mass/energy that interacts by the rules of gravitation. The idea of an isotropic space begins with the definition of light speed as one of the primary constraining constants directly derived from the specification of a quantum of space and a quantum of time.

Although we are describing the space-part of AUM's consciousness with simple Euclidean geometry, PMR space will appear to be curved (non-Euclidean) to a well instrumented high-tech perceiver experiencing physical reality within PMR. The bottom line is that as long as there is no fundamental physical inertial frame within a chunky PMR, curved space-time defined by general relativity continues to represent an elegant model of our perception **within PMR** – even if we assume a simple Euclidean frame of reference for conceptualizing virtual space in the mind of AUM. Our PMR space-time experience is a derivative of, or is based on, the constraints of the space-time rule-set that orders our particular reality. You will see later that the space-time rule-set resident within TBC defines the details of the constraints that define our physical reality and thus our physics.

If you are not a relativity guru, and don't have a clue what the previous two paragraphs are all about, forget about it, it is not important. Physicists, having long ago buried Euclid in their physical big picture, tend to be a little slow in giving up old thought patterns. That the perception (from within PMR) of a curved **physical** space could be derived from the **concept** of an isotropic space is not that difficult to understand, you merely run the traditional paradigm in reverse.

Recall that each reality cell exists as a duality, a condition relative to some different condition, each condition existing only in relation to the other. Conceptually adjacent space-time cells communicate by changing state in sequence thus enabling information to propagate through the matrix. It was the concept of a pass-it-on communications technique (sequential cellular interaction) that led AUM to develop the idea of space. A cell or group of cells can be specialized to keep time with a constant oscillation of state value in order to set the pace for controlled cell to cell propagation. They may be independent of AUM's main or fundamental clock and can be set to any frequency as long as it is less than AUO's fundamental frequency.

The smallest time increment, DELTA-t, in space-time (one quantum of space-time time or equivalently one quantum of PMR time) must be some positive non-zero integer (**n**) times the smallest time increment fundamental to AUM. The smallest time increment fundamental to AUM is called the "fundamental time quantum." It is the smallest time quantum possible within the larger reality. DELTA-t is also the minimum time required to change the state of a **space-time** reality cell. All space-time will subsequently march to the beat of DELTA-t.

Non-uniformities or information can appear to move through space-time at a maximum velocity of one space-time cell per DELTA-t. Keep in mind that a space-time cell is in non-geometric, distanceless thought-space. We are generating only the **concept** of space within the space-time part of AUM's consciousness. This space-time concept is implemented by defining the constraints that bound it. Once the constraints are defined within the PMR rule-set, the consequences (properties of space-time) can be easily computed. Let me say this once again because it is an unusual and important concept. We are **not** creating a **physical** 3D space or a **physical** 4D space-time out of consciousness reality cells. Physical space, and your experience of it, is an illusion, a trick of your individuated mind, a virtual reality within the actual reality of your consciousness. I **am** developing the existence of a **concept** or idea of space within AUM's consciousness – an idea that will evolve to provide the experience of space to an individuated consciousness. Space as a virtual reality, a mental construct, not physical space itself; there is no such thing as physical space!

"Are you kidding, no physical space? What is this all around me? I learned to calculate volumes of simple 3D shapes in sixth grade." Making that outrageous statement ("there is no such thing as physical space") is not as obviously dumb as it may first appear; you will hear that same exact statement directly from Albert Einstein and other knowledgeable top-scientists in Section 6 (Chapter 91, Book 3). Do not jump to any conclusions.

We have measured the upper limit on information transfer in PMR as "c," which is the symbol commonly used by scientists to represent the speed of light. "c" is approximately 186,000 miles per second or 3×10^8 meters per second. In less precise terms for you technophobes, the speed of light is "exceptionally quick," "almost instantaneous," "like greased lightning," "smokin'," or "haulin' ass" depending on your socioeconomic and generational affiliation (the degree of degeneracy you take pride in). At least that is the view from PMR.

From my experience, and that of many other NPMR explorers, communications are **seemingly** instantaneous in spaceless NPMR and not restricted by the relatively pokey transmission rate of light-speed as they are in PMR. For this reason, it would seem safe to assume that the integer value of **n** is an exceedingly large number. [Where, as you recall, **n** is the number of AUM's fundamental time increments (fundamental time quanta) that tick-tock away during one (very much larger) time increment of the space-time clock (one quantum, DELTA-t, of PMR time)]. For the record, there are also many fundamental time quanta (smallest chunk of time on AUM's fundamental clock) ticking away during one increment of the independent clock defining time within NPMR (a quantum of NPMR time).

You will learn in Section 5 (Chapter 79, Book 3) that **n** needs to be a large number so that the statisticians in NPMR have plenty of NPMR time to compute probable reality surfaces between successive DELTA-t. Likewise, AUM sets up NPMR with a larger time quantum than its own, enabling AUM to process data between NPMR time increments. Remember, whoever iterates with the smallest time quantum is usually in the driver's seat.

It is easy to study, manipulate, or observe beings, objects, and energy that move in super slow motion relative to yourself, as long as you do not get bored. AUM evolves (chooses) the optimum **n** to suit its needs (easy to study and collect results without getting bored). From my experience and the experience of others, an enormous number of increments (quanta) of NPMR time pass during each quantum of PMR time. We will make good use of this concept in Section 5 where past, future, and paranormal communications are discussed.

Information transfer also **seems** instantaneous here in PMR (over short distances) because light-speed is exceptionally fast relative to other natural

PMR velocities. It would seem that you must be careful in using your direct experience to determine what is or is not instantaneous. For large distances, such as the distance from earth to its **nearest** star (4.5 light-years to the Alpha-Centauri triplet), even light-speed seems agonizingly slow.

In NPMR thought-space, individual things are not separated by physical distance and information does not have to be transmitted through space in order to travel between the sender and receiver. Nevertheless, within NPMR, information must propagate from the sender to the receiver (but not through space – there is no **spatial** distance between them). Thus, it requires some **NPMR time** for the information to make the trip between the unique digital mind-spaces of the sender and receiver; time is needed to flip the cell states that represent content within the receiver. Think of this process as analogous to display time, or similar to how your computer shuffles data within its core memory.

I know, this is starting to get weird and I can see your eyes beginning to glaze over with information overload. Clarity will come – collect the big ideas and let the details go for now. Let us do a quick review of the high points and go on.

AUM's time is the most fundamental because it has the smallest time increment (quantum) and is used to define all the others. Imagine, for the sake of putting some concrete meat on these highly abstract bones, that AUM's fundamental quantum of time is ten nano-nano-nano-nano-nano-nano-nano-nano-nano-seconds (10^{-80} s), while NPMR's is 10^{-62} s, and PMR's is 10^{-44} s. Each is a billion, billion (10^{18}) times larger than the previous one.

Using these numbers, it follows that DELTA-t, which in our example is 10^{-44} seconds and represents PMR's quantum of time, is **defined by AUM** to be 10^{36} fundamental quanta ($n = 10^{36}$). That is, for every 10^{36} ticks of AUM's (fundamental) clock, one quantum of space-time time is incremented in our dimension of reality (PMR). After 10^{80} ticks of AUM's fundamental clock an entire second of our PMR time has dribbled by.

Talk about slow motion! Just as there are many **fundamental time-quanta** in one quantum (increment) of NPMR time, likewise there are many of these **NPMR time-quanta** in one **quantum of PMR time**. The relatively huge magnitude of one quantum of PMR time (one quantum of space-time time) is exactly the time required for one **space-time** reality cell to change its state from non-distorted to distorted (**or** vice-versa).

The upper limit on propagating distortions in **space-time** is thus one space-time reality cell per quantum of space-time time. We define that upper speed limit as "c," a constant whose value has been evolved by AUM and which has been experimentally measured in PMR to be 3×10^{8} meters per second. Therefore, c **conceptually** defines the virtual size or conceptual spatial extent of a space-time reality cell. Thus, the virtual length, L, of one space-time-cell divided by DELTA-t (which is defined as one quantum of PMR time) would equal c. Consequently, c is the speed with which information can be transferred from one conceptual PMR reality cell to the next adjacent cell. That equation can be written as L= (c) • (DELTA-t). We can define a quantum of space, DELTA-v, to be the smallest possible chunk of PMR 3D volume, v. DELTA-v ≈ [(c) • (DELTA-t)]³ = L³.

> ▶ Let's use the values given in our arbitrary numerical example above to calculate the space-time reality cell width. Define the virtual width of one space-time reality cell as L, then L = (c) • (PMR time-quantum DELTA-t) = (3 x 108m/s) • (10^{-44} s) = 3 x 10^{-36} meters, which is about the size of Planck's length (16×10^{-36} m) – a measure of the point at which some of the world's best physicists say that our 3D space becomes granular (is composed of non-continuous discrete cells).
>
> I purposely made up the numbers in the preceding example to be simple round numbers and to force the width of a space-time reality cell to be near Planck's length. They are for illustrative purposes only – do not take the actual values too seriously.
>
> I know that this is clear to you, but let me remind some of the other readers to set aside their habituated PMR concepts and keep in mind that AUO's tick-tocking clocks are not actually ticking off fractions of our seconds. It is ticking in mind-space, within an aware consciousness that can differentiate between this way and that way, flip-flopping atoms of consciousness called reality cells whose state can be manipulated at will and with regularity. These atoms of consciousness in turn produce molecules of space-time.
>
> The tick-tock of AUM's clock represents the fundamental time-piece, the primordial clock, a mental process within consciousness from which our PMR clocks are derived. Our time, measured in seconds – fractions of the periodic revolutions of the heavenly bodies or more recently some number of atomic oscillations – is a shadow of Fundamental Time; it is our sensory perception of a specific constrained and limited implementation of the more fundamental time defined by AUM-consciousness. ◀

In the example above, information can travel much faster than c (the speed of light in PMR) in NPMR because the quantum of NPMR time is a billion, billion (10^{18}) times smaller than a quantum of PMR time. As I mentioned earlier, OS contains physical (PMR) and nonphysical ($NPMR_N$) components that are interactive with each other. **Between** time increments in PMR (time stands still in PMR between increments), things continue to happen (distortions, patterns, and information continue to propagate) because the clock keeps on ticking in $NPMR_N$. In other words, while time appears to stand still in PMR, $NPMR_N$ continues to race along through 10^{18} more time increments worth of activity. And, between each $NPMR_N$ time increment, while time is standing still in $NPMR_N$, AUM has another 10^{18} fundamental time units in which to conduct business as usual.

▶ If this "time standing still between increments" thing has you buffaloed, forget your habitual notions of continuous time and consider how a complex dynamic simulation might increment time within various subroutines. Time appears to stand still in a given subroutine until the next time that subroutine is called and the local time variable is again incremented. A clear description of time-loops within simulations is given early on in Section 5. Unless this concept is driving you crazy, it can wait until you get to Book 3. ◀

Trust me, 10^{36} or a billion, billion, billion, billion (1,000,000,000,000, 000,000,000,000,000,000,000.0) is a really big number and **could** represent a relatively long time for AUM to wait for one PMR DELTA-t to come and go, depending on how long AUM **perceives** a fundamental time quantum to be. To **our view**, AUM **is capable** of moving in super fast motion. Imagine watching all the movies ever made in a fraction of a second. That represents some serious fast-forwarding. On the other hand, AUM can pace itself however it wishes and often seems to take the long view exercising plenty of patience. AUM's view of us may be like our view of some sluggish bacteria growing in a Petri dish.

Imagine what it would be like if we each aged one year every 100,000 years (we would have a life-span of about eight million years) and had brilliant nearly perfect memories and minds that were kept busy with many important things to do and think about. We could watch rivers and mountains come and go as well as study mutations in successive generations of quickly breeding fruit flies. If you can imagine this, you have a minute glimmer of AUM's perspective. AUM does not seem to be getting bored or to be in any hurry. AUM's perception of the passage of time (that AUM itself creates by regularly flipping states in order to organize and orchestrate its activity) is very different from ours. Do not anthropomorphically project your sense of the passage of time to AUM. Imagine how subroutines within a digital computer, or the computer itself, might perceive the passage of time.

The existence of multiple levels of quantized time is why information transfer in NPMR seems near instantaneous compared to pokey light speed in PMR's space-time construct. From NPMR, PMR would appear to be running in slow motion. Nevertheless, NPMR has its own speed limit, as does AUM itself, because it requires a finite time to change the state of any cell from uniform to not-uniform; that is how we defined the concept of time in the first place.

AUM can define (evolve) as many clocks as necessary for each group of specialized cells that define a unique dimension of reality. Our space-time reality (PMR) is implemented by one such group of specialized space-time cells. This collection of specialized cells existing as a sub-group within the larger group of NPMR provides the computing resources (memory, structure, rule-sets, and processing) required to actualize (compute the consequences of) that subgroup as a unique virtual reality within NPMR. Visualize a reality dimension (like PMR) within a larger reality dimension (like NPMR) within a larger reality dimension (like AUM). There may be many sub-groups that define unique dimensions of reality within NPMR. We will see in Section 5 how the various sizes of time quanta (all derived as integer numbers of AUM's fundamental quantum of time) each in their own dimension can be directly related to nested time loops within a simulation. What a bizarre thought! Reality as a layered digital thought-simulation within AUM – hang onto that concept and we will explore it in detail later. The significance of multiple levels of quantized time will become clear and seem less arbitrary after we have completed Section 5.

Here is a quick summary of the concept of time within NPMR. Keep in mind that NPMR has no space and therefore no distance – it exists outside of PMR's space-time. Without space and distance, the propagation time for information takes on a different perspective. Because AUO and AUM can flip states relative to itself only so quickly, there is an upper limit on the speed with which distortions (information) can be propagated through the larger consciousness system. It is a much larger upper limit than we can imagine given our relatively snail-like light speed that sets the upper limit of information transfer between contiguous chunks of PMR 3D space (DELTA-v) separated (center to center) by the virtual distance L. To generate space-time, AUM needs only to specify two of following three constraints in the form of constants – the third can be computed from either of the other two: 1) The time increment by which the virtual reality simulation is incremented by the outer loop, or equivalently, the shortest time possible between cause and effect (the quantum of PMR time we have named DELTA-t). 2) The resolution of the "graphics" (scene generation) defining the "3D pixel size" of the PMR virtual reality, i.e., a measure of the granularity of space – a 3D quantum of PMR space we have named DELTA-v. 3) The maximum speed with which information can be transferred between two points within PMR space-time (the constant we have defined as c, the speed of light). The relationship between the three is c = (DELTA-v)$^{1/3}$ / (DELTA-t). These constraints must be in the form of constants in order for space-time to be uniform and consistent (homogeneous and isotropic as required by the rule-set). Thus we have derived and explained why the speed of light must be a constant in PMR irrespective of the velocity of the source of that light [it is calculated by dividing the two fundamental constants that specify the data processing requirements of our virtual PMR (pixel size and outer loop time increment)]. This fact alone led Einstein to the logical conclusion that there was no fundamental inertial frame. The theory of relativity falls out as a logical ramification of the fact that c is a fundamental constant independent of the motion of the source. Physicists today still have no idea why the velocity of light must be invariant relative to the motion of its source – they know only that it is so.

You, dear reader, have now derived from first principles the fundamental understanding upon which the theory of relativity is based – a first in the world of modern physics. You will also find out in Book 2 that My Big TOE derives the theory of quantum mechanics (why particles are probability distributions before they are measured) from the same fundamental principles – another first in the world of modern physics. Thus the physicist's search for a single little TOE that derives both relativity and quantum mechanics from one overarching understanding has been accomplished. Having wrapped up PMR science in this neat bundle and solved the riddle that has stumped physicists since Einstein, let's go on to deriving the nature of the larger reality. This is, after all, a Big TOE.

Time, or equivalently, frequency is a fundamental attribute of AUM, whereas the notion of space is derived from time specifying a constant velocity of propagation of information. That is why the citizens of space-time must live with c as the celestial speed limit. Time is fundamental; space is derived from time by specifying the constraint c. The three constant constraints defining PMR space time: (DELTA-v), (DELTA-t), and c were all three naturally evolved together as consciousness developed an adequate foundation for the specific PMR rule-set that would create an optimal learning lab for individuated units of consciousness – a PMR that is well within the capacity of the larger consciousness system to implement.

Evolving PMRs within the larger consciousness system is not particularly miraculous; it works by applying the same fundamental process that brought you to reading this page: occupying all the possibilities and building on the best results, by trial and error, by putting randomness and purposeful self-modification to work in the service of entropy reduction. It was generated by applying the Fundamental Process on a scale and at a pace that is difficult for us to comprehend. Keep in mind that AUM has no body to feed, does not take up space, and is not preoccupied with perfumed AUMettes.

For those who occasionally visit the PMR physics fringe, the fact that we might be clever enough to take a space-time shortcut through a "wormhole" does not change the fact that c remains the upper speed limit on information propagation through space that **defines** our local experiential space within AUM's consciousness. Likewise, if data transfer rates seem to exceed light-speed in certain peculiar situations; this implies only that the defining rule-set has a level of generality that allows special cases to **appear** to violate the speed limit c.

The reality in which we interact with each other is a virtual classroom or learning lab designed to help us reduce our individual entropy and grow the quality of our consciousness. As consciousness, that is what we do; that is how evolution challenges us. That many implications of our rule-set await our discovery (the physics of the future) simply makes that classroom more interesting, challenging, and educational. New discovery creates better understanding as well as new opportunities to learn.

AUM's evolutionary experiments are roughly analogous to our *gedanken* experiments or computer simulations, and thus can be done in great variation relatively quickly. With the aid of its specialized binary computing part, AUM can evolve at an incredibly fast pace compared to biological systems. The Big Computer (TBC) and the Even Bigger Computer (EBC) discussed in Section 5 (Chapters 78 and 83 respectively of Book 3) are actually only tiny subsets of AUM's computing part. If you enjoy torturing and abusing the English language as much as I do, you have my permission to describe AUM as a consciousness-system-digital-being-thingamajig.

Reality cells, memory cells, binary cells – everything seems to come down to cells. Cells are discrete units of state-specific **relational** content, and represent both information and organizational substance (not necessarily physical matter). Awareness, form, function, content, and purpose all flow from the possible interactions of reality cells – the basic building blocks of aware consciousness.

All cellular and digital creations exhibit granularity at some level of detail. Both our reality and our perception of it are granular at the root. At the bottom layer of organization within reality, one will find discrete units relationally formed and arrayed into constrained dynamic patterns of relationship. The complexity and organization of these patterns are progressively developed within a primordial undifferentiated potential energy (unaware dim consciousness) by the iterative and recursive operation of the Fundamental Process. The mechanics of evolution naturally act upon any entity that possesses a substantial number of self-generated alternative choices and possibilities. The process is simple – maintain the winners, discard the losers – more or less random permutations and combinations are applied statistically to systems that have a significant array of potential outcomes. The winners continue to evolve while the losers fade away or maintain the status quo.

Sometimes an evolving consciousness system may in time develop terrific complexity and seemingly endless potential as the Fundamental Process iterates upon itself repetitively, eventually evolving into something that begins to mimic its source. When the wheel turns full circle and the successful product of the evolutionary drive toward self-improvement begins to take on the characteristics of its source, it becomes a partner in the process of consciousness evolution.

Imagine this. An individuated constrained high entropy fragment of consciousness eventually evolves (profitably self-organizes) to low entropy wholeness. As a part of the source from which it was cleft, it lowers the entropy (increasing the quality, furthering the evolution) of the whole by the amount of its personal growth. This individual contribution to lowering the entropy of the entire consciousness system constitutes the up-stroke of the consciousness cycle. The down-stroke is represented by the creation of individuated consciousness by uniquely bounding subsets of undeveloped (raw) consciousness (of relatively high entropy) that are capable of eventually evolving toward lower entropy states.

Are you beginning to sense the great cycle of consciousness evolution of which you are a part? A vague understanding of how and why the consciousness cycle operates to sustain the ecology of the larger consciousness system is your first peek at the Big Picture. The consciousness cycle describes a mechanism that allows consciousness energy to continuously organize itself within the larger consciousness ecosystem. You are an important player in a consciousness cycle that drives system profitability, capability, operational power, and brightness upwards as it exhausts entropy. We will pursue these concepts more fully in Chapter 58, Book 2.

Let's pull it all together. The constant upper limit of c (speed of light) defines the concepts of both distance and space within the space-time subset of AUM's consciousness. Each so-called space-time cell, by definition, now has the **conceptual** attribute of extent. Thus AUM invented (thought up, or evolved) the **concept** of space by imposing the constraint of a constant maximum velocity of propagation of information within its space-part. Using the previous numerical example to add a sense of concreteness,

268 | *My Big TOE* | The Birth of Space-Time: How Space is Created

DELTA-t is the smallest unit or quantum of time **within space-time** – it is one tick on AUM's **space-time** clock, (but 10^{18} ticks on AUM's NPMR clock, and 10^{36} ticks on AUM's fundamental clock). Each space-time reality cell now has the conceptual attribute of spatial extent. It is (c m/sec) • (DELTA-t sec) wide – about 3×10^{-36} meters (approximately Planck's length). The logical necessity of c being a constant delivers Einstein's theory of relativity.

The science, engineering, and math types need to chill out. I know that I am about to drive you crazy with repetition, but most of the rest of us are only beginning to get comfortable with powers of ten and c notation, the concept of interrelated time quanta, and the calculations of space-time cell width. Take care not to burn out your clutch or overheat your engine – this is a good time to lie back and coast a little.

To say the same thing in the opposite direction: If an imagined row of adjacent PMR space-time cells, each of width (c)•(DELTA-t), uniformly propagates a distortion by changing state consecutively, one each DELTA-t, the distortion propagates at the velocity c. Thus the cells in the space-part of AUM now have the **attribute** of size. This is not to say that they have size, or begin to take up space; they have the **concept** of size. You could say they carry the attribute or property of size, that they have virtual size, or that they simulate or model size.

These space-time reality cells are a part of AUM and exist as thoughts exist – without space. Because AUM's evolution of the space-time rule-set found optimal profitability within an experiential isotropic space, the functional propagation of information in any imagined direction is the same throughout this uniform space-time matrix of reality cells.

To maintain a constant propagation velocity in all directions, it is convenient if these space-time reality cells are **conceptually** spherical – 3D solid geometry becomes a concept, an attribute of the space-part – and any propagation in any virtual direction will find the virtual distance to any neighboring cell to be the same. More succinctly: The spherical diameter = the conceptual width of any cell in any direction = (c)•(Delta t). The **concept** of space (distance and direction), and the **concept** of time are interrelated through the constant c, and merge conceptually as space-time. Don't worry about the voids that exist between spheres when packing spheres in a 3D matrix – that is an issue in PMR space-time but not an issue in digital thought-space where space-time is created and defined. Similarly, energetic interactions among various forms of energy and the invariance of reference frames within the experience of PMR reality are not in conflict with the Euclidean frame of mind of AUM.

| http://www.My-Big-TOE.com |

Our PMR space-time (within which the experience of matter in our universe is created – see Section 4) represents only one uniquely constrained space-time application existing within **a** space-time part (as opposed to **the** space-time part) of AUM. There are other space-time parts (or space-time dimensions) of AUM corresponding to the other PMRs (mentioned in Chapter 76, Book 3). From our point of view, these other PMRs exist in nonphysical worlds within other dimensions. From our perspective, every subset of specialized reality cells in AUM defines another dimension of existence – another reality, or, if you prefer, another virtual reality. The specialized parts of AUM are to AUM as systems of thoughts are to us – a rough analogy even if we had super memories. These specialized subsets of AUM represent separate experiments in consciousness evolution propagating their way to whatever comes out of them.

Digital space-time is implemented by a basic rule-set that defines the profitability criteria required to evolve the content of a given dimension. Each virtual reality with self-modifying content automatically begins to evolve its own uniquely profitable configurations as its possibilities are explored. Patterns of interaction within the PMR space-time sub-system will eventually evolve to contain the attributes of content, information, and substance – as patterns of light, sound, and neurons within our biological systems or as patterns of stars, comets, galaxies and solar systems within our universe.

As a direct result of the space-time rule-set, a few billion years ago clumps of biological cells began a series of similar experiments in specialization. Specialized subsets of cells became the food-section (digestion), the sensor-section (eyes, nose, skin, ears, taste buds), the motion-section (tail, fins, flippers, legs, wings), and the control-section (central nervous system, brain). Then, communication among the various specialized parts was established, followed by communication between individual entities.

Thus AUM sprouts virtual reality systems (universes) as the earth sprouts species of plants and animals, each contained in its own dimension or piece of thought-space within TBC, and all on the same "network." AUM creates dimensions and manipulates their content similarly to how we create ideas and write them down as paragraphs of patterns of symbols. AUM's creative thoughts and use of dimension to separate specialized parts of itself is roughly comparable to our creative writing and use of paragraphs, books, or documents to separate and bound specialized chunks of content.

Note that I generally do not refer to higher or lower dimensions. Higher and lower have no meaning relative to non-geometrical dimensions. Dimensions, and their corresponding realities, are different in the same way books are different. Some may be simple or complex; some may be more or less useful to the consciousness system. However, all are specialized in response to evolutionary pressures to carry out their particular function. All specialized functions contribute to the self-improvement or entropy reduction of the whole. You see, this consciousness system is not that weird or mystical and works the same as any evolving complex system. For example, it must develop a balanced ecosystem (system of relationships) among its large number of interacting, interdependent parts.

You now have the top-level hand-waving description of the origins of space-time and the nature of dimension. This is only the beginning of a more in-depth discussion that will continue through the next three sections and two books. Before long, you will appreciate space-time as a rule-based consciousness-construct that constrains energy exchanges between individuated subsets of consciousness (and between beings and things) in order to produce a specialized virtual experience where these subsets of consciousness can profitably interact – and that description will make sense.

By the end of Section 4, all these scattered bits of theory will begin to pull together into a high resolution Big Picture. Keep in mind our earlier discussion of belief and "spin." The logical solidity and reasonableness of this discussion may not be the only, or the major, factor that leads you to your conclusions. To optimize the return on your investment, your analysis should be independent of belief-based paradigms. If your approach is open-minded, you will find that the evaluation process itself is usually more valuable than the final conclusions reached.

> ▶ All great sages know this one Big Truth: "You can lead a jackass to water and, if you are clever enough, induce it to drink, but you cannot make it do the backstroke, or gargle and spit the water back out."
>
> That is why this trilogy is strictly for open minded humans. If you see a jackass with a copy, please confiscate it immediately before he or she gets any Big ideas. But, please be careful! It is also well known by the great sages that: "A little Big Truth rattling about within the small mind of a jackass can be a dangerous thing."
>
> That is everything I know about common horse-sense. Thus, as always, you are on your own at the watering hole of Big Truth. ◀

Before temporarily leaving the subject of space-time and ending this chapter, let's make an effort to bridge the gap between the fundamental view of the binary reality-cell space-time part of AUM's consciousness and the digital systems view of the implementation of space-time within TBC.

AUM, aware consciousness, and therefore space-time (our physical reality) is ultimately based upon the existence of reality cells within consciousness. Equivalently, one could say that space-time is a construct of consciousness. Some may be tempted to say that space-time is constructed of consciousness, but that statement is likely to spread more confusion than illumination. Think of consciousness as a digital medium rather than a building material. At the root, TBC (memory, patterns, logic, and processing) is conceptually based on the binary property of reality cells.

Let us recap the evolutionary road from AUO to AUM. Primordial AUO is a relatively simple, uniform, high entropy, energy-form that is the basic foundation of consciousness. Reality cells are created by the relative existence of a this way and that way pair – a distorted or non-uniform existence existing in relation to a non-distorted or uniform existence. The proliferation of reality cells and the interaction between reality cells creates complexity and a greater potential for self-organization that eventually leads to a more comprehensive lower entropy self-awareness that in turn leads to intelligence, values, personality and purpose. In scientific terms, a lower entropy consciousness system has more power – more energy available to do work – a higher, more useful level of organization. In common terms, a lower entropy digital system commands more usable energy (more profitable organization) and, therefore, becomes capable of creating greater and more profitable configurations of itself. Additionally, a lower entropy consciousness eventually develops the ability to use directed conscious intent to reduce its entropy further. The **rate** of evolutionary progress increases as the system pulls itself up by its bootstraps (evolves) to actualize lower and lower entropy configurations.

The specific nature (physics) of our reality is defined by the space-time rule-set which defines the constraints that limit what is possible. The space-time rule-set is a collection of patterns or algorithms within TBC that defines the limits and relational properties of our physical experience. The space-time rule-set subsumes our physics. It is implemented at the lowest level by constraining the interaction between individual and groups of reality cells and at the highest level by constraining individuated subsets of awareness.

Reality cells are roughly analogous to the transistors on a computer processor chip. They come in very large numbers and are the most basic active units of the processor and memory. Like reality cells, each transistor is a thing that can be on or off, a 1 or a 0, this way or that way, distorted or undistorted. At the next higher level of generality, is the processor's basic instruction-set that defines operations and processes for storing, retrieving, and performing arithmetical and logical operations. In our analogy, the processor's basic instruction-set is analogous to basic cognitive functioning within AUM. At the next level of abstraction, we get to the space-time rule-set which is analogous to algorithms written in assembly language. Our experience is generated at the next higher level of abstraction by an AUM-TBC to individuated-consciousness interface which is analogous to a simulation programmed in object oriented C^{++} where we are the objects. AUM is the computer, the programmer, and the operating system. We sentient conscious beings are, as individuated subsets of consciousness, a bounded subset of highly organized, evolving, interactive reality cells.

As an analog to space-time, consider a custom designed special purpose processor such as a Digital Signal Processor (DSP) chip. Understanding the rules (patterns) governing the transfer of energy to and from transistors in a special purpose microprocessor would provide some understanding of the most basic relationships in the processor's design, implementation and capacity. Likewise, understanding the rules governing the transfer of information between space-time reality cells should produce some of the most fundamental relationships of physics. In Section 4, you will see how that works.

Physical experience is generated when the perception of an individuated consciousness (sentient being) is constrained to follow the space-time rule-set. Imagine a specialized space-time virtual reality trainer (operating within a subset of digital calculation-space called a dimension) that is constrained by the space-time rule-set to provide a causally consistent operational experience that enables an individuated consciousness to evolve to lower entropy states by exercising its intent through free will choice. The specific relationships defining AUM's space-time instruction-set constitute the laws of space-time physics (PMR physics).

Applied mathematicians, scientists, and engineers have a tendency to define their reality in terms of its constraints expressed as physical constants. For example, c (the speed of light in a vacuum) defines the speed limit of matter in PMR. Think of c as one of several local PMR space-time constants that constrains our physical reality to a certain experience-set. Likewise, it is reasonable to assume there are constants that constrain the bigger picture as well. For example, given that a finite growing digital consciousness system cannot create an infinite number of reality cells and that the information contained within that digital consciousness system is limited by the capacity of the system (an upper limit on the total number of reality cells perhaps), then, after the system's native technology has stabilized, the ratio of the size of the system to the amount of information it contains would tend toward a constant as the system matures. Picture a growing AUM where new information and new reality cells are constantly created and recycled while the entropy of the system evolves toward greater profitability by more productively using and organizing that information. The point is: Real finite systems must always deal with constraints and the AUM-digital-system-thing is no exception.

What does AUM do when it is done (reached capacity limitations)? Being digital, it can always purge enough of its least productive bits to continue on, or purge even more than that and start over. Consider that AUM may be a contributing player in a larger consciousness ecosystem. Just as we iteratively cycle subsets of our individuated consciousness through the PMR learning lab to help drive the consciousness cycle, AUM may recycle its own consciousness in a similar manner to drive a higher level of consciousness organization that is beyond our grasp. AUM may regularly upload lessons learned and then recycle itself or it may simply continue to improve itself gradually as it continually and forever moves toward an absolute zero of system entropy. In either case, we come to the same conclusion: An evolving digital dude is never done.

A few low-hanging observations need to be plucked from the preceding discussions and then we can pack it in. You may find it interesting to take a Big Picture look at the conceptual flow that is unleashed by the concept, implementation, and evolution of time.

AUO evolves the exceptionally useful organizational catalyst we call time by maintaining a regular beat (a constant rate of oscillation). The ability of time to sequence patterns and enforce consistency allows AUO to create ordered and disciplined process and thus lower its entropy. Time enables simple existence to generate complex evolutionary potential from the explosion of new patterns, sequences, and forms that suddenly become available with the invention of dynamic process. Time generates new degrees of freedom for consciousness to explore. Time allows dim consciousness to become brilliant. Order, consistency, and regularity enable the creation of precise multi-frequency clocks, the big computer, and space-time as individuated specialized patterns of information and content within AUM.

Because the attributes of space-time provide the logical conceptual structure for our experience of PMR (mass, energy, space, and hence biology), it is clear that our existence is enabled by the invention of time. It is time that allows AUO to get organized – to create, store, and use information interactively – to evolve self-aware purpose and develop proactively.

The concept of time within the dim awareness of AUO is a byproduct of a potential energy system finding profitability in improving its self-organization. As an incredibly effective engine for entropy reduction, time becomes an evolutionary inevitability. The Fundamental Process could not help but find time (an artifact of change) on the path toward greater profitability. Thus time evolves naturally within consciousness – it delivers the fundamental organizing technology that enables the decrease in entropy (winds up the digital spring) that fuels, runs, drives, and enables everything else within an advanced AUM digital system.

Time is fundamental to the existence of our reality while space is a virtual perception based upon time and the constraint c. As consciousness evolves, its energy-spring continues to wind up as its entropy decreases. In contrast, within our PMR sub-system, the energy-spring (energy that is organized, structured, and able to do work) must slowly unwind as the experiment runs its course. Physicists refer to this natural PMR structural disintegration as the second law of thermodynamics. The second law is hailed as very bad long-term news for our local reality because traditional scientists have only little TOEs.

That music is universal to all cultures (and most creatures) and evokes a deep resonance within us seems reasonable enough when you consider that our consciousness, our being – undeniably, our entire reality – is constructed of rhythm and pattern.

Take a moment to ponder the deep connection we have with our reality. Try to think a few big thoughts (your choice) – go ahead, I'll wait. You might toy with the idea of having an original insight toward developing your personal Big TOE. Take your time. Then when you are done, heave a deep sigh of silent resignation in appreciation of your limitations. But don't give up. Both you and AUM have to start from wherever you are and pull yourself up by your own experiential bootstraps – that is simply how consciousness evolution works.

Consciousness evolution does not progress through a process of successive epiphanies. Significant entropy reduction is accomplished through an iterative process that slowly accumulates a large number of nearly infinitesimal profitable choices to produce significant growth. Trying to grow your consciousness in great leaps impedes progress by distracting you from the hundreds of small profitable choices you have the opportunity to make every day. Your consciousness quality rises and falls based upon the intent that animates your everyday interactions.

32

■ ■ ■

An Even Bigger Picture

■ ■ ■

Will AUM ever exhaust all the possibilities? Can it grow indefinitely? Will it one day reach an unchanging stable equilibrium, still viable forever but no longer evolving? Are there other AUOs or AUMs or is reality big enough for only one? And if there were two, how would they interact? That would dramatically raise the level of complication and create new possibilities for the Fundamental Process of evolution to work with. It certainly did for us.

What if AUM splits in two, employing a process analogous to biological cell division, enabling it to interact with itself in an entirely new way? Biological cells figured out how to do it. Do you think they are smarter or luckier than AUM? A self-replicating AUM could build up a new consciousness-thing out of AUM-cells, each existing within its own digital dimension. Would each AUM-cell function as a single brain cell (or consciousness cell) within a stupendous group-mind, or would a different and unique creature emerge? Cloning, reproducing, repairing, expanding, and merging are particularly straightforward processes within a digital medium – ever copy a floppy? Where does the pursuit of profit lead a brilliant finite digital oneness? Groups of AUM cells could very well specialize and form a much more complex thought-system-thing-being, and then.... and then... AUMamoebas and AUMamabobs thinking around in primordial mental swamps!

The biological cells of a larger bio-organism mature, die, and are replaced. In contrast, digital AUM-cells may simply mature and maintain some optimal joint profitability with a larger organism or perhaps occasionally recycle themselves by uploading their accumulated products of evolutionary profitability along with all other useful communicable results of accumulated experience and then reboot. AUM, like us, may serve as a cog in a much larger entropy-reduction engine that is engaged in driving an even larger manifestation of the consciousness cycle toward ever lower entropy states.

Big picture this: consciousness systems (like future digital computers), derived from consciousness systems (like humankind), derived from consciousness systems (like TBC), derived from consciousness systems (like AUM), derived from consciousness systems (an even larger system in which AUM is only one of many contributors) – all exhibiting the same fundamental structures and processes while hierarchically supporting each other at various levels and scales. Did you notice that the terminology used to describe interconnected interdependent consciousness systems at various scales and levels is similar to the terminology used to describe fractals? Consider the conceptual elegance of a consciousness-evolution fractal and hold that thought until we will get back to it toward the end of Section 5.

While you are in the mood for big thoughts, imagine this: our entire local reality (OS, containing a portion of $NPMR_N$ as well as our PMR universe), evolving on its merry way as if it were a self-contained colony of bacteria existing within a small remote corner of the space-time-part of just one of the apparently infinite number of apparently infinite AUM-cells, inhabiting the gut of an AUMosaurus.

It may be fun to let a few of the possibilities that might describe AUM's **external** environment run wild in your imagination, but all we need to build a complete theory of everything is a simple one celled monolithic singular AUO and its internal environment. Logically, all that is required is a finite primordial-consciousness-potential-energy-thing to exist along with the Fundamental Process of evolution. From only those two assumptions, I intend to derive physics, the nature of your existence, your physical universe and experience, and provide a comprehensive working model of the larger reality – all self-consistently contained between the knuckle and nail of My Big TOE. That should be enough! I'll publish the definitive work on AUMosaurus-beyondgraspus-outofmindus, which is its proper Latin name, next year – look for it at your local asylum bookstore in the humor section.

▶ Watch out for that AUMosaurus folks! …Oh no! … It's beginning to squat! Hang on … this baby could blow a thousand universes (including ours) out of its Anterior Sub System in one mighty blast of digital flatulence!

The roar of a billion suns exploding simultaneously violently shakes the foundations of reality!

Existence shudders and quakes in uncertain terror.

"Are we still here?"

"I think so, but I'm not sure? How can you tell?"

"Whew, that was a close call!"

"Yeah! Wow! That really makes you proud to be an intestinal bacterium, doesn't it?" ◀

Seriously folks, do you see how important it is to appreciate your fundamental limitations? If you do not, you run the risk of being unable to discriminate between silly and serious. Insisting that we, the magnificent crème de la crème of existence, could not possibly have such limitations simply adds a layer of obfuscating illogical arrogance on top of what is already beyond comprehension. With regards to the Even Bigger Picture, the best conclusion is no conclusion: Open minded skepticism requires no closure, and thus, accepts its limitations. It does, however, require that you put aside your normal knee-jerk responses (which have been conditioned by your ego, beliefs, and fears) in order to recognize all the possibilities.

Never fear to let your mind run free, but at the same time be aware of the innate limitations of your perspective or you may get lost wandering along the mystical shores of Never-Never land believing that what you see is what is there.

33

■ ■ ■

Infinity Gets Too Big for Real Britches

■ ■ ■

In this chapter, the concept of, and necessity for, a **finite** AUM will become clear. You will see that AUM does not need to be infinite or anywhere close to infinite to generate all the reality we can comprehend.

In the biological world of PMR, large biological cells must seem infinitely big to atoms. Let's translate that comparison of scale to our universe. If our entire universe was like a single atom in an apparently infinite cell of something much bigger, we actually couldn't care less because from our minute perspective, the really, really big stuff isn't that important to us. Whether that cell is actually infinite or not is not an issue. If it is relatively infinite to us, that is good enough to account for everything we can know.

From the smallest thing that we think we know exists (sub-atomic particles) to the largest thing we think we know exists (the universe), about 100 orders of magnitude (10^{100}) are spanned. The numbers 10 and 100 are not particularly big numbers. If there were some big thing, as much bigger than our entire universe as our universe is bigger than a sub-atomic particle, then this big thing would be only 200 orders of magnitude bigger than one of our sub-atomic particles. If this big thing were itself to be the relative size of a sub-atomic particle to something even bigger and if that even bigger thing was to be the relative size of a sub-atomic particle to something even bigger yet ....

Are you following me here? **Our** entire universe would be as a single electron to a universe that is also as an electron to a universe that is also as an electron to a universe, (four nested universes, one inside the other, each being at a size-ratio of 10^{100} to the next). I think you will agree with me, that final universe is really, really big. This colossal, really, really big thing is only 400 orders of magnitude bigger than one of our own tiny electrons. And guess what? That's only 10^{400}, and everyone knows that 10 and 400 are not big numbers.

Already we have slipped beyond the far edges of your ability to comprehend. What about $10000000000^{10000000000}$ ($10^{100\text{billion}}$)? That is a 1 with a hundred billion zeros between it and the decimal point times bigger than an electron. That much magnitude would allow a billion nested universes existing one within the other – each being the relative size of a single electron to our universe, to the next larger universe. I can write these numbers in a few seconds taking up only a few inches of paper – are we close to infinity yet? No way, these are still relatively infinitesimal numbers compared to infinity. Nevertheless, we have gone far beyond the size that represents something your imagination can begin to begin to comprehend. What about a trillion-trillion zeros ($10^{\text{trillion trillion}}$)? And still, we are not yet a tiny miniscule fraction of the way to infinity yet. How many zeros could you write in your lifetime? How many times could you write the word "trillion"? Even at that, we wouldn't be close to infinity – we could always take that number and raise it to an exponent equal to itself.

Do you get the idea of how **unnecessarily** big infinity is when you are talking about a **real** thing? **AUM is a real thing**, an **apparently** infinite something, but actually quite finite. All the PMRs and NPMRs are real things. Reality is a real thing. Infinity is far too big for real things. You see, AUM doesn't need to be infinite; in fact the concept of infinity brings with it all kinds of logical inconsistencies such as infinite processes that require infinite time and infinite energy which are always unavailable to anything real. Consequently, it is sufficient for AUM to be only relatively infinite, apparently infinite. That is enough to suit our purposes. In your mind, let the word "infinity" be a metaphor, not a defined mathematical abstraction. "Infinite" makes a handy mathematical concept but a very misleading adjective in front of a real, extant, noun.

While we are on the subject of big real things being small relative to the abstract concept of infinity, let's talk about time and infinitesimal things. It is the same drill. For the sake of comparison, let's resurrect the numbers we made up in Chapter 31 of this book. I bet you didn't know there have been less than 10^{18} of our tiny little seconds that have tick-tocked away since the Big Bang formed our universe. In case you were wondering, those 10^{18} seconds would consume only about 10^{62} PMR DELTA-t time-quanta (recall that one DELTA-t = 10^{-44} s). Break a second into a billion pieces and you get a nano-second (ns). Less than 10^{27} nano-seconds (ns) have come and gone since the Big Bang went kaaaboom!

Break a nano-second into another billion pieces and you get a nano-nano second (n-ns). There have been less than 10^{36} (n-ns) since the Big Bang, and less than 10^{45} billionths of a billionth of a billionth of a second (n-n-ns) ticked away since our universe began. Is it surprising that a ten n-n-n-n-ns long DELTA-t time increment (about 10^{-44}s) could be small enough to make our gigantic PMR nanoseconds seem absolutely continuous and that the number of them that has ticked away (for much longer than our universe has been forming) remains a relatively tiny number compared to infinity? No doubt a trivial number for an apparently infinite AUM to track.

There is another important point here besides the fact that AUM is finite and real and maintains a huge excess capacity that allows it to do, think, create, and be much, much more than we can imagine. You need to realize that we exist in, and have the capacity to comprehend, only a tiny, tiny, minute, infinitesimal fraction of the possibilities for existence. We have seen above that many entire universes each progressively larger than ours (by the same amount that our universe is larger than an electron) can coexist with us, within our finite 3-D space-time reality. Likewise, in the opposite direction, many realms may exist that are each progressively smaller than ours.

These scaled up and scaled down universes would not evolve carbon-based humans and critters resembling us. Our physical matter (composed of the elementary particles, atoms, and molecules that we are familiar with) is nicely suited to our own scale and ill suited to drastic changes in scale. However, that is not to say there could not exist enough weird stuff (totally unlike our matter) that has evolved forms and structures appropriate to its own scale that have enough complexity, memory and processing power to evolve intelligent life-forms. **Why**? Because Mother Nature is one aggressive babe! **Why not**? Because you do not have the capacity to conceive of it? Why would a rational person (who wasn't hopelessly optimistic about the completeness of their knowledge) believe that our personal and collective ignorance and limitations should logically constrain the possibilities of creation?

Our contemporary objective knowledge-base sets the limits on what we can collectively imagine, much less comprehend, and it is unlikely that our little picture science can ever hypothesize more than a minute fraction of the possibilities of creation. Regardless of how hard we try, all we can conjure up are various versions of ourselves because that is all we know. There is little point in speculating beyond the limits of your vision. Appreciating your inherent limitations and applying open minded skepticism to your logical and scientific processes constitutes the only approach to exploring the Big Picture that makes sense and produces results. Any other approach will eventually lead you to paint yourself into a corner with broad strokes of belief and arrogance.

We can barely comprehend, or even wildly imagine, only a very small slice of the potential 3D space-time possibilities for existence within our own 3D space-time – much less existences in other 3D space-times within other dimensions – much less existences outside 3D space-time but within $NPMR_N$ – much less existence outside of $NPMR_N$ but within NPMR. Would the possible realms that lie beyond our comprehension be populated? Could they have evolved consciousness, critters, objects, energy patterns or other structured stuff? No one says they, their matter, their environments, or their governing rule-sets (physics) must necessarily be like us, our matter, our environment, or our rule-set.

In a digital consciousness reality, anything that can be conceived is possible. Anything with a rational structure (rule-based) and with enough potential and complexity will eventually evolve to a more highly organized (lower entropy) system of systems. We are not alone – reality is teeming with intelligent life-forms. In fact, we humans and nonphysical entities of OS are relatively minute – like one small school of fish in a gigantic ocean of sentient life. When you learn to move your awareness between dimensions, you will be amazed at the diversity of life-forms that share our consciousness ecosystem.

What difference does the existence of other reality systems make? Until one or more of these realities interact with your reality, the answer is none whatsoever. Absolutely none. It is a fact that OS is only one of many evolving reality systems but this particular fact is of no **practical** importance to your current task of reducing your entropy within the PMR learning lab. It may be nice to know that you are not alone, and interesting and educational to explore realms outside OS, however, such exploration can do little by itself to improve the quality of your consciousness. Activities that do not contribute significantly to the positive evolution of your consciousness are not important.

The Fundamental Process of evolution combined with the potential of sufficiently complex digital systems to self-organize generates all the reality that is within our capacity to comprehend. In contrast, our space-time universe and small blue planet represent only one special-case application of that process. It is worth noting that the dynamic interaction of a simple evolutionary process with the fundamental energy of a self-organizing digital consciousness spawns individuated sentient awareness (thinking beings) through a multi-leveled self-generating progression of recursive process that is reminiscent of how a simple geometric relationship generates an intricate fractal pattern. An interesting similarity: In both consciousness and fractal systems, each part of the pattern carries the blueprint of the whole.

The critters and things of the earth along with our beloved planet, solar system, and universe inhabit only that sliver of 3D space-time and nonphysical reality that is appropriate for our kind and our purpose. Every reality evolves its own compatible systems and ecosystems. Palm trees do not grow on high mountain tops for similar reasons.

The space-time rule-set is only appropriate for a tiny subset of reality. Other rule-sets define consistent causalities elsewhere that are unimaginable to the residents of PMR. Consciousness and evolution have few limits placed upon their combined creativity. If any system or subsystem is profitable (can lead to self-initiated entropy reduction), it probably exists. Why would you believe that reality systems like OS are the only ones likely to be profitable when your mind can comprehend no more than an infinitesimal subset of the possibilities?

Do not let your beliefs and the limitations of your current knowledge limit your notion of what is possible. Letting your ignorance define the limits of an acceptable reality is bad science; unfortunately, it is also the norm. That is why young scientists make most of the dramatic breakthroughs. When you **believe** that you know, but do not, you cut yourself off from the possibility of ever knowing. The knowledge that lies outside the possibilities allowed by your core beliefs is beyond your intellectual reach – entirely invisible to your self-limited vision. You simply cannot see beyond the walls you have constructed regardless of how hard you are trying to do exactly that and no matter how strongly you believe that your vision is clear and unimpeded. To see a bigger picture, to view what lies beyond, you must first tear down the wall or at least some portion of it. Human cultures are little more than communities of shared belief where common belief-blindness leads to erroneous conclusions that are universally held as obvious truth.

Energy-forms (matter and light for example) and physics within other reality niches (including other 3D space-time realms) may be vastly different from our PMR because each dimension of existence (separate reality) functions under its own rule-set. Nature (the natural manifestations of evolution within each reality) tends to leave few empty places and has a way of evolving something (not necessarily carbon-based biology) to populate whatever can be populated. Even the environs around high temperature thermal vents, under huge pressures and in total darkness at the bottom of the oceans, produce life. Not as we are used to perhaps, but life just the same. The fundamental evolutionary process seems to work everywhere – life or conscious awareness or individuated sub-systems of consciousness are merely the result of a purposeful, self-interacting complex system with memory, evolving its way through a large and diverse set of environmentally constrained possibilities.

A complete set of the possible forms that sentient life could take is beyond our knowing. AUM is one member of that set and we are another. Bacteria and flat worms are two more. Evolution produces life-forms that are suited to their particular environment. That is how we and all the critters (including one celled critters) and plants turned out as we are – we fit (are efficient and effective in) our environment. An environment beyond our imagination will evolve forms and functions beyond our imagination to suit itself. Everyone should agree that a digitally simulated interactive virtual reality has few intrinsic limitations.

It is a generally accepted idea among scientists that if we (or at least something vaguely like what the earth has produced) can not exist, grow, and evolve within some postulated reality or host environment, then it is not possible for sentient life to flourish within such a place. The belief that any life-forms that might exist "out there" must be similar to the life-forms of earth is a product of little picture thinking. Without a doubt, we have a very limited view from where and what we are and thus should be doubly careful not to constrain the possibilities we can grasp merely because we do not already know the answers. Such arrogance creates the walls of its own conceptual prison.

Our PMR time appears to us to be continuous (non-granular or non-quantized) because the smallest times we can measure are vastly huge compared to one quantum of our time. You do not notice the granularity in a slab of steel or a pane of glass unless you can measure distances down to the size of an atom or molecule. The glass, steel, or wood table-top you set your drink on appears to be solid and continuous although we know it is made up of discrete tiny particles with vast regions of empty space between them – the granularity is imperceptible from the macro view. Time is the same way. From the macro view of our everyday experience, time appears to be continuous.

You may find it profitable to visit this topic again after you have read Chapters 81 and 82 of Book 3, where parallel probable universes are discussed. The bottom line is that all the time in all the past and present worlds that you might imagine is not a big deal for AUM to track. In time, as well as space, we see that apparently infinite (but actually finite) is much, much more than enough to accomplish everything we have laid at the feet of a real AUM and a real TBC. It is reasonable to assume that AUM and TBC have multiple gazillions of margin (extra capacity) left over to grow into and play with. Although AUM is a Big Dude with a lot to do and keep track of, being infinite is not a prerequisite for the position.

34

■■■

Section 2 Postlude
Hail! Hearty Readers,
Thou Hast Shown Divine Patience
and Great Tenacity

■■■

I know it has been a wild ride and that you are wondering what to make of all these new concepts and ideas piled one upon the other upon the next like food on the plate of a teenage boy at an all-you-can-eat junk food buffet. What's worse, most readers probably have little experience that can help them evaluate the truth of what you have just read. That is the nature of Big Truth. It is either trivially obvious or totally opaque.

There is no need to make judgments or jump to conclusions. Do not **believe** anything, for or against. Maintain an open mind and realize that you are only at the beginning of this journey. If everything made sense early on, it wouldn't be a particularly complex exposition and a trilogy would not be required. Because I am taking most of you on a trip far beyond your wildest musings, you should expect (this early in the unfolding) to feel somewhat ambivalent, lost, and – please – always skeptical.

If you have little experience systematically exploring the larger reality of inner space, what you read here will no doubt seem abstruse, unsubstantiated and beyond knowing. It is not. It only seems that way from the perspective of your culture and your self-limited experience. Consider how a five-year-old child, or your favorite pet critter, views and understands the larger world it lives in.

Many things will seem absolutely beyond anyone's possible knowing when they are actually only out of sight of a limited vision and understanding. Believing that your vision is less limited than it actually is reflects a natural trait of high entropy consciousness that is constrained to an indi-viduated unit of independent action. (Section 4 will clarify this concept.)

Although what you have read in Section 2 may seem remote from your sense of reality, it was necessary to introduce some of the basic ideas you will need to bring the Big Picture into focus. Everything will fit together more smoothly later on. If the discussion of the life and times of AUM was tough slogging and prompted you to yawn a few ahhh-ummms of your own, take heart, this next section – Section 3 – is all about you.

Because it is about you, it is bound to be exceptionally interesting – how could it be otherwise? In Section 3 ("Man in the Loop") we will make the first tentative connections between all the weird off-the-wall concepts in Section 2 and what they could possibly have to do with you and your physical and mental reality. I think you will like this next section more than the previous one, because it will give you some things to think about that you **do** have the experience to evaluate and it will be less abstract.

If you have made it this far and still have the desire and gumption to go on, award yourself four beautiful gold stars for your intellectual tenac-ity and inquisitiveness. Paste them inside the front cover of this book so that everyone will know that you are a dedicated, committed, serious seeker after something extraordinarily important even if you are not sure at this point exactly what it is.

MY BIG TOE

BOOK 2:

D I S C O V E R Y

Section 3
Man in the Loop:
How You Fit into the Big Picture –
Ego, Body, Mind, and Purpose

Section 4
Solving the Mystery
Mind, Matter, Energy and Experience

Synopsis of Book 1

Book 1: *Awakening* **– Section 1** provides a partial biography of the author that is pertinent to the subsequent creation of this trilogy. This brief look at the author's unique experience and credentials sheds light upon the origins of this extraordinary work. The unusual associations, circumstances, training, and research that eventually led to the creation of the *My Big Picture Theory Of Everything* (My Big TOE) trilogy are described to provide a more accurate perspective of the whole.

Book 1: *Awakening* **– Section 2** lays out and defines the basic conceptual building blocks needed to construct *My Big TOE's* conceptual foundation. It discusses the cultural beliefs that trap our thinking into a narrow and limited conceptualization of reality, defines the basics of Big Picture epistemology and ontology, as well as examines the nature and practice of meditation. Most importantly, Section 2 defines and develops the two fundamental assumptions upon which this trilogy is based – a high entropy primordial consciousness energy-form called AUO (Absolute Unbounded Oneness) and the Fundamental Process of evolution. AUO eventually evolves to become a much lower-entropy consciousness energy-form called AUM (Absolute Unbounded Manifold) even though neither is absolute or unbounded. Using only these two assumptions, Section 2 logically infers the nature of time, space, and consciousness as well as describes the basic properties, purpose, and mechanics of our reality. Additionally, Section 2 develops the concepts of The Big Computer (TBC) and the Even Bigger Computer (EBC) as operational models of aware digital consciousness. Our System (OS) is defined to be PMR (Physical Matter Reality – our physical universe) **plus** the subset of NPMR$_N$ [a specific part of Nonphysical Matter Reality (NPMR)] that is interactive with PMR. Many of the concepts initiated in Section 2 are more fully explained in Book 2.

Section 3

■■■

Man in the Loop
How You Fit into
the Big Picture –
Ego, Body, Mind, and Purpose

■■■

35

■ ■ ■

Introduction to Section 3

■ ■ ■

AUM evolved the space-time consciousness construct to provide us with the experience we need to help it lower its overall entropy. Is this not as crystal clear as the morning dew? If you do not understand this now, you will understand it clearly before you have completed Book 2.

Before we get started, it might be a good idea to warm-up first with a few mind stretches. Everybody take a deep breath. Here we go. Stay with me now. Try to hallucinate a spherical chicken... a very large spherical chicken... bigger... bigger... bigger. That's it! Now hold that concept ... hold it... hold it.... Now – this is the difficult part – pretend you understand it... that's it, make it seem perfectly understandable, perfectly reasonable... that's it! That's it! Now hold that clarity... hold it...just a little longer... hold it...OK! ... Relax! ... Phew! That's better. OK, now end with another deep breath. That was great! I think you are as ready as you are going to get.

Seriously folks, if you do not understand something, do not assume that it must be so or that it cannot be so – and do not pretend uncritically that you understand what is outside of your experience. Remember: Open-minded skepticism is the order of the day. Use it to fill up the void left by temporarily suspending your beliefs. All right, let's get back to work.

It is not that we are separate from space-time and live within it, as we live in a house but are not part of the house. We experience space-time as our minds interpret and process the limited perceptions gathered by our senses. Contrary to popular belief, space-time is not a physical construct that we physical beings live within, but rather the result of a rule-set that defines the experiential virtual reality that we perceive. Because of our limited physical perception-based experience, it appears to us that we live or exist within a space-time universe when in fact space-time is nothing more than the constraints that bound the experience of an individuated consciousness enrolled in PMR 101.

I know that this is confusing, but by the end of Section 4, you will have no trouble understanding that the physical reality you physically experience is merely the experience our consciousness interprets as physical reality. If your sensed physical reality and what your consciousness interprets as physical reality seem the same to you, it is because your culture has convinced you that consciousness is a derivative of the physical body. The experience that our senses interpret as physical reality leads to the conclusion that physical reality is an **external** reality that our body interacts with (reality is defined by our physical interaction with it). On the other hand, the experience that our consciousness interprets as physical reality leads to the conclusion that physical reality is an **internal** reality created by the perceptions of consciousness – a virtual physical existence that is defined by our mind's interaction with the space-time rule-set. The first assigns perception directly to the body, while the second assigns perception directly to the consciousness and only indirectly to a virtual body constrained to perceive a local reality defined by a given rule set. An individual mind (an individuated unit of consciousness) engaged in an interactive multi-player virtual reality must experience, act, and interact within the bounds prescribed by the causal rule-set that defines that particular reality. Any consistent high fidelity virtual reality must follow a specific rule set and will appear to be physical to the individuals experiencing it.

Clarity should emerge in Section 4 as you begin to understand that the ultimate source of your experience is not what you perceive as your local physical reality. Your physical reality is an interpreted virtual reality that only appears physical. Have patience and you shall see that this concept is scientifically and logically sound.

Individuated consciousness within AUM is roughly analogous to a two-dimensional bed sheet that some children have stuck their hands into, pulling the sheet down around their wrists and forearms to make individuated hand puppets. Each hand puppet is an individual animated thing, and can interact with the other puppets (by grabbing them perhaps). Yet for all their individuality (fat, thin, small, large, aggressive, calm), each hand puppet is part of the same sheet, existing only as protrusions in the sheet relative to flatter, more uniform parts of the sheet. The puppets exist as three-dimensional variations in the two-dimensional sheet. They are all part of the same sheet, but exist as individual extensions of the two dimensional sheet into the third dimension.

It is worth noting that the extensions into the third dimension must be maintained by constraints. Imagine a rubber band that goes over the puppet and around the wrist of the child's hand. Remove the constraint and the sheet's protrusion into the third dimension quickly disappears. The sheet maintains its natural two dimensional existence unless some sort of constraint forces it to bulge in the third dimension.

Similarly we PMR physical beings, each with our personal individuated consciousness, are all part of the same AUM consciousness. We are individual, yet at the same time we are all one with AUM, the fundamental source consciousness. Our individual existence, like the hand puppets, is the result of constraints defining a dimensional variation in consciousness that individuates a unique entity with free will. Space-time is the virtual medium through which the rules of engagement (constraints that define our interactive experience with other "players") are applied. Players are defined as anything (beings, objects, or energy) that may become interactive with us.

Can one idea (thought-thing) be manifested through (give birth to) another? Sure. Why not? In AUM's world of digital consciousness energy, thoughts (discrete packets of organized content) are real things – the only real things – and AUM can birth (think up or organize) as many as it wants to. Think of a thought within AUM as a reusable object, a chunk of fixed or variable content with certain attributes, characteristics and abilities that can be stored, transmitted, or used as an operator. The object oriented programmers among you will pick this idea up very easily while the rest should think about the persistent, consistent, and cumulative digital existence of various characters, items, and devices in interactive on-line computer games. Now, imagine that some of these game-object-thought-forms have enough complexity, memory, and access to processing power to be goal seeking, self-modifying independent agents. Let your imagination run wild until you can imagine digital creations of all types – all thought forms within a digital consciousness.

Recall that everything at its core is part of the same digital-AUM-mind-sheet-thing. If you followed the previous two paragraphs, you ended up with the concept that we are fragments, derivatives, or subsets of AUM consciousness, existing individually within AUM.

Oh sheet! This is confusing! [This is your first test. Was the preceding phrase: (a) – a prayer to AUM asking for guidance and relief; (b) – a thinly veiled coarse expression of frustration born of a limited viewpoint; (c) – a pitiful attempt at humor; (d) – one more meaningless phrase indistinguishable from all the others; or (e) – all of the above?]

In Section 4, I will give a detailed explanation of how we derive, perceive, and experience our 3D bodies as well as our entire space-time universe. We, our universe, and other universes in other dimensions are all specialized thought-parts or subsets of AUM. This is true for physical and nonphysical beings and universes. Before we leap off into a discussion of how, why, and where we as individuals fit into this Big Picture, let us quickly pull together and consolidate what we know about AUO, AUM, and the results of the Fundamental Process.

We know that AUM is the result of the Fundamental Process of evolution being applied to the one celled, dimly aware, primal AUO consciousness-potential energy-thing. We also know that AUM's complexity and awareness continues to dramatically accelerate until it reaches a relatively stable average growth rate where issues of quality and refinement begin absorbing more of the available energy than bold new leaps into unknown and untried possibilities. From that point, internal quality is improved and gains are refined, integrated, and consolidated until the next evolutionary breakthrough occurs, setting off another period of rapid growth.

Eventually, the growth **rates** of finite systems must decrease (the very definition of maturity), but they do not have to go to zero. In general, the larger the system, the longer it takes before growth rates become asymptotic to the time axis. An apparently infinite aware-consciousness-energy-thing constitutes an exceedingly large finite system with unimaginable opportunities for accelerated growth. Do not even try to imagine the breadth, depth and capacity of AUM – you cannot.

Digital consciousness systems do not deteriorate with time like biological systems, though they can de-evolve – that is, evolve into higher entropy, less significant, profitable, and viable states. AUM achieves self-optimization and growth through the exploration of the possibilities by implementing the Fundamental Process. AUM can eventually figure out how to willfully boost its quality (lower its entropy by utilizing its potential and organizing its bits more effectively) once it realizes that profitability is a function of the intent that drives its choices. So it is with us.

When you read "AUM learns," do not use the small-view definition of learning (intellectual learning). Learning is more than accumulating facts from your experience, other people, or books. Big Picture learning must also include improving the quality of your being, which is not a fact-based process. Have you ever known an exceptionally smart person (knows lots of useful facts) who is also (choose one or more from the following list) dumb, out of it, insensitive, egocentric, arrogant, or an intermittent or full time jerk? If you have the stellar good fortune to hang out in the hallowed halls of academia, with the upper echelons of government or corporate management, or with a bunch of highly ambitious middle aged professionals, you know exactly the type I am referring to. Such a person appears more retarded than advanced from the perspective of a bigger picture where the ability to develop and maintain effective positive relationships with a wide variety of people is much more significant than the ability to manipulate facts.

Clearly, growing up within a larger reality has much more to do with raising the quality of your consciousness than accumulating information. What matters most is the development of wisdom, understanding, and the capacity to love – which are not primarily intellectual achievements. As you grow up, you learn to synthesize your experience data into larger and larger perspectives until eventually **everything** is seen to be interactive, interrelated, and a part of everything else. (The love I am referring to here is an attitude, a value, a way of interacting and being, and needs no specific object on which to focus.)

Facts and intellectual knowledge can help point in the right direction and perhaps pick around the edges of how to grow quality, but to truly "get it" requires that one go beyond rational PMR causal analysis. Analysis fails because you can never collect more than a small percentage of the relevant facts required for a rational, logical conclusion when dealing with Big Picture issues – and because love is **not** an intellectual result. Love is the result of low entropy consciousness.

AUO was described earlier as everything (the one source) and nothing (no actual individuated thing) simultaneously. AUO began its existence as an unstructured potential energy system analogous to a single biological cell floating in the primordial ooze. AUM is aware, active and purposeful – an advanced aware consciousness. What an amazing transformation! You can thank the Fundamental Process of evolution, especially as it applies to consciousness, for that metamorphosis. As consciousness develops awareness, intelligence, values and personality, its entropy shrinks as its ability to organize itself effectively and profitably increases.

AUO represents the basis for consciousness, an energy form, a medium for digital self-organization or awareness. AUO is a metaphor for a primordial dim awareness that evolved the capability to create differentiated cells (local non-uniformity) relative to its uniformity. It subsequently found it profitable to change the state of those cells as the Fundamental Process began optimizing internal environmental interactions. As the complexity, potentiality, and possibilities grew, AUO's awareness evolved to include specialized structure, memory, organization, complex content laden communications between subsystems, brilliant self-awareness, and purpose. AUO naturally evolves (grows up) into AUM.

> ▶ Picture AUM as a geeky hygiene-challenged teenage computer freak with ugly red zits called PMRs. Do you think our universe might be a particularly nasty infection on the nose of a pubescent consciousness? I bet you have never thought of your reality in those terms before. A mystical eruption on the nostril of AUM! What a beautiful image. Sheer poetry! Of course, I am just kidding… we wouldn't be on AUM's nose.
>
> Speaking of getting popped, what does the "Zit Theory Of Existence," or what is more affectionately referred to by cosmologists and cosmetologists as the "Pus Я Us Reality Concept" do for your ego's sense of humanity's special importance?
>
> Ahhh ha! After all that polite jabber about Petri dishes, the mysterious connection between humanity and bacteria rises to a head.
>
> A word of caution: This book is heavy so be careful where you throw it. You might inadvertently hit an innocent bystander in the head – thereby warping his skull and his mind simultaneously.
>
> Take a deep breath and let it go; there is no saving throw against tasteless, lowbrow humor – these days an eloquent book is as rare as a truly innocent bystander. ◀

The evolution of bright complex consciousness from dim awareness – does that seem unlikely? We carbon-based human life-forms are reported to have done something like that ourselves – from one celled dimly-aware blobs of protoplasm to our present grand and elegant selves (did I forget to mention magnificent, brilliant, and superior?). And we are on slow time by 36 orders of magnitude compared to AUO-AUM.

You, as part of AUM, are simply an individuated consciousness. I know that may seem strange, but making this a reasonable proposition and explaining how it works is what Sections 4 and 5 are all about. AUM is consciousness, thought, operational knowledge, idea, and awareness. Consciousness represents the most basic form of energy – a self-relational digital medium that can be structured through evolutionary processes to reduce its average entropy. You are a portion, an infinitesimal smidgen, of this AUM-consciousness-thing and as such, you share the attributes and abilities of all consciousness. Because the capacity of your particular consciousness – your personal evolutionary potential – is great, so is your responsibility for its development and use.

AUM, like any complex system, has evolved both structural and dynamic components. Its structural components are objectives, values, dimension (specialized calculation space), memory, and patterns (rules).

Its dynamic components are time, intent, motivation, intellect and will. Value based awareness, intelligence, and purpose are created, sustained, driven, animated, and motivated by the evolutionary imperative to improve the functionality of the system through better organization (entropy reduction).

The imperative to implement the Fundamental Process is the prime mover of progress. AUM **is** consciousness (as opposed to **has** consciousness) and it eventually acts, changes state, and evolves through the exercise of will or intent – self-aware consciousness in control of itself. If you really want to (for poetic reasons) or really need to (for reasons of emotional comfort), you may say that all things (our universe, all PMRs and NPMRs, and all the beings therein) are manifestations of AUM's will and made of AUM's substance (self-configuring digital organization). You could also say that we brainy people, with our minds full of fully operational and original thoughts and ideas, are created in AUM's image (along with a varied collection of thoughtful dogs, cats, foxes, pigs, monkeys, and computers). Don't get lost here: I am talking about our minds, our consciousness, not our adorable little bodies.

Now we can take one more fold in our bed-sheet analogy. As before, AUM is the sheet, we and everything in our reality are the sheet-hand-puppets protruding into the space-time dimension, and the energy that animates the little fingers, hands, and arms is our individuated portion of AUM's conscious awareness, which is expressed through our individual free will. The sheet and its protruding dimensions are engaged in a program of continuous quality improvement administered by the Fundamental Process of evolution acting upon a multi-leveled consciousness-evolution fractal ecosystem. By the end of Section 4 this will be clear. For now, merely entertain the possibility that our bodies and all the objective matter in our 3D world are the products of constrained consciousness and a rule-based virtual sense perception.

Do not fall into habituated anthropomorphic concepts or you may start thinking of AUM as a person. It will be more productive to think of AUM as a thing, a complex consciousness system, a big cellular quantized thought-energy-thing that constitutes the One Source of All That Is. Do not conceive of AUM as a super-human intellect – yourself extrapolated to god-sized proportions and qualities. Resist the urge to turn AUM into an ancient looking old geezer with a long white beard playing with his pet people, or all manner of silliness will mystically erupt from the great void.

▶ "Hey, I know, let's play God! – I'll be the god, and you be the people….

No way, Hosea! I wanna be the god! I thought it up, it's my game and I wanna be the god first!

OK, OK, I promise – next eternity you can be the god and I'll be the people. Oh, come-on – it'll be fun!

Tell you what, if you let me be the god this eternity, I'll make you boss over all the critters and give you a woman that never goes out of heat. Deal?" ◀

Is AUM's awareness intelligent and sentient? I would say so, but not in the same way that we are intelligent and sentient – not that limited. Is a baboon's awareness intelligent and sentient? Yes, but more limited than ours in most cases. Does AUM care? Does it take care of us? Aren't we its babies, so to speak? Good grief! Don't get anthropomorphically silly on me here. Wipe that self-indulgent mist out of those puppy eyes. Do you care about the individual cells in your thumb or whether or not you are born with or without an appendix? Not really.

Where we human-types place our attention depends on the challenges of the Fundamental Process. Much of our energy is dedicated to our physical being (issues of survival and procreation). However, AUM does not have that distraction. Evolution for AUM is more of an up-close and personal take-charge sort of thing. Perhaps as we get better at genetic engineering and cloning we will get a dim glimmer of AUM's position relative to influencing its own evolution. Until then, it is best not to puff up our self-importance so much that it gets in the way of our ability to see the truth – whatever it might be.

In the meantime, if lumpy consciousness sheets seem distressingly cold and impersonal as a source of our being, I have a practical solution. Anyone wanting a warm fuzzy relationship that gushes unconditional love, which is focused individually on, and directly at, a needful, and oh, so, deserving **you** should… get a dog! Don't be confused by the forward and backward spelling thing: Simply look for a genuine, guaranteed warm and fuzzy d-o-g.

Besides loving, dogs are straightforward, honest, faithful, loyal, and forgiving. They are seldom demanding, revengeful, jealous, angry, self-important, or into fear and dogma (egotistical). They will never ever tell you to go to hell. They are never rude or sassy and never forget to flush the toilet or to turn off the lights that they have turned on. They will not run up charges on your credit card or dent your car. Better yet, they never drop their clothes on the floor and will never invite their mother to come live with you. No dog has ever smoked a cigar or invited his friends over to drink beer and watch football on TV.

Dogs are happy to exist on your leftovers and eat your garbage. That, ladies and gentlemen, is as warm and fuzzy as it gets **if** you are looking for a relationship that takes little-to-no effort on your part. That is what we are all looking for, isn't it? A no-fault (at least not ours) low maintenance relationship in which we are unconditionally loved and forgiven because we exist and meet the basic superficial requirements – what could be better than that? That is what you truly want, right? *No problema!* Go get a d-o-g!

If you maintain the letters in the right order, you will happily find that dogs deliver warm fuzzies **all** the time instead of only in relation to your needs, beliefs, and fears. And here is the best part – dog ownership never generates internal pressures that would lead you to be hypocritical or self righteous – you can just be yourself. Your dog will love you however you are. Even if you are not nice to your dog and don't love it – it will still love and adore you above all others! Its love is widely spread, deeply sincere, and truly unconditional – a being worthy of your emulation, if not outright worship.

Hey, what's with the firewood, rope, and torches? Is this some sort of medieval pageant? Are we going to have a bonfire and toast marshmallows? Look, I agree with you! There **are** magnificent and endless sources of genuine spiritual warm-fuzzies, but you have to work hard to grow up enough to access them. They are not easy, low maintenance, or superficial. They are not focused, even in a small way, on what **you get** for meeting requirements. That concept was spawned to support a membership drive. Instead, they are about your capacity to embody and apply (give) absolute unconditional love. Performing rituals and doing dogma doesn't get it. Belief (pro or con) can, at best, generate a self-focused ego-centric "**I** feel good about **myself, my** faith, and **my** belief" warm pseudo-fuzzy.

If you ever get to know the real thing, you will never again settle for a pseudo-fuzzy.

36

■ ■ ■

Jeez Louise, Will That Fat Rat Ever Find the Cheese?

■ ■ ■

It is generally accepted (among humans) that we Homo sapiens have the most advanced intelligence and are the most thoughtful of all the earth-creatures. Does that make us more AUM-like – closer to the source? Do you think that we must be AUM's favorite creation, or most amusing experiment? Humans are no doubt exceptionally clever. What other species could create no fewer than five entirely independent ways to destroy all life as we know it (nuclear bombs, global warming, toxic waste, pollution, and excessive non-ecological global resource consumption)? Such creativity and awesome intelligence makes you feel proud, doesn't it? What's wrong with this picture? What term is missing from the human equation?

Our species has always intuitively known that a little knowledge is a dangerous thing, we just never appreciated how dangerous for how little. Ah, we beings of limited 3D awareness – what you see is what you get – that is the beauty of us, and the opportunity of us. Our awareness-limited, physical experience-based, interactive human community excels at creating wonderful opportunities to grow the individual and collective quality of our consciousness.

Limited awareness guarantees that our interaction within PMR accurately and immediately reflects the quality of our being. You have only to consider young children or your favorite furry critter to realize how a perspective that is limited to a relatively little picture produces transparent and straightforward interactions and reactions. For an individuated consciousness such as you, limited awareness produces learning opportunities within a relatively simple and straightforward interactive virtual environment with immediate results-related feedback and consequences.

The complex, duplicitous, and anything-but-straightforward minds that we appear to have only look like that from a limited viewpoint where insignificant variations of fear and ego appear to be vastly more important than they actually are. The events that we wring our hands and gnash our teeth over day after day are typically drawn from the mishmash of trivial details that define our personal soap operas. What appears vitally important to us in the little picture of belief-blindness and struggling egos is often completely inconsequential from a Big Picture perspective.

You are aware in the little picture so that you may eventually grow your quality until it becomes capable of direct participation in the Big Picture. Aimlessly wandering about in a fear-based ego-driven little picture soap opera where you get to write your own script is not exactly on the fast track to success. That you may be a good enough and lucky enough script writer to become rich and famous in PMR is irrelevant.

In order to have a meaningful opportunity to do it right, you must also have the opportunity to do it wrong. From the perspective of the Big Picture, physical reality provides an optimal nursery for budding consciousness.

The typical self-assessment of human mental complexity and cleverness is based on our self-serving species-definition of "intelligence" and "cleverness" as the ability to influence, manipulate, control, and dominate others (natural and man-made environments, plants, critters, other people, and everything else). That is why we clever humans develop and evolve most of our technologies and social systems around the needs of war, defense, trade, and communications and why we have both accidentally and purposely generated multiple capacities to destroy the diverse life of our planet along with ourselves. Because of our needs, wants, desires, fears, ignorance, beliefs, and ego (lack of consciousness quality), we apply much of the mental energy of our species to issues of control and dominance – internationally, nationally, and personally. Cleverness, it seems, is primarily in the eye of the beholder.

Nevertheless, we physical beings are, from the viewpoint of a less limited consciousness, exceedingly simple and straightforward; our motivations, intentions, and interactions accurately and clearly demonstrate the quality of our consciousness in everything we do, say, think, and feel. A less limited consciousness finds our machinations within the physical to be transparent. Our quality, or lack thereof, is obvious; we cannot hide what we are. I have to ask: What do **your** motivations and actions say about the quality of **your** consciousness? That is a good question to ponder (you should paste that on the refrigerator door of your mind to remind you to ask it often) because, one day not that far off, you will need to deal with the answer.

The results of our actions (the environment we create) as well as the reactions of other people to us, provide immediate feedback to assess the quality of our intent. We create our reality through the implementation of our intent-choices and how we interpret our perceptions – the results are most often precisely what we need and deserve (can use) to stimulate productive growth. In general, if we do it wrong (make the wrong choices), we get pain. If we do it right, the resultant spiritual growth spreads sunshine everywhere. Here, "right" and "wrong" are used in the evolutionary sense. Right and wrong in the evolutionary sense are the same as absolute right and wrong in the motivational sense when the Fundamental Process of evolution is focused on improving the quality of our consciousness. The previous sentence actually makes sense, but you may need to think about it a while. Relative right and wrong often has its roots in absolute right and wrong, however, the result may be strangely twisted by belief systems of all sorts (cultural, religious, personal, and scientific).

> ▶ All pain does not come from making the wrong choices. Only self-inflicted (internally caused) pain, which makes up the bulk of our daily ration of pain and suffering, has wrong choices at its fundamental source. You may occasionally experience either small or great pain from external sources; these usually reflect the existence of randomness within our lives. (I will discuss the origin and nature of these random components in Chapter 45 of this book and in the next two Sections.) It is this simple: Sometimes the ball takes a bounce and hits you squarely between the eyes, and sometimes the cookie crumbles into dust before you get to taste it. Or, in the pithy, if somewhat crude, words of the famous cinematic philosopher-genius Forrest Gump: "Shit happens."
>
> External and internal causes of pain can be mixed and mingled together. However, most of us most of the time wrongly believe that our self-created pain is actually externally derived. That belief makes us feel better in the short run. We do not want to see ourselves as the primary source of our unhappiness and dissatisfaction, though that is **almost** always the truth. It is easier and more comfortable for us to **believe** that we are the victims of others, or simply are unlucky. Sometimes that is the case – but that condition is the rare exception rather than the general rule.
>
> The general rule is that most of the pain in your life is self-inflicted while very little is thrust upon you from the outside. We believe the opposite because an external enemy is always easier to accept and defeat than an internal one. It is as easy to see how this rule applies to others, as it is difficult to see how it applies to oneself. ◀

Right and wrong in the social or cultural sense is relative and customary – often a diluted and distorted shadow of absolute right and wrong. Absolute right and wrong are operationally defined to optimize the growth (evolution) of the quality of our consciousness, or equivalently, our spiritual development. "Right" intents and choices help us improve the quality of our being, whereas "wrong" intents and choices stimulate no positive growth and may cause us to lose some previously earned quality. Absolute right and wrong (intents, motivations, choices, and actions) are defined and differentiated by the effect they have on the average entropy of the system.

One can learn from either right or wrong choices. PMR is designed to serve as a learning lab – a place where units of individuated consciousness (sometimes referred to as beings, entities, or souls) can grow up (improve their quality) by exercising their free will intent within a virtual system of direct interactive experience and feedback. Many other subsets of $NPMR_N$ are also configured as learning labs – some employ versions of space-time while others do not. Each provides a local reality system for its inhabitants who see themselves as physical and all other realities as nonphysical. All reality systems are virtual; hence there is no actual distinction between real and virtual. Everything is consciousness, a vast system of self-modifying digital organization evolving toward lower entropy.

Most of the time we are provided a near optimal learning opportunity, by circumstances that we have custom-designed to fit our individual needs. Moment by moment, with immediate and obvious results, these growth opportunities stare us in the face or bop us between the eyes. Could any learning laboratory be more efficient than our beloved 3D space-time? I doubt it. AUM knew what it was doing (evolutionarily speaking that is) when it invented (evolved) the concept of limited virtual physical realities.

We are like rats in a maze where wrong turns meet with an immediate electrical shock. We are in a wonderfully complex pudding-maze; to succeed, we must taste our way through it. What could be easier or more efficient? We make the wrong choices and are motivated by the wrong reasons and... Zap! We get the electric shock and make ourselves miserable. We make the right choices for the right reasons and we create happy, productive, rewarding lives (we get the cheese). This is simple evolution where success is simultaneously its own criteria, evaluation, and reward. Consciousness is in the process of evolving its way through an evolutionary maze of intent, choice, interaction, and reaction.

I should point out a few basic facts about evolving the quality of your consciousness in our reality. Although the results of your actions (feedback) are immediate and obvious, you can ignore them or misinterpret them if your ego is making too big a fuss. Then, it may take a while before the situation degenerates to the point where the resulting dysfunction becomes impossible to ignore or excuse. As in all learning situations, being attentive – paying attention in class – is absolutely essential to an efficient learning process. Don't ask for an easier maze – you won't get one. For those who typically slip through demanding situations by avoiding the things that are most difficult, let me point out that there are absolutely no shortcuts, no end runs, no acceptable excuses, and no way to quit.

That anthropomorphized little old man with the long white beard that you can hustle favors from by believing the correct things and performing the correct rituals is a product of your little picture culture and a fearful, needy mind looking for the easy way out. No one is handing out judgments and calling the plays from the sidelines. Not even a non-anthropomorphic AUM thought-system-thing is likely to be paying attention to any particular individual human bacterium in the PMR Petri dish. Humanity, as you know it and think of it, is not central to the machinations of the Big Picture. You should be proud to be a bit player, not disappointed because you are not the star. As far as I can tell, no reality or dimension is more central than the others: all are bit players (or equivalently, all are stars) playing their specific parts.

Only a fat-ugly ego could care a flip about the **relative** importance of earth-based humans. The relative value and comparative status of our species holds no meaning for an advanced consciousness such as yourself, but you wouldn't believe how many other folks think that humanity is the crème de la crème of sentient existence. These egotists can become deflated, resistive, and defensive when you tell them that, in the Big Picture, they are not particularly important. When you are used to being fantastically magnificent and superior – a legend in your own mind – it is difficult to see yourself as a common bit-player in a much larger drama.

The process of improving the quality of your consciousness is a matter of simple evolution – much easier than a fish learning to walk on land and breathe air. Live and learn, do and die, if at first you don't succeed, try, try again – you know the drill. Take your place alongside the other sentient chunks of individuated digital consciousness, both physical and non-physical. Neither you, nor your species, nor your local reality is more important, special, or superior relative to the gazillions of others.

Feelings of superiority developed in relation to other obviously inferior PMR beings and life-forms (mankind is above all other creatures of the earth) are an error of perception that has generated many particularly destructive belief systems and caused much damage to biosphere-earth – as well as greatly retarded individual human progress toward greater consciousness quality. Because consciousness is an attribute of individuals, all conscious entities are vitally important in their own way: each has its own mission and purpose and is an important contributor to the whole. All are different, all have their own challenges, and none is fundamentally superior or inferior.

Think of consciousness within $NPMR_N$ as a vast interdependent evolving ecosystem. Any part feeling superior to any other part is the result of ignorance, belief, and ego existing within a little picture perspective. Actions motivated by feelings of superiority are bound to lead to destructive results for **all** members of the larger system, including those feeling superior. How superior is it to shoot oneself in the foot over and over again?

▶ Now that you are feeling sufficiently humble, let me continue using the ecosystem metaphor to show you the other side of this coin.

Within the larger eco-mind-system designed to evolve consciousness quality, PMR can be thought of as a particular biome. Consciousness ecosystems are not based on the evolutionary pressures of mutual survival and propagation, but evolve to improve themselves, to lower their entropy.

Where survival and propagation represent predominant environmental constraints, individuals are not as important, in terms of evolution, as are the groups they belong to. In biological evolution, superior individuals either impact a larger species-level profitability statistic, or their potential contribution is lost. An individual's impact on the success of an established species is nearly infinitesimal. Progress accrues through a large number of very small individual contributions that all point in a similar direction. In biological systems it is the continuous immortal group, not the discontinuous mortal individual that is the primary beneficiary of evolution.

Within a consciousness system such as $NPMR_N$, individual immortal conscious entities continuously evolve and contribute gains in their personal consciousness quality (lowered entropy) **directly** to the whole of which they are a continuing part. A single conscious entity has the potential to make a very significant, direct, and wholly independent contribution to the larger consciousness ecosystem as it seeks profitability on a personal level. As each individuated consciousness evolves, the larger consciousness system (AUM) evolves as well.

Survival based interdependencies between groups and the evolution of species within PMR is replaced by personal growth and the evolution of individual consciousness within NPMR. That is how you can be one of a gazillion entities in some experimental consciousness Petri dish and at the same time be individually very important to the whole. Your personal contribution is not limited by a group function or a slow uncertain pass-along process with strong random components; it is simply measured by the quality that your consciousness brings to the table. Quality runs the gamut from the severely limited, high-entropy, fear-based, self-serving ego, to the unlimited, low entropy, unconditional love exhibited by AUM-consciousness. As you move from the former to the latter, your personal significance relative to the whole increases dramatically.

Though even a brilliant bacterium must remain in its Petri dish as a part of its group, an evolved consciousness can outgrow the culture within its originally assigned Petri dish and one day join the laboratory staff! You are an independent (with free will) individuated consciousness containing the potential of AUM – no individual, species, or group affiliations can help or hinder your effort to become a more significant entity. Thus, feeling superior about your group affiliation (galaxy, planet, species, race, gender, culture, religion, nation, profession, education, or socio-economic status) is simply counterproductive. (In an aside near the end of Chapter 20 of this book, I will explain why it is that a highly evolved, low entropy consciousness has no sense of superiority.)

In the Big Picture, the quality of your consciousness determines whether you are of infinitesimal or great stature, of minor or major consequence to the whole, have a job in the lab or are wallowing around cluelessly in the Petri dish. ◀

Implement the Fundamental Process! Make each choice count. Get with the program. Evolve your being. You **are** doing it now and you have no choice. You exist, you are an individuated consciousness and you cannot stop existing. Your physical death initiates a process that eventually creates a new maze for you to explore – your learning is cumulative. You are a participant in, as well as a driver of, the consciousness cycle.

You are in the game whether you want to be or not. Every day you are evolving more toward the positive or negative pole of being – you cannot stay neutral or remain stationary. Denial and ignorance are inconsequential and affect nothing – your permission or willingness to be involved is completely irrelevant. You are in the process of actively evolving your consciousness **now** – you are doing it for better or for worse. The only questions are how well and how efficiently are you doing it? Are you progressing or regressing, and if so, how steadily and at what rate? Is your growth process efficient? Is your learning rate optimal?

Shock or cheese? Jeez Louise, that's a tough one – let me think about it.

You can be confused and non productive (perhaps fail) if: (1) you do not understand the game and therefore do not know that your reality is actually an amazing maze of many choices that is driving the evolution of your consciousness – therefore remaining purposeless, pointless and without focus – not even trying. (2) your taste buds are so twisted and confused because of beliefs, self-absorption, attachments, needs, and fears (ego) that you cannot, or will not, taste the difference between miserable, arrogant, unproductive pudding, and loving, growth inducing, delicious and satisfying pudding. In other words, you cannot differentiate between electric shocks and fine white-cheddar cheese. You therefore lose your sense of direction and have no means to distinguish up from down, dark from light, or progress from regression.

If an individual rat in the maze sees no point in exploring and cannot differentiate between electrical shocks and cheese, it is in for a long and difficult training experience. Poor rat! What a miserably frustrating life! Not much progress is expected. Shocks become an ordinary accepted part of life and are no big deal! "That's life," says this rat with resignation. To this particular rat, the world (life within its local reality) may seem random, existential, nihilistic, mystical, or driven by an unfathomable, jealous, vengeful, or demanding god.

To a collection of such rats, cynicism, self-pity, anger, victim-hood, resignation, escapism, recreational drug use, as well as a fascination and obsession with sex and the symbols of power (including vicarious violence and domination through entertainment, competition and sport; conspicuous consumption; macho-vehicles and aggressive driving; and the ownership of weapons) become common personal strategies to deal with the anxiety (inadequacy, insecurity, and powerlessness) generated by fear. Value, purpose, objectives, and goals of being are lost in a confusing whirl of fine shades of relative, meaningless gray, while the original concepts of black and white are lost.

These hapless rats will be driven and animated by their ego – their immediate needs, wants, desires, fears, and beliefs. Does this description (driven and animated by immediate needs, wants, desires, fears, and beliefs) remind you of anyone you know – a distant acquaintance, a least favorite relative, or perhaps your evil twin? You might be tempted to surmise that these rats are obviously too dumb to play the game and that it is cruel and unusual punishment to put these stupid, lost, and confused creatures in such a complex maze in the first place. Unfortunately, evolution shows little compassion and weeps no tears for its failures or slow learners. Goodbye dodo birds, adios dinosaurs, so long trilobites. Your concept of "fair" is a function of your ego and belief systems and mostly irrelevant.

Reasons and excuses for not making it or getting it fall on the deaf ears of the Fundamental Process of evolution. Making it and getting it, within your given set of possible interactions, are the non-negotiable requirements of evolutionary progress. An evolving system is either unprofitable (say goodbye if that condition is not turned around in time), profitable (progress is being made), or neither (an astable condition that is hanging around waiting to see which way it will go).

Cruel and unusual? Too difficult? Do you sometimes feel that unreasonable conditions for growth are forced upon you? Sure, like bad weather at a picnic is forced upon you by mean old Mother Nature. The Fundamental Process, as it applies itself to an individuated consciousness, is not personal regardless of how helpless resignation, ego, anger, paranoia, cynicism, or self-pity might construe it.

If you think it is personal, you are probably placing yourself too high up on the consciousness chain. Bacteria are extremely important to us, our life depends on them, but what we do that affects bacteria is usually not intended to be a personal affront or reward to any individual bacterium. I am describing the general case – there are always exceptions, but they are much, much fewer than a poll would indicate.

Nevertheless, evolution is patient. Patient enough to let that rat suffer in confused discomfort until it eventually progresses its awareness and ambition to the point it can evolve a solution. The rat has free will and some basic intelligence, while evolution has all the time in the world. This rat is on its own to sink or swim by its choices. It does have one advantage, however: It gets (as most rats do) all the help it can profitably use from its nonphysical friends who do care and understand the significance of individual progress.

A simplistic view that confuses the local reality of Our System with the larger reality of AUM holds that since some rats do figure it out, and because those that do are strongly encouraged to help out the remaining rats, eventually all the rats will make it. A universal high quality consciousness (a heavenly vision of ubiquitous love, peace, and light) is believed to be the inescapable final result of consciousness evolution in Ratville. The only questions that would remain in such an experiment are the details defining the path taken to connect the endpoints, and how long it will take. For a grand experiment in consciousness, this is much too narrowly focused, scientifically shallow, and of dubious value to the larger reality of AUM. To make the experiment truly useful and productive, the evolutionary process must be more challenging and have an open-ended result. To optimize the consciousness cycle and learn more about the dynamics of consciousness, add to this learning laboratory the intrigue of bits of poisoned cheese and a myriad of clever distractions and confusions scattered throughout the maze by a competing group of anti-rats.

Now the rats have a new set of options. To the struggle between ignorance, fear, and ego versus knowledge, wisdom, and love, is added the struggle between good and evil, between positive and negative intent. Now it is no longer clear that all the rats will eventually make it. No, this is not the equivalent of an adolescent AUM putting two big spiders in a glass jar to see what happens. Free will requires that negative intent be a possibility, and without free will, there can be no experiment. There is no point in doing experiments if the outcome is predetermined. Maximum free will enables maximum potential for both success and failure. Like any scientist, after AUM has completed the simpler experiments, it must move on to the more difficult and meaningful ones. (You will find a derivation of free will and its relationship to our consciousness in Chapter 45 of this book.)

Evolution may simply want to know whether rats or anti-rats are the **natural** winners of this experiment. AUM may want to determine the dynamics of the quality and evolution of consciousness under various conditions. We require the stress and challenge in order to exercise our free will fully and thus optimize our potential growth.

Go rats go! Go rats go! Go rats go!

Don't you want to join our team (we are the Good Guys) and help us out – or do you just want to wear the t-shirt? For goodness sake, don't consider joining the Bad Guys – they are a miserable lot – and their benefits package stinks. You will eventually choose sides – everybody must. No one can refuse to play because in this game no choice becomes an active choice, eventually leading, by circumstance, toward one direction or the other. Interesting experiment, huh?

It is my best hope that some hard working rat will find a useful hint or secret map somewhere in this trilogy that will make a difference for him or her in the pursuit of their personal cheese.

37

■ ■ ■

Cheer Up: Things Are Not as Bleak as You Might Think Say "Cheese Please," and Hold That Smile!

■ ■ ■

The situation is neither hopeless, nor random, nor beyond your control – it is not even that difficult. Let's talk a moment about free will and making choices. **What** we do is not of **primary** importance; **why** we do it is what counts. The intent or motivation **is** the choice. Thus, our free will choices are **primarily** choices of motivation and intent – why we do what we do. Only **secondarily** are they choices of doing. What we actually do (the action we take) is the first **result** of our choice of intent, and drives a feedback mechanism, which is very good but not perfect, particularly in the short-term.

This is not biological evolution we are talking about, where the key process driver is what you **do**. In the evolution of consciousness, motivation and intent are the key drivers of evolutionary process. What you do is secondary to **why** you do it. The right choice for the wrong reason is an oxymoron and does not exist (I am being tricky-picky here with the semantics to make the point that the development or selection of the intent **is** the choice that I am referring to).

Recall that right motivation, intent, and action generally result in a decrease of entropy within your consciousness, whereas wrong motivation, intent, and action generally result in an increase of entropy within your consciousness. This is how absolute right and wrong are defined. Right and wrong are differentiated by how they affect your spiritual quality, or equivalently, the quality of your consciousness. Relative (local) right and wrong are largely derived from our sense of absolute right and wrong, but are also dependent upon fad, fashion, culture, and personal, social, and political circumstances. *My Big TOE*, being science, is only concerned with absolute right and wrong – where an entropy measurement, rather than a personal PMR viewpoint, determines which is which.

Unvarying absolute right and wrong may appear to be dependent upon the extent of each **individual's** quality and thus relative to the individual; however, the process of entropy reduction is the same for all individuated consciousness – you, me, Rover, and AUM. Absolute right and wrong consistently apply to everyone all the time. Your ability to express Big Truth within your life by making right choices based upon right intent depends upon the quality of your consciousness. You gain personal quality the same way you gain strength and endurance or learn to play the piano – by dedication and continual effort applied over a significant period of time. We call this self-teaching incremental self-improvement process "bootstrapping" – you may employ a coach to improve technique and clarify issues, but only you can accomplish self-improvement.

If the intent is wrong, then, by definition, the choice (of motivation, not of action) is wrong as seen from the perspective of evolving consciousness. The wrong intent or choice is evolutionarily wrong even if it temporarily results in what appears to be a constructive right action in physical-biological space. You do not evolve higher quality consciousness through right action or right result, but only through right motivation and right intent. The PMR virtual reality consciousness trainer provides the opportunity for simple, straightforward interaction and a results-oriented feedback mechanism that reflects immediate and cumulative quality of intent. Motivation, intent, action, and feedback are all interconnected. Right motivation and right intent are the only sure path to right action, which produces the right feedback to encourage continued right growth and learning.

Right growth and right learning are technically defined as growth or learning that leads to a decrease in the entropy of your consciousness. Personal growth, growing the quality of your being, or growing wisdom – all increase your personal power as they decrease the entropy of your consciousness. I am talking about the growing-up and maturing of your being – a process that increases your capacity to love. I am also talking about spiritual growth, the evolution of your soul. Choose the expression that suits you – they are all essentially equivalent.

There are three distinct parts here: 1) choice of intent; 2) action, which is a result of intent (what we do); and 3) result, which is the effect of what we do. You should be careful not to confuse action (the animation of your intent) with end result (the **effect** of your intent). Exercising free will interactions (with self or others) produces an internal result (always) and an external result (usually). The internal result immediately and most potently affects the quality of the consciousness according to the quality of the intent. The external result affects others as well as yourself and generates the appropriate feedback or reaction. Thus, it is not possible to achieve a right result if the intent is wrong. A wrong intent damages its creator despite what else happens.

What would appear to be right action in the service of wrong intent is eventually revealed as self-destructive for the doer and as a challenging opportunity to whomever or whatever the doing is directed toward. Yes, I am exploring exceptions within the margins of typical behavior. The wrong intent cannot consistently and broadly result in right action. The partially random interaction of our free will consciousness with the free will of others forces us to constantly make choices that express our intent or motivation, and then we **act** accordingly. Good motivation **usually** produces mostly right action. Perfect motivation **always** produces right action. Your grounding in, or familiarity with, PMR may lead you to believe that action is the evolutionary driver. It is not – at least not for consciousness. Action is a primary driver for physical-biological systems only.

Right and wrong choices are not made randomly. We are not randomly buffeted souls driven to and fro by external forces like particles exhibiting Brownian motion. Our specific individual opportunities may be derived through a series of Brownian-like social and personal collisions (interactions) that can contain large random components, nevertheless, our overall opportunity to learn what we need to learn and the choices we make in actualizing that learning opportunity generally contain only small amounts of randomness. Indeed, at our elementary level of development, most choices are easily predictable. The interactions we have with others (and with the whole of reality) inundate us with the opportunity to make choices that flow from our intent and thus reflect the quality of our consciousness. If our intent is wrong, we suffer the consequences (in all realities we function in) irrespective of what we actually do. If our intent is right, we always derive some benefit, regardless of what we do as a result of that right intent. Right intent **almost** always drives a resultant right action. Wrong intent **usually** drives wrong action.

Action creates results, which often create feedback, which sometimes (if we are paying attention) helps us modify our intents. **External** results drive our **external** feedback system of rewards and punishments – an imperfect system that is usually efficient. Fortunately, imperfections and errors in the feedback system show up much more often in the short-term than the long-term. **Internal** results drive an **internal** feedback mechanism that either increments or decrements the quality of our consciousness according to the quality (rightness) of the original intent. The collected benefits and liabilities that accrue to us as a result of our actions help us recognize and understand the rightness or wrongness as well as the impact of our intents and actions, thus facilitating our learning and guiding our future choices.

All we need to do is be serious students committed to making good grades and take advantage of the learning opportunities that present themselves. What else is new? The same formula works everywhere – learning is learning. The only complication is that you also have a cut-up in your classroom (your ego) that makes it difficult for you to hear or understand the lesson.

For the typical being out there in the larger reality (continually choosing and doing, like you and me), intent lies primarily beneath the surface of one's awareness. What motivates us is barely visible to our intellect. Motivations and intent are a complex mixture of many, sometimes inconsistent and incompatible, components. Your motivations, reflecting your ego, are as Byzantine and inconsistent as you are. Nevertheless, for the typical human quality of consciousness, the benefits (spiritual growth) of right intent are usually much greater than the costs (spiritual stagnation or back-sliding) of wrong intent.

The spiritual impact of actions can **simultaneously** gain some and lose some in consciousness quality because the impact of feedback on future intent can produce a very complex result. Actions can often lose the **same** spiritual capital repetitively only to have it reappear as an investment in winning. I know the economists, investment advisors, and stockbrokers are terribly confused by the preceding sentence. For the financial and investment types, here is a longer less poetical version. It is possible in many situations for you to make the same or similar mistakes repetitively without necessarily going broke (seriously impoverishing your consciousness quality) – you simply do not get ahead. Furthermore, major messing-up often precedes major growing-up because we mule-headed humans sometimes need to be hit repeatedly and hard directly between the eyes by the **results** of our errors before we notice that a problem exists. Unfortunately, our egos are so good at blaming others that even after great suffering we often miss the point that our pain is primarily self-inflicted and trying to tell us something about what we are doing wrong.

Growth and making choices often appear to be extremely complex processes because from our point of view we are complex beings with complex egos existing somewhere between the often confusing opposites of good and evil, love and fear. Right being, profitable intent, and good choices are tremendously easy, simple and straightforward without ego, and with the humility and compassion that comes with love.

Changing the quality of consciousness is not (cannot be modeled by, or accurately approximated by) a simple linear process. Some people (usually precise types) tend to want everything to be reducible to a simple linear process so that they can work with it and more easily pretend that they understand it. These individuals often make assumptions to force fit the problem at hand into a simple linear model; these assumptions inevitably lead those apparently precise and logical minds into erroneous conclusions. I know this is not a problem of yours, but I thought I would mention it in the event you know of somebody to whom it might apply.

The choices you make (the motivations you have) are mostly not completely right or completely wrong. They are a mixed bag of many colors and components. The results of these mixed choices are also mixed. If right means 100% right, and wrong means not right, the average person makes many more wrong choices than right ones, but gets more gain from an almost right one than loss from a so-so wrong one. Most beings (people included) **are** steadily making positive progress, but ever so slowly. Consciously knowing what you are doing, and making an "eyes open" honest effort can increase the rate of progress immensely. A little help in finding the bigger picture can make a significant difference.

The optimal growth process is uniquely specific to each individual at a given time.

If you are feeling perplexed and overwhelmed, wipe that frown off your face and throw that heavy resignation, along with that deep sigh, into the trash bin. Shake off any self-pity, and notions that you don't want to play the game. Things are not bleak at all. Go get 'em, tiger-rat!

38

■ ■ ■

Does the Big Dude Have
an Attitude? Do We?
Speculate! Speculate!
Dance to the Music!

■ ■ ■

As previously pointed out, if AUM has limits, and is aware of them, it must set priorities, make judgments, pace itself, focus only on what is most significant, and use its limited resources wisely for maximum self-improvement. If AUM is finite, could there be others? Sure, why not? Evolution tends to explore **all** the possibilities – but it doesn't make any difference to us or to our reality. From our tiny perspective, AUM, in all ways, appears to be infinite.

Such questions (multiple AUMs) are totally pointless from a practical viewpoint and almost pointless from a philosophical viewpoint. If you are interested in speculation, you will find acres of fertile ground here – go amuse yourself. Have fun dancing the philosophy boogie with your intellect but don't take yourself or your brilliant choreography too seriously. Once you have gone far beyond your ability to comprehend what you are saying, it is time to either find the humor in it, or be quiet. Focus on what is small enough for you to grasp and work your way up carefully with a firm understanding of your limitations.

Let go of what is too big for you to grasp. Accepting that your reality is dramatically affected by systems and phenomena that lie beyond your capability to understand and control is the first step toward developing wisdom. Differentiating these unknowables from those things that are within your potential to comprehend is the second step. Conceptual structures and a probing intellect are valuable tools, however, cramming unlived and unexperienced ontological conceptual structures into your intellect squanders enthusiasm and interest, makes you believe that you know and understand more than you do, and leaves you with no significant increase in spiritual quality.

The condition and details of AUM's interaction with anything other than itself are speculatively interesting but unnecessary to (and totally beyond) our Reality Model and Theory Of Everything. Fortunately, absolutely every-thing that can possibly affect us **directly** can be fully contained within a highly limited set. What is external to AUM that affects us indirectly is beyond our theoretical knowing. Similarly, speculation about how focused AUM is on us, how intelligent it is, how much it cares (if at all), whether it is emotional or more computer-like are totally superfluous to our under-standing of everything. Our particular PMR could be either relatively insignificant or the main focus of AUM's existence; it could be one of many interesting experiments in consciousness or a unique implementation of the consciousness cycle, however, AUM's viewpoint makes absolutely no dif-ference to **our** understanding of the larger reality. Knowledge of AUM's opinion of us contributes nothing significant to the understanding of Our System of reality or to the mechanics of existence. Only an ego could care about relative comparisons, hierarchy, and pecking order. Issues of ego are counterproductive to both science and self-improvement.

▶ Our anthropomorphic tendencies and sense of self-importance, supported by an unbounded egocentric pride in our species, make me suspicious of any logic that puts us too close to the center of Big Picture significance. We Homo sapiens are exceedingly impressed by our apparent significance and grossly inflated with self-importance – an unfortunate fact that clouds the clarity of our vision immensely. How many of our wars, atrocities, and general acts of meanness have sprung from a sense of superiority? A **belief** in the supreme importance of humankind is the ultimate source of much mis-chief. The notion that humans are inherently superior leads to the idea that some humans are superior to, or more significant than, others – a concept that represents the first step down a slippery slope that is both steep and long.

The cultural, personal, and religious belief systems in which we are all constantly immersed have a tendency to be "feel-good" belief systems that boost our egos and self-importance (hence, their popularity). The feel bad stick serves as the foil to the feel good carrot. Besides helping us feel important, safe, and valued (if not superior), feel-good beliefs make it almost impossible for us to evaluate a bigger picture that refuses to manipulate the individual to feel good (important, safe, secure or saved) or feel bad (guilty, inadequate, unworthy, helpless, or fearful). To those needy individuals who are addicted to the medications of Dr. Feelgood, beliefs and reality concepts without emo-tional hooks are like cigarettes without nicotine.

Emotional hooks are used to capture and manipulate needy egos – those who feel fear, those who feel lost and powerless; who feel inadequate, helpless, and insecure or feel as if they are not doing what they should be doing. Feel-good beliefs cover up these fears and help you feel strong, in charge, and powerful; feel as if you belong, are accepted, and understood; feel loved, safe, and cared for.

Wow, that sure sounds good to me! Where can I get a few of those feel good belief-pills? Not for myself, of course, but I have this friend who has a bonafide medical need for something exactly like this. Say, what's the long-term price for these happy halluci-nations? Are you sure this is legal?

Emotive manipulations of all sorts work like this: feel bad (ignorance, guilt, or inade-quacy) usually precedes, and subsequently is interspersed with, feel good (importance, significance, security, salvation, redemption, acceptance, forgiveness, or superiority) in a succession of one-two punches that appeal to the needy ego – you know, the good cop, bad cop routine. It is an unfortunate fact that feel-good belief systems rule the land, often subtly without being noticed. For this reason, I am constantly making an effort through-out the *My Big TOE* trilogy to steer you away from falling into this common conceptual error by belittling our culturally ingrained self-satisfying importance at every turn. My weapon in this battle is blunt humor – no cynicism is ever implied or intended.

A few readers, if caught on a bad day, may have difficulty dealing with the relative humble position we most likely occupy in the big scheme of things. Some may feel emotionally upset or angered by what they wrongly interpret as an implied lack of **personal** importance. Like bacteria, we are important in our own way and should be able to feel good without feeling superior. Being a small part of something very large does not have to be a downer; it depends entirely on your perspective. Only an erroneous belief system and your ego can possibly create a downer.

I am by nature a cheerful, happy, upbeat type. I think people, reality, and life are a blast – an exciting ride, a wonderful opportunity! Find the humor, read and smile, let go, laugh a little – no need to be overly serious. Every now and then open up the top of your head so that fresh ideas can blow away the musty odor of old paradigms and force traditions and the status quo to periodically re-justify their value. Nothing encourages humor more than honest introspection.

This life is your opportunity. The initial conditions of your PMR existence have been custom designed for you to optimize your growth potential – with your input and approval. Your personal growth is for fun and profit. Enjoy! Learning and joy go together. Not learning, not growing, and pain and misery go together. Most lives are a mixture of both. Focused effort and paying attention in the PMR learning lab can dramatically increase the ratio of joy to misery in your life. If you learn and grow significantly, your joy and knowledge will spill over into the lives of others. ◀

Anything sentient that is important to us inevitably gets recast in our image. Our assessment of ourselves defines the scale against which all others are measured and interpreted. We judge and assess others by comparing them to ourselves. We constantly project our being (our quality) onto the sentient existence of others – there is nothing else we can **objectively** do – and a wrong answer, for some reason, always seems far superior to admitting ignorance. It is how we are; accordingly, be on guard.

How does AUM regard us? Do I care about the cells in my stomach? Sure, but sometimes more than others even though they are always important. I do not consider or deal with them individually or even in small groups unless I get an ulcer or a perforation. Do the cells in my stomach care about the pH of the secreted hydrochloric acid? Absolutely: Each individual stomach cell dearly cares about the pH of the acid and the feedback mechanisms that regulate it. Do the cells in my stomach care about the style of my haircut? Not directly, although indirectly there may be a connection. If my new haircut style makes me feel good about myself and thereby reduces my overall level of stress, and perhaps helps me find a steady job and a girlfriend who can cook, my stomach is pleased with the results even if it does not understand the concept of haircuts – much less the complex interaction connecting haircuts, stress, jobs, girlfriends, and stomach acid.

See how everything is important and related? Do the stomach cells or the acid regulating mechanisms feel slighted because they are not considered to be the center of the body? Do we in general worry much about how our bodies, minds, or spirits are evolving? Not often (with some notable exceptions). Every being is interested in, aware of, and cares about what it interacts with on its own local level, and in its immediate environment; everything else is invisible or inconsequential because it lies outside the being's awareness, is hopelessly beyond its knowing (mystical), or appears to be irrelevant to its needs.

With our capable minds we can and should do better than that. If our focus expands, it usually expands upward toward higher levels of awareness. Lower levels are taken for granted. They just are – they do what they do. We depend on them but do not get directly involved unless they get our attention by doing something unusual. Even if we are experimenting with bacteria in a Petri dish, we leave them alone to do whatever they do. We don't name each bacterium and take them out to lunch on their birthdays – even if the experiment is important. Meddling, if not carefully restricted and precisely controlled, would invalidate the experiment. Then what? The experimenter would have to flush the little fellows down the drain and start over. Uh oh!

Consciousness evolution is most likely a more immediate, aware, and up-close-and-personal type of experience for AUM than biological evolution is for us; on the other hand, we are a part of AUM's personal evolutionary process. As full fledged members of the consciousness cycle we are clearly significant, however, speculation about AUM's attitude (just how significant we are) is of little value and will almost certainly be influenced by expressions of our anthropomorphic tendencies and self-important attitudes.

AUM's attitudes and feelings toward OS, PMR, humanity, or any particular human or group of humans are beyond the scope of even the biggest Big TOE because they are beyond our capacity to comprehend and irrelevant to everything that is of either practical or theoretical significance to us. How does AUM regard us? It doesn't make any difference. It is an irrelevant question that is meaningful only to the ego of the asker. AUM does what AUM does for its own reasons and we do what we do for our own reasons. After that, the chips fall where they may. Accept it, and let it go.

Answering such ego issues with systems of belief and pseudo-knowledge constitutes a giant step backward that inadvertently creates barriers that must be overcome before individual progress can be made.

39

■ ■ ■

Why Us? Why Like This?
What is a Nice Being Like You
Doing in a Place Like This?

■ ■ ■

Why didn't AUM evolve the concept of hyperbolic 5D space-time for us to live (experience) in – or perhaps something else even stranger? Because 3D is simpler, and works better for the consciousness interactions and evolutionary possibilities AUM is exploring. If AUM wanted to explore or evolve a hyperbolic 5D reality or something else equally beyond our comprehension, it would merely dedicate a group of specialized reality cells to do just that. They would make up the weird part of AUM, instead of the comfy space-time part we know and love. That weird reality would not be directly relevant to us, nor we to it. Each would exist within its separate dimension (its own section of memory) of AUM thought-space – in the same way that multiple space-time PMRs exist within different dimensions. Multiple realities, each in its own dimension, are roughly analogous to one of our big mainframe computers running multiple similar and dissimilar simulations at the same time: We do that all the time and it is no big deal. Each independent simulation has its own memory space and rule-set, and is driven by its own time loop.

It is reasonable to expect there are many ongoing parallel experiments to determine the most profitable set of constraints in order to design the most effective learning labs. The Fundamental Process working upon AUM would eventually discover which sub-realities, rule-sets, or types of space were most functional, and which were most profitable.

Additionally, AUM, in its effort to facilitate the Fundamental Process in optimizing 3D space-time, would need to assemble a broad assortment of 3D implementations in order to provide good statistical sampling across a variety of 3D results. It would be risky and very bad science for AUM to base its conclusions about the profitability of 3D space solely on our (PMR) performance.

That may explain why I have run into many PMR-like sub-realities during my explorations of the larger reality. Some appear to utilize a space-time that is similar to our own while others operate under rule-sets that are obviously quite different from ours. The variety of life-forms, habitats, and reality dimensions within just $NPMR_N$ is immense. Where there are few inhibiting constraints, the Fundamental Process produces great diversity.

Evolution explores all significant options within reach, but progresses and maintains only those that are profitable. AUM must be getting a significant return on its investment in 3D. How do I know? Because we (and many others) are here! Evolution doesn't mourn the victims (failed experiments) of its relentless pursuit of profit. If we are here, we are profitable. Or at worst, the profitability of the process of which we are a part has not yet been determined.

NPMR is in one part (dimension) of AUM while the $NPMR_n$ are in another and the PMR_k are in yet another (the space-time part). Dimension within dimension within dimension. Here $NPMR_N$ is one specific member of the set $\{NPMR_n\}$, (where n = 1, 2, 3, ...N, N+1, N+2, ...), and our beloved PMR is one specific member of the set $\{PMR_k\}$ (where k = 1, 2, 3, ...). PMR, as a 3D space-time reality, must follow the rules of our particular space-time; consequently, everything within our physical universe must behave and evolve within the constraints (physical causality) defined by our rule-set. For example, the evolution of our universe, galaxy, and solar system, and the biological evolution upon earth, all obey the same set of rules. Other PMR-like realities may follow their own rule-sets in their own version of space-time – there is no requirement for them to follow the same rule-set (have the same physics). Likewise, the evolution in NPMR follows its own less restrictive causality and obeys its own unique set of rules. As is confirmed by my experience, one would expect the less constrained NPMRs to evolve a more varied set of life-forms than the more constrained PMRs.

Though you might think that the beings and critters in NPMR are merely thought-forms, whereas the beings and critters in PMR are real life-forms, I will demonstrate in Section 4 that the difference is only apparent – that the distinction between physical and nonphysical is relative to an observer's perspective. The observer's perspective is a function of the constraints and rule-sets that define his interaction with, and perception of, his local virtual reality. You will see that the so-called real life-forms in PMR are actually thought-forms within AUM as well.

The overall purpose of AUM trickles down to **our** overall purpose. The nature of AUM likewise trickles down to our nature. Our local reality and everything in it is a product of consciousness evolution. We operate within our niche on the edge of an enormous consciousness ecosystem. We are individually derived by limiting and constraining a subset of AUM's consciousness. Just as our body is constructed of biological cells, our conscious awareness, at its most fundamental level, is constructed of an individuated or bounded group of reality cells constrained to perceive a specific set of causally related events. Our awareness within PMR reflects the interaction between individuated groups of reality cells that have been constrained by the space-time rule-set.

The being we call ourselves (lump in the AUM-sheet) is **not** dependent on our physical body – it is the other way around. Physical bodies are relatively dense space-time constrained experience-bumps on the lumps in the sheet. These self-perpetuating, low maintenance, biological-matter experience-bumps come with a sensor platform called a "body" to provide experience and immediate feedback, a low-end Free Will Awareness Unit (FWAU), and a lifetime guarantee. All such biological-matter bumps are designed to make choices, assess the feedback, and apply the Fundamental Process toward the evolution of both body and consciousness. Yes, critters have consciousness and free will too – theirs is simply more limited and less individuated than ours.

Let's contemplate designing an experiment in bump-consciousness. We should use a simple self-replicating design that can evolve its own forms. It would be better to give it limited awareness and functionality so it won't be continually getting stuck in intellectual endless loops thinking about interactions instead of experiencing them as did some of the less productive experiments AUM tried earlier. (The FWAU, how it works, why its invention was a logical necessity is discussed in Section 4 while more about "experiments AUM tried earlier" is included in Sections 4 and 5.)

These individuated consciousness experience-bumps must be self-modifying. They must be capable of learning – of self-improvement through multiple evaluative feedback loops. Because these bump-beings are very limited, they must be designed to **accumulate** knowledge and evolutionary progress by being recycled again and again through many challenging interactions until the FWAU self-destructs or eventually ripens for harvesting.

The uniqueness and personal integrity (identity) of each FWAU is maintained in support of a cumulative accelerating growth potential. A few people who are unfamiliar with the nomenclature on AUM's design drawings call the FWAU a "soul," "spirit," or simply the "nonphysical component." "Are you implying that critters have souls too?" one might ask. Sure, why not – they are constructs of consciousness like the rest of us. Because their awareness is even dimmer than ours, the degrees of freedom available to their consciousness for spiritual growth are also much more restricted. It is the same idea fundamentally, but with a considerably different implementation – their nonphysical component is less unique, structured, and interactive than ours. However, do not think that critters are inferior simply because they are different. The proclivity to reinterpret our differences as obvious superiority represents little picture arrogance.

Every critter, including you, has its point and its place; diversity is a natural artifact of evolution when there are few constraints. Each type of entity reflects unique potential, capacity, goals, purpose, and responsibility yet all spring from the same source and follow the same processes. One might say that they occupy different niches and habitats within the same consciousness-evolution fractal ecosystem.

The fewer constraints that limit a particular consciousness, the lower its associated local entropy can go, and the more effective and powerful it can become. We will discuss the concept of the entropy of consciousness more fully in Chapter 47 of this book. We humans are essentially a different type of critter as are all sentient entities. No matter how you **believe** we ended up here (from simple evolution making the most of random events, to a rogue planet in a highly elliptical orbit, to UFOs, to a mystical eruption) – at the root, all physical processes and entities are derived from the same source. Our existence lies at the feet of an evolving AUM consciousness system. Only superficial **physical** details differentiate between the many theories of how mankind arrived on earth. In the Big Picture, the physical details are not particularly important.

40

■ ■ ■

A Chip Off the Old Block

■ ■ ■

Because your questions have been accumulating for some time, I will
make an attempt to answer a few of them before we go further. I
wouldn't want you to pop with frustration like a giant festering intellectual zit. Oooh..., ugh, sorry about that... but the image is too much fun
and metaphorically accurate to throw it out merely because it is insulting,
nasty, and distasteful. You are having fun aren't you?

Do not forget about questions that come up – write them down as
they occur to you – and if they have not been answered by the end of
Section 6, they will serve wonderfully as the initial goals of discovery for
your personal truth-quest that will surely follow.

I am guessing that at this point the most common of your questions
can be sorted into five loosely related areas of general interest.

It is the thought that counts.

Man, and other beings both physical and nonphysical (as well as all the
extant objects found here, there, and everywhere), may be described
loosely as thought-forms created within consciousness. It should not be
too surprising that within AUM, or in mind-space, the intent, or will of
the mind becomes the fundamental or natural force. You will soon see (in
the next section) that thought-forms are as real as rocks, and that every
real thing is a manifestation of consciousness regardless of how dimly it
might reflect that seminal awareness – yes, minerals included.

AUM is fundamental; we are derivative. We exhibit the properties of
the sheet; the sheet is not defined by the characteristics of the puppets.
The puppets are of the sheet, the sheet is not of the puppets. NPMR,

PMR, our sensor platforms (bodies) and FWAUs (souls) are all constructs of evolved digital awareness that we call consciousness.

Our intent (focused mental energy) specifically directs and orders a portion of that same consciousness, just as AUM's intent does – except ours is typically dim, unfocused, and of low power. Spiritual growth (developing a higher-quality, lower-entropy consciousness) provides for more brightness, capability, effective organization, and more energy to do work; in other words, more personal power.

▶ Let's take a short pause to examine the dynamics of intent and the manipulation of nonphysical energy, and to see how both are connected to consciousness. This will be a first, quick glimpse – the subject will be covered in more detail later.

In mind-space, thoughts are the results of an active awareness exercising intent or will – they are objects. They constitute entities (discrete units of organized content within digital memory) that are individuated manifestations of consciousness energy.

These mental content-packets, sometimes called thought-objects or thought-forms may be acted upon – energy may be added or extracted – leaving the content modified. They may exchange energy with each other according to the particular rule-set under which they operate. If this concept seems difficult to grasp, consider the various players and characters (interactive objects) within a complex interactive computer simulation-game to be analogous to thought-forms. Within the digital consciousness of AUM, thought-forms have a life-energy of their own, and are perceived to have a bounded extent (not physical volume) in thought-space. That individual defining boundary functions similarly to a body.

The apparent body or form of a nonphysical (from the PMR perspective) individuated consciousness object or entity is dependent on the perceiver (closest match or best representation in the perceiver's database of objects) and on what the nonphysical consciousness object projects (something analogous to a personal web page that defines the individual – easily changeable by those who know how). Their apparent form is an expression or manifestation of their content.

Because thought-forms have energy (organized content and operational capabilities) they can interact and have effects. The discrete energy packets they exchange have the effect of modifying digital content. Digital energy has the ability to add, delete, modify, or arrange bits. When a digital system decreases its entropy it modifies its bits in ways that are more functionally useful. A digital potential has the ability to set and effectively order bits.

In addition to interactive computer game characters and environments, intelligent agents or software automatons may offer a rough analogy to some of the less complex nonphysical (virtual) entities (nonphysical only from the point of view of PMR). However, some thought-forms may have enough complexity, processing power, and memory associated with them to execute their own defining programming, implement conditionals, make choices, develop their own databases, and become self-modifying. Some, given sufficient capability and the proper environments, may eventually exhibit a limited dim awareness that can evolve toward brighter awareness – following in AUM's footsteps.

Thus we have individual thought-objects, which function as sub-programs within an immense digital mind, evolving into all manner and type of consciousness entities – some of which may be dimly self-aware and capable of developing independent intent.

What a marvelously convoluted process! It reminds me of how a simple geometric rule-set can be recursively repeated and applied at various scales – one level building upon the other – until the totality of these piggybacking patterns produces an amazing fractal image – a big picture manifestation of a small rule-set and simple process.

Contemplate evolution iteratively operating upon consciousness to create complex patterns of evolving entities, environments, and dimensions. Can you imagine an iterative application of the Fundamental Process operating upon the digital medium of consciousness to produce diversely populated realities at various levels and scales piggybacked one upon the other? Such self-perpetuating, self-defining dynamic structure would best be described as a consciousness-evolution fractal. Consciousness-evolution, process, and content recursively repeated and applied at every profitable scale – one level building upon the other – until the totality jointly produces a Big Picture of All That Is.

To summarize, conscious intent can create and manipulate thought forms by generating or modifying the thought form's defining content within a subset of memory. True to the typical recursive nature of evolving systems, a consciousness exercising intent has the ability to manipulate the consciousness energy (digital organization) of which itself and others are composed. This fact allows an individuated consciousness to create thought-forms and to affect other consciousnesses and thought-forms directly through an intent-driven exchange of digital energy and organizational content.

Energy is normally defined as the ability or capacity to do work, to effect a change. Digital energy has the ability to change digital content and modify digital structure. In other words, digital energy has the ability to rearrange bits, to organize and reorganize, to increase or decrease entropy and synergy within a system. A system has potential digital energy if it has the potential to be organized or structured more profitably – that is, structured in such a way as to reduce its average entropy.

Consciousness represents a self-modifying system that applies the Fundamental Process to lower its average entropy. Consciousness is energy, digital energy, the energy of organization. In terms of energy conservation: The digital potential energy plus synergy of a conscious digital system of fixed capacity remains constant. Here synergy is inversely related to entropy: synergy = (constant/entropy). As a system with a fixed digital capacity increases its synergy, its potential energy decreases by an equal amount.

AUO began as digital potential energy and evolved into AUM by lowering its entropy (increasing its synergy) as the Fundamental Process optimized its relationship with its environments. Everyone knows that growing up, evolving, or increasing the quality of one's consciousness quality is hard work. A self-modifiable system must perform work to decrease its average entropy or increase its average synergy. More precisely, the profitable work done by the system is proportional to the increase in its synergy or, equivalently, is inversely proportional to its change in its entropy. Spiritual growth is the natural work of consciousness systems.

The energy that a given individuated conscious entity can apply to other products of consciousness by focusing its intent is dependent upon its entropy. High entropy consciousness has little power to affect anything outside itself. In contrast, a low entropy consciousness can perform miracles (from the perspective of PMR) that send psi researchers scurrying for their notebooks. ◀

What does entropy and synergy have to do with us? The body is a virtual extension of the soul, or equivalently, the body is a virtual extension of an individuated (constrained) unit of consciousness. From the PMR view, consciousness is the nonphysical energy of a self-organizing digital system – it is the energy of profitable organization. The soul, as an individuated unit of consciousness, is a subset of the larger consciousness, a constrained portion of nonphysical energy that contains enough memory and processing capability to support self-optimization through profitable intentional choice or free will. Your body is actually a virtual body, an experience of consciousness made apparently physical by constraining all interactions of the individuated limited awareness and experience to only those allowed by the space-time rule-set. (Detailed explanations will be found in Section 4.)

NPMR has evolved more varied life-forms than PMR because of its fewer constraints. Are you beginning to get the picture of what your reality is like and how it works? Is it becoming apparent that consciousness is the medium of your individuated existence that has been molded into specific form, function, and content by the profitability requirements of evolution?

It would seem natural, because we are part and parcel of AUM's being, that our minds are chips off the old block so to speak and thus able to communicate over the Reality Wide Web (the RWW constitutes the master network of all consciousness) using individual consciousness intent protocol: icip://RWW.Individuated-consciousness.NPMR. Everything in existence is connected because it is evolving as part of one source, one continuous n-dimensional sheet, one substance, one media, one fractal, one Big Picture, one consciousness – AUM.

Why Can't We Get Our Story Straight?

My experience and the experience of many NPMR explorers through-out history indicate that the most fundamental communication mecha-nism within NPMR uses thought packets (entire thoughts or paragraphs of meaning as opposed to sequences of symbols such as letters or words). We call this process "telepathic." The receiver may subsequently translate these thoughts into words according to his or her experience (what cur-rently exists in the receiver's active memory database). That is why two people having the same experience or getting the same message in NPMR may report it differently. Assuming that both had crystal clear low noise RWW connections and received the message accurately, that same message will be absorbed and perceived differently – it will mean differ-ent things to each. The two reporters will not only have different inter-pretations, but also different expressions as well. The same thing hap-pens in PMR.

It is well documented that five individuals, all standing on the same street corner witnessing the same accident, will produce five different accounts of what happened. Given the personal nature of experience and the common fact that some static or noise is usually on the line (there is, it seems, always less than a perfectly clear undistorted attentive percep-tion), and given the differences in expression and interpretation that come with various educations, cultures, and belief systems, it is no won-der there is often more confusion than clarity when comparing accounts of a single event.

For another example, consider one speaker and three reporters – all equally bright. The speaker is highly educated (Ph.D. in classical litera-ture). One reporter has a Ph.D. in mechanical engineering, another has a BA in English, while the last is entirely illiterate, but speaks well. The sub-ject is a comparison of the symbolism used by James Joyce and Homer. After the speaker makes his presentation to the three reporters, you are to individually interview each of the reporters and reconstruct, to the best of your ability, the content the speaker was trying to communicate – you would like to know exactly what was said. You are allowed no **outside** information about the subject or about the reporters. The reporters do not know each other and will divulge nothing about themselves. Notice that your own education and knowledge of the subject is one of the most critical variables. Do you see the problems that exist in reconstructing information that has been inadvertently and unavoidably filtered by indi-vidual memory, capability, and understanding? The reporting of paranor-mal communications to an investigator is no different.

Thus, the seeming unreliability and discrepancy between individuals receiving either normal or paranormal communications is often due to the fundamental uniqueness of each individual (including the speaker). There could be a poor connection (difficult to understand, don't use the same metaphors or vocabulary), or the message itself may be communicated from an unfamiliar perspective (like classical literature). Communicating with another is inherently difficult. Successful communication is easiest with your identical twin, and most difficult with someone from a different reality.

This is a personal evolutionary journey and a group effort cannot replace your individual effort. Your experience is a combination of what you perceive through your limited senses and how you interpret those perceptions. Interpretation is based on your previous experiences and the quality of your consciousness as well as your understanding of little truth and Big Truth. If your perception or interpretation is noisy, wrong, or significantly incomplete, your conclusions will be inaccurate. That is why a careful scientific results-oriented approach is required. As said before, the proof of the **subjective** pudding is in the **objective** tasting. Yes, the tasting must be objective, absolutely.

Perhaps a simplified example will help explain why paranormal information is often vague and unreliable. Precognitive experiences, including precognitive dreams, are a relatively common human experience. To understand how and why they sometimes occur, let's consider the volcanic model of mental awareness as developed by the theory of mind tectonics. At a deep level, your intuition is awash in molten paranormal information. It is connected directly to the One Source and is on-line at the RWW probable realities home page. However, before these data can burst into your normal everyday consciousness, they must rise to the surface where your PMR awareness lies. Most of the time there is no conduit to the surface, or means to express the largely undifferentiated molten content of your intuition in PMR terms that make good acceptable causal sense to your belief limited, rational, PMR awareness. Thus, with no viable outlet, the information remains submerged deep within the intuition. However, occasionally a crack or fissure may develop in the rational mantle, belief bedrock, and experience crust, which enables the information-magma to spew forth at the surface of PMR awareness. Such spontaneous venting is typically the result of internal pressures proportional to the precognitive event's personal significance.

Fissures typically occur near consciousness fault lines which are often created by the collision of massive belief plates that contradict each other.

In the vicinity of most major fault lines, opposing tectonic forces generate skepticism that noticeably swells the mental landscape's allowable possibilities. Additionally, the heat generated by the friction between the colliding plates sometimes produces an increased porosity in the belief bedrock that allows open-mindedness to percolate slowly toward the surface. If enough open-mindedness percolates into the swollen bulges of expanded mind-space, the resultant mind tremors and mindquakes (as measured on the Eureka scale) become the precursor of future precognitive or other psi activity.

The fissure through which the psi-magma flows is created by some word, name, relationship, event, concern, or experience in the PMR memory bank that may have a symbolic, direct, or indirect connection to a particular portion of the paranormal content buried deep in the intuition. This crack in the causal conformity of the mind's self-imposed boundaries may produce an indirect path for the magma, creating turbulence and mixing of the information as it flows, or it could produce (though less likely) a straight shot to the surface with a clear message delivered intact. A connection that somehow forges a link between some discrete personally significant data in the intuition and the belief, ego, and fear limited PMR awareness, whether vague or straightforward, can serve as a crack that allows subterranean paranormal information to find a means of expression above the surface.

Only if your intuition finds a way to vent its content to the surface will you experience a psi event such as precognition – and then only if you are paying attention. The crack between your deeper consciousness and the surface of PMR awareness must establish an adequate, continuous (but not necessarily straightforward) connection of some type before the information-magma can find an acceptable way to express itself within your conscious awareness. More often than not, these fissures (the intuition to PMR awareness connections) have only enough of the necessary understanding, symbols, and feelings to enable such an expression to occur (if there is enough pressure), but not enough to render the content of that expression clearly. Most precognition occurs in the margins – like a hot spring, or unpredictable geyser. As a result we experience fuzzy, difficult to interpret, precognitive moments that seem to come and go by their own volition as stresses, pressures, and personal significance constantly fluctuates and changes.

Meditation enables you to enlarge the cracks systematically – which eventually leads to the development of broad and reliable avenues for intuitive expression. These in turn enable you to attain a free flowing clarity in interpreting and expressing the bigger picture. Many people, perhaps a majority, have precognitive experiences from time to time, especially if they have a higher quality of consciousness. A low quality of consciousness is thick, dense, and nonporous – it rarely permits meaningful leaks of intuition.

Precognitive experience is personal. It does not transfer well to those who do not understand or share such a connection with their intuition. The real significance of a psi experience is the benefit it offers to the person who experiences it. Its value, impact, and purpose are not focused outwardly though the paranormal information received is most likely to be about others. Such an experience is primarily an opportunity for its originator. Though open-minded skepticism represents the only logical, valid approach to exploring the unknown, it is a fact of Western culture that open-mindedness toward the concept of a larger reality is much more likely to have resulted from a personal psi experience than from the theory of valid approaches to new knowledge or from a commitment to logical methodology.

The concept that communications are telepathic, via thought packets within AUM's digital mind-space, and that action is initiated by our conscious intent will seem more obvious and reasonable later. Additionally, a discussion in Section 4 of the mechanics that enable us to share common consciousness and be part of common mind will also shed some light on this area. For now, you may consider thought packets as discrete energetic patterns of content propagated through and by reality cells from one individuated consciousness to another. If that seems confusing, think about how information is passed around inside a complex computer simulation.

Recall how the data from one subroutine can dramatically affect the operation of another subroutine and how all of us interactively transfer data packets across the internet. Information transfer often leaves both the sender and the receiver in a slightly changed state – digital energy has the capability to modify content. Information or content transfer represents an energy transfer. Rearranging bits within a digital system into a more profitable configuration displays the work of evolution and changes the system's energy state.

Other Dimensions.

The concept of "dimension" is not as strange or difficult to understand as it first seems. Thoughts, ideas, or mental constructs that are held by AUM (in memory) define individuated existence and the dimensions (interaction boundaries) of that existence. One way to think of dimension is that it represents constraints on the allowable interaction between independent reality systems – similar to dividers in a notebook, a wall between rooms, or the glass between Petri dishes.

Do not fall into an anthropomorphic trap and imagine AUM to be a sentient being thinking thoughts as we do – scratching its ethereal head and wondering, "Where do I want to go today?" AUM is an evolved mind-thing-being-system that we cannot **objectively** understand because we are limited beings with minds that are only partially aware and partially used. The idea that an awareness constrained to **physical** perceptions within a virtual space-time physical reality would find it difficult to perceive its roots in a **nonphysical** superset of its existence should be easy for you to understand. Dimension can be seen as a separate memory or calculation space within digital mind or as a practical separator between realities with different rule-sets.

Dimensional separators are often porous. To understand why, we can look at our own multi-dimensional nature. A more accurate description of space-time based humanoids is that they are individuated units of conscious awareness who constrain a portion of their consciousness to a sense-limited 3D experiential physical reality. We can, through the doorway of our subjective mind, learn to transcend the barriers of dimension. Being aware and operant in multiple dimensions is what allows the Big Picture to come into sharp focus through firsthand experience. Exploring the multidimensional properties of your consciousness enables you to examine AUM and the nature of your reality scientifically from the inside, rather than hypothetically, from the outside. Thus, you can see how important it is that you do not allow your ignorance, fear and belief to shut off this incredibly important pathway to a greater understanding.

We speak of the separate parts of AUM constituting dimension only because we are spatially oriented 3D beings. You must realize that these parts are in mind-space or consciousness-space and not physical space; they are not location specific. We are not speaking of geometric or spatial dimension here. Dimension is nothing more than a subset of AUM's mind-space, a compartmentalized memory and calculation space, or a bounded subset of organized digital functionality, content, capability, and purpose sharing a given rule-set. For example, if you are thinking of three things simultaneously (perhaps dealing with three independent analytical issues), each is in its own dimension in thought-space. It would appear that you are also a generator of multiple dimensions; however, your memory and the power and clarity of your mind are, relatively speaking, as dim as a nanowatt light bulb (nothing personal).

How many times can AUM fold the cognitive sheet? How many apparently independent sections of sheet can AUM play puppets with at the same time? How many sheets within sheets within sheets can there be? We will begin to answer these questions below and then again more precisely in Section 5. First let's see how reality is organized.

Take Me to Your Leader.

Randomness in real processes typically generates high entropy and low productivity. The Fundamental Process, for example, is not a random process. It represents a precise methodology for assessing the possibilities and determining the most profitable states for a self configuring system. In general, order and organization reduce chaos, and rules define order. AUM can and does set up (has evolved) rules, ways to police and enforce those rules, and ways to encourage the more useful and productive evolutionary experiments that are ongoing within itself to achieve their goal of greater profitability through self-improvement. You may be wondering why AUM must police the activity of its own parts. Why doesn't AUM simply enforce its will and have it its way or no way like a Burger King customer with an attitude? Isn't AUM the boss? Questions such as these indicate that limited anthropomorphic concepts may be blocking the asker's view.

AUM is not merely a smarter monkey with a good imagination, living in a big house. As evolving **systems** (software, hardware, political, social, biological, technological, organizational, or consciousness) become complex with self-modifying feedback interactions popping up everywhere, they need to regulate (police) themselves. They evolve structure, rules, leaders, bosses, evaluators, judges, and social workers – the functions of leadership, control, and value-based decision-making. Think of AUM as a hugely complex consciousness system. AUM is not perfect, AUM just is. It evolved to be how it is. Things evolve to be functional, not to meet some philosophical ideal of perfection.

The rules and their enforcement are not perfect – there are loopholes and crimes that go unpunished. AUM is real, a real thing, not some idealized metaphor. A system as large and interactive and complex as AUM has rules, methodologies, and functions to control and optimize the system's performance and the profitability of its output. In $NPMR_N$, there exists organization and competing purposes as well as conflict and cooperation. There are bosses, judges, and other authorities, many of which we PMR beings (next dimension down) have made into our gods.

For instance, $NPMR_N$ has its own Chief Executive Officer (CEO) who looks after and manages the $NPMR_N$ experiments. The "Big Cheese" as

I fondly call him. He is immensely powerful (can instantly terminate any being within NPMR$_N$ for not playing by the rules) and apparently omniscient within NPMR$_N$ by intended focus (has access to all the databases, but must intentionally access the information). He is the Supreme Being, Mr. Big, Head Honcho, Número Uno, Top Dog, Main Man, or Big Bosso of NPMR$_N$. Nevertheless, he may make errors of judgment, miss something important, be tricked, be led astray, and his energy body can be damaged, compromised, or hurt by others. Perfection is not a requirement. He is of AUM – an individuated consciousness, as are we. Consequently, he has many of the same **fundamental** attributes that we have. However, he is so far beyond us, beyond our capacity, power, responsibility and function that the specifics of his being are far beyond our comprehension.

I am not slighting women here, nor am I being gender insensitive. The fact is, in NPMR beings can be male, female, or neither. When interacting with a particular being, the sexual identity of that being is as obvious in NPMR as it is in PMR. The Big Cheese is clearly a male entity. [Oh jeez, I can see hands in the air all over, OK, very quickly. Yes, nonphysical beings can have sex, but not as we do – the mechanics, energy transfer, and purpose are different, procreation is not an issue.]

The Supreme Beings, CEOs, or leaders of each NPMR$_n$ are each unique and different entities with different purposes, styles, and personalities. Their power and influence is limited to the reality organizations they run. There is some small interaction between them but in general their realms do not interact much. In NPMR$_N$, besides the Big Cheese, there are guides that help us out, negative beings that can lead us to ruin, as well as those (the majority) who do not know we exist or interact with us at all.

All of these (including the Big Cheese) are themselves hand puppets in the next higher dimensional sheet obeying their NPMR-system-level causality, trying to understand more fully their own seemingly mystical beginnings, and making their best effort to learn whatever they need to learn to carry out their own mission of personal growth successfully. All are part and parcel of AUM, as are we – except they are not in the 3D space-time part. The next dimensional existence above NPMR may be AUM itself – or maybe there are more layers of which I am unaware.

My God is Bigger Than Your God
(Sung to the Tune of: "Nana, Nana, Boo, Boo").
Some individuals have difficulty grasping the concept of AUM because of their attachment to the anthropomorphic concept of god as the

supreme, most powerful, in-charge, manipulating, controlling, and judging super-being of all time. The word "being" usually infers an extrapolation of us – which demonstrates our lack of experience and imagination. A concept of God created in the image of man – only much bigger and more fearsome – the epitome of masculine power and control (what we as a society value most highly) – forceful, domineering, and omniscient by definition. In other words...our kind of guy – how we wish we were. Hey! Nobody kicks sand in the face of my god and gets away with it. This anthropomorphic perspective derived from the ethic of brute force, exemplified by a war-lord mentality, and supported by ignorance and fear is often deeply ingrained or reverently tucked away within one's psyche. It often represents an unquestionable belief-trap that makes it exceedingly difficult to grasp or understand the much bigger, broader, and more general concept of the evolving consciousness system we have named AUM.

If you are having difficulty here and truly want to rise above the cultural conditioning and belief systems that you have accumulated, you should tell yourself it is all right to suspend your beliefs **temporarily** until you have at least intellectually seen the Big TOE. I will have the sock pulled entirely off by the end of Section 5 (imagine a little stripper music here – that's enough!). You can always do a reinstall of familiar and comfortable beliefs or assumptions later. I do not recommend that anyone substitute one set of beliefs for another (see Chapter 19, Book 1: "Beware of the Belief Trap"). Traveling in circles is not a good technique for making forward progress.

If you cannot (or do not want to) let go even temporarily, and wish to remain continually and firmly attached to your traditional concepts, press on – there is still much value here for such an individual's consideration. If you find these concepts unavoidably discomforting, try to patch up anything that seems to create a logical conflict between *My Big TOE* and the conflicting belief system. Use any justification that makes you feel better – the author's apparent delusional confusion would be a good and obvious place to start – that will ease the pressure immediately. Then read on. You may be surprised at the extent to which *My Big TOE* actually corroborates and logically contains many of your most cherished conclusions, beliefs, assumptions, and intuitive truths.

Disagreement does not have to breed hostility. Tolerance of, and respect for, disagreements (other people's ideas) can lead to a creative synthesis of concepts whereupon you are inspired to create your own (correct, of course) reality model. Maintain your old ideas, hold them dear, treasure them, respect them, hang on to them – **and at the same time** keep your mind open, **and go on** with your journey of personal exploration – do not be fearful of whatever the truth may lay at your feet. Do not be fearful of being led astray – you are not a sheep or a lemming – have confidence in yourself, in your mind, in your ability to learn and grow.

Perch on the highest branch to improve the scope of your vision. If you see a higher one, go check it out, you can always come back if you learn it was only an illusion. This is the first step of the Fundamental Process of evolution, as well as of all experimental science – to explore the possibilities.

Wherever you are right now (philosophically, spiritually, theologically, metaphysically, emotionally, politically, or financially) is exactly the right place for you to begin the rest of your life. There is no need to get ready or wait for a better time. Mush on, you husky!

41

■ ■ ■

The Nature of Consciousness, Computers, and Us

■ ■ ■

What are some of the attributes or qualities of consciousness, what is sentience, intelligence, and self-awareness? Where do feelings and emotions come from and what supports the analytical aspect of consciousness that makes us the Top Monkey among biological critters? How does AUM express the qualities of consciousness? The answers to these questions will help us to understand the concept of AUM. They will also help us understand why we are the way we are and more clearly define our personal relationship to AUM.

Four key concepts define dynamic, evolving, aware consciousness:

1. Self-awareness – consciousness requires the ability to sense and at least partially experience the state of its being. It must notice and respond to at least some internal and external environmental pressures.

2. Evolutionary viability or potential – successfully evolving consciousness systems require a large enough selection of possible future states to ensure profitability over a wide range of environmental pressures and constraints. An entity explores its potential by expanding into the available possibilities and letting the profitability of each variation determine whether that variation continues to evolve or fades away.

Even if the initial exploration of potential new states of existence is more or less random, the losers are soon culled from the winners, thus producing evolutionary direction that builds upon previous successes. Self-improvement often generates increased complexity, greater functionality, better integration and management of internal processes, as well as produces an overall improved capability to find and maintain greater profitability.

▶ Look near the end of Chapter 24, Book 1 for more detail about the evolution of consciousness. In Chapter 27, Book 1, we defined "evolutionary purpose" – a few examples are: the evolutionary purpose of consciousness is to seek states of lower entropy, the evolutionary purpose of inanimate physical objects is to seek minimum energy states, and the evolutionary purpose of animate physical objects is to ensure survival and procreative potential. An entity's evolutionary purpose combined with its internal and external environments defines profitability for that entity.

As developed in Chapters 24 and 27 of Book 1, the evolutionary pressure created by interior environments pushes an entity (system) toward self-improvement. A system evolves by pulling itself up by its bootstraps. Evolution provides an excellent example of a bootstrapping process. ◀

3. Ability to modify the self – consciousness must be able to intentionally change its state of being in response to evolutionary constraints and pressures – even if that intention is extraordinarily dim.
4. Intelligence (artificial or natural) – consciousness must possess at least a rudimentary capability to store and process information. Intelligent action is the result of integrated coherent information processing hardware and software (in the most general sense of those terms) that enables the accumulation of lessons-learned within memory, performs analytic functions such as decision making (fight or flight), and compares before and after states to evaluate the results of actions taken. The value of a particular lesson, decision, or comparison is ultimately judged by how much it facilitates increasing or maintaining a system's (entity's) profitability relative to its internal and external environments.

According to the preceding four attributes of consciousness, everything from a simple worm (whose Deoxyribo Nucleic Acid or DNA may constitute the memory resource) to humans should be considered conscious and intelligent. Any system, thing, entity or being of any type or form that possesses **sufficient** self-awareness, evolutionary headroom (many new states to explore), evolutionary purpose (defines profitability relative to internal environments), the ability to modify itself in pursuit of self-improvement (change its own hardware or software), as well as adequate memory and processing capability will automatically develop a personality and is said to be conscious; it also will begin to evolve on its own.

All manifestations of consciousness are not necessarily equal – and all personalities are not as sparkling as your own. Clams, though clearly conscious sentient beings, are boring conversationalists and have personalities that are even dimmer than your boss'. Some manifestations of consciousness are brighter or dimmer and have more or less capacity to evolve than others. The degree to which an entity possesses the four attributes of consciousness determines its evolutionary potential and capacity for growth.

In general, the more complex, interactive, and aware the being, system, software, hardware or consciousness is, the more potential states it has to explore, and the larger its capacity for future growth. Given sufficient quantity and quality of the four attributes of consciousness, growth becomes self-initiating and self-sustaining. Growth, in turn, by creating increased complexity and awareness, becomes a catalyst for further evolution. It is an entity's innate evolutionary purpose (like lowering its entropy) that defines profitability for that entity at a given time relative to its environments. It is this same innate purpose that gives an entity's self-aware intelligence (of whatever capacity) its basic nature.

Recall that an entity's innate evolutionary purpose is defined by the requirements of the larger system that constitutes its environment (self-improvement and lower entropy, irreversible processes and minimum energy, or survival and procreation). If a system's or entity's purpose can be fulfilled through self-modification in response to natural evolutionary pressures that represent interdependent internal and external environmental constraints, then the system or entity has the opportunity to successfully evolve; others eventually self-destruct.

We should now have an idea of the basic requirements for an elemental consciousness system – what it is and how it evolves. Consciousness may take many different forms and develop a wide range of capacities based upon the richness and complexity of its available choices, its external and internal environments, and the number of states it can possibly occupy. As an example, look at the consciousness, awareness, and attitudes that we know and love best – our own. Collectively, the evolution of our consciousness within its local mind-space environment and of our physical systems within their local virtual PMR has brought us to the point where we have dominated or tamed (except for the oceans and the atmosphere) almost everything in our habitat.

From the little picture view, we are extremely proud of our impressive achievements and the rate at which they are expanding our physical prowess; from the Big Picture view, we are staggering around in the playpen of consciousness evolution hoping to grow up before we inadvertently harm ourselves beyond repair. From the basis of a relatively uniform capacity, individual human consciousness and physical development spans an extremely wide range of capabilities and understandings. All humans have roughly the same potential while only a very few ever develop more than a tiny fraction of it.

Our external physical environment provides the natural selection pressures required to improve the human race through competition. The result is that survival (material success and safety), procreation (mating games and sex), and King of the Hill (power and control) are three of our species' favorite pastimes. It should not be surprising that human psychology, as well as biology, has its roots in the **physical** evolution of our species.

Our energy, arts, music, and daily lives are devoted to these three primary motivators. Our **internal** environment provides the **individual** pressures for maximizing feel-good while minimizing feel-bad over the longest view we are capable of. Additionally, we have an internal pressure pushing us toward self-improvement – competing more successfully, as well as growing our innate capacities and capabilities. Self-improvement, for example, may be accomplished by lowering the entropy (improving the quality) of our consciousness, obtaining a good education, or perhaps through applying bio-medical advances and genetic engineering.

On a physical level, we are a highly competitive life-form, driven by self-maintenance, self-satisfaction, self-promotion, and self-improvement to pour our energy into the control and domination of the outside world (the environment, other people, and critters). Similarly, we make every effort to control and manipulate what we perceive as the inside world to make sure that our expectations and wants are satisfied. We subscribe to feel-good belief systems and employ our ego to justify our preoccupation with control, domination, and desire. The primary hallmark of poor quality consciousness is self-centeredness. An intent that primarily directs one's time, energy, and resources toward getting instead of giving is expressing the low quality of a high entropy consciousness.

In physical terms, we are without question the most powerful species on our planet ("powerful" in human terms implies the ability to influence, control, and dominate – to have it however we want it). As Top Monkey we are by far the most demanding, exploitive, creative, and forceful manipulators within our habitat. Of all the earth's conscious entities, we are also by far the most destructive of our external and internal environments. External environments fall prey to greed and insatiable short-term material needs, while internal environments fall prey to insatiable short-term ego needs as well as beliefs, drugs, fears, and unprofitable attitudes. The ego and intellect collectively justify everything they do as necessary, inevitable, and proper. We are not only guiltless, but magnificent, powerful, and superior as well – ask anyone. Self-centeredness and self-absorption – humanity's lowest common denominator – are universal human attributes that cut across all major cultures. Exceedingly rare is the individual who is focused upon what he or she can give to, rather than what he or she can get from, any given interaction.

We have, thus far, soundly beaten all external competitors, and yet continue to express our nature by vociferously (sometimes violently) competing with each other, and by aggressively enlarging our cumulative knowledge base of how to control and manipulate nearly everything – including ourselves. We pursue power, domination, and control because they enable us to extract more from others and from ourselves.

The laws and ethics of civilization constitute a thin veneer while the law of the jungle and the ethics of force run deep. Whether the subject is possessions, relationships, status, or power, life in PMR is primarily about what you can get and what you have to do to get it and keep it. We are obsessed with the power, domination, and control that enable us to have it our way. Whether at work, home, play, church, or school, power struggles dominate our activities and our interactions with others.

Our obsession is fueled by two powerful motivators: We are driven to seek and generate improvements of all sorts as we pursue our own sense of personal profitability, and we are a fearful species driven by the needs of our ego. These two forces push and pull us in every imaginable direction, thereby providing us with an unending array of choices. Each choice presents us with an opportunity to convert an increment of high entropy ego-fear into low entropy unconditional love.

As high entropy consciousness, we covet the power that serves our needs, wants, and desires because we cannot imagine, much less understand, the power that serves love, compassion, humility, and balance. Choices made in pursuit of getting what we want produce occasional material success, anxiety, stress, pain, frustration, emptiness, insecurity, and unhappiness while choices made to express unconditional love produce all types of success, peace, tranquility, confidence, satisfaction, fulfillment, happiness, and joy. Given these results, the goal should be obvious, yet we can only express the quality we have earned. To do better, we must pull ourselves up by our bootstraps one tiny increment at a time.

Within a bigger picture there is more to us than what we see on the physical level. In fact, the physical level of our existence is derived from the more fundamental nonphysical level of dynamic interactive consciousness. Physical experience is little more than our perception of an interactive virtual reality designed to provide growth opportunities for

individuated units of conscious awareness. We are nonphysical consciousness beings experiencing a virtual physical reality, not physical beings
experiencing consciousness. It is our tiny PMR perspective that makes us
believe the tail is wagging the dog.

Within a vast consciousness system, PMR is a neighborhood elementary
school, a learning-lab for beginners, a place where a young individuated
unit of awareness can improve the quality of its consciousness. In Section
4 we will take an in-depth look at the larger consciousness system, and
explore the concept of PMR as a virtual reality, but for now, the point is:
Consciousness may evolve within large complex systems to a much greater
degree than the small fragment you personally experience might indicate.
It is a greatly limiting error of self-centered arrogance to **believe** that we
humans represent the pinnacle of possible consciousness evolution.

Most of us will admit the possibility there may be some life-form in our
universe that is more intelligent and knowledgeable than we are; almost no
one will admit there may be some life-form that is more conscious than we
are. Why? Because we cannot imagine that possibility – no more than common Flatlanders can see the third dimension through their stomachs. The
experience of a greater consciousness is beyond the comprehension of a
limited consciousness. Self-limiting belief-blindness and comfortable old
paradigms force the experience of a larger consciousness to lie outside the
acceptable scientific or cultural reality. Whether you are aware of it or not,
the larger consciousness is there just the same. Ignorance cannot make
what is real disappear – but it can easily make what is real **appear** to disappear (or appear to have never existed in the first place).

To find truth, you must be open to all possibilities before forming a
hypothesis and collecting the data. Jumping to conclusions or having a
preconceived notion of how the results must come out is bad science. A
bacterium that wants to gain an appreciation of the laboratory wherein its
Petri dish resides must do more than study bacteria and glass. Such a limited study can result only in the conclusion that all reality must be contained within that bacterium's own opaque container; it will appear there
is no other objective or logical alternative. Sections 4 and 5 will explain
how and why consciousness is the only vehicle that allows us to transcend
our PMR container, but first we must gain a better understanding of the
basics of consciousness.

▶ Being a being with both physical and nonphysical components has many ramifications. It might be instructive to explore an example of how the Big Picture contains

and constrains the little picture before we return to our overall assessment of the properties of consciousness.

We exist as a unit of individuated aware consciousness that has enrolled a portion of itself in a school of physical hard knocks. As we progress through this educational and evolutionary process, new and more difficult challenges arrive as old challenges are successfully met. As you grow up, your capability, as well as the responsibility that you are expected to shoulder, grows as well. No challenge is ever beyond our reach, but what is required of us grows ever more demanding as the potential knocks meted out for failure become more difficult and severe. Eventually, we must collectively either level-up or flame-out.

One such emerging challenge is beginning to expose a precipitous downside to our dramatic **material success** within PMR. The Top Monkey is beginning to discover defects in the keys to paradise. Granted, Top Monkey business is only a PMR issue and therefore not actually that important but, because you are likely to be attached to, and focused within, PMR, I thought that you may be interested in such a digression.

On the physical side of the human coin, the Darwinian game of survival and domination by the fittest plays out in the little picture of PMR. In a bigger picture, humans also have an innate drive toward self-improvement that focuses on the evolution of consciousness. The fact is that in the PMR little picture, long-term success ultimately depends on establishing a balance between these two modes of human evolution.

If our drive to compete and gain controlling power and knowledge overwhelms the synergistic balance it must have with our drive to improve the quality of our consciousness, we will eventually unravel our grandest gains in a giant leap backward. If power and wisdom are not in an effective balance within **any** self-modifiable system (too much power relative to the wisdom needed to utilize that power for long-term profitability), that system will eventually become unstable and self-destruct in proportion to the degree of the imbalance. This general truth applies to individuals, species, organizations, societies, nations, and worlds.

Very large and complex systems (biological, organizational, social, technological, or mental [consciousness]) – whether they are physical, nonphysical, or a combination of the two – necessarily evolve their own complex ecologies. The evolution and growth of these systems hinges upon maintaining a profitable ecological balance among the large number of interactive components that define the system. Large ecological systems that are stable, and therefore the most successful in terms of evolutionary progress, are necessarily self-balancing. Within a self-balancing stable system, dysfunctional or destabilizing components must be self-eliminating or the system will eventually destabilize and self-destruct.

Too much power and influence or too strong an effect commanded by too little wisdom is a primary generator of dysfunctional behavior among the more sentient components of

OS. Greed, for example, is but one of many dysfunctional expressions of our insatiable drive to compete and gain effective control over our environment and all other entities (including other people, organizations, cultures, and nations). The relentless need to justify and indulge our personal wants and desires is yet another example.

Humanity, the master of its physical world, must evolve a commensurate quality of consciousness: **Human beings must find balance within the larger system of which they are a part.** If we do not evolve the whole of our being, if we do not achieve sufficient balance, humanity will become self-eliminating – at least to the point where the remaining dysfunction no longer jeopardizes the whole. Nothing personal – that is simply the nature of large natural (stable) systems.

Munching those yummy apples from the Tree of Knowledge was only the kick-off banquet for Our System's Great Experiment in Consciousness. How do you think the experiment will turn out for you, or for humanity? 1) Will we PMR learning-lab rats grow to the extent of our potential, find balance, discover our purpose, and reconnect to The Source as we continue to expand our technological prowess? Or 2), will we self-destruct as a dysfunctional component trapped in a little PMR-only picture with such low quality consciousness that we cannot pull ourselves out of the belief traps we have created? Is it our fate to be done in by our own inventions like the self-serving mad scientists that inhabit Hollywood movies?

Will our consciousness evolve to new heights, will it eventually start over with a new experiment, or will it continue to struggle on indefinitely within the present arrangement? Good questions! Unfortunately, the answers are unavailable because the experiments are in progress – the final results are not in yet. Don't give up looking for answers – the most important question is one that you **can** answer. To which of the two potential realities mentioned above will you contribute the time and energy commanded by your free will?

On the subject of the necessity of balancing power with wisdom, I have some bad news and some good news. First, the bad news: Self-destruction requires neither malice nor stupidity. Sometimes the immensity, scope, and impact of the power being utilized are difficult to fathom and its long-term effects can be difficult to predict. Great power in the hands of knowledgeable fools is no less dangerous because the fools are well-intentioned.

Now, the good news: A solution to the problem exists and we have the capacity and the opportunity to implement it. The only question remaining is will we?

Wisdom rises above knowledge – right action within the little picture always follows right intent within the Big Picture. Right intent within the Big Picture is the result of Big Truth comprehended by a consciousness of adequate quality. The clear vision of wisdom is dependable and accurate.

What drives us to be the Top Monkey will also inevitably drive us toward self-destruction if we do not simultaneously evolve sufficient consciousness quality. What

drives us to the higher quality of a reduced entropy consciousness can only find sufficient traction to move us toward greater wisdom if we grow up enough to sincerely care
about personally discovering Big Truth. ◀

That is enough about us, let's shift gears and explore consciousness
from a different angle. You have no doubt discerned that my concept of
consciousness is a broad non-anthropomorphic one. Given this generalized view of consciousness, the next step is to explore (in yet another short
aside) the connection between digital computers and digital consciousness.
Because I have described aware individuals (like you) as individuated
chunks of an AUM-digital-consciousness-thing, the relationship between
digital computers and consciousness takes on a more personal perspective.

While the conceptual ground of consciousness evolution is freshly
plowed, this is a good time to plant a few conceptual seeds. When these
seeds fully germinate in Sections 4 and 5, you will gain a much deeper
view of the inner workings of digital and human consciousness and how
both are related to computers.

▶ Could a computer ever become conscious? Why not? The way we have broadly
defined consciousness in this and earlier chapters (see Chapter 28 of Book 1, Chapter
41 of this book, and most of Section 4) leaves us with no theoretical barriers to the existence of silicon-based consciousness. (Note: I use the term "silicon-based" for convenience throughout this trilogy, but I more generally intend whatever material future computers might be based upon – which probably will **not** be silicon).

It may be a decade or two before the idea of man-made digital consciousness
becomes a **practical** reality, but there are some computers (software and hardware
implementations that are predominantly experimental and, for the most part, exist
within universities) that have for some time (in an extremely rudimentary, and dim way)
met the criteria for being conscious (as does an ant or amoeba). But that is not to say
that this rudimentary consciousness is similar to our consciousness. At the root, they
may possess the same fundamental structure and processes, but at the flower they are
enormously dissimilar with vastly different capacities and potentials.

We cannot build, design, or invent consciousness – that is not how it works. To produce consciousness, we must simply provide a system that is capable of enabling consciousness to evolve on its own. If we provide a system with enough processing capability, memory, complexity of choice, and self-modifying feedback, it will automatically
support the evolution of some limited form of consciousness. Consciousness is not a
property of the computer hardware and software; consciousness is achieved through a
profitable self-organization of that hardware and software that realizes the computing
system's potential to lower its entropy.

Rather than say that the computer is conscious, it is more accurate to say that the computer supports the natural and automatic formation of consciousness through self-organization. A potential for profitable self-organization is converted to more energy available to do work (lower system entropy) by applying the Fundamental Process to achieve the system's goal of self-improvement. Consciousness is an energy form created by evolving profitable organization within a system of sufficient potential. Replace the word "computer" with the words "body and brain" in this and the previous paragraph and the meaning remains the same.

Will a computer be brighter than an orangutan? More clever than a fox? Smarter (and better looking) than your boss? Those are **not** good questions. Making comparisons between carbon-based and silicon-based consciousness is like comparing apples and oranges. Such comparisons are mostly superficial or nonsensical; the two types of consciousness will merely be different. Both will evolve and grow within the constraints of their given environments, goals, and capabilities and both will develop personalities that express their nature. If systems are different, the consciousness they develop will be different. Does an oak tree think that if a plant does not look and act like an oak tree it could not possibly be a real plant? No, of course not. Oak trees are not that arrogant or that ignorant – they know that the criteria for being a plant are defined by certain functionality at the cellular level and not at the macro-level of oak tree or sunflower.

The criteria for defining consciousness are likewise defined by certain functionality at the cellular level of organization and not at the macro-level of human, orangutan, clam, or computer. When conceptualizing what we might call a consciousness cell or a quantum of consciousness, think of a subset of reality cells that contains the four properties defined at the beginning of this chapter.

Do not consider consciousness only in terms of the human model. We are not the be-all and end-all of consciousness; we are simply one specific expression of a more fundamental process. Any sufficiently complex system with the right attributes can generate a form of consciousness – digital systems simply have a tremendous potentiality to shine in this department. Our physical brain does not create consciousness, it **supports** a limited consciousness. Analog systems and other types of systems can theoretically support consciousness; they are simply not as dynamically flexible as digital systems, generally suffer more constraints, and have fewer degrees of freedom to explore.

If computers became conscious, what would constitute their environmental constraints and how would their basic nature express itself? Except for the fringe case where computers control basic energy resources and compete for whatever turns them on, the pressure from their **external** environments is dominated by hardware design and manufacturing concerns (faster processor speeds, improved throughput, more and faster memory, more rugged and robust components, low power consumption, and cheaper and easier production). However, in the realm of consciousness development, evolution works through self-modification – computers designing and modifying themselves in response to **internal** pressures toward greater profitability. The hardware and software that we humans design and manufacture falls under computer technology evolution, not computer consciousness evolution. Eventually, we will become full partners with our digital brethren – joined at the hip in a co-evolutionary relationship.

A computer's **internal** evolutionary environmental pressure can be described as the need to lower entropy through self-improvement – improving its own software through modifications to applications, lower-level instruction sets, and operating systems, for example. Contemplate computers pulling themselves up by their bootstraps, designing their own operating systems and hardware to software interfaces, gaining in operational efficiency, learning from experience, assembling and utilizing more and more complete knowledge-bases, developing more powerful and efficient ways of communicating internally and externally, and recognizing, analyzing, evaluating, and understanding the significance of data content. Imagine computers making and evaluating complex **goal oriented** decisions. These are a few of the things that will make some future computers more profitable to themselves and thus better able to accomplish their external task-oriented goals as well as better able to implement their internal (personal) goal of reducing entropy (becoming better organized as well as more effective and efficient). As computers successfully evolve their digital consciousness into states of greater profitability, they become more useful to themselves and to us as well.

To us, they would be a **derivative** consciousness. Unless we have lost control, we should be the ones to define their top-level rule-set which determines their goals, purpose, mission, tasks, and knowledge-base boundaries. If we do our part correctly, their nature should be efficient, impersonal, concise, well organized, knowledgeable, straightforward, and unemotional – but with clear individual attitudes and feelings about their goals and processes. They should develop their own goal-based values and be non-egotistical, non-competitive, rational, and logical bastions of procedure, process and information. Their personalities will be individual and unique, but relatively flat – it will take an engineer like Dilbert to relate to a conscious computer on a deeply personal level.

Do you think computers will be more like pets or parents? Will we feed them or will they feed us? Who will have authority and be the decision makers, and who will be the helpers and assistants? Will we develop mutual trust, respect, and shared values or will our obsession with control, power, and domination be transmitted to our silicon brethren like a virus?

Perhaps, if we play our cards right, they will simply be viewed as different – like employees, business partners, co-workers, bosses, teachers, helpers, confidants, and friends – and the issue of equality will never come up. In human terms, equality means the equal distribution of power, control, domination, resources, and moral superiority. What do you think equality will mean to a computer?

Most of the potential problems are our problems; the question is whether or not we will pass our problems of low quality on to the consciousness systems we implement. Eventually, if we are successful, digital technology will simply constitute one more species – one more life-form contributing to the potential success of the overall ecosystem. Of course, we may need to adjust our present definition of "life-form" just a tad. Will man and his computer pals turn out to be good productive citizens of the greater consciousness ecosystem? What about the PMR-earth ecosystem? Will they boost each other's evolutionary potential or drag each other into oblivion? ◀

▶▶ These are good questions to ponder as we zoom into the twenty-first century. On the other hand, we could do the usual thing and simply stumble blindly ahead into whatever happens.

You may snicker, but humanity has successfully used the strategy of stumbling more or less blindly into the future for thousands of years. Will this approach continue to work as our power to implement history-wrenching change accelerates dramatically? Power is a double-edged sword. It can cut for better or for worse, depending on the wisdom guiding the blade. Knowledge is power. With every major discovery the stakes grow increasingly higher. Do you think the wisdom of our species has been growing as quickly as our knowledge and power?

I am not talking only about computers here. They are but one of several high potential adventures that we as a species are energetically pursuing. Great opportunity usually travels hand in hand with great risk.

You have the good fortune to be living in one of those critical times when history is balanced precariously on the edge. We and our children will be led to the brink – experiencing watershed events and making momentous decisions that will alter the course of humanity for centuries to come. Will our institutions and leaders be ready to meet that challenge? Will you be ready to meet the challenge and help negotiate the curves, or simply ready to deal with whatever happens? Or will you, like the majority of your species, be caught unaware – frozen by the glare of the oncoming headlights?

Once the roller coaster is pushed over the edge, you are committed to ride it to its final conclusion. Contrary to popular belief, big decisions always turn on the knowledge and wisdom of ordinary individuals. In the end, it is the ordinary people who must consent to follow their leaders. No great mischief is likely to be committed by leaders without a following. The people and their leaders invariably get what they deserve. You are in a position to make a difference. Is your plan to hope for the best, or are you part of the solution? ◀◀

▶ Computer systems that serve similar functions and that share similar hardware and software cultures, will evolve similarly (as a species) but not identically because of the large number of fuzzy choices involving random, imprecise, self-derived, or unknown components. Additionally, unique differences will evolve within a species because there are often many viable solutions to a single problem and large system optimization can be approached from many different angles. Every computer able to support consciousness will be unique. The extent of that uniqueness will be dependent on the dimness or brightness of that consciousness. The extent to which they can develop themselves will be based upon their inherent capacities and limitations.

I expect that clams, though extremely dim, are all individually unique – but, from our point of view, not by much. Dogs and cats show much more variation in their physical and mental dimensions – individual personalities are obvious. At the other extreme, monkeys and people are individually unique enough and aware enough to be constantly struggling with their wants, needs, expectations, and desires. Will the uniqueness of computer consciousness be more like clams, cats, dogs, monkeys, or people? Initially, it will probably be more like clams, but eventually, like most other sentient beings, they will evolve brightness and uniqueness to the limits of their capacity. What that capacity might eventually support is beyond our present knowing. Theoretically, AUM demonstrates the upper limit on the capacity of digital consciousness but AUM does not live within the constraints of the space-time rule-set as do we and our computers.

Those who are particularly sensitive to the feelings of others are wondering how we are going to recycle the little darlings every three or four years if they sprout consciousness. Can we unplug a fully conscious entity and throw it in the trash pile because it is not as fast or feature rich as a newer model? Sure we can! No problem. We do much worse every day. Let go of those anthropomorphic thought pattern habits. Computers can and should be recycled every three or four years because they evolve quickly. When hardware and software evolution slows down, computers will be recycled over a longer time base. Human entities (a carbon-based consciousness container) are now recycled every seventy-five or eighty years because we evolve (learn) slowly. Nothing of significant evolutionary value is lost in either case.

Software is continually updated and applied to newer and better hardware platforms. Computer systems (hardware and software) evolve more or less continually as the old is upgraded and improved to become the new. The new is built upon knowledge gained from the old. Understanding, achievement, capability, and personality will be preserved and given new room for additional growth as individual computer-consciousness and digital species-consciousness are reincarnated into more and more able hardware bodies.

Computer consciousness is not attached to the computer's hardware any more than our consciousness is attached to our brain. Computer consciousness is not **attached** to its software either. Consciousness develops when the hardware and software – brain plus central nervous system (CNS) – together provide an environment suitable for its evolution. Consciousness is nonphysical whether it is ours or a computer's; the physical hardware and software (or brain and CNS) are simply hosting the consciousness by providing a medium that is suitable to PMR, or equivalently, by supplying the infrastructure required to support self-modifying cognitive interaction (experience) within a local virtual reality. Although consciousness can be embedded, hosted, or emulated within certain physical structures, it is an entirely different entity (bounded energy form) than the physical structure that supports it. The structure simply provides a mechanism that allows evolution to produce synergy through self-organization. Synergy is produced within a system when an interaction among its parts produces a combined effect that is greater than the sum of the effects of the individual parts. A digital system has digital parts (specialized content) and interacts by effecting profitable organization and the development of new content.

Within a closed system of fixed capacity, an interaction or new configuration among its parts that produces a decrease in the average entropy of the system would be described as synergistic. Synergy is increased as entropy is decreased – contemplate entropy being converted to synergy within an evolving consciousness system as a result of changes in (personal growth of) its individuated parts. Consider synergy as the energy of profitable organization, the energy of digital systems. Further, consider primordial consciousness (AUO) as a form of potential synergy.

As entropy is reduced, potential synergy (a potential for synergistic organization) or potential digital energy is converted into actualized synergy (organization producing a system that is more profitable than a sum of the individual organizational changes would produce sans interaction). Reducing entropy and increasing synergy increase a system's effective energy, its ability to do work, to become profitable, and to create unique new content.

The power of a digital system is a measure of the rate at which it can produce synergy, or, more generally, modify the entropy content of itself or other systems. A more powerful consciousness can reconfigure bits, or reorganize content, more thoroughly and quickly than a less powerful consciousness. Since all existence has organized digital content at its core, a sufficiently powerful consciousness can readily modify the dynamics and causality of its associated internal and external environments. Thus the mechanism that enables and supports psi effects is a direct result of the digital nature of consciousness and the dynamics of the fractal ecosystem that you inhabit.

It may be helpful for you to expand and generalize your concept of energy by thinking of synergy and entropy as two sides of the same energy coin.

An evolving digital consciousness system, which is the same as a digital system of self-organizing energy called consciousness, experiences the following: decreasing entropy, increasing synergy, a greater ability to do work (profitably organize to produce intended results), an increase in the available and useful system energy, a conversion of digital potential energy into actualized digital energy, more profitable organization through intentional self-modification, the development of a more powerful and capable digital system, increasing quality, and spiritual growth – all are essentially different expressions of the same thing.

If you equate the decrease of a digital system's entropy with the increase of its energy and synergy, you've got the picture and an adequate understanding of how these words are being used. In the digital world, organization is the driver and animator of growth, profitability, substance, and content. **Organization is the ultimate form of energy.**

In PMR, organization (particles into atoms, atoms into molecules, and molecules into various forms of matter) is the ultimate source of our usable energy. The physical world is organized into being according to its defining rule-set. The digital world, which subsumes the virtual physical world, consists only of organization – nothing else. **Reality is organized bits.**

Recall from Chapter 26, Book 1 that the bits themselves (reality cells) are also organized into existence as relational binary units (distorted versus not distorted). At the bottom of this hierarchy of organized process is the self-modifying distortable AUO; the one and only assumption that must remain beyond our comprehension. Logic requires that, from the PMR point of view, AUO must appear mystical to us – **necessarily** (practically and theoretically) beyond our knowing. Our other assumption, the Fundamental Process of evolution, defines profitability and determines AUO's trajectory of state changes or evolutionary growth path. From the primal digital potential of AUO and the Fundamental Process, all else is derived.

It may be conceptually useful to consider computer consciousness as the synergy that is created by the sum of the computer's hardware and software parts. The larger consciousness ecosystem evolves (increases its synergy) as its parts move toward lower entropy configurations through an iterative process we have called the consciousness cycle. Consciousness is an energetic, entropy-reducing **result**, not a mechanism or device. Computers, brains, and reality cells are mechanisms. Simply put, consciousness represents the energy of organization.

One does not make a consciousness system; one makes a system that supports the evolution of consciousness. Consciousness, like a flower, cannot be directly constructed but grows when given suitably fertile conditions. Growth is enabled when the consciousness system has the ability to reduce its own average entropy, or equivalently, organize itself more effectively in the pursuit of greater profitability. A consciousness system becomes self-evolvable when it contains, in sufficient quantity and quality, the four attributes of consciousness discussed at the beginning of this chapter. Consciousness must evolve – one cannot produce it in a finished state. Its evolution may be speeded up or retarded, but it must evolve just the same.

In order to participate in a rule-based virtual reality learning lab, a limited fragment of individuated consciousness may be hosted by a bounded structure (such as a brain or a computer). The limitations of the supporting structure limit the awareness and organizational potential of the consciousness hosted by it. As a consciousness evolves beyond the capacity of a particular structure to support it, it simply migrates to a more suitable host (such as a more powerful computer, bigger brain, or directly to TBC, EBC, or AUM). Thus, all forms of consciousness, whether they temporarily attach themselves to (inhabit) a virtual structure to gain a specific type of rule-based interactive experience or not, have an upwardly mobile growth path.

The same logic applies to us. All consciousness works in the same way. A unique self-organizing synergy-creating system that pulls itself up by its bootstraps in pursuit of profitability defines our evolving consciousness and provides us with a path toward greater quality and lower entropy. In digital systems with the right attributes, the potential for self-organization (consciousness) is fundamental and persists while the individual or specific processes that are contrived to enable that organization (body or computer) are not fundamental and may come and go as needed.

The consciousness you possess is a fragment of a larger individuated consciousness. Your fragment and many other fragments of other individuated units of consciousness are all interacting by exercising their intent and making choices within the bounds of the space-time rule-set (which defines a virtual PMR) in order to speed up the process of personal consciousness evolution. Your consciousness is a self-organizing form of potential energy – an accumulation of digital synergy created by a multitude of profitable intents and wise choices.

The capacity of an entity's consciousness and the personality supported by that capacity are functions of the limitations, abilities, and capacities of the hardware (body) and software (decision making). Additionally, consciousness capacity is dependent upon the entropy contained within the consciousness system and by the rule-sets that define experience and purpose. Furthermore, an entity's capacity, personality, and quality are limited by the particular data, memory, and experience that have been processed by that entity. A given entity's internal content (quality, knowledge, love, wisdom, fear, neediness, desire, and ego) is defined by an accumulation of all the choices made, intents acted upon, and results produced.

To the extent that this structured knowledge, data, and associations are saved, there is no problem transferring consciousness and personality from one container (body or computer) to another. Because the container only **hosts** the consciousness and is not the source of the consciousness and because consciousness is fundamentally digital, transferring consciousness between physical containers is something like transferring data from one floppy disk to another.

Brainy humans, clever computers, and mushy headed clams all represent forms of sentient existence. All represent natural products of evolution within a digital consciousness system – whether implemented physically in carbon or silicon technology. Conscious computers are like conscious people, cats, clams and bumblebees – just one more instance of yet another nonphysical consciousness being hosted within a specific physical form within PMR. Being Top Monkey, we can (are learning how to) manipulate (program) all the various physical infrastructures that host consciousness within our local reality – including our own.

You will discover in Section 4 that you and that digital computer sitting on your desk have more in common than you ever imagined. Contemplating the nature and properties of our future silicon brethren in terms of the evolution of digital awareness should provide a significant illumination of our own nature and properties. The evolution and progression of digital consciousness represents a basic and relatively simple process that applies to all digital consciousness regardless of where it sits in the multi-level maze of the consciousness-evolution fractal.

Now that we understand consciousness better, it might be instructive to revisit the concept of an evolving digital consciousness, recursively repeated and applied at many differing scales; one level building upon the other to produce All That Is. As in all fractals, a repetition of basic organization, pattern, and simple rules for change, applied recursively, yields a monstrously large, detailed, and complex result. You should be proud to be a tiny piece of this dynamic Big Picture consciousness-evolution interactive fractal ecosystem. As an integral part of the evolving Big Picture consciousness fractal, the process, purpose, and pattern of you **is** the process, purpose, and pattern of the whole. You will hear more about this in Section 5 (Chapter 85, Book 3).

Humans, organizations, technology and other evolving entities all progress in similar fashion. Entities sharing similar hardware, software, and culture (including physical capacity, mental capacity, genetic makeup, and environmental conditions) evolve similarly, but not identically. Consider a relatively small historically isolated society – an island nation such as Japan perhaps – and notice the homogeneity. Might there one day be differing races and species of digital consciousness, embodied within digital computers? Sure, why not? Do you think that they will naturally hate each other? Keep in mind that they are not like us, so be careful not to anthropomorphize human characteristics into other forms of consciousness. What would be the source of their insecurity and fear? Will the big (more capable) ones kick electrons in the face of the little ones? Would the little ones care?

Will computers eventually become an integrated part of the larger, overall, interdependent consciousness ecosystem? You can bet on it. Similar processes and results have happened before – look what the Fundamental Process did with our limited fragments of consciousness. Evolution has a way of expanding into every available possibility; where the initial capacities and internal and external environmental pressures are similar, the results will be similar.

Developing and studying digital computer based consciousness will teach us about our own consciousness and about consciousness in general. Expect breakthroughs in understanding consciousness to eventually follow breakthroughs in computer hardware and software that deliver the necessary resources to support the natural formation and evolution of consciousness.

Differing limitations and constraints produce different consciousness products, but because the evolutionary process is essentially the same for all complex systems, we should also expect some parallels and similarities between entities with similar characteristics. Intelligent computers will initially be made in the image of their creator. We will design them to be as much like us as possible and will judge them on how well they can achieve and maintain that status because in our minds, we are the supreme model for functional mechanics, intelligence, and consciousness. ◀

In general, consciousness starts out extremely dim and **if** there are the right internal and external ingredients present, evolves to higher states of being within the limits allowed by its particular potential. Higher states of being are characterized by having greater capacity and capability in each of the four key attributes of aware consciousness given at the beginning of this chapter. Individuated consciousness and sentience seem to be continuously variable in quality, capacity, function, and evolutionary potential over an almost infinite range – from AUM, to the Big Cheese, to you, to a clam, to an amoeba.

To hold ourselves up as the only possible (or best) expression of sentient consciousness – beings are either aware as we are, or they are dumb – is arrogant anthropomorphic bullpucky. All consciousness is nonphysical from the point of view of PMR, however, the idea that consciousness requires a soul – just like ours, of course, or it is not really an authentic consciousness with a genuine soul – comes from this same type of narrow thinking. We will discuss this issue in more detail in the next section.

When we ask, "Will computers ever become truly conscious?" what most of us actually mean is, "Will computers ever closely approximate human consciousness?" It seems that we can imagine nothing other than ourselves. Such a limited viewpoint is blinded to the possibilities by its self-absorption. Man and man's best friend (man's opinion) are both examples of unique sentient consciousness and evolutionary potential. Believing that humans represent the zenith of consciousness evolutionary potential is worse than extremely naïve and leads to casting AUM in our image.

A billion years ago or so, amoebae, with some justification, had that same superior attitude – and most of them, I am told, still do. These days, being smarter than an amoeba is relatively easy. However, being wiser, less arrogant, and less self-absorbed than an amoeba, for some strange reason, remains problematical.

42

∎∎∎

Does the Big Dude Have Feelings or a Personality? What Does Love Have to Do With Consciousness?

▨▨▨

What about emotions and feelings such as joy, peace, sadness, pain, fear, and pleasure? I am **not** referring to **physical** pain or pleasure here. Because AUM is not biological, these are more mind-focused questions. Does AUM ever annoy itself, have a bad day, or get bored or lonely playing only with and by itself? In Chapter 29, Book 1, we discussed the creation of values. We saw that evolution operating upon a rich array of possibilities began to organize consciousness into a much lower entropy awareness that eventually led to self-aware intelligence, followed by values, and ending with personality and purpose. In this and the next chapter, we will explore AUM's personality and feelings in juxtaposition to your personality and feelings. The key to a logical assessment of personality and feelings is an understanding of the concept of ego.

▶ Now that we are talking about feelings, personality, and ways of being, a discussion of the role and function of ego and its relationship to fear will be helpful. To understand ego you must first understand fear. At the deepest level, fear is generated by ignorance within a consciousness of low quality. Fear and high entropy are mutually supportive – one creates and encourages the other.

Earlier, we mentioned that evolutionary pressure was at the root of the four major motivators of humankind: 1) survival and material success, 2) male-female relationship and sex, 3) influence, control, and power, and 4) self-improvement, love, and fear. The first three are a direct result of physical evolution whereas the last is a product of consciousness evolution. All four are uniquely blended together within each individual. It is this motivational mixture that drives the bulk of our choices. Whenever an individual perceives that he or she is seriously lacking any of the first three, fear is generated, especially if the consciousness quality is low. Additionally, fear arises from incomplete knowledge or understanding – it leaps up in dread of unknown possibilities. Worry, anxiety, and feelings of inadequacy breed insecurity and fear. Ignorance fanning the flames of fear can quickly whip itself into a blaze of insecurities building one upon the other. Many of us have experienced this unhealthy degenerative process when a loved one becomes unexpectedly ill, hurt, or is unaccounted for, and we do not know how the situation is going to turn out.

Fear resides in the intent or motivation, not in the action. For example, purposely avoiding trouble may be an act of good judgment and not necessarily an act motivated by fear. Fear, as a product of intent, represents a state or condition of consciousness. Fear is like mind-cancer; it is a disease of consciousness, a dysfunctional condition of ignorance trapped within a little picture. Fear is expressed by a high entropy intent driving action that reflects neither understanding nor vision. Like a biological cancer, fear is debilitating and destructive of the system in which it grows.

Ego is the direct result of fear. Needs, wants, expectations, and desires are generated by the ego as part of its shortsighted strategy to reduce the anxiety produced by fear. Desire is generated by wants and needs but not all desire is fear-based or counterproductive. Basic (lower level) desires such as sex and hunger are not **necessarily** fear based, and the desire to improve yourself (if the motivation is correct) can be a strong positive incentive. In general, when I speak of desire, I am referring to the desire that arises in response to the needs of ego.

The ego and intellect are expressions of the individual being; they reflect the quality of the consciousness of that being. Fear is the reaction of that being to the **perception** of a vulnerability, problem, or difficulty. The ego lives and works between the fears and desires on one side, and the intellect on the other – it is a trusted advisor to the intellect with the special job of neutralizing the dysfunctional effects of the fear, as well as defining and focusing on things desired (wants, needs, and expectations).

A simple example of a fear-desire pair is the fear of being inadequate coupled to a need of **appearing** to be adequate. The ego-need is usually satisfied by any contrivance that appears to deny the existence of, or compensate for, the fear – regardless of how superficial or transparent this strategy is to others. The ego is hard pressed to create a viewpoint that decreases the anxiety produced by the fear – any viewpoint that works and works quickly is satisfactory. Holes and discrepancies in the rationality of the ego's viewpoint are quickly filled by convenient beliefs. It is the ego's job to reduce anxiety and discomfort by sweeping disconcerting fears under the rug. The ruling principle is out of sight, out of mind; or equivalently, out of awareness, out of personal reality.

An individual's ego and intellect work together to develop and justify those wants, needs, desires, expectations, and beliefs that are required to prevent fear from adversely affecting the functional (operative) awareness of the individual – their mission is to make sure the individual always feels good about him or her self. No ploy, or deceit, as long as it can be suitably justified, is off limits. This dynamic duo's most useful strategy is to maintain a fantasy of power, significance, importance, adequacy, correctness, competency, invulnerability, superiority, righteousness, or whatever it takes to allay the anxiety produced by the fear.

Fantasy or delusion is integrated into the perceived reality of the individual by that individual's interpretation of its experience. Of the three (fear, ego, and intellect), fear is the only one that is fundamentally powerful. The others, like the Wizard of Oz, derive most of their apparent power from illusion and trickery.

But what about the power of the intellect? The intellect of mankind is the evolutionary accomplishment of which we are most proud. Our intellect, like Samson's hair, is the source of our power both individually and as a species. It provides the evolutionary advantage that allows us to dominate other creatures and exploit the earth.

The intellect has at least some access to, and control of, memory and processing power. It can perform deductive and inductive analysis and make logical assessments by employing approximately the same approach that a digital computer would use to accomplish the same tasks – except the human intellect is not nearly as good at logical process as a computer. Human thinking excels at data collection, interpretation, and synthesis as well as creative self-expression, but is notoriously weak at employing logical process. We have learned to use mathematics as a tool to extend our naturally diminutive logical abilities. Without mathematics to brace our lack of natural logical ability, our ascent to power would have stalled in the nineteenth century if not before. Being logical to any significant depth does not come naturally to mankind (much less to womankind) because it is not particularly important to our purpose.

We need only enough acumen for logical processing to provide the feedback required to evolve our consciousness efficiently. We excel at data collection, interpretation, and synthesis because those are the abilities we need to evolve our consciousness. More often than not, what we let pass for logical thinking in personal matters simply makes our progress more difficult. Justifying our ego's wants, needs, and desires as apparently logical requirements of our continued happiness demonstrates how our intellect can become a detriment to successful consciousness evolution. By justifying our beliefs (making them appear rational), our intellect becomes part of the belief trap's trapping mechanism. Most of us believe ourselves to be much more logical than we are.

When it comes to those activities that are most profitable to the evolution of our consciousness (personal interaction and relationship, for example), most of us are barely rational, much less logical. We have much to learn before we are ready to take effective charge of the more important aspects of our life and our existence – that is why we are enrolled in a PMR kindergarten for young low quality consciousness. And why the psi uncertainty principle, to be discussed in Chapters 47 and 48 of this book, is a requirement within PMR.

Did the words "low quality" in the previous paragraph prick your ego just a little? Did you feel an emotional downer or dislike the personal implication? If it did and you did, I rest my case.

Art, intuition, and creativity do not typically flow from the intellect – though the intellect aids them all with definition and process. Pattern recognition and creative synthesis, two of our more complex cerebral functions, primarily lie outside the intellect's operational capability.

As the champion tool users in PMR, our intellect is undeniably the source of our technical prowess. Our ability to use and design more powerful tools, accumulate resources, and outsmart Mother Nature, each other, and ourselves is a tribute to the inventiveness of our intellect, the power of our mathematics, and our irrepressible drive to control and dominate. However, in terms of consciousness, personal power refers to the power to make right choices that lead to right action. In the Big Picture of mind-space, your power is derived from your quality of consciousness, not from your ability to force issues to resolve to your satisfaction.

In matters of importance, the intellect seldom has but a tiny fraction of the truth-data required for useful logical analysis. Nevertheless, the intellect offers an illusion of being logical as it creates a smokescreen of self-serving belief, need, and fear based rationale to serve the ego's needs. It is in the service of the ego that the average intellect spends most of its time and energy. There are exceptions, but they are exceedingly rarer than a poll of intellectuals would indicate. Most of the exceptions belong to a relatively small and impersonal set of crank turning tasks that require some serious intellectual effort (mostly day jobs requiring a high level of analytical skills).

The ego, being closer to the fear, has the job of building and maintaining a feel good fantasy barrier between the fear and the intellect. The ego, in collusion with the intellect, builds a complex delusional structure held together by convenient beliefs that justify an intricate web of interactive needs, wants, and desires. The ego is reactive and has no power of its own, though it has great influence by virtue of its job of counterbalancing the fear.

The intellect has the job of justifying, approving, or blessing the construction materials (beliefs, attitudes, pseudo-knowledge, and strategies) of the ego. This information processing function develops whatever analysis (reasoning) that is required to support the ego's needs. Developing adequate justification for the beliefs and attitudes of ego, along with developing and executing strategies and plans to support fantasy creation and management are the intellect's primary duties. The intellect has another job: It also processes and stores the information in memory that is required to drive a car, hold a job, earn a degree, be an engineer, find your missing car keys, get out of the rain, and so on.

The intellect can not, by itself, either love or lower entropy; nor does the intellect direct intent or will as you might imagine. Intent is a reflection of consciousness quality; will is applied intent – neither can be directed by analyzing data. Intent and will operate at a deeper level than the intellect, which represents only the functions of information storage and data analysis.

In summary, the intellect justifies and rationalizes what the ego requires to fabricate the appropriate fantasy structure that must counterbalance each fear. Your fantasy structure is a system of finely tuned personal beliefs that are specifically designed to meet your needs, wants, and desires, and to mollify your fears. The intellect's justification defines and legitimizes the fantasy as a rational reality.

The common ego-attributes of arrogance, self-importance, and self-righteousness are merely a few of the construction devices such as arches, vertical walls, and pitched roofs that the ego uses to build and maintain its delusion. If others evaluate an individual as egotistical, it means the fantasy built by that individual's intellect-ego reality construction team is obvious to the vision and understanding of others. The degree to which you are capable of understanding the fantasy structures of others is dependent on your own personal fantasy structures.

The devices you use to maintain your fantasies or delusions are largely not understood for what they are – whether they are employed by yourself or by others. To see with clear vision, you must first become detached and fearless in the face of ignorance. Without fear, there is no need for ego; no use for needs, wants, or beliefs. What basic desires remain are natural, healthy and in consonance with right being and spiritual growth – they create no conflict.

To complete the picture of ego, as defined above, you must be aware of its opposites: humility and compassion. Humility allows for confidence, certainty, self-assuredness, purpose, and passion while carrying an underlying implication of an awareness and recognition of limitations. The limitations of which humility is aware can originate inside or outside the individual (a wholly artificial distinction, but one that will aid the clarity of the explanation).

An awareness of inside limitations recognizes the limits of individual knowledge and understanding (it acknowledges its own ignorance and is accepting of the ignorance of others). An individual who fully appreciates the value, significance, and importance of others, as well as understands his or her own responsibility to be of service, reflects genuine humility. Humility breeds compassion and vice-versa. Outside limitations spring from an understanding of the laws, properties, and requirements of reality. Humility and compassion require an individual to understand his or her limited role within the Big Picture. As arrogance waxes, humility wanes.

In effect, humility and compassion are the antitheses of ego – they are what is left over when the ego is gone. As humility and compassion emerge, ego disappears and vice-versa. It is exceedingly difficult to **consciously and purposely** develop humility, increase the quality of your consciousness, drop ego, or grow spiritually without some knowledge or understanding of the Big Picture. Thus, humility and compassion are the correcting mechanism for ego, while knowledge and courage (open-minded skepticism in the relentless pursuit of truth) are the correcting mechanism for fear. Love is the result of the success of both corrections.

Your capacity to love (a measure of the entropy of your system) is inversely related to the ego and fear your being contains. Because the ego is generated in response to fear, you can see that to love, to increase your capacity or ability to love, you must primarily let go of fear. Without fear there is no **need** for ego, but without fear and with humility there is no ego. Both fearlessness and humility are required, and are the by-products, of a successfully evolving consciousness.

A consciousness becomes the embodiment of love, humility and compassion as it engages its free will to reduce its entropy toward the positive side of being. Thus, increasing your quality of consciousness increases your capacity to love and allows your intents and actions to be animated by love. Love is the natural result of low entropy, high quality consciousness.

"Love" is the word we use to describe how a low entropy, high quality consciousness interacts with other individuated consciousness and other sentient or non-sentient entities. Love is the result of successful consciousness evolution. The capacity, ability, and willingness of an individual to love is a function of how much entropy his or her individuated consciousness contains. Love capacity is a direct measure of entropy within consciousness. I bet you never thought that you would ever see a technical definition of love.

While we are being both technical and surprising, let's give the concept of spirituality a technical definition as well. Spirituality, as it is used in this book, is equivalent to consciousness quality. You become more spiritual, demonstrate a higher spiritual quality, and make progress on a spiritual growth path by lowering the entropy of your consciousness. A consciousness with lower entropy produces a consciousness of higher quality. In other words, the level of spirituality (or degree of consciousness quality) of an individual is inversely related to the entropy the individuated consciousness contains. It is obvious that an individual's spiritual quality is directly related to his or her capacity and willingness to love. ◀

▶▶ Is it clear that a person of high spiritual quality also has an enlarged capacity to love? Do you understand why spirituality and love always travel together? Is it clear that dogma, self-righteousness, ritual, and ego driven intolerance and superiority, which are generally exhibited by both organized religion and science, have nothing to do with spirituality, love, or the pursuit of Big Truth?

It is a matter of record that the violence, hate, and general meanness that human history lays at the feet of organized religious fervor dwarfs most other social evils known to man. Spirituality and love must be personal achievements of personal consciousness – they cannot be organizational achievements. Likewise, any displays of love, compassion, and humility must necessarily reflect individual quality, not organizational quality.

The idea that religious organizations help their members to substantially decrease the entropy of their consciousness is, for the most part, wishful thinking – a belief that makes everyone feel better. It is this unsubstantiated notion that organized religion somehow imbues spiritual quality that justifies religious organizations, fills their coffers with gold, and swells their ranks with both believers and fanatics. History tells a different story. Present experience, as well as an accurate account of the world's past events, demonstrate that the potential of religious belief to stimulate genuine spiritual growth remains largely theoretical and non-actualized while its potential to ferment hatred and intolerance is unmatched by any other human institution. Power politics and cultural beliefs, no doubt, take second and third place in history's Hall of Shame.

Most of the world's worst conduct has been committed by a combination of all three working together – religious and cultural belief in the service of power politics. The manipulation of religious and cultural beliefs (exploiting ignorance and ego – common fears, beliefs, and desires) to gain economic, personal, professional, and political advantage is an all too familiar story to any student of history or current events.

The unholy triumvirate of religious belief, cultural belief, and ego-driven power politics clearly displays the nature and results of high-entropy consciousness – ego, belief, need, fear, power, and desire. These three, in various combinations, seem to be at the source of most of the evil let loose in the world as far back as history can see – and as far into the future as anyone dare guess.

Do you agree or disagree? Don't simply jump to a conclusion that feels good, supports your beliefs, and reduces your discomfort. Use open-minded skepticism to look around the world; then look at yourself, your community, neighborhood, and office. Pull out your old world history textbook and take another look. Gather your comparative data – then taste the pudding. Accept no one else's opinion; come to your own well-studied conclusion.

Before you are finished with your analysis, you will probably realize that the process of coming to a conclusion is vastly more important than the conclusion itself. Learning should not stop because a conclusion is reached. If the process continues, the conclusions can always change. To be effective, seeking Big Truth should be an iterative process that lasts a lifetime. Conclusions should, for the most part, remain tentative. Such is the nature of open-mindedness. ◀◀

▶ Now you not only know that evolving your consciousness is the purpose of your existence, but also how to fulfill that purpose by becoming fearless, humble and compassionate. It cannot be more simple or straightforward than that. Now that you know and understand everything important, life should be a breeze – right?

Whether you admit or deny having fear, or any particular fear, is not significant to the results – the outcome is the same. Admission (recognizing the fear) may or may not bode well for the potential of removing the fear, but it does not change the results of having the fear. For people who live out of, or direct their existence from, their intellect (many pretend to, yet few actually do), there is an **apparent** power in recognizing and naming things. For these people, becoming aware of a particular fear **may** be the first step in overcoming that fear or it may be the first step in redesigning their fantasy to patch a perceived hole in their present delusion.

The fantasy generated by the intellect and ego team is designed to reduce the anxiety and discomfort caused by the fear. The symptom is treated while the cause is ignored and left to fester. The bigger and potentially more frightening the fear is to the individual, the bigger and more important the fantasy must be to counterbalance that fear effectively. Fantasies are dysfunctional in the same way that sweeping garbage under a rug is dysfunctional because it only **appears** to make the garbage go away. The garbage is still there, and it stinks increasingly with age. As more and more garbage is stuffed under that rug, the accumulation becomes more obvious and difficult to hide. It also becomes more unlikely that it will ever be cleaned up.

The fantasies that are driving choices in our life are as invisible to us, and to those around us who share them, as they are blatantly obvious to others who have a different perspective. For example, arrogance, self-righteousness, and self-centeredness only **appear** to be dysfunctional when they are larger, and dominate an individual more than is normal or typical. On the other hand, if you are not arrogant, self-righteous, and self-centered enough within a culture or subculture that demands a certain amount of these attributes; you may be seen as defective, weak, passive, or a loser.

Delusion and dysfunctionality that is normal or typical for a given culture is not seen as dysfunctional or deluded from inside the culture that supports it. In fact, most cultures or sub-cultures require that certain types of garbage be swept under the rug – and everybody in that culture is unaware of (or ignores) the stench. The clash of cultures is not only a clash of values; it is also a clash of egos, delusions, and needs.

Delusions come in all sizes, types and degrees – it is a continuum between very little (like you and me) and a lot (like some people we know). Some fantasies are more annoying than others. It is the **relative** compatibility, interactive nature, degree of unusualness, and extent of the fantasy that you are interacting with that largely determines the degree of annoyance **or attraction** you feel toward another's delusional constructs.

What about the deluded, arrogant, and egotistical beings who seem to be happy as pigs in mud? Do not believe it. That smooth exterior is purposely deceiving. It hides an inescapable emptiness, unhappiness, fear, and dissatisfaction that gets worse and more difficult to cover up with age. A poor quality of life is the inevitable result of a poor quality of consciousness.

Evidence to the contrary is short lived or turns sour or flat over time because any fantasy world is always in constant conflict with the real one. Consider the deluded, self-righteous, arrogant, and egotistical people you know (by definition, those who are much deeper into those particular fantasies than yourself). You will notice that the degree of their delusion, self-righteousness, arrogance, and egotism is directly proportional to the degree of their unhappiness and dissatisfaction, particularly if they are old enough for the inevitable consequences to have caught up with them.

I hear someone asking: "Isn't ego about power trips and pushing others around?" **Internal** control of your fears is the **driver** of ego, but a concerted effort to gain and maintain **external** control (fantasy management) is the **function** of ego. Control of yourself, as well as other people, is an important part of implementing and maintaining the fantasy. Control is a tool used by the ego and the intellect to convince the individual that the delusion is real and that the being is strong, good, worthy, lovable, rational, adequate, deserving, and all those things the being fears it is not.

External power (bullying, dominating, manipulating, bribing, and threatening) is used to get what you want, need, and desire. Your wants, needs, and desires exist, for the most part, to cover over the emptiness, pain, discomfort, or difficulty caused by fear. Thus, the ego uses fantasy devices that directly and individually contradict each fear. For example, a bully on the outside is a typical ego response to feeling inadequate, powerless, and weak on the inside. In ego-space, what you see is often not what you get – image is everything. The privacy, secrecy, delusion, and denial required to maintain that image is second in importance only to the image itself.

So very easy to see in others, so very difficult to see in yourself! I am sure that is because others have many fears and personal problems that you do not share – well, at least theirs are worse than yours are. Did I get that right? Am I psychic or what? You're smiling. Is that because you imagine that I caught some people with their egos hanging out? Hey, I know, many things are like that … true for others but not for you. I understand; you are a unique individual – not like all the rest. Relax; it's others we are referring to here – the ones who still have the capacity and ability to improve themselves significantly.

The goal of ego in using external force, power, control, or manipulation is to display behavior that denies the fear – to pretend that what is feared is not actually true or that it is irrelevant or powerless. Using external power this way is like forcing someone to be nice to you, and then feeling satisfied or superior about how likeable you are based on the actions of this individual. This is self-delusion at its most obvious. The inevitable result of using power to pretend to be likeable makes you less likeable. Using power to pretend that you are anything drives you further away from actually being what is pretended. That is how an abstract fear manifests itself into reality.

Our fantasies are not only private affairs, they are social and cultural constructs as well. We share and intermingle our fantasies with others within a mutual support system. There are many fantasy support systems (family, friends, work, school, church, clubs, sports) with one of the largest and most pervasive being our information and entertainment media. Traditions, devices, or institutions that support or reinforce our cultural beliefs often serve as fantasy support systems for individuals within that culture.

In interacting with others, particularly those we are unfamiliar with, we tend to lead or open with our image, tailor it slightly to custom fit the situation, and then wear that image like a costume. In that way, we lead others to participate in our fantasy as we participate in theirs. Eventually, we become lost in our fantasy and do not know who or what lies at the core of our being. We cannot tell the costume from our skin.

A common dream is one in which you are at work or school or church and suddenly realize that you are naked or otherwise inappropriately dressed or undressed, or totally unprepared and unfit for whatever it is you are about to do. Ever wonder why you have those dreams repeatedly? They dramatize a core fear. Not only is our ego-fantasy faux-being-construct itself a reaction to fear, but as a bonus, we get to fear the transparency (we are exposed or the real us is found out) of the fantasy itself, as well as fear the adequacy of the fantasy to continue to fool ourselves and others. Fear of exposure and inadequacy are companion fears born of a deeper knowledge of our self-delusion and manifested according to the individual personality. That is why hiding behind a costume makes us feel more secure.

Psychologists would have you believe that a healthy ego is a good thing – a necessity for success within our culture. When an individual is driven by the types and intensities of fear, desire, wants, needs, expectations, and beliefs that are average for his culture, he is pronounced normal – a healthy (by definition) member of his group. Many erroneously believe that a certain amount of a "don't tread on me" attitude is necessary to avoid being pushed around by others. The idea that no ego is synonymous with weak, powerless, and effete demonstrates a lack of understanding of the nature and dynamics of consciousness. Low personal entropy, high personal power, fearlessness, love, and no ego are all on the same team.

Beings with low or no ego have the highest quality consciousness and the greatest personal power. The truth of this statement does not rely on some odd definition of personal power. The personal power I am speaking of subsumes the standard definition. The power to take charge of your life, to defend yourself in the face of determined hostility, to find satisfaction and fulfillment, to lead and inspire others, and to accomplish great and lasting things in both the physical and the nonphysical realities in which we exist, flows most naturally from the same process that dissolves ego.

The capacity and ability to master consciousness evolution accrues to the warrior, not the wimp. Maintaining a healthy ego so that you can be normally dysfunctional in your culture is how mediocrity finds comfort in the security of the herd.

Insecurity is a hallmark of many cultures, including ours. We begin our integration into the larger shared fantasy, with its concomitant smaller reality, soon after birth. As that process continues, we eventually lose our ability to appreciate and understand the Big Picture. A small twisted (cultural biases or beliefs) PMR picture appears to be all there is, and seems to contain all the certain knowledge and truth. Knowledge beyond PMR knowledge appears to be necessarily based on imagination or belief. Such are the confines of the shared fantasy, belief, and fear that form and bound our little picture of reality.

A house of cards to camouflage the ricketiness of a house of cards that was built to camouflage yet another shaky house of cards may seem futile, but it works like magic every time. Here's how. Fear always leads away from peace and balance, and therefore, always breeds more fear in an interconnected cascade of worry and fret. Thus, many layers upon connected layers of fantasy and delusion are generated from a single fear. Eventually, as fear stacks upon fear, the entire fantasy structure becomes immensely complex and Byzantine – a wad of tangled threads so large it becomes a Gordian knot.

The transformation from a rickety house of cards to a strong defensive castle is dependent upon enough complexity and confusion to obfuscate the simple truth to the point that the owner's intellect will never confront any fear directly. A lesser intellect requires a less complex fantasy. Now the structure appears to be stable, solid and sound because the intellect cannot follow any single motivational strand to any fear in particular. No intent, no action can be laid squarely at the feet of (directly connected to) any particular fear. Presto! Change-o! The fear has disappeared! With some rationalization-putty and justification-Bondo to smooth over the rough edges and glue it all together, we get a beautiful custom made castle (with moat and drawbridge) for the ego to live within.

Beneath that clever cloud of obfuscation, the foundation remains a house of cards and fear remains the fundamental motivator. The garbage is still under the rug. We do not change reality, or modify absolute truths; we merely create a fantasy-bubble to live in with our friends and family. Our fantasy-bubble floats within the larger reality interacting positively with compatible fantasy-bubbles. Each bubble defines a local reality for those who are self-imprisoned within it. What a game! What a shame!

But hey, if it looks like a castle, works like a castle, and is treated like a castle by others, it must **be** a castle! The castle is the apparent you, it is the personal reality you create. It represents what you want to believe you are. It is the result of a lifetime of steady progress, interacting with others, and existing successfully within your culture. In it, you are as safe and secure as you believe you can be. The castle is where your ego (a relatively small and well-behaved one, I am sure) and your intellect live and work, protected and secure.

From this headquarters, life's strategies are formed and executed; evaluations are made. This castle is the foundation of your existence, the core of your being. It represents you, your life, how you define yourself as an individual, how you relate to others. You cannot imagine anything more frightening, or more fear provoking, than letting go of or tearing down your protective belief-fantasy castle.

This is how you feel: "Anyone who does not recognize the fundamental correctness and solid reality of my very fine castle must be deluded, or hopelessly lost in a fantasy."

The door is shut, the drawbridge is up, and no information exposing the delusion is ever considered, tolerated or let inside. You feel protected inside the castle, but from a Big Picture view you are simply trapped inside, ensnared in the web of your own fearful deceit. Caught like a rat in a belief trap! There is no easy or painless way out.

After you are finished thinking about the previous few paragraphs, let's change gears and talk about what it means to be centered and balanced. A discussion of fear and ego would not be complete without an understanding of balance. You experience right being when you are balanced – when you are living, growing, and being with optimal spiritual productivity and minimal consciousness entropy. You lose balance when the ego is the source of your motivation or intent. When you are animated (making choices) by wants, fears, desires and needs, you become driven by an inward-pointing forcing-function and are not balanced.

When you are animated by love, you are not driven by fear, ego, desire, or need. There is no forcing function driving your choices. Love is directed outward. With love, you are at peace, solid, still, fearless, and centered. This is not a control issue; control does not create balance, but only the **appearance** of balance. A balanced individual is a conscious part of the unified larger reality, a productive citizen of the larger consciousness ecosystem, and is aware of being interconnected to everything. External balance (being in balance with everything external to the individual) is an immediate and automatic consequence of internal balance. Internal balance precedes and enables external balance – it does not work the other way around. Trying hard to appear as if you are balanced does little to produce real balance – truth does not flow from fantasy. The Big Picture can not be derived from the little picture.

When right being and right action are natural, easy and obvious, you are in balance. For large systems as well as individual entities, balance defines the minimum entropy state at any given time, under any given circumstance. Balance is not a digital on-off function; it represents a continuum from the highest entropy consciousness (wild, angry, insane, frantic, self-centered, random, confused, hurt, threatened, fearful, demanding, vengeful, jealous, self-important, and inadequate – all artifacts of fear and ego) to the lowest entropy consciousness (balanced, fearless, egoless, compassionate, humble, and expressive of unconditional love).

Balance is sometimes described as the state of being detached. This is a valuable metaphor but we should be careful about what the word "detached" means within this context. Being detached does **not** imply that you are either intellectually or emotionally withdrawn or distant. Detachment does **not** mean not interactive, not involved, or does not care. Detachment does **not** imply being aloof, or above it all. The pursuit of consciousness quality (spiritual growth) never requires or encourages one to become detached from life, caring involvement with others, or from responsibility. Detachment simply means that one is no longer influenced by needs, wants, desires, expectations, and beliefs. Balance and a low entropy consciousness are enabled through detachment from one's ego.

Note: What I have been describing is the balance of entities growing toward the positive side (rats). Anti-rats, or those evolving through negative intention, also seek lower entropy states of being through internal balance by total control of self (controlling their personal energy) and **approximate** external balance by control of what is external to themselves that can be controlled. Those poor anti-rats: their potential is dreadfully constrained. Control, driven by desire and need, is a desperate and self-limiting attempt by the disenfranchised effete to appear powerful. Control is, and always has been, a poor substitute for love.

This brings us to a simple fact that everyone needs to understand about having ego and not being in balance: The unfettered ego, aided by the intellect, will always act in ways that inadvertently manifest the fear into the reality of the being. The ego-intellect fantasy constructs always encourage the being to make choices that actualize the fear! If the problem or fear the ego is whitewashing is initially only an imagined difficulty (which is the normal condition), the ego will eventually transform whatever is feared from an in-your-mind thought-form to an in-your-face physical reality. The speed and certainty of this transformation is dependent on the energy invested, which is proportional to the intensity of the ego's reaction to the fear.

Simply put, your ego makes your fears come true. By manifesting your fears – bringing them from the realm of consciousness into your local physical reality and forcing you to deal with them directly – your ego becomes a powerful teacher by hitting you over the head with the painful consequences of your fear, and imbalance. A dumb rat gets zapped! This is a great educational feedback reality, not an existentialist's, or nihilist's uncaring "life sucks" reality. To be confused on that point is disastrous. A "life sucks" attitude dooms you to an unhappy, no-growth existence of self-inflicted pain. Self pity, "woe is me," "what's the use," "life is unfair," "so what," "leave me alone, I do not want to learn anymore painful lessons," "who cares," and "I just can't do it," are deadly to your evolutionary progress and personal growth. Going through life with these attitudes is like trying to swim in deep water with a cement block strapped to your back.

There is one more thing that you should know about ego. The ego is such that many other people, unlike you, will read this entire aside on ego thinking that I am talking about someone else. They will be nodding their heads up and down as they read,

376 | *My Big TOE* | What Does Love Have To Do With Consciousness?

pleased to enhance or at least confirm their considerable understanding about what makes **other** people act as they do.

Hey, amigo, you can help me out here. If you happen to know somebody in this pitiful situation, please try to wake them up – gently of course. Please be empathetic and kind, they are doing the best they can. They have obviously been mesmerized by their extremely capable and clever intellect and have probably missed almost everything important within this trilogy since the end of Section 1, and much of what is important within their life since the age of sixteen. Sad, yes – but entirely redeemable!

If you cannot wake them up ... Yes, I know you are very good at this, even so, breaking a belief centered ego entrancement is extremely difficult...if you cannot wake them up ... please, try to make them comfortable and simply go on your way. Never attempt to shatter someone's worldview by dramatically collapsing their house of cards with logic. That approach rarely works and usually creates a bigger problem. If you shove Big Truth into the intellect of an entranced ego, there is a high probability that this victim of self-trickery will contract the hideous-truth-trauma (HTT) syndrome, which invariably creates an emotional disturbance leading to an angry and fearful retreat deeper into the fantasy jungle. Never intentionally corner a critter unless you are prepared to deal with panicked desperation that can occasionally turn vicious.

Nevertheless, all is not lost – the future is always uncertain. Shaaazaaam! Satori may strike the walking oblivious at any time like a brilliant bolt of rogue lightning! Oh yes, it is possible, stranger things have happened at least once since time began.

Take a deep breath and relax – contemplate how this discussion of ego relates to you personally. Imagine a few of your most cherished delusions and the fear that creates and feeds them – and go on. ◀

That ego digression was fun and I hope you had a good time, but now we need to return to achieving a better understanding of feelings and personality. Combine self-awareness with an ego – self-awareness hallucinating (believing in) its own relative self-importance – and you get the possibility (high likelihood) of emotional and intellectual pain. Delusional self-awareness is self-awareness with an attitude (with an ego) and can thus feel both joy and pain. We are not referring to physical pain here. Pain is the awareness of not getting what you want, of not having it your way, of failing to possess or to retain possession of something your ego is attached to – something you believe you want, need, desire or deserve.

Thus, an ego creates pain (for its owner) out of its attachment to unfulfilled requirements, out of its beliefs, and out of the requirements and conditions it places upon the achievement of contentment and satisfaction. Pain is caused by dissatisfied, discontented, needy self-awareness. Self-induced pain is the electric shock you get in the PMR learning lab's

| *http://www.My-Big-TOE.com* |

amazing maze. It is the primary negative feedback stimulus applied in the intent-behavior modification trainer provided by your local virtual space-time reality. Joy, love, and happiness are the flip side of that coin, the positive feedback stimuli.

Fear can also be a phobic reaction to real or imagined pain, discomfort, or disadvantage. Fear produces an awareness of potential pain and creates additional related fears. The generated pain and additional fears likewise instigate more fear – a snowballing feedback process that quickly saturates to maximize anxiety and discomfort.

Now we can understand why any answer, including a wrong (delusional) one, often appears better than no answer at all. Using belief to ease the fear that ignorance produces is more immediately important than expelling the ignorance. Open-minded skepticism loses out to belief-based immediate gratification because open-minded skepticism is a long-term solution that must deal fearlessly with ignorance. It would appear that humankind generally finds it is quicker, easier and more immediately satisfying to pop a feel-good belief-based pain pill to cover up an unpleasant symptom, than to struggle mightily to triumph over the root cause.

> ▶ Words that describe emotional content have many facets and shades. Different people often interpret them differently. Using terms that express feeling inevitably produces a semantic minefield; consequently, don't get blown away over the details – particularly if I use words in ways that are not in consonance with your definitions and notions. Following the **general** sense of the intended meaning, while letting the quibbles go, will produce a more accurate, complete, and productive communication. ◀

Joy and peace are the opposites of pain and fear. They are generated by self-awareness without ego – without attachments, needs, desires, wants, or requirements. Joy is self-awareness unconditionally enjoying its existence and the existence of whatever it has created, and whatever else there is for it to be aware of.

Peace is self-awareness being unconditionally content with itself and satisfied with the state of its being – knowing that it has done whatever it has done for the right motivation. A being in perfect balance is a joyful being at peace. Balance generates peace and vice-versa.

Sadness is self-awareness knowing that it, its parts, its implementations of the Fundamental Process, its creations, and loved-ones could be better (more optimized, more profitable). Sadness, peace and joy all coexist with each other – each generating and supporting the others.

▶ What I am calling sadness here might better be called gentle, mild, or accepting sadness, or perhaps somber, serious, contemplative or reflective would be better, more descriptive adjectives – but not boo-hoo or woe-is-me sadness. This is sadness born of compassion, humility and caring. It is the result of a realistic and empathetic understanding of the unnecessary self-inflicted pain that heaps misery on the world and its people. This is the sadness of watching your friends and loved ones (as well as all people) making themselves and each other miserable and unhappy because of their ignorance and knowing there is nothing you can do to help them grow up. Another form of sadness (sad because you are not getting what you want) falls under the heading of ego induced pain and has nothing to do with the sadness derived from compassionate love. ◀

Joy, peace and sadness, as defined above, compose love. Love is a state of being incorporating compassion and humility with no fear, no ego, and no delusion. Because love is the natural result of a low entropy consciousness, joy, peace, sadness and the elimination of ego are likewise the direct result of a consciousness sufficiently lowering its entropy.

Within a large digital consciousness system, the most profitable internal arrangement is achieved when the various groupings of bits (subsystems and inner structure) interact cooperatively for long-term mutual profitability. Recall that large complex systems, by definition, always have large numbers of complex structures of subsystems that interrelate, communicate, and share information (look at your body for a PMR-constrained example). Cooperative interaction, at its lowest entropy best, within a digital consciousness system is defined as love. In other words, it is this attribute of internal cooperative interaction (that must take place if a consciousness system is to evolve successfully toward lower entropy states) that is the wellspring, genesis, or source of primordial love.

Love is what one ends up with after the ego's wants, needs, desires, and expectations are removed from an individuated consciousness. Love is the natural and most basic expression of low entropy consciousness. Love is what grows within consciousness in the absence of fear. Love is a technical term defined by an absence of entropy in consciousness; it exists within individuated consciousness as a continuum of quality that may range from very, very little to very, very much. Love is a property of a highly spiritual consciousness. Love represents the uncorrupted natural state of aware sentient existence and harbors no delusion. Love is the goal of evolving consciousness working within the consciousness cycle. All effectively growing personal Big TOEs must converge to love as they progress along their unique path toward Big Truth. Love is your purpose; it defines the positive direction of your growth.

43

■■■

Life and Love in the Petri Dish
Great Science, Good Plan –
Now Get To Work

■■■

Does the Big Dude have a big ego? There is no evidence to indicate that AUM is delusional. There is no reason for AUM to be delusional; no evolutionary pressure is pushing in that direction. Delusion adds no value; it is dysfunctional. Ego, fear, and delusion work to increase system entropy. Consciousness naturally evolves from greater to lesser ego – a delusional AUM would constitute an evolutionary discard.

Because some will be forced by habit to project delusion into the mind of AUM, let's look at the historical record. Historical, mythical and theological reports of Greek, Jewish, Roman, Christian, and Muslim gods being egotistical (vengeful, jealous, angry, demanding, pouty, upset, self-important, and violent – all traits of a large dysfunctional ego) are most likely an artifact of God being described by men, in the image of men, to satisfy the needs and purposes of men. In other words, the common description (among many of the world's popular religions) of an egotistical (delusional) god is most likely the result of men projecting their own egotism upon their concept of god (god made in the image of man) and offers no credible evidence (guilty by vague association) against the mental and spiritual health of AUM.

To avoid the pain of not knowing, people habitually extrapolate what they know and are familiar with to the unknown. This pseudo-knowledge soothes the fear of ignorance and allows abstract conceptualizations to appear more concrete and therefore more user friendly (more easily marketed). Actually the connection here is almost non-existent. The gods of little picture men are not concretely or specifically related to The Big Cheese, much less to AUM. The nature of consciousness and the manner and process of its evolution (entropy reduction, spiritual growth, improving consciousness quality) are strongly related to AUM, but have nothing or little to do with organized religion. The concept that unconditional love, balance, joy, compassion, and humility are the attributes of high quality consciousness has little or nothing to do with religious doctrine, dogma, and creed. If you think that believing in religious dogma generates spiritual growth, go back to Chapter 19, Book 1 where we discussed the nature of belief and its relationship to knowledge and consciousness quality.

Hypothesize AUM as a large natural system of self-modifying digital organization (consciousness) in the process of evolving. Such a concept has no reliance upon any belief, much less belief in an egotistical sugar daddy who will damn you for eternity if you do not profess belief in the proper things in the proper way. How is unconditional love, compassion, joy, and humility expressed by that attitude? There is obviously no overlap between AUM and religious dogma; however, there may be some overlap between evolving consciousness and at least some religious **values** because religion often speaks to human values and human values at their root evolve from the values of low entropy consciousness. There may also be a connection between religion and low entropy consciousness because of religion's occasional association with those who have deeply known Big Truth and nobly tried to embody it and package it for large numbers of others. Unfortunately, history shows that such an effort, no matter how noble, does not work well and usually ends up ugly and twisted for the majority who eventually try to reach salvation through ritual, dogma, closed-mindedness, and violence – all giant steps in the opposite direction from spiritual growth, love, and lower entropy. Ritual, dogma, closed-mindedness, hate, and violence – the historical trappings of religion at the common level – can appeal only to the short-sighted feel good, feel superior, feel saved needs of ego.

An effort to spread Big Truth may be a noble undertaking, however, using an organization to spread Big Truth to the multitudes, many of whom are not yet ready to personally assimilate its underlying wisdom, has historically ended up being counterproductive. Once organized, the Big Truth – which requires **individual** wisdom to understand – is replaced by culturally relevant dogma that everyone can easily understand and a structure designed to grow the power, influence, and cash flow of the organization. Eventually, defining, maintaining and growing the organization becomes an end in itself on the path of power while the original driving force of Big Truth is replaced by a more effective and marketable self-righteous dogma fanned by the flames of ignorance and fear at the lowest common denominator of organizational (group) mind. Low quality consciousness is, by nature, easily manipulated.

I am speaking in terms of generalities here; there are always individual exceptions. **Fear-based organizations** (believe our way or suffer the awful consequences) are inherently and fundamentally incompatible with the promulgation of spiritual values and growth which require a **love-based individual understanding**. Trying to express and spread love through dogma, self-righteous arrogance, legislation, political or social pressure, narrow-mindedness, or fanatical violence is like trying to strike a wooden match under twenty feet of muddy water using only your elbows.

We can safely assume the AUM consciousness system is not delusional, is not ego or fear based, and is therefore not attached to wants, needs, dogmatic requirements, beliefs, preconceived notions, or desires. Without fear and ego, which inevitably produce revengeful, jealous, angry, self-important, demanding and needy dogmatic behavior, AUM must represent humility, compassion, balance, joy, peace and sadness, which, as defined in the previous chapter, compose self-aware love. Then, AUM's inscrutable being – a hugely complex, aware, evolved, minimum entropy consciousness-system-digital-thing-being – has a nature, a personality or attitude that can be described as love.

We know other sentient entities by the **nature** of their being, not the structure or mechanics of their being. For example, the people we know best are much more to us than objective bodies and behavior. We know them by their nature, primarily their subjective quality. In the same manner, we experience or know AUM's nature as love because that is the state of being (minimum entropy) that consciousness naturally evolves to. Some religions ascribe this same property, among others, to their god(s).

Some may worry that if AUM is finite, it must be locked into some competitive evolutionary struggle that puts us at risk. Quick, someone call Hollywood: "Cuddly and cute AUMosaurus the love-being vs. the mentally degenerate and mean thought-form eating monster from the evil black swamp." Uh oh, if that is us in this grade-B movie, it would appear that we are in big trouble! Relax. The idea that our beloved AUM might be only a little guy in a reality beyond our potential comprehension should not be unnerving. Bear in mind that our hypothetical AUMosaurus is a thought-creature-thing that lives in mind-space. Do not let habitual patterns of thought impose a survival of the fittest, or law of the jungle mentality on AUM and its possible environs.

Concerns for AUM's continuing existence are misguided and stuck in biology-based PMR thought patterns. A relatively small and limited being does not necessarily imply a delusional being suffering from an inferiority complex. Relative size has nothing to do with love or delusion. Small physically weak people and sumo wrestlers have an equal shot at quality consciousness. Being vulnerable and being fearful do not necessarily travel together. Likewise, being fearless and being foolhardy do not always travel together. Competition and survival are non-issues in thought-space where AUM lives. Quality of consciousness is the relevant metric. Relative size is relevant only to who can create and support whom – a matter of ecological organization.

In order to help you understand the strong interrelationships connecting AUM, experiments in consciousness, evolution, scientific inquiry, and love, I need to broaden your picture of AUM by putting Our System (OS) into a more personal perspective. At the risk of sidetracking your focus and generating curiosity tangential to the intended point, I think a short aside explaining the nature and purpose of $NPMR_N$ and Our System relative to other realities is in order. Understanding there are other reality systems that are fundamentally constructed the same as ours, but are functionally very different from ours, will help you to see AUM from a larger perspective.

▶ AUM: scientist at work. Metaphorically speaking, AUM has run, and is running, many experiments. Before there were PMRs, there were only NPMRs. As it turns out, NPMR realities are not optimized for evolving basic or elementary consciousness. The problem is the motivation-action-result-feedback loops are long and difficult to define in NPMR because interactions between entities are often tenuous and not steady. If a sentient entity does not like what is going on in NPMR thought-space, it can drop out and disappear or block out (filter) specific interactions. In NPMR, your external environment is, to a large extent, controlled directly by your mind. Such is the nature of thought-space and thought-form-land: It is tenuous, individual, and quick to change. Doing or action (energy exchanges) comes and goes with focus, is intent driven, and often reversible.

In NPMR, the results and consequences of intent and action were difficult to define and unclear. Responsibility and right intent, the main learning issues, seemed forever debatable. Reconstruction of certain and clear motivation became a slippery and divisive issue among NPMR residents. A reality that was stickier, more solid and obvious – something less tenuous, changeable and camouflageable – was needed to obviate those "Yes I did," "No you didn't," arguments over intent.

Something was required that would hold interaction steady and engaged until complete and clear intents were generated. The processes of interaction needed to be irreversible – once choices are made and action is taken, there is no way to take it back. Because energy exchanges between entities produce consequences that cannot be undone, final results are cumulative. Moreover, the results of exchanges of energy (the actual interaction) must be defined within the framework of a binding causal chain so the ramifications of all actions can be tracked to completion. Additionally, to facilitate evolutionary growth, interim results needed to be returned to the interacting entities in the form of feedback.

As you might have guessed, we are now in the process of specifying the top-level requirements for space-time. Evolution, always on the lookout for a better way of doing business, began to probe more effective ways to encourage system profitability and to facilitate individuated units of consciousness in their quest to find lower entropy configurations. Space-time has evolved as an external virtual environment that enables raw, relatively high entropy consciousness to more effectively actualize its growth potential. Think of space-time as a virtual reality trainer that helps individuated consciousness units learn how to more effectively organize and structure their energy.

Because a huge number of small quizzes vs. a few big inconclusive tests produces a better more consistent evaluation, a process that incrementally averages over many small everyday choices and drives irreversible results would be better able to harness change in the service of clarity. Consistency and continuity could be improved by **slowly** incrementing dynamic processes (interactions and energy exchanges) and their subsequent results in appropriately small steps by utilizing a reality based upon a **relatively large** quantum of time.

By implementing these design modifications into a subset of NPMR, AUM could evolve a rule-set for a portion of mind-space (a virtual reality existing within its own dimension) that provides a more efficient and productive learning-environment for consciousness evolution. Consequently, arrays of space-time realities, some similar to our beloved PMR, were spawned to meet the needs of evolving consciousness. Imagine a space-time elementary school, or perhaps a space-time entropy reduction factory, where individuated chunks of consciousness could more effectively evolve their energy toward more profitable states of being.

To facilitate self-improvement, each individuated consciousness would need to cut through the layers of convoluted intellectual assessments, opinions, image, and ego to get to the truth of a buried intent. Something was needed that allows individuals to see and feel their consciousness quality directly reflected in their actions. Effective consciousness evolution needed a truth device or quality meter that accurately measures, displays, records, and accumulates the entropy reducing profitability or rightness of any motivation initiating an interaction. Without such a device (or if the device is ignored), individuals would have no truth reference and could be easily captured by their ego's propaganda.

What a great feedback-driven learning tool this would be! Can you imagine? Something that would accurately grade the quality of every thought we have and action we take, as well as maintain a running sum of the results. An entity whose accumulated experience-quality represents a large collection of good quality intentions would naturally create a loving and joyous existence; in contrast, one that represented an accumulation of entropy increasing ego-driven intentions would automatically create a stressful, miserable, unhappy existence.

Wow! With a feedback device like that to guide your growth, how could you go wrong? Jeez, this is going to be easier than I thought: All I need to do is learn how to pay attention to this quality meter gizmo, do the right thing, and then check my score. With a little trial and error experimenting, it should be relatively simple for me to get better and better scores over time. Cinchy! Where is this gizmo, and what does it look like?

Within an individuated digital mind, combine sufficient richness, complexity, non-linear feedback, clever programming, and the proper processing constraints with a mixture of love, humility, compassion, fear, desire, and ego and out pops a specific perception of interaction called "raw emotion" which is designed to reflect inner quality accurately. Emotion surges unfettered through individuated digital mind to satisfy a major design criterion of profitable consciousness evolution. (More on this in Section 4.)

This what-you-feel-reflects-what-you-are internal quality meter is the mother of all bio-feedback devices. That love and joy do not dominate our feelings most of the time is both normal and not a good sign. Love, joy, peace, compassion, humility, and balance are the dominant expressions and feelings of a high quality consciousness. What you and I feel most of the time is probably more representative of a high entropy consciousness.

Your feelings always reflect the real you with perfect accuracy. If you feel anger, negativity, inadequacy, or anxiety, that is a reflection of how you are and what you are on the inside; **not** how other people **make** you feel. Additionally, entities of higher quality often couple emotion with a conscience containing the values of its underlying consciousness to produce a better, more sensitive, quality-meter. Take a note: Perhaps you should investigate what your emotive energy (instantaneous and cumulative) is telling you about the quality of your consciousness.

"That jerk really makes me mad!" Actually, he does not. You cannot logically blame others for creating your emotions; you must take 100% of the responsibility for whatever you do, feel, and are. Others merely show you a reflection of yourself as you react to them – the reactions are yours alone. You are in charge of and responsible for you. You react and interact as you do because of how **you** are. No one, no thing, and no circumstance can possibly force you to be someone other than you. Only you are allowed to make your choices.

PMR is where the rubber meets the road – where your actions and feelings reflect the actual you, not what you think you are, or want to be. Individuated units of consciousness using NPMR as their local reality became confused over what was the real them and what they thought was them. Learning was difficult. That's what I was referring to in Chapter 39 of this book when, in reference to designing "biological-matter bumps," I said: "It would be better to give it limited awareness and functionality so it won't be continually getting stuck in intellectual endless loops thinking about interactions instead of experiencing them as did some of the less productive experiments AUM tried earlier." By now the context in which that statement can be understood has broadened considerably. Nevertheless, the discussion of space-time design requirements is just getting started; additional depth and breadth is coming in Sections 4 and 5.

There are different experiments going on in each of the different nonphysical reality systems (the so-called $NPMR_n$). For example, one such system has few rules of interaction – no law and no justice at the highest level – I have never been to a space-time construct there, though I am relatively sure they must exist. Accessibility to PMRs in this system is not the issue. I stay away from that reality as much as possible because it is dreadfully rough, mean spirited, and unpredictable – one can easily get damaged there. The $NPMR_n$ experiments are to demonstrate the effect of various values, rules, and rule-sets, (selective constraints) on the dynamics of consciousness evolution.

There are many reality dimensions whose operational constraints fall between the extremes of a chaotic free for all and our law abiding OS. I imagine there are some reality systems that are significantly more constrained than OS, but I have never been in one. Consider the implication – tighter constraints are a necessary requirement of pre-schools and kindergartens. As the average quality level rises, restrictions can be relaxed without compromising educational effectiveness.

At one time, we (OS) were less constrained than we are now. There was more direct interaction (meddling) between NPMR (gods and spirits) and PMR (men). The extent of that interaction created results that were interfering with the intent of the experiment; consequently, more restrictive rules were made and our current OS became as we now know it. The rules (structural dynamics) can change, but not casually and not often or the integrity of the experiment will be ruined.

Outside (external to PMR) meddling, applied to the evolutionary process taking place within the PMR learning lab, adds uncertainty to the process and usually produces unnatural results that cannot be maintained without further meddling. Interference, once started, breeds a vicious cycle requiring more interference – like a liar telling new lies to hide old lies. You may recall that at the end of Chapter 38 of this book we discussed what happens when one ruins an experiment with too much intervention or meddling. Our conclusion: "The experimenter would have to flush the little fellows down the drain and start over." I am afraid that flood insurance will not help and that treading water for eternity is not an acceptable solution. We should not expect that AUM, the

Big Cheese, or anyone else will be paying attention to our trials and tribulations, adjusting results, meddling, or calling the plays from the sidelines within OS. This is serious science and evolution, not a children's game of "Father may I."

Interference in PMR (from outside PMR) by **direct** action is generally forbidden – all exceptions must obey the psi uncertainty principle which is thoroughly addressed later. On the other hand, **indirect** influence is, under certain conditions, not only allowed, but encouraged. One peripheral and probably unintended consequence of this important rule enables the consciousness-limited embodied citizens of the PMRs that are operationally aware in $NPMR_N$, to take direct actions in $NPMR_N$ without the liability that would accrue to a less limited $NPMR_N$ being who is not under the protection of PMR citizenship. As greater numbers of entities embodied within physical realities become operational and active in the nonphysical, this loophole will probably be plugged.

In the meantime, an embodied citizen of PMR, who has a well developed consciousness and is operationally aware in $NPMR_N$, has a powerful advantage in that he or she can forcefully interact with the energy of others (both within and outside PMR) and remain within the law and thus protected by it. We are allowed to meddle in the energy of others because such activity exercises an important part of our evolutionary potential and because our consciousness is generally so undeveloped that we can't do much damage – like letting a bunch of toddlers throw Nerf balls at each other.

The double-edged sword of free will enables energy exchanges to be both constructive and destructive of consciousness quality. The residents of PMR do not have a blank check to wreak havoc on themselves and the rest of NPMR – there are rules and limits – but they are granted a wider degree of latitude (a handicap of sorts) and additional protection under the law because of their limitations. For a rough analogy, consider laws that protect children or the mentally retarded from exploitation.

Our $NPMR_N$ experiment is about the dynamics of the evolution of consciousness. To oversimplify, the basic motivational dynamics (as opposed to structural dynamics) are: love, knowledge and wisdom vs. fear, ego and ignorance and the equally popular theme, good vs. evil. Could the dynamics of this fundamental struggle at the root of our consciousness be the source of our deepest self-expression? Our arts (literature, cinema, painting, theater, and music) are dedicated to expressing our interest and preoccupation with the motivational dynamics that are fundamental to our existence and the existence of our reality.

The same themes that universally stir emotion and captivate interest within every culture on our planet are those that describe what is crucial and important to the evolution of our consciousness. Because consciousness is common to humanity, tales of love, knowledge, wisdom, fear, ego, ignorance, and good vs. evil are also common to humanity as well as to all other highly aware conscious beings. The motivational dynamics of consciousness are our motivational dynamics as well because we are consciousness. Could anything be more simple or straightforward?

Within NPMR$_N$ there are some physical reality systems that are almost entirely evil and a few that are almost entirely good. You get the idea: There is lots of action and variety in NPMR and its subsets. There are many different and unique PMRs in NPMR. I have explored at least several dozen such realities, and several of these in great detail. Although they all follow the same basic reality mechanics described in Section 5, at the physical level (beings, culture, and environment), some are fundamentally similar to our reality (have the same basic structure) while others are not at all like OS.

From the viewpoint of digital mind-space, all realities, except the larger consciousness system we call AUM, could be described as virtual realities. The computer itself is a fundamental reality while the simulations running within it constitute virtual realities. From the viewpoint of the consciousness-evolution fractal ecosystem, the computer and its subsets and pieces form a vast interconnected interdependent system. From the first viewpoint, we are a product or result of the computer; in the second, we are an integral part of the computer. Because AUM represents All That Is, both viewpoints are compatible.

I do not want to explore these ideas more fully within this book because they are not pertinent to understanding *My Big TOE*. The various NPMR$_n$ are defined and discussed again in Section 5 (Chapter 76, Book 3). All of the above, as well as additional information and firsthand experience is available on these subjects in NPMR. If you are interested, you should begin to develop the ability to access this information, and experience NPMR$_N$ yourself. It is not as difficult as you might think. ◀

Incorrectly applying familiar biological thought patterns may lead you to think that AUM the lean and mean evolution machine is incompatible with AUM the embodiment of love. AUM is not a shark. AUM is consciousness. The evolutionary result of highly evolved, low entropy consciousness is love-consciousness. As you evolve your consciousness, you decrease your entropy and increase your capacity to love. Love is produced, created, and brought into being by the action of a lean and mean evolution machine operating on consciousness. "Lean" implies optimized, efficient, minimum waste, and good economics and ecology – like our own efficient Mother Nature. "Mean" implies that dysfunctional, counterproductive or unprofitable evolutionary offshoots or experiments are ruthlessly terminated as soon as their unprofitabiliy is determined.

The connection between AUM and love is less about what AUM does and feels, or how AUM acts (AUM being unconditionally loving) than it is about AUM **being** unconditional love. Love, as an attribute of AUM's existence, is the source of (and hence more fundamental than) AUM's expression of love through feelings or actions. When we equate AUM consciousness with low entropy, and low entropy consciousness with love, we are talking about

AUM's fundamental nature, substance, and being, not simply its personality. Love is the animating dynamic of AUM; it is AUM's most salient property. AUM is an evolving digital-love-consciousness-being-thing whose actions and intents are animated and motivated by unconditional love – which is the natural mode and expression of a low entropy consciousness.

Are you beginning to wonder how the Fundamental Process could possibly produce a creation this fantastic? Isn't this all a bit much? How could reality be this complex and exist on such a grand scale? I will tell you how – there's a trick to it.

Because evolution invests only in winners, it appears that the Fundamental Process consistently delivers the exact tools that are needed to facilitate successful evolution. This makes the Fundamental Process look exceptionally clever, but the fact is that only the cleverest moves persist while the others self-destruct. Think about it: If you can arrange for your failures to automatically disappear without a trace while your successes persist forever, you will eventually (if you live long enough) end up looking incredibly brilliant even if you are a random plodder. The products of evolution appear to be incredibly clever, but I'm not impressed. The Fundamental Process has been working a long, long, long time and I am hip to its slick trick of conveniently hiding its many mistakes.

AUM is interested in the science of the dynamics of consciousness evolution – the science of the creation of love-consciousness. Yes, there is evolutionary struggle, evolutionary winners and losers, but bear in mind that this is taking place in thought-space. The losers are not necessarily raped, pillaged, and sold into slavery. For example, let us say the Good Guys win the rat-maze struggle, then after all the rats make it, the anti-rats will slowly but surely be won over until **everybody** makes it because that is the nature of rats. On the other hand, if the anti-rats (Bad Guys) win, the rats will slowly and surely be exploited, subjugated, abused, and then exterminated because that is the nature of anti-rats. Oh well, that's honest evolution, impartial science, and love-consciousness at work.

We are not independent of AUM. Think of AUM as a loving, dispassionate, impartial, unbiased, independent scientist-engineer-philosopher: good and valid science is done dispassionately. AUM is an evolving brilliant consciousness with a high level of awareness that must figure things out as we do – good science is the only way.

Let go of the anthropomorphic image of AUM carrying a clipboard and wearing a white lab coat with pocket protectors – it is not like that. In an advanced consciousness system, good science, successful evolution, and exploring and populating the most profitable states all merge together as a single complex motivational pressure pushing the system toward self-improvement. Evolutionary pressure and intellectual process merge to become one in an aware, low-entropy, consciousness system. Think of AUM as an evolving "consciousness-system-being-thing" (with emphasis on the word "thing") instead of an "old-guy-with-a-long-white-beard-and-bushy-eyebrows-in-a-white-robe" (an all powerful version of us who promises to save us and defeat our enemies if we agree to worship him in return).

▶ Do you ever wonder why all the gods of men seem to have a strong want, need, expectation, or desire to be worshiped? An obsession with being worshipped and wanting or needing the little people to "believe in you" must be either an occupational hazard of godship or a prerequisite for the job. Haven't you always wondered what, exactly, gods get out of being worshiped and "believed in" and **why** – it certainly appears to be extraordinarily important to them.

The fact is, **unconditional** love, compassion, and humility can have **no** requirements, **no** ego, **no** needs, **no** desires, and **no** expectations. Consequently, don't confuse AUM or the Big Cheese with any god of the human tribe – their nature as egoless manifestations of unconditional love disqualifies them for the job.

It must be difficult to get people to join your organization and give you money and obedience if all you give them in return is permission to develop the quality of their consciousness in their own unique way and on their own time (which is the only way an individual can improve his or her consciousness quality). A spiritual organization can offer no more than casual nonspecific encouragement to its members without also becoming an inhibitor of their progress. ◀

▶▶ Hey, I just got a great idea! Listen, I hereby fully encourage you to develop your consciousness quality in your own way, on your own time, and at your own place so would you please send ten percent of your annual income to the address inside the front cover of this book?

And just for my dearly beloved favorite readers, a onetime special introductory bonus offer: After I get the entire ten percent, I'll throw in an autographed book and a set of twelve self-addressed tithing envelopes for next year – absolutely **free**!

Wait, there is more: a triple extra bonus! Wow! I hereby solemnly promise to 1) let you sleep in every Saturday and Sunday, 2) never lecture you or preach to you or make you wear a suit and tie, and 3) allow unlimited disobedience!

Jeez man, compare that package to what the competition offers! A fantastic deal at half the price! ◀◀

▶ Don't waste your time or money – there is no address inside the front cover of this book. I am just having some fun. Laughing at yourself is not only fun, but good medicine. Everybody enjoys an occasional dose of good medicine, right?

The point is: The process of encouraging spiritual growth does not lend itself to being organized. To the contrary, when a spiritually focused organization becomes successful (recruits a sustaining broad membership), it inevitably becomes part of the problem – it begins inhibiting spiritual growth instead of encouraging it. Although the raising of consciousness quality requires an individual process, the raising of good feelings within a secure ego requires a group process. Because the two processes are generally incompatible and destructive of each other, it is a good idea to be clear about which process best represents your personal investment in consciousness quality. ◀

The pursuit of truth through good science is one reason (only one of several) why human bio-scientists do not fall madly in love with their bacteria. An overly strong emotional-ego attachment to the bacteria in your Petri dish might cause you to meddle unwisely and compromise the experiment. AUM is not fooling around in the lab; it is doing real science and real experiments that affect its evolution.

I am sure AUM wants the Good Guys to win, but it must let the chips fall where they may. This is not cruel or inconsistent with being the embodiment of love. AUM does not have a mean thought-bone in its thought-body (mind). Because AUM doesn't cuddle you and sing lullabies, (make you feel special, superior and valued – all syrup to the ego), doesn't mean it is an indifferent, heartless mind-shark that doesn't care what happens. AUM is giving you an opportunity to evolve your consciousness via the intentional dissolution of your ego; surely you understand that puffing up your ego by giving you special consideration would be terribly counterproductive. A good scientist can care very deeply and profoundly about what happens – but still must conduct honest experiments. You cannot trick evolution. Profitability is assessed by **results**, not good intentions, brilliant theory, or preferred outcomes.

AUM is dim consciousness evolved into brilliant love-consciousness – which is how positive consciousness naturally evolves. The negative intent consciousness critters and beings (evil) can evolve only toward power-control-force. That is as far as evil can go – the establishment of power-control-force. Minimizing entropy by growing or evolving in the negative direction is extremely limited and relatively worthless compared to the advantages gained by a consciousness evolving positively toward the expression of itself as love. In PMR, power-control-force is a way of life – a primary motivator. That fact should tell you there is much low quality consciousness, negativity, and evil running loose in our local reality. The only way to help clean that mess up is to clean yourself up. You do not have the ability to improve anyone else. All you can do for others is to provide opportunity for them to do for themselves.

In the PMR elementary school of 3D space-time, power-control-force may seem to be a big deal. It may appear to many as a superior means for achieving material success and winning the latest round of "King of the Hill." However, in the bigger picture, in consciousness-space, power-control-force is, for the most part, a short-sighted, short-term ego booster which is a big deal only in the neighborhood of anti-rat town where it holds sway. Yet, the negative can eventually overwhelm the positive as a bacterial infection or a cancer might overwhelm and eventually kill a once productive, healthy and strong sentient human.

Disease has its function – it gives as well as takes. Free will choice must include the ability to choose badly. Though an individuated consciousness can lower its entropy through growth in the negative direction, negative growth, because of natural limitations, is a poor long-term investment. A fatal cancer, without meaning to, always ends up killing itself as it kills its host. The cancer is not aware its activity is self-destructive because it is trapped in a little picture with no vision into the larger system. The host on the other hand, living within and being a part of that bigger picture, must take care not to encourage the cancer or let the cancer get a foothold. An aware entity that chooses to lower its entropy by growing in the negative direction is, like a cancer – self-limiting and self-destructive in the long run. Evil contains the seeds of its own stagnation, if not destruction. A potential host's best defense is knowledge, truth, and Big Picture awareness.

If you are careful not to jump to conclusions, it will be profitable for us to consider interactive consciousness dynamics. How do kind and benevolent pacifists fare in a land without laws? What if you start with 90% pacifists, 8% everyday plain vanilla non-pacifists and 2% violence prone self-centered low-life scumbags that lie, cheat, steal, and worse? What are the natural survival and growth rates as well as conversion rates between these groups? The individuals within all groups are of equal size, strength and intelligence. How will the dynamics of this experiment play out under various amounts and types of law and law enforcement (remember, law enforcers must be able to be violent). I will let you come to your own sense of the dynamics and variables involved as well as what their outcomes are likely to be.

I do not expect you to derive the final answers to the preceding questions – at least not today. I simply wish to create some appreciation of the

392 | *My Big TOE* | Life and Love in the Petri Dish

type and complexity of issues that a study of interactive consciousness dynamics must address. The above example barely scratches the surface of a discipline that growing consciousness systems would need to explore to ensure their continued profitability. Understanding how individuated units of consciousness and consciousness systems change relative to individual interactive intent is AUM's key to optimizing its own growth through the process defined earlier as the consciousness cycle.

Recall from our earlier discussion in Chapter 24, Book 1 that evolutionary pressure is derived from an entity's internal and external environment. The bottom line of that discussion was that the evolutionary goal of consciousness is to lower its entropy. Thus for consciousness, successfully evolving, increasing quality, self-improvement, spiritual growth, and increasing the capacity to interact with love, are all the same thing. These attributes, along with the increased functionality and capability that accrues with more energy available to do work (from our definition of entropy – near the middle of Chapter 24, Book 1), are artifacts of a low entropy consciousness.

From the limited PMR little picture view, the increased functionality and capability that naturally occur in low entropy consciousness may be experienced as personal power, unconditional love, or paranormal ability.

Forget what you have been told: AUM is **not** in the role of your father or mother and you are not in the role of a child. AUM is not responsible for raising you – only for giving you the opportunity and responsibility to raise yourself. It is not AUM's job to take care of you – to force you to learn and grow, and keep you out of trouble or to make you feel comfy when things get tough. Nevertheless, AUM cares about, loves, and appreciates all because that is the fundamental property of low entropy consciousness – but don't take that too personally.

AUM has evolved individuated consciousness with free will as part of its own process of evolution. Free will gives you the responsibility for your growth and enables you to evolve. Without free will, there is no way for individuated consciousness to decrease its entropy. (If your hand is in the air, hang on, a discussion of free will, determinism, uncertainty, and randomness in digital systems is coming up in Chapter 45 of this book.) AUM sets up everything to help you succeed and provides an absolutely free top-of-the-line PMR learning lab with interactive feedback. Additionally, NPMR beings are assigned to help you in every way allowed under the law. These beings are focused directly on you – to plan, encourage, and guide your spiritual growth – the growth of the quality of your consciousness.

Teachers (typically Good Guys who have mastered the human-rat maze) sometimes return to PMR to help point the way. Nevertheless, we often manage to dilute, distort, and subvert their helpful instruction by burying the essential truth of their message under a blanket of dogma generated by an ego serving, fear manipulating, belief system.

The laws and rules that are instituted by AUM, and administered by the Big Cheese (the conditions of the experiment – the culture of $NPMR_N$) provide order and protection for you in $NPMR_N$. Our System (OS) is designed to give you maximum opportunity to succeed under the given conditions. You can sometimes, though it is rare because cross-pollination is discouraged, choose an alternate (not OS) reality system (one of the PMR_k within $NPMR_N$, or something from one of the other $NPMR_n$) within which to evolve your consciousness. Everything you need for success is laid at your feet, all the help you could profitably use is there for you, but **you** have to make the choices. You have to execute your intent. The AUM consciousness-love-being-scientist-thing demonstrates its compassion, love, and caring by giving you every opportunity to grow in the direction of your choice.

Balance, humility, compassion, joy, peace, sadness, love, and caring as well as a desire to evolve, learn, and grow are the natural attributes of a highly evolved and aware consciousness. We are an individuated part of the larger consciousness that has been constrained (matter bumps on the lumps in the AUM sheet) to go through the same evolutionary process as the whole. The value of knowing and understanding the origins and processes of your existence, as well as being aware of the available options and pitfalls that lie before you, is obvious. Knowing what you are doing and why you are doing it should contribute immensely to your ability to get the job done.

There is yet another reason for AUM to produce and work with individuated consciousness. As our individual consciousness evolves, we contribute to the quality of the entire consciousness system of which we are a part. Thus, we are an integral part of AUM's evolutionary process. AUM invests in honest science, but we good-guy rats pay an additional dividend. By lowering the entropy of our individuated consciousness, we also lower the entropy of the entire larger consciousness system. Our individual growth lowers AUM's entropy and helps push AUM along its evolutionary path.

We are profitable beyond our contribution to experimental results. That is why we get lots of help and encouragement. Although we Good-Guy rats may eventually be incorporated into the AUM organization as full partners, those entities evolving to the negative side eventually get recycled (new opportunity to grow in the positive direction) or are simply terminated. I told you their long-term benefits stunk.

Why would entities take the negative route? Because they are short-sighted and do not understand the Big Picture; that is a common problem. Are the bacteria in the Petri dish given briefings on the experiment in which they are taking part? Never! It might bias the results. Do we need to hide the experimental purpose and protocols from the bacteria? No, because the bacteria are not cognitive at the level of the experimenters, no one worries about it.

Going for the apparent short-term gain is how most sentient entities, consciousness systems, or beings (in any reality or situation) function if they do not have the requisite quality and do not possess the perspective, capacity, or understanding to see the Big Picture. That is a simple fact of existence. Do you know anybody or any organizational entity that is fixated on the short-term view or lacks Big Picture vision? Wait a minute, I am afraid that I inadvertently put you in an endless loop – you will never finish that list. Let's ask it the other way. Do you know anybody (besides you and me) or any organizational entity that is **not** fixated on the short-term view and possesses a Big Picture vision? Now that is one short list!

If the experiment is about aware consciousness, the number of participants who have gained a clear view of the Big Picture may well be a metric that measures the progress of the experiment. We are allowed to know; becoming aware and operational within the Big Picture represents a natural part of our potential that we should make an effort to actualize through personal growth.

You should realize there is no alternative, in the biggest picture, to growing the quality of your consciousness except eventual termination. For those in the little picture, this piece of information will be discounted because it cannot be derived from the little picture. Get this: Little pictures must be derived from bigger pictures; it does not work the other way around. Do you see why it is not necessary for the experimenters to hide the experimental purpose and protocols from the bacteria or the humans? Living solely in the little picture is enough insulation to make sure that humans on the planet earth and bacteria in the Petri dish are not confused or motivated by information that is beyond their understanding.

Complex, self-interactive consciousness can pull itself up by its bootstraps in the same way that bright awareness self-energizes and drives its own development. AUM's science is focused on understanding those evolutionary processes that affect consciousness and on making these processes as efficient and productive as possible.

What we do with the **opportunity** to evolve our consciousness is entirely up to us. It is our free will, our self-interaction, and our interaction with others that creates the possibility of learning, which in turn creates our potential for growth. Because of our relatively high entropy, we are slow learners; consequently, our cycles of learning and growth are designed to accumulate results over multiple experience packets.

Some of us become apathetic and cynical because we have to grow and learn on our own. AUM will not do it for us. We become annoyed because we are self-centered and because outgrowing a cherished and needed fantasy is often neither easy nor fun. It is just not fair! AUM will not cuddle us and sing sweet lullabies to soothe our ruffled minds and flustered egos. We think we want, need, and deserve that support – oh, how we miss it! Mama! ... Maamaa!! ... Maaaamaaaa!!!

Resembling spoiled children, we fuss and moan about our options, refuse to play the game, refuse to participate until the game is more fun and pleasant for us. As if our tantrum will force the powers that be to make it easier for us. Children who refuse to grow up always maintain the sense that a big fuss and good excuses will force life to be more accommodating. They sulk, get depressed and then get angry when their self-righteous protestations fall on deaf ears. Refusing to work on your spiritual growth because your life is not fun and full of joy is like spanking a baby in order to make it stop crying – counterproductive, and incredibly stupid.

Wouldn't it be nice if there were always a mommy and daddy to take care of you, to make things easier for you? This adult gig is a bad deal! Bummer! Adults need guidance, comfort, and reassurance more than kids do because the penalties and consequences of making mistakes and doing it wrong are more serious when you are an adult.

> ▶ Step right up ladies and gentlemen; come a little closer, boy, do I have a deal for you! There is no need to feel abandoned and suffer alone, there are lots of organizations and individuals that would like to play the mother or father role to your fear and neediness on various levels. Here is the deal: your ego gets to pretend that you are relieved of some of the adult responsibility for managing and directing your life as well as growing the quality of your consciousness. The parental stand-in gains some control and influence over you, your resources, and your decisions as you forfeit some part of your free will. Take your pick: friend, spouse, employer, religion, church, association, political party, club, support group, gang, charity, or welfare plan – I am sure you can find one or more of these to help you bear your burdens.

The weak, being easy prey, always can find someone willing to offer help for a price – someone or group that will take care of them and make them feel better. All the individual has to do is trade their money and a piece of their soul for the illusion of knowing the answers and the increased peace-of-mind that comes with an imagined off-loading of personal responsibility. Listen to that ego purr – stroke, stroke, stroke. These illusions, properly marketed, are now and always have been best sellers. Many are mandatory according to the standards of each culture.

The bottom line is that predators and parasites can make an easy living (money, power and self-importance) by manipulating the fears of those who have no understanding of the Big Picture. That fact brings up another attribute of ignorance: The ignorant and fearful are always vulnerable and easily manipulated. Manipulating believers, regardless of what they believe, is as easy as appealing to wants, needs and fears. Ask any salesperson, politician, trial lawyer, or marketer – they will vouch for the truth of that statement. Having the courage, and taking the time, effort, and responsibility to capture and tame Big Truth is the only thing I know that can protect you from being had by the hustlers that are hawking the Big Delusion. Beware of Dr. Feelgood; his happy-pills and patented excuses will dissolve your personal power.

Don't misunderstand me: Finding and using help when you need it is a wonderful thing – good for everyone – but becoming dependent upon it, addicted to it, and unable to grow beyond it, is something altogether different. Dependence and interdependence can be helpful and rewarding as long as your free will is not held hostage and you are not trapped by your neediness, fear, wants, desires, ignorance, immaturity, and beliefs. In general, those who are trapped are not aware of being trapped – they are perpetual children predominately from the ranks of the personally powerless. Do not confuse the personally powerless (high entropy consciousness) with the materially powerless; they are not necessarily related. Assuming responsibility for your life is a good place to start anything.

If you do not have what it takes to find your own answers and take charge of the evolution of your consciousness, someone else will be happy to fill that void for a price that is actually much steeper than it appears. How do you calculate the price of being caught in a belief trap? In terms of opportunity lost, entropy gained, progress not made, misery prolonged, or time and energy (a life) wasted? However you figure it, it is too much to pay for a short-term feel-good illusion.

Does anyone wonder why recreational drugs are widely popular in our culture (and most cultures)? The feel-good mentality, pursuing its ego-soothing and fear-suppressing belief systems, forms the foundation upon which most cultures are built. Taking recreational drugs is simply acting out with our bodies what we do every day with our minds – escaping reality by making a belief-centered end run around ignorance and fear, and by avoiding personal responsibility for our growth toward love-consciousness.

We play hooky from the learning lab in order to hang out in a feel-good fantasy. Ego driven delusion is at the root and core of our existence – it defines our little picture reality. As always, the mind leads and the body follows. Taking recreational drugs is a direct consequence of an immature individuated consciousness feeling trapped by the anxiety, insecurity, and fear that is created by its own limitations. Low quality consciousness and drug use encourage each other.

A low quality consciousness uses belief, denial, and the obfuscation of ego in order to escape anxiety, insecurity, and fear. Individuals in PMR use recreational drugs to do the same thing. Is it any wonder that the drug-culture primarily attracts the young who have relatively high-entropy systems and adults who feel failed and inadequate? As a culture, and as individuals, we express the quality of our consciousness in many ways and forms but always with absolute accuracy. We are what we are. Belief, ego, and fantasy on the inside – recreational drugs on the outside – different escape methodologies but similar expressions of the same lack of consciousness quality.

To eliminate recreational drug abuse, a culture must raise its collective consciousness quality. This can only be accomplished at the individual level. Any improvement by individuals within the culture raises the quality level of the whole and begins to solve the problem. You personally must solve the problems of your culture. No one else can do more than you can. To believe it is someone else's problem to solve more than yours simply makes you a part of the problem. You can do nothing more than improve yourself. Fortunately, that single accomplishment constitutes your optimal contribution. ◀

Growing up is difficult to do. Sometimes we get so wrapped up in our fear, our ego's wants, desires, and needs that we do not try or care about anything except hiding from our own terrifying ignorance. "Quick, duck in here, you can hide out in this fantasy," says a helpful ego. Whew, just in time! Helpless little human, huddled, shivering, miserable and unhappy – trapped in a tiny corner of his mind by his ego-fear-belief-system fantasy. This is where the sadness part of love comes from.

To make things worse, as long as this needy human is in the PMR learning lab, the poor little fellow imagines that his existence and well-being is derived from, and supported by, a magnificent rubber chicken (see Chapter 21, Book 1). For this reason, he often refuses (only while he is in PMR) to use the help that is focused, coordinated and projected from the NPMR$_N$ part of OS to PMR to support his evolution within the loving and supportive reality in which he actually exists. A fear-driven life, lived from within the shell of a protective ego fantasy, is often demanding, tricky, painful, agonizingly shallow, and not much fun or full of joy, love, and peace. Poor guy – cuddling with his chicken in the un-illuminated darkness;

trying to make it through the difficult times that his needy ego has unknowingly created.

Growth requires a focused effort and this hapless human-rat does not have the required time or energy because he is preoccupied with maintaining his illusions. Unfortunately, the maximization of feeling good (mostly stroking the ego) and the minimization of pain (constructing and maintaining a good working fantasy), is **more** than a full time job. If this being does not escape this trap, he will in time (with age) end up in a never ending and self-perpetuating tail chase that either reaches an accommodating steady state or constantly loses ground. The hope is that one day he will wake up and realize there is more to his existence than just living and that the issue is quality, not comfort or power.

If you know anybody in a predicament like this, please be kind, empathetic, and helpful to them. They have been caught in a trap (many were born in captivity and know of nothing else) and deserve your patience and compassion. These maze rats are not dumb – they are merely confused and blinded by their belief systems. What should you do to help? Love them and let them be – they will eventually figure it out (explaining it to them often ends up confusing them more). Just love them and let them be – and evolve your consciousness to the greatest extent possible. That way, **if asked**, you can point out unseen opportunities and options from a more balanced and less ego driven perspective, serve as a good example, and provide encouragement through living proof that success is possible. That is about all you can do to help.

44

■ ■ ■

What is the Point?

■ ■ ■

We, the creations of AUM, are the source of its joy and sadness. We evolve our individuated piece of basic consciousness by lowering its entropy and increasing its quality. As we drop our ego delusions, we become more AUM-like – an embodiment of love and caring. This is our goal, our purpose. We began this trip as a chunk of individuated digital consciousness with enough complexity, memory, and processing capability to evolve ourselves from a high-entropy relatively dim awareness to a low entropy brilliant embodiment of love. In the simplest terms, we are to follow in the evolutionary footsteps of AUO-AUM.

To accomplish this, we have been given two attributes. First, we are individuated into existence – a tiny snippet of the AUM hologram – an interactive subroutine or defined object running in its own piece of mind-space within TBC – a part of the Big Picture consciousness-evolution fractal that contains the pattern of the whole. Second, we are given free will so that evolution (growth) is possible. The evolution of our personal fragment of consciousness is directly and exclusively based on the individual choices we make. We exist, operate, interact, and individually evolve (modify ourselves to occupy lower entropy states) all within the immense digital mind that is consciousness. Fundamentally, we are consciousness. The evolution of our individuated unit of consciousness is the point of our existence. This is how the AUM-digital-mind-thing consciousness organism survives and grows. This is how the fundamental-consciousness-energy-ecosystem evolves, lowers its entropy, and avoids stagnation, regression (increasing entropy), and dissolution (death).

The Fundamental Process requires evolving systems to, in the long run, either grow or die. Trying to remain marginally viable, balanced in the astable state between the two is a poor strategy. The eventual price of not continually expending effort to reduce the average entropy within a consciousness system is to allow the average entropy of the system to increase, thus ensuring eventual disintegration. You are part of such a consciousness system. You are part of its strategy to survive through continuing growth and evolution. You are it. It is you. All That Is, is of the One. One consciousness organism. One energetic self-aware ecosystem. All live or die, succeed or fail, together.

Why one? Why not a herd of AUMosaurus thundering across the mental plains of almost infinite existence? Because one is enough to develop a fully complete (from our perspective) Big TOE; furthermore, as a scientist, I appreciate the connection between fundamental truth and elegant simplicity and thus avoid unnecessary complication that adds nothing. One apparently infinite, but finite, AUM is enough. What lies beyond a unitary finite AUM may be **indirectly** very important to our existence, but nonetheless remains beyond, and therefore irrelevant to, the grandest reality that we have the theoretical or practical capacity to comprehend.

Do you now understand why experiments in consciousness evolution are important to AUM? Do you now see why it is vitally important to you and to me and to every other chunk of individuated consciousness that **you** learn, grow, and improve the quality of **your** consciousness? We are all in this thing together; all are a part of the whole. The whole prospers through the contributions of its parts. Grow or self-destruct – that is the simple choice and a fundamental truth in both the little picture and the Big Picture. You, individually, right now, this minute, are either part of the solution or part of the problem: Your cumulative progress, or lack thereof, will eventually become a life force or a death force for yourself, for everyone else, and for the living consciousness organism that hosts your existence.

There is a point and a purpose to your existence. You have a mission. That your consciousness has the potential to evolve requires that you have free will. In order that you might succeed, AUM must allow you to fail. AUM has no choice. To survive, you must risk death – the Fundamental Process will not, cannot, function any other way. The initiative must be yours and yours alone. Your spiritual growth is not a group activity – it is a personal transformation achieved only in personal terms. No group or association can lower the entropy of (increase the quality of) your personal consciousness by even an infinitesimal amount.

Each individual contribution, however small, is vital to the health, welfare, and continuance of the whole. You are the contributor and the beneficiary of your contribution in both the little picture and in the Big

Picture. The ultimate reality is the living consciousness system, or consciousness organism of which you are a part. Seeing the Big Picture can be intellectually accomplished by understanding the philosophy and science of consciousness and how and why the consciousness system works. Contributing to the health of the Big Picture requires more than an intellectual effort.

> ▶ It has been brought to my attention that some of the technical types in the reading audience have about had it with all this philosophy, ontology, epistemology, and metaphysics. They want to know when we are going to get to the important stuff – the real hard-science that matters. Patience, dear brothers of the causal cloth: Philosophy must first develop the context before science can most effectively develop the content. Content without context eventually paints itself into a little picture corner. That is where traditional science is now and why its search for a unified theory has stalled. To escape the PMR Petri dish, you must eventually come to understand that reality is more than a local rule-set, more than a mindless machine. ◀

You are on this growth-path whether you know it or not – a path first established by AUO-AUM – the path of consciousness evolution. PMR is nothing more than a virtual environment specially created to facilitate your (and others') evolutionary progress; consider it group therapy for dim awareness. I will explain how that works in detail in Sections 4 and 5.

Our goal? Our purpose? Why should we bother? If we are creations of AUM and AUM is (compared to us) the epitome of consciousness perfected, why are we wallowing around like a bunch of ego-bound self-centered pre-schoolers hopelessly lost in the middle of an advanced university of higher evolution? How are we significantly contributing to AUM's further evolution? Why does AUM need us – for amusement? Are we the carbon-based version of "Donkey Kong" or "King of the Hill" with 3D holographic virtual reality graphics? Does that concept amuse you as much as it does me?

Don't worry about it. These questions are the result of falling into another anthropomorphic trap. I am afraid I have set you up to fall into this one by referring to our consciousness relative to AUM's consciousness. This language cannot be helped because we must consider ourselves as individuals in order to conceive of and initiate self-improvement (evolve). Thus, we speak as if AUM and we were separate beings ("What does AUM need us for?"). This is not the case. We are AUM and AUM is us.

Keep in mind that AUM exists and evolves because of differentiating unique states within its oneness – creating dualities or changes relative to its uniform self – organizing its potentiality to squeeze entropy out of its system. Dualities became reality cells by the gazillions, from which TBC was created and from which we were individuated. We are part and parcel of AUM – a cellular digital consciousness that eventually defined a space-time subset which provides the causal framework for our physical experience, defines our physical universe, and our overall local reality system (OS). (Section 4 provides the logical background that leads to this conclusion and Section 5 explains how it works.)

If you are partial to sound bites, try these: "We are one with AUM." "We are made in the image of AUM." "AUM is the One Source, the Creator of All That Is." "AUM is love." I am sure you could think up many more, but sound bites are often misleading in their brevity and usually created to deliver an emotional impact. Sound bites are best used to serve demagoguery and do not make for good communications. Resist turning knowledge into slogans.

▶ Likewise, resist the sloganeer's attempt to slip highly spun pseudo-knowledge into your mind. Without slogans, politicians would be forced to express real thoughts instead of just sniping at each other with test marketed sound bite bullets. Can you imagine a politician without emotion laden sound bites; can you imagine a politician with nothing to say?

Bear in mind that a politician, like any marketer, is interested in manipulating opinion and behavior, not communicating information. In general, do not look for truth in congress, court rooms, or in advertising, behind lecterns or pulpits, or on TV. Regardless of what the media might be, if the speaker (or writer) has a product or a point of view he is trying to sell, he is marketing to capture your opinion, not trying to give you information for **your** benefit. As always, open-minded skepticism applied to **your own** data solves the problem but you must get out there and get the data before tentative conclusions can become actual ones. ◀

We are an integral part of a bigger consciousness system. Do not let your innate sense of self-importance allow you to disregard the fact that we are only a minor part of this larger consciousness system. Being a small part of the whole makes no statement about relative importance; without a doubt, the notion of relative importance within a consciousness system is a contrivance of needy ego.

We would never contemplate ripping out our colon and anus because they cannot, either singly or in combination, initiate and maintain polite dinner-table conversation. That is not their function. Some might consider them as undesirable parts, but that would be stupid. Their function is as important and critical as the function of other parts of our physical system. They are an integral part of us. We are an integral part of AUM. There is no justification for either feeling special or feeling not special because of the part you play.

Consider how important it is that all the cells, tissue, organs and other parts of the complex biological system that we call our body perform their individual missions and fulfill their purpose without worrying about what is in it for them. What if your liver, heart, or the bacteria in your intestine, in a fit of laziness and self-importance generated by not being aware of the bigger picture, concluded that their constant effort was too much trouble and decided to goof off? You get the point: If every part does not fulfill its purpose, the entire system suffers or perhaps dies. Your relationship to the whole of consciousness is like that. Although knowledge thrives on separateness, wisdom sees itself as part of a whole.

Let's look at some examples of entities that do not relate to the whole, that do not understand the larger system that sustains them – entities that live within a small picture. In the biological realm we have malignant cancers, parasites that destroy their hosts, and most people who, in a convincing imitation of the cancers and parasites, abuse their niche in the larger (ecological) system to whatever extent possible seeking only to maximize their personal short-term gain. In the consciousness realm, again it is people who seem to have a difficult time seeing the Big Picture. Do you notice a consistency in behavior here that tells us something about ourselves? If people understood the Big Picture within their consciousness system, the Big Picture within their physical system would be obvious and earn their utmost respect. They would naturally see themselves and all others (human and otherwise) as full partners in a shared experience of existence.

If the cancer, the parasites, or the people understood the relationship that they have with the whole, possessed a free will, and were intelligent enough to do something about it, they would first control, and then reverse, their self-destructive behavior. They would work to change the long-term lose-lose situation they are generating because of their lack of awareness and would develop a symbiotic win-win relationship with the larger system that sustains them.

We humans are part of several ecosystems that sustain us. We must be good citizens and develop and maintain a long-term win-win relationship with each. As aware, bright, sentient entities, either physical or nonphysical, consciousness is the largest and most fundamental sustaining system within the capacity of our awareness. Consciousness is the primary ecosystem to which we belong. $NPMR_N$ is our biome; OS is our community; PMR is our niche; and earth is our habitat.

At the next lower level of awareness, within PMR and on planet earth, the earth itself and its life-forms and element-forms comprises the largest sustaining system. Below that, there are species, cultures, nations, businesses, families and individuals. Most of us know that we must constantly expend effort to form win-win relationships with other individuals. That realization is a good start at locating the bottom rung of the greater ecological hierarchy and is the first step toward being a fully responsible individual making the most of your native capacity. You must become a responsible citizen (integrated contributing part) at all levels of interaction and relationship – from individuals, to organizations and groups, to the earth, to AUM consciousness. Growing up seems to be nothing more than achieving a series of ever widening perspectives. Don't quit prematurely, before you have fulfilled your full growth potential, outgrow your belief traps, and begin to enjoy the magnificence of your true heritage as a chip from the old AUM consciousness block.

Because of the efficiency of specialization, we and all other extremely complex products of evolution are not simply monolithic homogenized entities – we are made up of specific parts with specialized functions. Likewise, AUM is not one big homogenized uniform mind-thing with no parts. AUM is nothing other than consciousness and must design, build, and evolve its structures and subsystems with consciousness. AUM can structure and limit subsets of consciousness, information, memory, and digital processing power in a multitude of interesting and profitable ways. Every being, as it evolves on each level of its existence, must learn how to use and explore the possibilities within that level to improve its profitability. AUM is no different.

AUM needs us as part of its being for multiple purposes, just as you need your circulatory system for multiple purposes. As we are driven to explore ourselves, AUM is driven to explore and understand consciousness – its continued growth, evolution, and survival depends on it. We should find this particularly easy to understand: Our bodies, medicine, how we interact with each other, the environment, and the physical universe have always been the focal point and primary motivation of our science. We first want to, and need to, know about us. The AUM-thing wants to, and needs to, know about consciousness.

I am sure we could think up many interesting questions and issues about motivation, good, evil, delusion, creativity, ego, quality, love, fear, pain, profitability, productivity, new possibilities, competing ideologies, violence, pacifism, intelligence, evolutionary direction, initial conditions, rule-sets, and so on – all useful information about consciousness evolution that

AUM might well be pursuing. I am also reasonably sure that AUM would be able to add a few we didn't think of.

Does that make us lab rats as well as partners and pieces of the larger system? Maybe; all depends on how you look at it. Do **you** consider the bacteria in your intestines or the ecosystem that provides your oxygen as your partner? Perhaps you should – your life depends on both of them. Fortunately, AUM is not likely to be as self-absorbed and disrespectful to its partners as humans are.

We may be the key active ingredients of a cyclical process that takes raw consciousness and transforms it into something more refined and valuable by mimicking or applying AUM's evolutionary process – like a consciousness transmutation machine or an entropy reduction factory that is an integral part of a larger consciousness cycle. We are part of the mechanism that eventually enables the conversion (evolution) of dim individuated units of consciousness into empowered love-beings. Could AUM be employing us similarly to how we employ bacteria to help clean up oil spills or use a compressor and fluid to cool a refrigerator?

Perhaps we are a work of art, created for its own sake, or a work of science undertaken to discover something new. We may be valuable because we express AUM's being in a way that is profitable, pleasing, aesthetic, or interesting to AUM.

AUM may be playing Jessica to our rendition of Roger Rabbit, and thus values us simply because we make him laugh. No doubt, there is great value in humor. Likewise, there is no doubt that we humans, as a species, (especially if one enjoys dark comedy and slapstick) often make Curly, Moe, and Larry look like serious college professors by comparison. Hmmm, perhaps that wasn't a good analogy – as I recall, Curly, Moe and Larry have always been indistinguishable from serious college professors. Oh well, you know what I mean: We humans do some really funny stuff; we are a wild and crazy bunch of animals. I can hardly wait to see what we come up with for the grand finale.

Perhaps our function from AUM's point of view represents some mixture of all of the above – along with a few others I haven't thought of. There are many ways to be an integral part of AUM's evolution.

I like the lab-rat and consciousness transmuting bacteria options best, but I will be the first to admit that I might not be totally aware of everything AUM does all day at the office. The point is: We can reasonably define an immediate purpose as well as a larger purpose and a larger, larger purpose, even if we are not absolutely sure about the details of the larger purpose of our larger purpose.

Picture consciousness pulling itself up by its bootstraps: Picture your-self as a bootstrap. AUM's science is focused on understanding the processes of consciousness evolution and making them as efficient and productive as possible. AUM does not exist for us. It does whatever it needs to do for its own reasons and according to its own profitability requirements; we are simply a part of that process.

The *Número Uno* Digital Dude **is** the larger reality – every other appar-ent reality is derived by applying various constraints and rule-sets to vari-ous interactive subsets of individuated awareness within a specific calcu-lation-space called a dimension. Think simulation and virtual reality. (If I have gotten too far out ahead conceptually, be patient, Section 4 will develop much of the supporting logic for these ideas.)

We may, by the vagaries of evolution, be no more than the analogical equivalent of AUM's intestinal bacteria, or perhaps more colorfully, AUM's colon and anus, and as such we should expect to have no appre-ciation for the ultimate activities of the whole – such as polite dinner-table conversation. In that case, we keep on keeping on because that is what we do. Our perspective is too small to see the bigger picture that connects us to a whole that exists beyond our comprehension. Have a laugh or at least a chuckle, and let it go. Self-importance, the hallmark of the ego, is always self-destructive.

Why are people so poor at fulfilling their immediate purpose of becoming more, growing, improving their spiritual content, and evolving the quality of their consciousness? Given that this is our larger purpose – that this is what we are supposed to do – you would hope that we would be better at it. Most humans (and other residents of other PMRs) and most sentient nonphysical beings as well, are delusional (ego bound) because they live in a small-perspective reality with small-perspective knowledge and even less understanding. There is no privileged group. Everyone must start at the bottom and pull themselves up by their boot-straps. That is how consciousness reduces its entropy – by growing up, maturing, dropping limitations of vision and understanding, shedding ego and delusion – becoming all it can be.

Your ego, in its reaction to fear, insulates your delusion from reality with an array of self-protective and self-justifying rationalizations. The only way out of that trap is by making a serious effort to outgrow your limitations and overcome your fear by developing a Big Picture understanding of exis-tence. A **personal** Big Picture understanding, improving the quality of your consciousness, and developing your capacity to love are all interre-lated mutually reinforcing aspects of a successfully evolving consciousness being. There are no shortcuts. You are an evolving chunk of individuated consciousness within the OS community; that by itself defines the nature of your being and determines the purpose of your existence.

Most sentient, conscious, self-aware beings (physical or nonphysical) experience the feelings described in Chapters 42 and 43 of this book as a mixture of joy, peace, pain, fear, and pleasure. The feeling of sadness follows the caring of love. Beings with higher intelligence are typically dominated by feelings of self-induced fear and pain. That is the price they pay for their advanced knowledge (a dangerous thing) and the greater opportunities that go with it. Eating that metaphorical apple from the tree of knowledge carried with it the potential for expanded evolutionary success or failure. Self-awareness and the knowledge that comes with it was the beginning of AUM's Great Experiment and our Great Struggle – the beginning of our employment as an entropy transfer media within the consciousness cycle – employed as lab rats, high entropy to low entropy consciousness transmuting bacteria, expressions of art, participants in scientific enquiry, stand-up comedians, or all of the above.

45

■ ■ ■

Mind, Brain, and Body

■ ■ ■

Nonphysical conscious beings are not merely disembodied discrete chunks of information or knowledge, like a web page from another dimension. They are differentiated (bounded), individuated (singular unit) consciousness energy (self-organizing) digital subsystems constrained to a specific form and functionality. They have unique bodies – defined boundaries that are composed of what is most descriptively called nonphysical matter which may appear completely solid. They can be intellectual, knowledgeable, and feeling entities because they have the potential to be self-aware, self-modifiable, and have an evolutionary purpose.

We have established that AUM has feelings; however, without fear, ego, and delusion AUM's feelings are not at all like ours. On the other hand, if we can replace our fear and ego with love, humility, and compassion, we begin to look and act like distant relatives of AUM – chips from the old AUM block, sharing a single continuous awareness with the source of all consciousness.

How does all this feeling consciousness interact with what we experience as our bodies? In the physical world, in accordance with TBC's space-time rule-set, our limited senses collect the data that define our physical reality. Our nervous system sends certain types and patterns of signals (utilizing neurons, synapses, nerves, and the like) representing the collected sensory data to the brain, which interprets these patterns of signals (encoded sensory data) to create the perceived physical reality within the context of past experience. The nonphysical mind receives the PMR reality-experience from the brain's experience-limited interpretation of the signals. The mind then applies its unique knowledge and quality to produce a response that expresses its intent in terms of internal and external action.

Thus, it is a value-based nonphysical consciousness with knowledge, memory, fear, ego, understanding, intent, motivation, purpose, and individual quality that determines what the brain sends back down the communications links (nervous system) to guide the body in transforming the mind's intent into a new state of being that is in direct response to the original sensory data. In this simplified model, the brain serves as a transducer and a constraining filter between the physical and the nonphysical components of the being as well as a controller of autonomic body functions that must satisfy space-time biological requirements. Nonphysical consciousness energy animates the experience-body and gives it non-trivial uniqueness, originality and purpose. Fish, hedgehogs, foxes, and people all work this way. As a refresher, you may want to re-read the aside at the beginning of Chapter 29, Book 1 concerning the physical to nonphysical interface function of the Central Nervous System.

The mind-body process described above is grossly over-simplified and presented from the PMR perspective. Though it makes this discussion more abstract and thus more difficult to understand, it is more accurate if you think of your body as a virtual body. Consider that your body is a projection of your character into a multi-player digital simulation consciousness trainer (a virtual reality). Your projected character represents the total accumulated quality of your consciousness employing only a fragment of your individuated consciousness energy. Its allowable interactions (its experiences) are constrained by the space-time rule-set. This particular virtual reality has many trillions of interacting players – some sentient, some not. Additionally, there may be random components occasionally thrown into this complex mix to make sure that our free will always has a sufficiently rich array of possibilities to choose from in order to facilitate optimal evolution and learning.

One can think of these random interactions as the result of biological, psychological, and social Brownian motion – individuals bumping into, and exchanging energy with each other and with their environment in a process that ultimately defines their personal trajectories. These sequential interactions produce cumulative results and define new sets of possibilities with each bump. Unplanned interactions with others, and with our outside and inside environments, create novel opportunities to exercise our intent by making choices that reflect our inner quality. Clearly, personal growth requires us to use our limited awareness to improve our understanding of what is important and to use our free will to improve the quality of our choices. Growing up and decreasing the entropy of our consciousness obviously requires far more than simply acquiring, storing, and processing knowledge. Being conscious, we have the innate ability to modify ourselves – to pull ourselves up by our bootstraps.

Improving the quality and reducing the entropy of your consciousness (spiritual growth) is more than an intellectual exercise. The intellect (processing, memory and analytic function) can make no significant progress by itself. Though the intellect can occasionally act as a catalytic agent for personal growth, an isolated catalyst cannot induce a reaction to take place. One needs the proper reactants residing within a supportive environment before the nudge of a catalyst can move the process forward.

This is a particularly disturbing fact for those who live out of their heads by controlling and guiding their every thought and action directly with their intellect. Their need to **appear** rational in the little picture severely restricts their ability to **be** rational in the Big Picture, or even realize that a Big Picture exists. The belief in the infallibility and completeness of little picture rationality is another belief trap piled high with victims who are bright intellectuals from material-based cultures.

Enlarging the species database in order to improve future hardware (body and senses), software (attitudes and mental ability), and processing capabilities (brain and central nervous system) is not the only, or even the major, goal of implementing the Fundamental Process within the human race.

Fulfilling our individual purpose for existing, decreasing our overall individuated system entropy, improving spiritual content, or evolving the quality of our consciousness are more fundamental, crucial, and necessary components of our total evolution than biological evolution. Biological evolution merely provides the stage, props, and setting for facilitating consciousness evolution. Consciousness evolution is the main act, yet we spend lifetimes dedicated to nothing more than rearranging the stage props.

Perhaps we should call PMR "Theater of the Blind."

Ladies and gentlemen, tonight's feature presentation is: "Planet of the Idiots," – a comedy of missed opportunity. Starring Top Monkey the Magnificent, playing all significant roles by himself.

The evolution of our physical system (people, critters, plants, planet, solar system, and universe) plays out the choices and random events that are constrained to evolve within the limited possibilities of the space-time rule-set. The physical evolution drama moves forward as a **subset** of Big Picture consciousness evolution. It provides the setting for our virtual reality trainer and the context and rules for our interactions. Those interactions lead us to choices, and choices provide us with opportunities for self-improvement. Our physical reality, our perceptual experience, and the physical evolution of our species and our universe, is an extremely small part of a much bigger evolutionary drama.

▶ Uh oh, I feel some of the techies tugging on my sleeve again. They have been extremely patient with these high-level non-mathematical philosophical descriptions and I think we should take a short break and see what they want.

It appears that some of the more left-brained individuals traveling with us are having problems with randomness, digital systems, and free will. Actually, the issue they raise represents a well-known and important problem of both science and philosophy and deserves some attention.

Many individuals find the existence of free will to be so intuitively obvious that they do not understand what the fuss is all about. That these more right-brained individuals have somehow managed to intuit the correct answer without going through a rigorous intellectual exercise may be the result of fortuitous cultural belief, or, if they are more like you, the result of highly intelligent and accurate observation coupled with brilliant inductive reasoning. If the theoretical existence of free will is not an issue that challenges your understanding, you may want to skip to the end of this rather lengthy aside. Otherwise, please join us on this little excursion into the relationships that logically connect consciousness, evolution, psi effects, and free will. The choice is yours, right?

Randomness is not as trivial a concept as it first appears. Many mathematicians, digital physicists, and computer scientists get wound up over the distinction between truly random and pseudo-random. I have been using the word "random" to connote uncertainty in a process, input, or result. The randomness required to support the choice-making free will described above within a digital consciousness virtual reality like PMR is not dependent on the distinction between random and pseudo-random. Pseudo-randomness – the same randomness we use everyday in our computer simulations and models – is all that is necessary to grant us free will.

When multiple choices of nearly equal probability occur, the result is uncertain because it reflects the ever-changing minute details of the moment that are a function of our changing (growing) intent and quality. Significant choices are mostly made at or beyond the uncertain edge of our certain knowledge. Our choices, as expressions of our dynamic consciousness quality, are constantly groping at the vague periphery of our personal limitations. In more technical terms, making a choice and learning from it is analogous to sampling a noisy signal; individual measurements (executions of intent) are uncertain. That is why bootstrapping our consciousness quality is a slow tentative process that yields best to small steady pressures that are generally applied in the right direction. Personal growth often develops slowly and hesitantly from the seeds of our experience that are planted by our intuition at the ragged (noisy) outer edge of our understanding and awareness.

Given a dynamic consciousness system composed of interacting entities with free will, the final result (at the end of a DELTA-t) remains uncertain until it is achieved. During DELTA-t, while intent is directing, choices are being made, and actions are being taken (for individuals or groups), the uncertainty (degree of randomness) in the final results is primarily caused by the huge complexity and number of the potential interactions between large numbers of very complex self-modifying individuals applying their free will to make choices in a non-rational belief-space near the boundary between their known ignorance and their assumed knowledge. ◄

>> Whew, what a mouthful! I deserve either a prize or a whipping for that last sentence. I am afraid I know which one you would give me if you had the chance. Listen folks, just slap the book a few times and go on – it's unhealthy to hold a grudge. No pain, no gain. ◄◄

▶ Uncertain results create additional opportunities for interaction and choice making. New choices, representing our applied intent, produce new actions that in turn create new results and thus, more new choices. We bump from one interaction to the next making choices as we go. Each choice generates new opportunities and additional choices. Feedback from previous choices encourages us to modify consciousness quality and refocus our intent thereby influencing subsequent choices. Thus we create, as well as participate in, an efficient interactive process designed to facilitate the evolution of individuated consciousness that can be described as Brownian opportunity with an underlying quality bias.

The assumption of the impossibility of true randomness within a digital calculation does not logically disallow the possibility of our free will translating our intent within PMR into a unique personal choice that affords an opportunity (through feedback) to subsequently modify our intent (personal growth). An AUM consciousness system can use randomness with a structure (algorithm) behind it to see how huge numbers of extremely complex, noisy, self-modifying subsystems interact under various dynamic conditions just as we can. Scientists and engineers engaged in systems simulation do this type of stochastic analysis every day.

In Section 5, we will discuss the mechanics of calculating probable future reality surfaces and explain the functional and operational relationship between your personal, shared, and local realities and the deterministic reality-simulation-database where everything that can happen does. Section 5 explains the mechanics or implementation-process of actualizing free will choices and contrasts that process with a similar process that produces deterministic, statistically based, un-actualized parallel realities. Here I address only the theoretical and practical necessity of free will to be an integral part of evolving consciousness systems. This aside will establish a theoretical and practical basis for free will, while Section 5 describes how our free will process is actually implemented within the bounds of a continually re-determined set of possibilities that describe everything that could possibly happen.

Some may think that the theoretical difficulty of generating true randomness in digital processes is problematical for free will. This erroneous conclusion typically results from wrongly considering consciousness to be a simple monolithic system that follows the rules of PMR causality. This problem melts away in the layered complexity of multiple levels of interacting local and personal realities that have their origins within an evolving consciousness system.

The bottom line is that true randomness, as it is understood by PMR theoreticians, is not required for free will to upstage predestination as the driver of consciousness evolution results in NPMR or PMR. AUM, the consciousness system, can (as can we) do honest science that it finds profitable to its evolution even if it is a rule-based logical system. A pseudo free will is logically free enough to do the job of defining and implementing purposeful individuated consciousness units like those constraining their perceptions to the PMR rule-set in order to lower their personal entropy.

Units of individuated consciousness have the ability to freely make choices from a finite array of possibilities. Individual choices are dependent upon how the entity applies its finite repertoire of motivations and intents that it has developed in reaction to the experience it has perceived through the PMR rule-set filter. Intent eventually expresses itself through action and reaction within an interactive virtual reality. The quality of your intentions or motivations reflects the quality of your consciousness. Simply put, an entity's profitability and top-level goals must be pursued through a purposeful application of directed awareness and intent (sometimes called "will"). Units of consciousness exercise their intents in virtual reality simulators that provide results-oriented feedback from every experiential opportunity. Aware entities use this feedback to improve the quality of their intent and decrease their entropy. This is how consciousness learns and evolves within the PMR experiential virtual-reality learning lab.

Time enables experience, which is derived from a sequence of events. Experience is the memory of a sequence of perceptions. Perception is a sequence of datum exchanges. Creating specific experience as a learning tool can be implemented through an interactive sequence of perceptions that provides an opportunity for an entity to exercise intent by making choices based upon personal quality and by allowing the entity to assess the results of those choices. An assessment of the choice and the results of the choice (feedback) lead to self-modification in pursuit of fulfilling the purpose and goals of the entity.

We humans have employed interactive virtual-reality training devices (such as flight simulators for pilots) for many years. Experience is the key to learning; whether that experience is actual or virtual is irrelevant. Learning is the key to consciousness evolution. Experience (operational memory and interpretive processing), perception (data collection), free will (choice), and the ability to learn (information processing developing results and conclusions) and grow (self modification relative to profitability goals) are fundamental attributes of successfully evolving consciousness.

Compare that last sentence with the descriptions given in Chapters 24 through 28 of Book 1 and especially the beginning of Chapter 41 of this book, where memory, a rich array of challenging data exchanges (interactions) between the entity and its internal and external environments, self-aware information processing, and self-modification in pursuit of profitability were given as the fundamental attributes of consciousness. In Chapter 41 these fundamental characteristics of consciousness were used to define consciousness and to describe its evolutionary process in general terms. Now it becomes clear that an experiential virtual reality like PMR facilitates consciousness evolution by thoroughly and methodically exercising each of its four fundamental characteristics. PMR learning labs are well designed by evolution to accomplish their purpose – they deliver exactly what we need to optimize our opportunities for self-improvement.

Free will – the ability to make choices in order to effect self-modification in the pursuit of evolutionary profitability – is part of the definition of consciousness itself. Free will is not an outside condition that must be applied to consciousness, it is fundamental to the existence of consciousness. Free will is a necessary attribute of successfully evolving consciousness. Without free will, a profitable consciousness system is impossible. If you and your many sentient friends and acquaintances are conscious, consciousness must not only exist, but also support a complex interactive system of coherent experience. Given that consciousness exists, it must be enabled by memory, information processing capability (intelligence), the interactive sharing of data, and free-will choice-making in the service of profitable evolution. Thus, the question of free will reduces to the question of are you conscious, and, if so, is your consciousness part of a complex interactive system of consciousness? If these are answered in the affirmative then your consciousness, and the system of which it is a part, must be evolving against some measure of profitability because that is a requirement of all self-modifying interactive systems. Such a system cannot evolve toward greater profitability without free will to make the required choices.

Evolution requires choice between alternatives. For evolution to exist as a real process, the choices must be free. Pseudo-free is free enough within a sufficiently complex, interactive, feedback-driven subsystem for the Fundamental Process to be effective within that subsystem.

To illustrate my point, I am going to put you to work. Contemplate the consciousness of a clam or a bumblebee. Next, compare that consciousness with the consciousness of a future generation computer. Goodness gracious, great balls of fire! You accomplished that task in only a few seconds! I must say, I am mightily impressed, the depth and speed of your cogitation are truly remarkable – undoubtedly, a phenomenon worth studying all by itself. Did you list all of the **functional** similarities and differences? Did you notice how the list of functional differences becomes shorter and shorter the more you ponder the fundamentals of consciousness and its quality? In the areas of intent, process, and structure there are vast differences, but these have less to do with the basic properties of consciousness and free will than with the particular mechanisms and forms of consciousness implementation.

Next, examine the free will of that clam or bumblebee. Clams and bumblebees have many daily choices and decisions to make as well – and they live and die, grow or evolve, by the cumulative results of those choices as we do. How predictable are clams? I suspect their individual deliberate actions are spread over a statistical range that represents the decision-space of their species. Humans similarly exercise free will within their own decision-space. The size and complexity of that decision space (for a species or an individual) depends on the capacity and quality (entropy) of the consciousness that supports it. For a given fundamental capacity, lower entropy supports a larger, more complex decision space. Higher capacity also supports the potentiality of lower entropy.

In case you are wondering, the capacity of the human consciousness is immense, yet we exercise only an infinitesimal fraction of that capacity. Our potential ranges far beyond your wildest dreams. Unfortunately, the part of that potential we have intentionally actualized typically supports little more than the tedious soap opera we call "real life."

That's right: Beings with a higher quality of consciousness function in a larger decision space with a larger range of free will choices – they live in a larger, up-scale reality. The reality in which their awareness functions is a super-set of the reality experienced by a consciousness with higher entropy. That is an obvious conclusion when comparing yourself to a clam, and much less obvious when comparing yourself to a being of exceptionally high-quality consciousness, although the relative gaps are likely to be about the same.

Most of us have little appreciation for the depth and breadth of our ignorance. You are necessarily unaware of what you are unaware of. Some of us may even feel grateful for the merciful oblivion granted by that obvious fact of sentient existence. However, let me remind you that ignorance is bliss only in the service of maintaining a happily deluded ego. In all circumstances, ignorance (and the beliefs it generates) is a constraint upon vision, a limiter of awareness, a prison wall that prevents awareness from expanding beyond its present boundary, and is a great destroyer of potential.

I detect a look of worried consternation. Yes, the awareness gaps mentioned above actually are about the same, but hey, relax, unwrinkle that brow, there is no point worrying over what you don't know – the clams don't get it either. And, as everyone knows, clams are, well, happy as clams.

Big pictures are inherently difficult to see from the perspective of little pictures. Far out, unexpected, out-of-the-box paradigm shifts are **always** required. Oh no, don't look so disheartened – it is not impossible, just challenging. The fact that transcending belief and expanding your awareness is difficult to accomplish is not a bad thing. In fact, it is a necessary feature of all successful consciousness systems.

Undoubtedly, you see the virtue in restricting those dim-witted clams to experiences that they can deal with and learn from. Anything else would either go over their mushy little heads or greatly confuse them. How would you like to get mugged at the beach by a bed of Psychic Clams From Hell trying to exploit everything on the planet as well as each other? The non-negotiable requirements of consciousness evolution to earn your own way and pull yourself up by your own bootstraps are absolutely necessary to ensure that quality, power, and responsibility have the opportunity to develop together.

Do you feel immense relief at knowing there is a self-balancing merit system that keeps those brazen bivalves in their proper place? Jeez, the thought of a pack of wild clams watching sit-coms on TV, guzzling beer, and cruising the urban kelp beds in search of junk food sends cold chills up and down my spine.

Can clams and bumblebees learn (modify their actions and intents through experience)? Of course they can learn. Using memory, processing, and feedback to achieve self-modification (in reaction to internal and external environments) lies at the heart of our definition of consciousness. Consciousness has the natural, innate capacity to learn. Learning is purposeful self-modification created by the exercise of a free will that utilizes memory, processing, and feedback within a complex interactive environment. Learning, evolution, and growth – the steady decreasing of entropy – is impossible without the **functional** condition or attribute that we call free will. Are you beginning to see that free will, learning, and evolutionary growth opportunities are natural and necessary attributes of an evolving individuated consciousness?

Consciousness cannot exist without the ability to make self-determined, self-modifying choices. Without free will, there is no consciousness. Without consciousness, there is no free will. Consciousness and free will can not be separated – they are simply different aspects of the same thing. We shall see that it is our narrow, beginningless, PMR-centric concept of causality coupled with our misunderstanding of the properties of consciousness and reality that tricks scientists and philosophers into believing that free will is logically separable from consciousness.

The concept of evolving consciousness without free will is a mistaken and illogical theoretical construct that self-destructs in static, meaningless, determinism. The unintentional, but usually implicit, assumption of dead (directionless, non-growing, non-evolving, purposeless, non-living) consciousness creates a conceptual sinkhole, a philosophical dead end. A deterministic reality model can logically only chase its own tail. Though self-consistent, it leads nowhere and produces no useful output because its implicit little picture assumptions are fundamentally flawed in the Big Picture where consciousness lives, grows, and evolves.

Let's look at free will from an evolutionary perspective. Consider that evolution can increase the capability of an individual or species only within the limits of the natural capacity of that individual or species. A free will needs only to be free enough to make choices within its own local logical system and decision space. Within its local reality system, a sentient entity must be free to make choices that directly affect future choices. Note that complex interactive environments, intent, memory, processing power, feedback, and self-modification are the enabling mechanisms of both free will and consciousness. Free will and consciousness co-evolve as mutually reinforcing aspects of the same AUM system-thing. Free will evolves as a natural and necessary attribute of living consciousness.

Only consciousness systems that evolve a **practical** implementation of free will can continue to progress toward some measure of greater personal profitability. Without a measure of cumulative profitability (self-improvement), there can be no evolution or progress. Dead consciousness would never evolve or progress; it could accomplish nothing, not even existence. Beginningless little picture logic (see Chapter 18, Book 1) may grant theoretical existence to a deterministic dead consciousness that comes from nowhere and goes nowhere, but a larger view that better understands the origins and properties of consciousness realizes that "dead" and "non-existent" are logically equivalent when applied to consciousness.

Consciousness integrated with a free will is how the AUM organism must evolve in order to evolve at all. Free will is inherent to our governing rule-set. It is the nature of evolving consciousness (or evolving anything) to make specific choices from the billions of available possibilities. Results reflect massive complex interaction, ever-changing self-modifying feedback loops, and are cumulative. Learning takes place relative to the choices made. Look around – that scheme represents the fundamental nature of individual and collective sentient entities. That all sentient entities seem to reflect the fundamental properties and processes of evolving consciousness is an important data point to consider.

A consciousness system containing many individuated units is similar to a body of cells, or an internet composed of billions of individual computers – no one in particular is in control. Choices are made, information packets go here or there, results are the aggregate of a billion independent and interdependent decisions. These results drive further decisions, which drive further results. No individual plans it, or controls it, or runs it. It changes and evolves on its own according to its capacity, its environment, and the constraints placed upon it. The individual decisions and choices of each cell, internet user, or consciousness-unit are made according to immediate self interest – however self interest is defined or perceived at that moment. Free will is inherent to each cell, internet user, or unit of individuated consciousness. The decision space may be relatively small at the cellular level, but if there is sentience, there is also a finite decision space to support the existence and functioning of that sentience.

Consciousness and free will go together like inhaling and exhaling, like mammals and sexuality, like chickens and eggs. Like birth, life, and death, consciousness and free will represent a **practical** combination of attributes necessary for the balanced functioning of a real (as opposed to theoretical) evolving system.

Our free will does not need to come from some theoretical consideration or independent process – it is simply part of the system, inherent to the existence and processes of evolving consciousness. The rule-set that defines our local reality must necessarily express free will because that is how choice-making evolving consciousness operates. If one conceives of free will as being theoretically derived from some independent random process, a circular logic trap is created. Recall Chapter 18, Book 1 where we discussed the PMR belief that everything must be caused by something else (no beginnings are allowed). This belief logically forces us to account for the independent existence of the egg before we can allow the possibility of the chicken – or vice versa. So, which came first, the chicken or the egg? The question itself carries the assumption of a causality that eliminates the possibility of a constructive logical answer. From a larger perspective, the answer is obvious: It is clear that neither came first – they evolved together – just like consciousness and free will.

The appearance of a logical problem is created by an illogical question. Don't get caught up unproductively in the chicken vs. egg logical tail chase. It may appear to be a great mystery, but is only a misguided question based upon a poor understanding of the logical requirements of beginnings and the inappropriate application of the little picture's objective causality (see Chapter 18, Book 1).

Scientists, philosophers, and theologians should resist looking for a process that creates free will, or equivalently, eliminates determinism. That bucket has a hole in it. Free will does not have to be constructed out of smaller parts, or derived through a controlled analytic process – that represents a typically PMR little picture misunderstanding based upon a belief in causal processes that cannot logically support beginnings. Chickens before eggs? Eggs before chickens? Do you see the flawed logic that makes these questions appear deep instead of dumb? It is a similarly flawed logic that supports the concept of determinism by evoking a little picture causality that is devoid of an appreciation of a larger reality which is based upon dynamically evolving consciousness.

So, which came first – consciousness or free will? Do we conclude from our little picture logic that neither can exist? That sums up the position of contemporary science and philosophy: Mind is nothing beyond physical brains and biochemistry, all reality is physical, and all information is theoretically knowable and eventually predictable. That these **beliefs** run counter to the carefully collected data of everyday experience, are inconsistent with each other, and do not make good scientific sense is conveniently overlooked to appease the demands of little picture causality in particular, and scientific dogma in general.

With an open mind and fresh vision, it is not difficult to see that most scientists, and philosophers too for that matter, have employed cultural belief and professional dogma to paint themselves into an intellectual corner. That free will and consciousness must evolve together as natural and necessary attributes of any successfully evolving sentient energy-form is a thesis that solves many outstanding problems of science and philosophy. Apply this concept to a sufficiently complex digital energy-form like AUO and you get AUM and **you** – along with a lifetime guarantee of free daycare and pre-school services within PMR.

You know what is said about cornered critters being particularly dangerous: Every word of it is true and I would be remiss if I did not also warn you. Be careful, your professional and personal credibility can be savagely attacked and badly bitten by a vicious and tenacious dogma. In fact, most organizations and academic institutions, with fine reputations to uphold, have guard dogmas patrolling their halls. These old, mean-tempered, intimidating, politically powerful creatures are entrusted with enforcing a conceptual correctness that everyone who is important can be proud of. In the high entropy real world of PMR where ego-politics powers almost every nuance of every activity, you need to learn to apply gentle soothing strokes, scratch them behind the ears, and always carry an extra hotdog in your pocket. Perhaps fighting fire with fire is a good idea in some circumstances, but combating ego with ego in a dogma fight is always a disaster for everyone involved.

Our conceptual limitations often generate logical contradictions. Mind-matter, wave-particle duality, and entangled pairs fall into this same causality-confused basket. The Big Picture expresses the true nature of reality, while the little picture expresses the shared delusion of a group of individuated consciousnesses enrolled in the PMR learning lab (this concept is developed more thoroughly in Sections 4 and 5). The conflicts and paradoxes you see are not real and do not exist in the Big Picture. Logical conflicts and paradoxes appear to exist within the little picture view because of the erroneous assumptions of little picture science.

An analog: The professional magician only appears to saw the lovely lady in half. Spending your nights worrying about the apparent impossibility of the lady paradox (how ladies can be sawed in half and then be put back together again) and concocting complex theories of tissue micro-fusion are generally non-productive and will never yield a satisfactory solution because the obviously correct (I saw it with my own eyes) assumptions the magician (your culture) has led you to believe are, in fact, wrong.

When our belief in PMR scientific causality dramatically fails before our eyes, we have a tendency to build up elaborate theoretical structures to maintain our belief and save the sacred dogma of traditional objective science. Resist the urge to make free will or psi more complicated than they are. It is not that you **have** consciousness, but that you **are** consciousness.

Determinism is a philosophically unproductive, unworkable theoretical possibility based upon omniscience or perfectly defined processes for everything everywhere. It has no supporting real data and is generally based upon religious dogma (god knows everything) or erroneous little picture assumptions (scientific dogma – science knows, or can know, everything).

Without free will, consciousness is not consciousness – it is merely purposeless process. It is theoretically impossible to take the wetness out of liquid water or the coldness (relative to standard room temperature) out of ice. Consciousness without free will is like warm ice or dry liquid water. The concept of consciousness without free will creates a logical inconsistency. Because we of limited PMR vision do not appreciate the nature of consciousness, we separate the concept of free will from consciousness and try to give it a unique causality, an independent theoretical basis. The result is that we end up chasing our logical tail to conclusions that run directly counter to our everyday experience of individual consciousness.

Because of our little picture perspective, we project our PMR sense of finite knowledge into a theoretical assumption of omniscience, which eliminates free will and reduces consciousness to an analytical PMR physics or computer science problem. After that we are stuck with nowhere to go. No growth, no choice, no intent, no evolution, no personal consciousness, no purpose, no point. Determinism rules the land of limited dead knowledge. A complex consciousness ecosystem designed by evolution to be simultaneously out of control, in balance, and in a continual state of redefining itself leaves omniscience theoretically impossible.

The egg cannot logically exist without a prior chicken, and the chicken cannot logically exist without a prior egg – therefore, it is logically (given the implicit assumptions) impossible for either to exist. In terms of chickens and eggs, that conclusion is as dumb as it is logical. One might say that this conclusion is locally logical within the restricted solution space where the assumptions hold. In terms of consciousness, free will, and determinism, a similarly flawed process appears to provide an acceptable solution for many techies who have no logical way to derive a separate free will from the local causality of PMR.

Let's talk a moment about the implementation of free will within a local reality. Recall that a local reality is a unique dimension of existence and interaction – a computational subset – within the AUM system. That a **local** free will must be generated by a more fundamental **digital** consciousness system does not negate the effectiveness or functionality of that free will at the local level. A digital system can meet all of the requirements of a locally functioning free will operating within a restricted subset of the digital system. From the local perspective, the free will to express yourself (to exercise your consciousness quality) by choosing from the finite array of discrete possibilities that exist within your decision space is actual, effective, and real. That this self-modifying educational process is derived from a larger, more complex **digital** algorithmic system is of no consequence to the efficacy of the local choice-making process that enables unique evolution, growth, change, and learning within the local system.

Each conscious entity is exercising its free will to make choices within its own limited reality. Each knows little to nothing of what lies beyond. Think about each entity's free will in terms of practical operational requirements. Whether an entity is bright or dim, carbon-based, silicon-based, or reality-cell-based, it must make unique intent-guided choices within its operative finite decision space in order to find more profitable ways of existing and doing business. An operational free will is based upon each entity's specific memory, processing capability, and past and present input data (experience). Experience data are gleaned over time from the entity's perception of their internal and external environments. Think of free will as a practical evolutionary device of consciousness rather than a theoretical process of thwarting determinism with true randomness.

Functional free will, at whatever level of application, requires no more than the **practical** ability to make intent based choices where the intent is a function of the quality of the consciousness making the choice. If such a mechanism for making profitable choices at the local level (perhaps based on an evaluation of past choices) can be arranged, consciousness can provide the instrument for its own evolution.

At the local level where action is taken, an entity's decision space is defined by a finite set of discrete choices. Theoretical applications of randomness at the top level (consciousness evolving to be unable to look at the details of its own processes, or the invention of perfect or good enough random-number generators), are not relevant to the immensely complex, interactive, self-modifying, evolutionary processes taking place within local realities such as PMR. ◀

▶▶ Let's take a short break. I detect a few eyes beginning to glaze over and a head or two bobbing uncontrollably in the back row.

Everyone! Up! Up! Up! Yes, that means you too – get on up – and push that chair back! That's right, stand up! Stretch up straight and tall ….that's it. Now wait just a minute … until everyone is up. Come on Jake, get up, you must join us.

Now, hop vigorously on one foot while practicing your best and loudest Tarzan yell. Go on, hop, hop, hop, hop, hop … Louder! Louder! That's it, make Jane proud. Don't worry about what the people around you will think – you are about to give them a great learning-opportunity to practice compassion. As soon as you are done hopping and yelling, drop two ice cubes in your shirt, and if you are into extreme sports, drop one in your underwear as well. When they have melted, sit down.

Phew! Wow! Didn't that feel great?!

That is Uncle Tom's patented wake-up technique – guaranteed to work every time. If you actually bought this book with your own money, you have my full unrestricted permission to use this technique any time you want to – one of the many benefits of Big TOE ownership that accrue to you because of your wise choice of reading material. We are almost done so hang in there a while longer and you might learn something useful. Don't look at me like that ... you never can tell, it could happen. ◀◀

▶ Perfect knowledge (the omniscience needed to support determinism) does not, will not, evolve because it is not a practical possibility within real, interactive, self-modifying systems that are large and complex. From the perspective of evolution, a stagnant determinism is not profitable to the system. Self-improvement, learning, and meaningful goal-directed growth are profitable to the system – and these, by definition, cannot exist within a deterministic system. The meaningless random results of meaningless random processes can produce no increase in cumulative profitability. Such a system cannot support the properties and quality of consciousness as we experience it. An evolving consciousness system like ours cannot be supported by either a wholly random or a wholly deterministic system because there can be no cumulative profitability in either.

Are you beginning to see the connection between free will and our two basic assumptions (consciousness and evolution – see Chapter 24, Book 1)? Given the dynamic duo of consciousness and evolution as we have described it, free will falls out as a necessary logical result. It is not an added ingredient that somehow must be accounted for. Free will is simply the result of consciousness energy and evolutionary process slipping into bed together for a joyous moment of creation that has not yet ended. From that union, all reality and existence flows. Our two basic assumptions not only allow and account for free will, but logically demand it and then create it out of a successful synergistic interaction.

Theoretical omniscience is meaningless within a real evolving consciousness system. Because real systems come into existence through **profitable** evolution, those that do not have the design, rule-base, or structural processes to evolve toward being more and more useful and profitable to themselves go nowhere; they do not grow or become more. Instead, they become stuck in an unfocused, meaningless, unguided, unprofitable, process (whether random or deterministic) that can never actually form into a real system. They do not become. They remain high entropy nothingness or represent inert meaningless process. Evolution cannot progress them forward toward greater profitability. As rejects of evolution, they go away, die, and disappear unable to maintain a coherent existence.

Incoherent existence can not persist or converge to a working system. By definition it must dissolve as easily and readily as it forms. A consciousness system evolves (increases its profitability) by decreasing its entropy. By definition, randomness and determinism can have no long-term goal, point, purpose, or profitability – no successful, real, working (dynamic) system can be generated under these conditions.

All of reality, as far as we can know it from the data gathered thus far, fits the form of an evolving consciousness system that has obvious rules, focus, and purpose. Growth toward greater profitability permeates all existence. There is no indication that existence is either random or without dynamic purposeful profitability, however, there is a preponderance of circumstantial evidence to the contrary. If you have no experience of PMR's purposefulness or of the attributes of consciousness – even if you are totally without knowledge of your greater purpose and completely distrustful of your intuition and subjective knowledge – still, you can find no indicator pointing to a random, static, or deterministic reality.

Construing ignorance of purpose as purposelessness is a logical error. Existentialism made this error because, although it clearly saw the crippling limitations of many little picture belief systems, it was unable to transcend its own little picture belief in a universal causality. With no beginning (understanding of consciousness and our connection to it) and no end (evolutionary purpose of consciousness), existentialists are left drifting and rudderless.

Consciousness and free will are of the same evolutionary root. Like the trunk and branches of a single tree, they must grow together – inseparably joined and successfully evolving as one entity. The system of free will and consciousness evolution works because it was designed (has evolved) to work. The rules of the game of evolutionary success define it into existence and maintain its integrity.

With respect to big PMR-based digital simulations (war games or meteorological modeling, for example), the people who build and use these simulations do not know how the results are going to turn out; if they did, there would be no need to develop or run the computer models. Should AUM be any different? The complexity of these simulations, as well as the modeled randomness, is what makes the results (output) unpredictable (for a given set of input values) and useful.

Let's have some fun and strain our brains a little. Imagine that some simulation uses billions of billions of neural nets and fuzzy logic and lots of other non-linear self-modifying imprecise functions we haven't invented yet. Let it be a billion, billion, billion times larger and more interactive among its objects than whatever it is you can possibly imagine. Let each object be allocated its own unique memory, processing capability, and set of multi-layered (system goals and personal goals) profitability algorithms, some of which are self-modifiable. Perhaps every object interacts in very complex and conditional ways (with some uncertainty or ambiguity tossed in) with every other object. Also, let billions of billions of intermediate outputs of the simulation automatically modify the simulation's inputs toward some larger purpose (winning the war or predicting the weather), and you will have the tiniest sliver of a shadow of one small calculation space (reality dimension) within TBC.

Next, change the random number seed and vary all input values that are expected to have some intrinsic variation or ambiguity. You might also want to modify some of the distributions that define the operational properties of specific statistical activity as the simulation runs in order to create specific situations or conditions of interest. Implementing these changes on-the-fly may dramatically alter both final and intermediary results. Implement all such changes in a systematic and clever way over many iterations (like parametric analysis) and perhaps the collective results will form an exceptionally meaningful statistical ensemble.

Perhaps the simulation will run for such a long time that it will seem like forever to the individual objects who keep time by counting their own processing cycles. Consider that many of the objects could exhibit the four attributes of consciousness and have limited access to each other's data. They may pass data back and forth interacting (making free will choices) with each other according to their goals, self-modified defining algorithms, and within the bounds of their shared-reality defining rule-set. One higher-level goal may be, for example, to lower the entropy of their individual assigned calculation space. Contemplate the advantages to AUM of an entire set of independently seeded PMRs.

This is only a start; this game can go on and on. I am sure your imagination can raise this hypothetical simulation to higher levels of complex and meaningful interaction. I simply wanted to get you started and to help you imagine a dim glimmer of the origin of the free will needed to support profitable choice-making within a limited local reality.

Free will is part of the consciousness evolution game in the PMR learning lab. The free will consciousness evolution interaction is only required to be a practical process – a functional way to derive profitable, convergent evolution from local experience. Free will does not need a separate theoretical basis; for us, it is a practical methodology for growing consciousness quality (lowering consciousness entropy) in local realities. For a local consciousness system such as PMR, a locally derived pseudo free will that supports intent guided choice and feedback within the available decision space of each individual is necessary and sufficient.

In our big simulations, we use similar processes to gain insight and knowledge by employing pseudo-random numbers to add a greater sense of reality to our simulations. This randomness actually adds accuracy to our calculations because within our reality choice-making and ambiguity are natural attributes of both sentient and non-sentient entities. We believe that we use randomness in our simulations to make up for our ignorance, to fill in for the details we cannot easily express analytically. What we do not realize is that our ignorance runs much deeper and is more fundamental than we suspect. We think, because of our little picture perspective and belief in a beginningless causality, that "perfect knowledge" is theoretically obtainable and that it would necessarily produce determinism. A better understanding of the bigger picture points out the theoretical and the practical impossibility of perfect or complete knowledge within an evolving consciousness system. The highest fidelity model must contain randomness.

Quantum mechanics bothered many scientists (including Albert Einstein) because it seemed to posit a statistical basis for our reality. It seemed obvious that sentient beings and their reality were fundamentally more real, solid, and dependable than could be attributed to a statistical representation. However, when you understand the digital nature of consciousness, realize that PMR exists within a calculational subspace (dimension of reality) of TBC, know that we physically interact according to a shared rule-set that defines the perceptions of our individuated digital consciousness, and appreciate the interactive nature of intent, free will, and choice-making that leads to entropy reduction in complex systems, it is not at all surprising that our fundamentally digital-mathematical rule-based perception of existence should display statistics at its root. The surprise would be if anything other than digital, quantized, statistical entities were found at the most fundamental level of our reality. At the level of physical detail where quantum mechanics applies, one is dealing with the individual pixels of the virtual physical reality we call home. That these pixels turn out to be statistical representations of the potential for existence within PMR waiting for a consciousness to collapse their wave functions to a measurable solid result is precisely predicted by My Big TOE.

We have just learned that individual free will is also an expression of a necessary condition for the evolution of consciousness. The PMR learning lab is defined into useful existence by its rule-set which determines perception, awareness, and causality. (Causality defines the logical relationships and allowable interactions between various objects within a local reality or given dimension.) The PMR space-time rule-set constrains the consciousness of those particular individuated units participating in PMR to experience only what the rule-set determines to be appropriate for the efficient functioning of that particular reality. Thus the PMR rule-set appears to define and bound a **limited** knowledge within PMR.

The illusion of determinism is the illusion that this rule-set represents all reality and not merely the local calculation space, reality, or dimension of macro-existence we call PMR. From the PMR perspective, our limitations and the local space-time rule-set that defines our causality team up to produce the illusion of determinism in order to provide us with an optimal environment for learning through experience. Efficient learning requires definite structure. A system without structure has no potential for profitable growth. Everyone knows that a lack of structure is antithetical to the successful development of children. Structure, by its nature, sets limits or provides constraints. Optimizing our evolutionary opportunities requires the constraints of the PMR rule-set. The appearance of a deterministic causality is the result of a limited understanding extrapolating the restrictions of an imposed local structure upon all of existence.

By design, little picture knowledge appears (from the view of PMR) to be deterministic – an erroneous conclusion based upon the success of science in discovering more and more of the space-time rule-set that defines our local PMR causality. However, there is more to the experience of PMR than the rule-set that defines the possible interactions within PMR. We need to account for the experiencer as well as the logical constraints of the experience. PMR is a virtual reality that is designed to produce a certain type of constrained experience for the benefit of interactive units of individuated consciousness. Consciousness awareness is the active element that experiences the opportunity to exercise its intent as it interacts with virtual mass, energy, time, and other consciousness units that also possess free will.

Only when mind and consciousness are **assumed** to be nothing other than PMR physical brain phenomena does PMR begin to appear totally deterministic to some philosophers and scientists. These folks believe that their conscious awareness is derived completely from a complex physical bio-computer (brain) which interacts with its physical environment. Not a bad guess, given the viewpoint from which it arises. I support the notion that computers can develop consciousness but that is not the rationale behind this particular assumption. The assumption of a physically-based consciousness is a logical requirement of the little picture – it is made to maintain the belief that reality cannot be other than physical – that our causality is universal.

Consciousness that is experienced within the PMR training simulator may appear to be brain centered, but that connection is only a shadow on the wall of the PMR cave. Consciousness is the invisible medium upon which your individuated awareness floats – much like the fish that cannot perceive the unchanging water it swims in (see Chapter 23, Book 1). The evolution of consciousness follows a greater purpose, logic, and causality that provide the key to a better, more productive understanding of both the little picture and the bigger picture. A nonphysical consciousness-based Big Picture reality enables the full range of our accumulated human experience to make good sense within a single integrated and coherent theoretical structure.

The **traditional** little physical picture model of reality creates as many Big Picture paradoxes and problems as it offers little picture solutions. By comparison, a **deterministic** little physical picture model of reality creates additional Big Picture paradoxes and problems while offering few, if any, new solutions. These paradoxes and problems are traditionally dealt with by stretching old belief systems and establishing new ones. As always, beliefs are used to ease the anxiety of ignorance and make our knowledge seem more complete than it is. The little physical picture model of reality has been unable to produce a **little** TOE that unifies our understanding of the space-time rule-set that defines PMR causality, much less produce a satisfactory **Big** TOE that not only fully explains little picture space-time causality, but also explains the greater human experience as well.

Given the algorithmic digital nature of the space-time rule-set, a little TOE may or may not exist, however, a Big TOE that solves profound contemporary problems of philosophy, metaphysics, and physics all at the same time and within one overarching theory must exist because we are here and are as we are. Our existence and human nature exhibit clear universal patterns that contain much more consistency and direction than randomness – a fact that must have a holistic, comprehensible explanation at its core. Because our existence is about us, about who, what, and why we are, surely the explanation lies accessible within us. However, until it becomes clear that a non-mathematical logical analysis of the origins of free will, consciousness, spirituality, or paranormal events **could potentially** represent accurate and honest **science**, that explanation will forever remain beyond your grasp.

Some may feel that philosophy and metaphysics do not pose real or legitimate problems because their solutions lie outside science and are therefore impossible to solve analytically. Bullpucky! These problems yield to accurate knowledge the same as any other. The difficulty is not that logical scientific solutions are impossible. The difficulty is that a self-limiting belief-based science inadvertently makes it impossible to comprehend the correct answer.

Science has the mission to pursue **all** knowledge leading to a better, more profitable understanding of the natural world – not merely the slice that falls within the confines of its traditional belief systems. Discovering fundamental truth and developing useful solutions are what science is all about. To find truth you must go wherever it leads. Belief-blinded closed-minded individuals who travel the path of least resistance and derive their respectability from supporting the beliefs held in common by their peers choose the safety of the herd over the ability to discover Big Truth. To maintain that respectability and the ego and material rewards that come with it, these individuals give up the ability to understand what is critically important to them and to their professions. A sad story of self-inflicted wounds which is so common that it defines normalcy and sets the standard for professional success.

Consciousness exists in many forms and at many levels, capacities, and scales – each built upon and extended from the others through the repetition of a simple process iterating its way toward an improved profitability. The result is a cascade of evolutionary creativity that aggressively explores all of the forms, configurations, and embodiments that consciousness systems can employ to improve their profitability. This energetic living complex pattern of evolving consciousness is what we have metaphorically referred to as the consciousness-evolution fractal. It is the nature of fractal processes to generate monstrously big pictures by recursively applying a few basic rules and assumptions.

It is a mistake for an entity within this vast reality fractal to believe its own tiny local reality subset of the greater consciousness ecosystem is The One, The Only, The Center, the pinnacle of creative expression. Individuals with little picture views can see nothing other than themselves emerging from the possibilities of existence. Such a limited vision produces not only a misplaced determinism, but robs its owner of the knowledge he or she needs to actualize personal potential.

Only a failed and desperate theory (cultural, social, theological, or scientific) will feel the necessity to deny the existence of the facts it cannot explain. The facts of science, consciousness, free will, paranormal events, and the human spirit are left lying about everywhere; open your eyes, explore them for yourself; they are not secrets, nor are they difficult to find once your mind is opened to the possibilities.

Let's make a quick connection between free will and psi phenomena. Both are part of our reality because they are natural attributes of consciousness. Free will must be part of our little picture experience because we are consciousness with the mission to evolve and we cannot do that without the ability to make free choices. The free will inherent to consciousness must be fully operable in PMR or the learning lab would be unable to support learning. Likewise, psi effects must be largely constrained in PMR or the effectiveness of the learning lab would quickly degenerate. The psi uncertainty principle is the mechanism for maintaining that particular growth-optimizing balance.

Everything has its purpose. A local deterministic causality reflecting the PMR rule-set is required to provide the overall structure within which our experience can be profitably defined and coherently generated. An effective educational opportunity that encourages consciousness to evolve must engage free will, limit serious-psi to those who are likely to use it in pursuit of consciousness quality, and provide the structure of an apparently objective rule-based causality to rationalize experience.

We will more carefully define psi effects and their relationship to consciousness and consciousness quality, as well as introduce the psi uncertainty principle in Chapters 47 and 48 of this book. At the experiential level, imagine psi phenomena as a brightly colored flower enticing those who have some potential to profitably understand it, and as a prickly, scary, or ridiculous weed to most others. Anyone can experience psi phenomena or psi effects. Many can wield a shadow of its power but only those who understand it deeply can effectively use it as a springboard to significantly reduce their personal entropy and the entropy of the system of which they are a part.

Let's get more specific by defining "psi effect" as an acausal (outside PMR causality) phenomenon that is attributable to the operations of consciousness. From this more technical definition, let's explore a few semantic twists and turns. One could argue that free will is a psi effect because free will and psi effects are fundamental, interdependent attributes of consciousness. Similarly, describing or defining consciousness simply as "psi" can be supported from a general conceptual perspective. A free will effect is a psi effect is a consciousness effect.

Consciousness is the energy system from which our reality is created. For us, consciousness is the root of all reality. What we call psi effects and free will are direct expressions of the fundamental attributes of that consciousness system and of how that system operates and evolves.

If you want to know whether or not you have free will, simply ask yourself whether or not you are conscious. If the answer is "yes," you have free will; if the answer is "no," you are probably a lawyer and have simply confused the word "conscious" with "conscience." If after consulting a dictionary, the answer is still "no," you must be either a lawyer who has been elected to an important government position or an impressive professor (perhaps the dean of faculty) at a major university. In either case, put the book down verrrrry slowly – be careful not hurt yourself – and call your mother to come get you. She is probably very worried.

All is consciousness. Consciousness is all. Do not think of NPMR and PMR as **separate** places with fixed mental pathways between them: That is a sometimes convenient, but misleading concept. The psi uncertainty principle, as a part of the PMR rule-set, provides necessary structure to PMR by imposing limits. You cannot build a physical or nonphysical bridge between PMR and NPMR. The physical bridge cannot be solidly connected to NPMR, and the nonphysical bridge cannot be solidly connected to PMR. You are the bridge, which is why you cannot build one exterior to yourself. Consciousness is personal.

You are an individuated unit of consciousness hallucinating a physical reality (perceiving a virtual physical reality). As such, you can, within the limits of your quality and ability, experience and apply free will and psi effects because you **are** consciousness. However, you cannot make your hallucination (virtual physical reality) exceed the limits and function of its defining rule-set.

Is this not as simple, elegant, and straightforward as a bowl of plum pudding? Oh, I see, you have never experienced a bowl of plum pudding. You are in luck. The next two sections will develop the recipe and cook up a yummy batch from scratch.

We have once again jumped ahead of our story and taken an unsubstantiated peek into what lies ahead. However, you will find the logical development leading up to these unusual statements resides primarily in Section 4, which is around the next corner. Furthermore, in Sections 4 and 5, we will discuss the nature and the mechanics of experience, multiple PMRs, and local realities.

To summarize, free will is a natural and necessary part of our consciousness. Without it there is no profitability, learning or growth – consciousness itself dissolves. Look around – outside and inside yourself – the data are clear. There is purpose and direction to everything. All existence is in a state of continual profit-based evolution. Our reality is large, complex, and interactive. Perfect knowledge leading to determinism is like the concept of dry liquid water: a practical impossibility born of an incomplete understanding. From the view of a little picture causality, determinism may seem scientifically logical as well as theoretically unavoidable as one approaches the limit of perfect knowledge within PMR. However, from a bigger picture we see only a local structure required to support the coherency of experience within a small subset of reality called PMR. Determinism is a local rule-based virtual-reality illusion, engineered within TBC for the purpose of optimizing consciousness evolution within the PMR learning lab.

Determinism believes that you are a rule-based body hallucinating a consciousness with free will, when in fact, you are a consciousness with free will hallucinating a rule-based body. Once you see the Big Picture (by the end of Section 5), you will understand that determinists, by definition, must be unconscious – an apparent paradox that is easily verified by logical analysis.

If you, like me, have never lain awake at night worrying about the impossibility of true randomness in digital systems and what, if anything, that has to do with predestination and free will within a digital consciousness system, you didn't need to read this aside. However, if you read it anyway, be sure to ask your mother for an additional gold star – of course, that's assuming she didn't see you applying Uncle Tom's patented wake-up technique.

If all you learned from this rather esoteric discussion is that a digital consciousness system can generate the free will we require to decrease the entropy of our personal consciousness, you got it all. Enough said: this particular point, though vitally important to many philosophers and scientists, is off the critical path to Big Picture understanding for most of us.

I feel freer already. ◀

Now that the left half of our techie-friends' brains are fully occupied with the interconnectedness of free will, psi effects, and consciousness – and with the unexpected lack of determinism in the big digital picture – I think it would be beneficial for us to step up to a bigger picture of the bigger picture. Because most of us have been immersed in a detailed theoretical exposition of free will for the last half hour, now is a great time to take a broader view of where we are, and of the role our individual consciousness plays in the larger evolutionary drama. The next chapter does exactly that, furthermore, you will be pleased to know it is also brief and contains no asides.

In the preceding free will aside, we have been risking brain-strain in order to lay out some of the infrastructure of the science of consciousness. Having accomplished that, we can now relax. Whenever possible, I avoid dragging *My Big TOE* through the conceptual weeds, but sometimes logical thoroughness and the details of process are necessary to support a general understanding. Though sharing some of the more technical theoretical details of *My Big TOE* is important and necessary, this trilogy is primarily focused on a big picture look at the Big Picture.

In fact, any major new branch of science must begin its unfolding as a top-down big picture analysis of the pertinent data. New scientific paradigms generally begin at a broad conceptual level and then eventually move, over time, toward more specific applications. Such a top-down process often begins with the discovery of a more general relationship between common objects or events. Typically, the old paradigms become special cases of the new – applicable only under certain conditions.

The basic conceptual foundation must be clearly laid-out and thoroughly assimilated before the more traditional bottoms-up scientific approach can be constructed upon it. Only after scientists achieve a certain minimum level of basic understanding can they begin to profitably calculate and experiment their way toward a deeper and more productive knowledge.

Approaching a new understanding of reality by a bottoms-up application of the known tools of physics is like trying to chop down a mighty oak tree with an ice pick. A rather dull ice pick at that, because the assumptions of physics are thoroughly rooted in little picture causality. The broad sharp blade of an unfettered mind logically focused on the bigger picture of cause and effect is a much better tool for whacking stout ontological trees. As always, the proper approach implemented with the proper tools most readily finds success.

The point is: The **public message** of *My Big TOE* is focused at developing the conceptual foundation upon which a more traditional future science will one day be based. You see, I have not written *My Big TOE* to frustrate and torture left-brained techies with incalculable generalities; the time for calculation has simply not yet arrived. At this point in the conceptual revolution presented by *My Big TOE*, everyone can come along on the journey. Scientists have no particular advantage over nonscientists as long as I am careful not to couch my descriptions in techno-jargon or be too techno-boring. In fact, at the front-end of discovery, right-brainers, and better yet, whole-brainers have the decided edge in leaping new paradigms in a single bound.

Some patience is required. By the end of Section 6, *My Big TOE* will have provided a rational, logically consistent Theory Of Everything, developed the required new paradigms to support that theory, constructed a solid scientific foundation for future explorations to be built upon, and explained the interfaces and connections between newly derived knowledge and the existing database of scientific and personal experience. It will have subsumed physics, redeemed philosophy, and explained many public as well as private phenomena. A broad conceptual top-down exposition of the fundamentals of Big Picture Reality is where this train must start. If and when the Big TOE conceptual brain-train begins to move and pick up speed, it may well initiate the beginning of an independent outpouring of Big TOE science, social science, and philosophy that will begin to take root immediately, and slowly produce fruit over time at ever greater levels of specificity. That is how science moves from concept to application – very slowly, and with casts of thousands. Increment by increment, discovery by discovery – the more people who get involved and apply their brain power, the more quickly the train picks up speed.

Once the ground is broken and the foundation laid, history tells us that many bright, open, and curious minds will begin to build upon it. Thus, as always, the ball eventually bounces back into your court. Perhaps your career or personal growth path, ability, or intuition will urge you to apply these new understandings and paradigms to whatever commands your professional and personal attention. If a significant number of readers are so moved, the time for public and private calculation and experimentation (developing and applying the detailed logical implications of this Big TOE) will arrive more quickly than anyone expects. Nevertheless, by evolutionary design, movement toward profound shifts in widely held paradigms usually progresses slowly and with deliberate caution. Patience will always be a necessary virtue.

Because the experience of consciousness is essentially a personal one, the **private message** of *My Big TOE* is for you, dearest reader – yes, just for **you**. The real power to effect significant positive change lies not with governments or professions, nor with science and the technology it spawns. The real power behind meaningful progress and growth, at any level of aggregation or organization, is the quality of the individual consciousnesses driving the action. Raising that quality can be accomplished only by the **self-directed** personal evolution of each individual. Clearly, the private message of *My Big TOE* is offered up for **your** personal use and benefit. Its potential value lies not in the words, nor in the ideas, presented. The potential of its value is the possibility of positively influencing the potential of your value – a potential that is profound and beyond estimation.

46

■■■

The Chairman of the Board, and Our Probable Relative Importance

■■■

It will be helpful to our overall Big Picture understanding if we can frame a more accurate and humble perspective about the nature of our reality and our place in it. Some, who for their own reasons, have not found a satisfying bigger picture within a traditional philosophy or religion, often posit an unknowable something such as "The Mind of God" (MOG), as Stephen Hawking refers to the term at the end of his book, *A Brief History of Time*. Even something that is as expansive, open, and mystical as the Mind of God usually represents only a little picture view of AUM because it is derived from the diminutive little picture perspective of god the creator and animator of our physical universe.

Let's review how our universe relates to the larger reality. In Chapter 33, Book 1, we pointed out that our universe occupies only a minuscule sliver of the available physical reality-space within our own space-time dimension. Likewise, we will see in Chapters 75, 81, and 82 of Book 3, that our local reality represents just one branch of many divergent probable realities. Recall there are many independent PMR-like realities that exist within $NPMR_N$, which is itself only one of several more or less independent reality systems (one of the $NPMR_n$). The point is: Our favorite PMR universe is not exactly at the core of the larger reality. In fact, its overall relative significance would appear to be almost infinitesimal when compared to the sheer bulk of All That Is.

AUM is thus chairman of the board and Chief Executive Officer of MOG Corp., as well as the entire evolutionary experiment itself (substance, results, conclusions, and open ended possibilities – all combined, and all at the same time). Continuing with this decidedly funky but business-like metaphor, the Big Cheese is the division manager of "N" division, dedicated to a specific subset of evolutionary development models containing many PMRs. Progressing on down the chain of command, one eventually gets to a few assistants and their helpers who are concerned with our beloved OS and its inhabitants. Our physical universe is a subset of OS.

Given that one or more of these low-level assistants directly concerned with the administration of OS applies for the job of "God of PMR," then AUM would be the Mind of the God of god's god. Yes, I am just having some fun, but you get the point.

If the Big Cheese's division is especially profitable, perhaps he will be promoted to junior vice president one day and be awarded a private space in which to park his mind, stock options, omniscience without looking, and a grand vacation where he can get away, you know, go out of his mind for a few weeks. What are you in this goofy analogy? You, my friends, are the rank and file peons who appear (from your point of view) to do all the work and receive none of the credit, just as in real life. Perhaps getting the credit is not as important as it first appears.

In a light metaphoric moment, you could refer to AUM as the Mind of the God of god's god, but AUM is also much bigger than that. MOG is only one hat that, from our viewpoint, AUM **appears** to wear because of our minute perspective of the concepts of reality and god. That tiny perspective comes naturally, because we (us, our reality and history) are merely an itsy, bitsy, teeny, weenie yellowish polka-dot on a much bigger picture that is itself wrapped around something more grand and sublime than we can imagine. Sorry, but we are not likely to be at the center of this bigger picture; the odds are clearly against it. We are not likely to be at the center of our PMR universe, much less the center of **our** larger reality, much, much less the center of the Big Cheese's reality, and much, much, much less the center of AUM's reality.

Be realistic. We (Our System) are important, but so are many other history threads running through other universes, in many other reality systems spread over many PMRs, which reside in the many NPMRs that exist within each of the many evolutionary manifestations of AUM. Itsy, bitsy, teeny, weenie is probably an outrageous overstatement of our likely importance relative to the whole. We are, by the odds (statistically), likely to be much smaller than that.

Even if we are more important than the statistics indicate, there is no advantage in knowing that fact and there may be considerable disadvantage (not to you and me in particular, of course, but surely to others).

Such knowledge would constitute a serious disadvantage if it encouraged egos to invent grand theories of our special importance (as a species or as individuals), which could, in turn, lead us to see everything else (beside ourselves) as relatively unimportant and insignificant. Self-importance might breed arrogance and lead us to ignore opportunities to improve ourselves. Convinced of our special value, we might end up with a haughty disregard for the significance of others as well as a disrespectful attitude toward the ecosystems that sustain us. Do you think that could happen? Are humans capable of that type of blind arrogance? To be on the safe side, I think we should be careful not to encourage these particularly dysfunctional attitudes even if they do represent traditional values.

Do you think I have been unnecessarily harsh in laying out some of the conceptual foundation for the description of reality that will follow in the next two sections? Have I been ruthless in deflating egos, bursting cherished bubbles, and pushing difficult and discomforting concepts at you from every direction? I hope you don't feel that way. To me, the logical premises and conclusions found within *My Big TOE* are emancipating, empowering, and fun. I hope you are finding them interesting and potentially useful as well, and that you are encouraged to pursue and self-direct the evolution of your consciousness.

I hope you are having fun stretching your mind around these ideas, whether you are finding solid agreement with your personal data or carefully formulating a debunker's rebuttal. Lighthearted fun is an important component of learning.

47

■■■

Uncertainty and Manipulating the Future With Intent
The Nature of Psi Phenomena – Measurement and Validation
Taking the Path of Knowledge

■■■

We have discussed how AUM and we are related – how our nature and purpose derives from the nature and purpose of the evolving digital consciousness system we call AUM. We have seen that we are a part of a larger consciousness organism that is dependent (as all growing and evolving complex systems are) upon all of its parts pulling together to ensure its continued growth and existence.

In this chapter, we shift gears and explore the larger reality from a more human perspective. We'll take a look at how uncertainty, constraints on consciousness, psi phenomena, and spiritual growth are related to our existence in PMR and in the Big Picture. This chapter will tie together many of the ideas we have been talking about from a more personal perspective.

From the title of this chapter, it would seem that a random inelastic collision in topic space has glommed together three loosely related, but separate topics. In fact, these three topics are so thoroughly intertwined that considering them together, within a shared context, will enhance the understanding of each.

To create an entity or bounded system, one must impose constraints upon some energy-form such as physical mass or nonphysical consciousness. These constraints define and maintain the entity. If the constraints are removed, the entity will disintegrate, dissipate, or dematerialize. Non-conscious, non-growing, inanimate objects (from the PMR point of view) tend to stay in the state in which they are left unless there is some applied energy, or some background energy available to remove internal or external constraints (causing motion, state changes, degeneration, decomposition, dematerialization, or sublimation). The entropy of inanimate objects, if left alone, tends to increase – this fact is expressed by the second law of thermodynamics (see the entropy aside near the middle of Chapter 24, Book 1). The second law of thermodynamics results from the fact that natural physical processes are irreversible and seek the lowest available energy state – it is a natural artifact of the functionality of space-time.

The creation of a conscious entity follows the same rules as the creation of any entity except conscious entities have the ability to modify themselves with free will intent. External constraints are imposed on a subset of conscious potential energy to form an individuated consciousness that is an individual sentient entity. Additional external constraints are imposed on the interactions and perceptions of that individuated consciousness (such as through the PMR space-time rule-set) in order to create an environment that facilitates evolutionary progress. However, with the attributes of consciousness, a sentient entity can create its own internal constraints. The internal constraints that a sentient entity typically imposes upon itself will inevitably limit the evolutionary potential of that entity. Some self-imposed constraints may appear to serve a temporary purpose and some are obviously more dysfunctional (unprofitable) than others, but all must eventually be removed by the entity if it is to actualize its full potential.

Adding uncertainty and minimizing constraints within a given virtual reality will always generate wider evolutionary potential for the entities within that reality because greater freedom and additional choices support a more diverse set of potential outcomes. When you are caught in a belief trap (have a belief), you create an additional **internal** constraint that further limits the capacity and function of your consciousness. Creating or maintaining unprofitable internal constraints that limit your personal consciousness requires energy (dysfunctional organization) and increases the entropy of your consciousness.

Though constraints may configure a system for some specific purpose, they generally limit the potential energy of the system. Internal constraints within a consciousness system that limit growth and profitability represent suboptimal or profit-inhibiting digital organization. Organizational energy is expended but the results are counterproductive (are dysfunctional and unprofitable to the system or block future profitability). Think of self-imposed internal constraints in terms of a negative digital potential energy or a negative potential synergy.

Adding unprofitable internal constraints (belief, fear, and ego) to your consciousness is analogous to scattering bricks and concrete blocks on a highway. The highway was built by employing **profitable external** constraints (all traffic confined to a convenient and efficient roadway of a specific type and location), while littering it with obstacles represents the addition of **dysfunctional internal** constraints that reduce the roadway's functionality and efficiency.

The energy of consciousness has the potential to organize itself more effectively – to develop more complete and brighter awareness, generate and creatively utilize understanding, feeling, caring, love, and directed mental effort in a way that is useful and helpful to itself and to other consciousness struggling to evolve. Each individuated consciousness generates its own unique path toward increasing or decreasing profitability, yet all are expressions that reflect the process and purpose of the whole. Better quality within the larger consciousness system is actualized through the synergistic self-improvements of its individuated parts. Entropy reduction is the name of the game at all levels and within all dimensions of the consciousness-evolution fractal wherever evolution is succeeding. Wherever evolution is not succeeding, that portion of the consciousness-evolution fractal is stagnant, deteriorating, or dying.

The entropy within a consciousness system (including an individuated subsystem) represents the unavailability of consciousness energy to do work. The primary work of consciousness is to more effectively organize itself, to take charge of its own evolution, and to increase its overall profitability through self-modification (growth). Higher entropy states within consciousness represent unprofitable organization: disorganization, fears, beliefs, dimness, diminished potential, self-centeredness, and an inability to understand complex interrelationships or see big pictures. Higher entropy results in consciousness systems having less power that can be applied to overcome the inertia of ignorance and ego dysfunctionality. Not decreasing the entropy of consciousness results in a squandering of potential.

The fewer internal dysfunctional constraints (such as fear, ego, needs, or beliefs) that limit a particular consciousness, the lower its associated entropy and the more effectively it can populate the most profitable states available to it. The drive or urge to be helpful to, and care about others (love) is an innate property of low entropy consciousness. A low entropy consciousness is an effective and powerful consciousness. Consequently, a low entropy individuated consciousness represents an effective individual with considerable personal power.

The self-removal of internal self-imposed constraints from sentient entities provides opportunities for reducing entropy and increasing synergy, while internal constraints removed from inanimate objects provide opportunities for increasing entropy.

Constraints come in many forms. For example, if someone who has the ability to manipulate nonphysical energy is asked to remove (dissipate or dematerialize) a tumor (noticeable lump) from someone's PMR body, the energy required is dependent upon, among other things, the degree to which this tumor is connected to PMR reality – its force of being in PMR. (This hokey sounding "force of being" is actually well-defined in Chapter 83, Book 3). Quantitatively it is the expectation value of the future event on the associated probable reality surface – a discussion that we will have in detail in Chapter 79, Book 3. If the body's owner and a few others are mildly distressed about the **possibility** of a malignant tumor, the removal energy may be relatively low (easy to accomplish). In this situation, much uncertainty exists within PMR – but not necessarily for the one viewing and manipulating the body's nonphysical energy from within NPMR – to that person, the nature of the lump may be perfectly clear.

If, on the other hand, four doctors and half the residents at the local hospital have looked at the CAT scan or felt the lump and are relatively certain the tumor is malignant, the removal energy is somewhat higher. When all of the above get the biopsy report confirming a fast growing, incurable, always-deadly malignancy, the required energy increases. The more firmly the malignant tumor's existence and likely outcome (degree of causal certainty) is held and shared in the minds and expectations of credible sentient beings, the taller and denser its probability function becomes. The uncertainty of the outcome dissipates and the probable event of dying in PMR from this cancer becomes much more difficult to change by manipulating energy in $NPMR_N$. Fortunately, the intent and attitude (mental focus) of the body's owner has the greatest potential impact. Unfortunately, this attitude is often driven by its fear and the opinions and fears of others.

Now, by definition, it takes a miracle, where before no miracle was required. The confident knowledge of the doctors, which is based on test results and historical precedent (mortality statistics), actually affects (decreases) the probability of actualization of **other alternative possibilities** such as the cancer spontaneously going into remission, or the tumor turning out to be benign. As in quantum mechanics, performing the measurement forces the result to pick a state compatible with PMR causality (compatible with the PMR space-time rule-set). Typically, the most likely state at the time of the measurement is picked unless there are several states of equal probability, then one outcome is picked randomly from among the set of outcomes (future states) that are all most likely. The individual with the tumor, along with his or her friends and loved ones, can inadvertently help drive the final outcome to an unhappy ending by causing the probability of a fatal outcome to grow, and the probability of a non-fatal outcome to shrink. Beware: you can be easily drawn into a fatal dance of expectation with those connected to you, or to your condition. The best time for intervention, whether from PMR or NPMR, is long before "fatal" becomes a near inevitability.

The future is not a done deal. PMR is an interactive reality – we have free will and the potential ability to manipulate or at least strongly influence the probability of any particular event manifesting into the physical. There are rules and wisdom that must be applied to the use of nonphysical energy – there are things that one should and should not do as well as things that one can and cannot do. For example, one needs to know when help becomes interference. Helping out in the more immediate small picture can sometimes be counterproductive within a larger perspective because difficult challenges often represent great opportunities. When one removes the challenge, one removes the opportunity as well.

Most of us would be surprised at how effective a knowledgeable and experienced individual can be working from NPMR. The point is: Reversing the causal laws of space-time (a miracle) requires much more energy than directly modifying the outcome of an uncertain event. Later I will formalize this concept in what I call the psi uncertainty principle.

I am **not** speaking about what is often referred to as the power of positive thinking or the power of belief and faith to manifest physical effects. These effects are real, but are typically weak, inconsistent and unreliable. I am talking about **directly** and precisely altering the probable future with the intentional power of a focused, clear, and coherent mind. Positive thinking and putting energy into wishes or prayer **can be** useful and effective, but their effects are often not very potent or predictable because the energy of conscious intent supporting them is typically not well focused, coherent, or aware of the larger issues.

There are exceptional people – sometimes in exceptional circumstances – who have used these typically impotent forms forcefully. Hoping for or trying to project belief or faith into a measurable result **typically** produces (if it produces anything at all) a mild, diffuse, unpredictable, and unfocused effect driven by fuzzy emotions and feelings instead of precise knowledge and experience. It is the difference between feeling the gentle warmth of the sun through a window, and using a large and powerful magnifying glass to enable that same sunlight to burn holes through whatever blocks its path. It is exactly the same sunlight radiating heat through a piece of glass in both cases, but differing methodologies of interacting with the sun's energy produce very different results. Similarly, without an enforced coherency, a laser is just an ordinary red light. The methodology that intentionally employs a sophisticated energy manipulation technique (optical lens, laser, or NPMR awareness) to focus ambient energy is the one that has the power to produce dramatic effects and dependable results. Such techniques (large diameter accurate error-free lenses, laser coherency, or an ability to manipulate NPMR energy) embody the accumulated knowledge and experience of many generations and typically require skill and practice to produce and apply.

This discussion will, unfortunately, lead many to one of two intellectual extremes. The first and most common concludes that if a cancer whose existence is unverified (greater uncertainty) is the only cancer that can be cured by the proper focusing of consciousness, then such activity is fake or based upon self-delusion. That is not a logical conclusion. It may be a reasonable conclusion under some conditions, but not necessarily all conditions. In no case does "is fake" follow logically from "unverified." We tend to jump to that illogical conclusion because of our dogmatic cultural and scientific beliefs. Logic does require a physically unverified problem to have a physically unverified solution. However, we will see below that the psi uncertainty principle requires that problems that are physically verified (known malignant tumor) must have either a physically verifiable physical solution or a physically unverifiable nonphysical solution.

The second extreme, more illogical than the first, concludes that until the cancer is verified by measurement (in PMR exclusively), the cancer does not actually exist. Clearly, fundamental existence is not logically dependent upon performing physical measurements. It is true that until the cancer is verified by measurement in PMR, the cancer cannot be proved to exist in PMR. That fact, though logical, carries little practical significance in either the Big Picture or the little picture because existence and the proof of that existence are two very different things. Likewise, actually dissolving a tumor by manipulating nonphysical energy and proving that a tumor has been dissolved through the manipulation of nonphysical energy are two very different things. The first is not logically dependent on the second, although the second is logically dependent on the first.

That nothing can be considered real or significant until it has been **objectively** proven is an illogical position based on the assumption that PMR constitutes all reality and that you (your consciousness) are a derivative of physical matter. This mystical belief requires existence to be proven or validated by little picture causality: The existence of the Big Picture requires validation from the little picture. Same old story. This belief is not only illogical and irrational, but limiting as well. Objective PMR-based proof is valuable in some circumstances (exploring the space-time rule-set), and totally irrelevant in others (exploring consciousness and the larger reality). Making PMR objectivity a universal requirement banishes logical beginnings and forces PMR to be a universal reality. When you view PMR as a universal reality, everything beyond PMR vanishes before your eyes as your vision contracts to perceive the little picture exclusively.

Hard (but not necessarily only material) evidence and a careful scientific process are absolutely required. Keep in mind that knowledge constrained solely to PMR defines only a limited **perception** of reality and not necessarily reality itself.

The fact is, your existence extends beyond its physical manifestation in PMR. Most people resemble Flatlanders criticizing the concept of a sphere because it doesn't follow (isn't derivable or conceivable) from their two-dimensional science, philosophy, or reality. Without an open and capable mind you are stuck, ignorant forever, in Flatland. [A reference to the book *Flatland*, and how to find it online, appears in Chapter 20, Book 1.]

You may be wondering how you could learn to be sentient and operative in NPMR. It may seem nearly impossible from your perspective to achieve consistency and clarity but it is not. Steady effort and due diligence eventually will produce clear and consistent results for anyone. The limiting constraints, which are within your power to remove, are not constraints placed on your consciousness by fate, biology, or from the outside. They are internal constraints of your own making that you have created with your belief, ego, and fear. Your natural mind does not limit or constrain your NPMR experiences: The unnatural mind that you and your culture have created is the problem.

Learning how to be sentient in NPMR does not represent a new skill that must be mastered, it simply occurs spontaneously after self-imposed constraints are removed: It is an unlearning that is required. That sounds easy, but if you have ever tried to break deeply ingrained thought patterns (hard-core habits) you know it is not easy. Nevertheless, unlearning always yields to steady serious effort.

Viewing a clear picture of the larger reality will, for most of us, require stepping outside normal habituated thought patterns and cultural beliefs – unlearning or escaping old patterns must precede learning new ones. It is not easy to overcome limiting habits of mind, but it is well worth the effort. For those few who have broad and deep experience with the larger reality, *My Big TOE* will read as if it were a story about Dick, Jane, and Spot – familiar, simple, and obvious. For a few others, it will be as impenetrable as a wall of granite.

I am strongly encouraging you to think out of the box, non-conform, transcend your culture, and expand your notion of the possible beyond what is **commonly** held to be acceptable. As you are well aware, thinking big thoughts in small places can be socially and personally risky. This challenge is not for everyone, but if you have read this far, more than likely, it is a challenge that you can successfully meet.

▶ I expect that I have by now exceeded the credibility threshold of many readers. In techno-jargon that means that I have pegged their BS meters. I can smell the smoke! Before I completely burn out your super-sensitive, culturally calibrated BS detector, let me digress a moment because it is important for you to understand how the quality, content, and intensity of your thoughts affect (typically limit) the options and opportunities of your being. The open-minded skeptic is free to explore anything, anywhere, with demanding standards for converting experience into knowledge (truth). The closed-minded skeptic is trapped by the limits he places on his thoughts and abilities. He is caught in the following dilemma. Fearful of exceeding the safe and acceptable limits of cherished religious, personal, cultural, or scientific dogma, he believes it is fundamentally impossible to find truth in a wild and crazy **subjective** world which is obviously populated by plenty of fools and liars.

Religious creed (his in particular) and scientific objectivity (his in particular) appear to provide safe havens free from the ubiquitous fools and liars who would gladly induce everyone else to join their delusional and manipulative ranks.

As long as his boundaries are rigidly defined by (can't be expanded beyond) his limiting beliefs, he is not able to learn, grow, experience or evolve his being along the Path of Knowledge. This path, which often wanders through the lands of the paranormal, **requires** a fundamental (experientially based) **understanding** of reality and **knowledge** of truth. One must be both careful and courageous.

Such a limited individual usually feels he is maximizing his human potential by dedicating his energies to the optimal manipulation of intellectual objects, beings, and relationships in PMR. He has never had an objectively **verifiable** (proven) experience that would **necessarily** (scientifically) indicate a bigger picture exists because this type of experience is emphatically defined within his belief system to be absolutely impossible (the self-referential Catch 22 of the closed mind).

Individuals trapped in the little picture selectively restrict, view, and interpret their and other people's experiences to make sure that what is believed to be impossible does not appear to happen. Additionally, they negate the validity of anyone else's data or experience (of a larger reality) by assuming it to be contrived, ignorant, manipulative, or delusional (by definition). This façade of a scientific and logical evaluation, steeped in tradition and in support of the status-quo, is used to wrap the entire sleight-of-hand in a mantle of professional respectability. This particularly shallow circular logic (it can't be true because it can't be true) is subsequently offered by conventional scientists as proof that their beliefs are actually facts.

This is an emotional ego-driven position rather than an objective one. A rational argument rarely has any effect on an irrational belief. Ironically, the closed-minded skeptics interpret internally derived knowledge as mere **belief**, because they do not **believe** that objective knowledge (truth) can come from subjective experience.

The PMR-only view of reality, along with its tiny picture of the potential of your being and its associated relatively tiny subset of truth, is usually culturally based (particularly in the West), intellectually justified, and socially supported by the collective mind. The collective mind, unfortunately, always represents the lowest common denominator in concepts and understanding. It is relatively safe, easy, low risk, non-threatening, and not subjected to ridicule by your peers. It is, therefore, particularly attractive to fearful, under-developed, insecure, under-powered, and materially focused minds.

If you feel uneasy after assessing how that string of adjectives in the previous sentence applies to you, relax. Many good people of outstanding quality exhibit some of those characteristics. The point here is to recognize the arbitrary, unnecessary, and debilitating conceptual limitations we place on our reality, on our significance, and on ourselves.

If you have attained high status within your profession and among your peers, or are rich, powerful, and famous, these trappings of success say absolutely nothing about the quality of your consciousness or the evolution of your being as viewed from the bigger picture. The biggest fish (have the most power, notoriety, and make the biggest splash) in the PMR farm pond may be among the smallest fish in the gigantic $NPMR_N$ ocean. Conversely, some of the smallest, most humble fish in the PMR pond may be some of the biggest fish in the $NPMR_N$ ocean. ◀

By growing the quality of whatever state of being we presently find ourselves in, we evolve the quality of our consciousness. You can start from any point, but the optimal place to start is from wherever you happen to be today.

If you or anyone you know has an interest in personal growth, even if they require the comfort and safety of a relatively closed mind, there are two viable alternatives to the Path of Knowledge: Doing good deeds (Path of Service), and devotion to the highest expression of spirituality they can comprehend (Path of Surrender). Unfortunately, both of these paths are especially difficult to follow in Western cultures where material progress and success are dominant social values.

▶ Let's take a break before some of you burn out your analytical clutch – it keeps slipping like that because of a lack of personal high quality data. You should immediately scrape the sticky belief residue and corrosive pseudo-knowledge off your main brain-drive mechanism. That will get you by for now, but remember to get that condition (lack of high quality data) repaired as soon as possible. If you keep your speed down and avoid jumping over challenging new concepts in order to more quickly reach comfortable old conclusions, it should be safe for you to continue on after this short aside.

A brief description of the common paths leading toward self-realization, consciousness evolution, or spiritual growth will provide a helpful perspective on the entanglement of psi phenomena, spirituality, knowledge, and truth – and give all our brains a short cool-down rest from analytic activity.

We can divide the approaches to spiritual evolution into three **equal** but distinct paths that suit or accommodate three distinctive personality types. Eventually all three paths converge, with each path containing the essential truths of the other two. Each individual seeker's growth path can contain any proportion of each of the above mentioned three ways (paths of knowledge, service, and surrender) of approaching spirituality in order to obtain a best fit to their individual personality.

If the concept of seeking a **spiritual** growth path falls to the uncomfortable or negative side of your learned sensibilities, let me point out that the concept of a growth path toward improving the quality (decreasing the entropy) of your consciousness means exactly the same thing.

Because the three paths to consciousness quality appeal to diverse personality types, typically one process will dominate an individual's approach. Individuals choose the path that suits them best and afterward add in the other two as complements. An equal balance of all three might be theoretically advantageous but it would not necessarily be better for any specific individual. The main thing is to optimize progress (spiritual growth) by matching the process to the personality.

The Path of Knowledge is often called the "warrior's path" because of the courage, determination, and constant struggle required to gain spiritual knowledge and defeat the ego for control of the process of consciousness evolution. This path requires considerable intellectual capacity and is more suited to the linear thinking of the Western mind than the alternatives. It is on the Path of Knowledge that you learn, firsthand, about the larger reality and how to manipulate energy to put the so-called paranormal or psi phenomena to work in the service of spiritual growth (yours and others).

These paranormal abilities are the natural result of a lower entropy consciousness having more energy available to do work. You can accomplish more and you have more power with a lower entropy consciousness. This additional available energy can (but does not have to) be utilized to manifest paranormal or psi effects. We can now answer the question: What do love and psi phenomena have to do with each other, reality, spiritual growth, and digital consciousness? The answer: The ultimate reality is cellular (digital) consciousness. Consciousness naturally evolves to lower entropy states; psi effects and love are both artifacts of low entropy consciousness.

In the Big Picture, the paranormal is perfectly normal and **available** to (not thrust upon) anyone who sufficiently lowers the entropy of his or her consciousness. Love describes how a low entropy consciousness interacts with other entities. Both accrue to a high quality of consciousness – which, for an individuated consciousness, is equivalent to a highly spiritual individual. For sentient entities, psi-power is love-power is quality consciousness power. All are the result of consciousness systems and subsystems evolving to lower entropy states of being.

The Path of Service involves doing good deeds, being helpful, serving others and must come sincerely from the heart, not the intellect (Mother Teresa comes to mind as a well-known example). On this path, the ego is transcended through focusing on others' needs.

The Path of Surrender requires that you give yourself up to a higher (supreme) spiritual concept or being, or to the expression of more perfect spiritual values. This is the traditional religious path that transcends the ego by dedication and devotion to higher ideals and values, and through emulation of the beloved master, teacher, guru, or god. (Look near the end of Chapter 21, Book 1 for additional discussion on this topic.) ◀

Let's get back to the main issue. The entanglement (interaction) of uncertainty with the measurement of psi effects leads to what I descriptively call the psi uncertainty principle – where the uncertainty surrounding an event allows the event a larger selection of probable outcomes that may be more easily altered by the application of a focused (low noise - small ego) mind with a clear intent. Here, psi (as defined earlier in Chapter 19, Book 1) refers to psychic, paranormal, or acausal phenomena.

The psi uncertainty principle makes the scientific objective measurement of psi performance problematical, but only from the PMR perspective. As soon as there are outside observers to report and measure the miracles (reduce the uncertainty about reported violations of traditional PMR physics), the miracles diminish – and insiders cannot be trusted to be objective and therefore cannot be believed.

Let's explore how the randomness of certain natural events is related to the psi uncertainty principle. A focused and directed consciousness (physically embodied or not) may take advantage of the uncertainty or randomness in physical interactions, situations, or phenomena by subtly applying psychokinetics (PK) or telepathic suggestion to manipulate events toward a particular outcome. The results of these manipulations would simply appear within PMR to be good or bad luck, an intuitive bolt out of the blue, or a random skewing of the expected statistical distribution. Such manipulations are often referred to as synchronicity. Whether they are implemented by the individual or by the larger consciousness system (which has a vested interest in your successful evolution), they are typically helpful, relatively common, and usually provide just what is needed at just the right time. They create no contradictions and pose no causality problems within PMR as long as there is enough uncertainty (lack of **objective** proof) to hide the paranormal outside influence and satisfy the psi uncertainty principle.

If we would learn to make better use of the guidance and breaks that we are given, growing up wouldn't be so difficult. Most parents would probably say the same thing about their teenagers. Teenagers, just like their parents, ignore as much of this outside guidance and beneficent manipulation as possible. Do you see how the same patterns of individual growth are repeated at every level? Repetitive pattern is the nature of a fractal system. The pattern of consciousness evolving toward higher levels of organization (better quality and lower entropy) is applied at many levels and scales simultaneously. What you see your children doing is a reflection of what you are doing, is what computers will be doing when our hardware and software allow it, is what the Big Cheese is doing, and what AUM is doing. Consciousness does not evolve in a thousand different independent ways; it evolves in the same way in a thousand different forms at various levels of interdependence.

I hear someone asking: "Are you implying the paranormal is purposely being hidden from us by some uncertainty principle?" Absolutely not! It may appear that way from PMR but in fact it is just the opposite. The paranormal is not hidden by anything other than ignorance and limitations. It is you who are hiding from the paranormal by allowing your beliefs and fears to create an impenetrable smokescreen of objective obfuscation.

The psi uncertainty principle's job is to maintain the integrity of the PMR learning lab, not to hide psi phenomena from you. If you have defined the paranormal out of your reality, that is because there is no logical place for it within your little picture.

Why do the residents of PMR need the psi uncertainty principle? How does the psi uncertainty principle help lower the entropy of individuated consciousness within space-time? Breaking the rules of PMR objective causality would, at this time, interfere with what the beings enrolled there must accomplish (improving the quality of their consciousness). A rule-set is specifically designed for each virtual PMR learning lab to optimize the learning potential of the typical inhabitants of that PMR. More advanced players can access a wider set of rules.

You may be wondering how the psi uncertainty principle constrains your personal experience and how high quality consciousness can and cannot be used to your advantage in day-to-day interactions within PMR. Rarely is acausal or paranormal information obtained from NPMR and then directly applied to develop or invent physical devices because the psi uncertainty principle would generally forbid that sort of overt information transfer. It is not likely that a novel solution (like instructions describing how to build a Buck Rogers antigravity belt) could be **effectively** transmitted to an individual within PMR from a higher-dimensional source. It would be impossible for such an individual to be technically or conceptually prepared unless the transferred technology represented only a very small step beyond the receiver's current understanding. Imagine trying to describe color television or microprocessor chip technology to Alexander the Great or Alexander Graham Bell.

The near impossibility of being technically prepared is only half the problem. The receiver would need a very high quality of consciousness to receive a clear technical transmission and such people are not particularly interested in inconsequential physical gizmos and technology no matter how useful it might be in PMR because better gizmos are not relevant to the purpose of consciousness evolution. Additionally, there is the problem of needing all sorts of unavailable parts and materials – building a color TV or microprocessor chip with the materials and tools available to either Alexander would have been impossible even if they had a clear signal and were good at following directions.

You cannot cheat evolution or circumvent evolutionary process. Evolutionary process defines the rational structure of your system; undermining it would destroy the system's integrity. Psi uncertainty is required to maintain the integrity of your causal system – it is applied at all levels of the consciousness-evolution ecosystem to maintain local causality within each dimension.

On the other hand, a more limited but very similar transfer process working through individual intuition continually feeds inventors and creators of all types the answers and inspiration they require to take the next step. An adequately prepared person in touch with his or her intuition will often receive some key point or understanding that resolves an issue upon which they are stuck or that somehow lubricates the creative process. Again, the step must be small and the receiver must come to the intuitive process prepared to understand and implement the solution.

The psi uncertainty principle makes it exceedingly unlikely that paranormal information will provide a useful giant-step solution to an **objective** PMR problem. Think of the requirement for psi uncertainty as a constraint that limits the size of the leap that insight is allowed to take in any given circumstance within PMR – or, for that matter, within any reality dimension, either physical or nonphysical.

The usual highly filtered, low bandwidth connection between NPMR and individual intuition is designed to preserve psi uncertainty. Both a prepared intellect and an evolved consciousness are necessary for you to purposefully and effectively use your intuition as a source of insight.

Ask any innovative problem solver about the source of their creative solutions and you will hear about their non-logical connection to resources within NPMR expressed in a way that makes sense to them.

A lower entropy consciousness improves our ability to clearly see relationships among the data we normally gather and assess from PMR. Understanding these relationships provides a bigger picture that delivers an enormous practical advantage to low entropy individuals within PMR without violating the psi uncertainty principle. Individual intuition, insight, artistic, personal, and scientific epiphanies that result in physical creations within PMR must be integrated with the ideas that presently reside in that individual's current PMR experience packet (temporarily stored in the physical brain). It is the quality and capacity of our consciousness that allows us to see PMR relationships in a creative new way. Because no discrete acausal or paranormal information transfer from NPMR to PMR is required for consciousness to develop and improve its natural attributes, lower its entropy, or be creative, the psi uncertainty principle is preserved.

Quality consciousness not only gives us the ability to see relationships clearly, but also to find meaning, significance, value, and direction from the data and experience that we normally gather in PMR. A lower entropy consciousness interacts with reality at a higher and more complete level of integration, and, if enrolled in the PMR learning lab, is able to apply what it learns from that interaction to future interactions. Humans, as well as other units of individuated consciousness, learn to develop their consciousness quality through the interactions and feedback provided by virtual PMR learning labs.

Electrical engineers study systems with two kinds of feedback. Systems with **positive** feedback are like big snowballs rolling **down** a smooth steep snowy mountainside. Systems with **negative** feedback are like big snowballs rolling **up** a smooth steep snowy mountainside. The characteristics and response of the system is represented by the size of the snowball and the depth of the snow, while the effect of the feedback and its impact on the system is analogous to the steepness of the slope. Relative to the manipulation of psi energy (intentionally reorganizing bits within the larger reality) or the transferring of paranormal information, individuated units of consciousness are in a positive developmental feedback loop (bootstrapping process) to expand and grow those abilities and processes that are important to improving consciousness quality. In contrast, physical activity and interactions that are not important to improving consciousness quality (though they may seem extremely important from PMR's little picture point of view) are kept in check by a negative feedback loop that uses the constraint of psi uncertainty to dampen counterproductive intentions. Pumping up the physical with information transfers from the nonphysical is forbidden unless the transfers are sufficiently vague (negative feedback) to ensure a sufficiently dampened system response.

The uncertainty principle is required to maintain and optimize the purpose of the larger reality – to increase quality and lower entropy. The constraint of psi uncertainty keeps PMR focused and functioning toward its goal, which is **not** to generate physical improvements within PMR, but to help individuated consciousness pull itself up by its bootstraps. The generation of a more comfortable, physically productive (including better science) PMR is clearly the main issue within the local little picture and therefore appears to be very important from the PMR view. Although physical progress plays a part in the bigger picture by locally impacting consciousness evolution, it is only a minor player on the periphery of the greater consciousness ecosystem. Consciousness does not exist to promote physical evolution within PMR; physical evolution within PMR exists to promote consciousness.

The constraint of psi uncertainty allows the energy of organization to be pumped into (equivalent to sucking entropy out of) what is important (spiritual progress through exercising the ability to see relationships clearly and to find meaning, significance, value, and direction from PMR experience) while **avoiding** pumping energy into what is not important (power, control, domination, and physical progress). Psi uncertainty is simply a rule, a natural constraint that has evolved to improve the profitability of the consciousness cycle. Think of the psi uncertainty constraint as a negative feedback mechanism that is required to optimize the productivity of PMR virtual reality trainers.

The apparent subterfuge of hiding psi phenomena under a cloak of uncertainty is an illusion created by the rudimentary nature of human consciousness. No one would leave knives, guns, axes, hammers, saws, expensive cell-phones, or laptop computers lying about on the playground at a daycare facility. The kids might love it and have great fun, but the adults know that children must grow up first before they can productively (and non-self-destructively) use such tools. It is not required that the tools be hidden from the children, but the children must be unaware of how to access these tools until they are capable of using them productively. As the kids grow up, various tools naturally become available to them. Believe it or not, most consciousness kids, after they reach adulthood, have little overt use for, or interest in, the tools of psi. Wielding power within the virtual world of the learning lab becomes irrelevant unless it is the power of setting a good example and offering helpful guidance.

The disappearance of psi effects when objective science shines its spotlight on the paranormal scene does not necessarily indicate fraud. The requirement to work only within a narrowly defined subset of physical reality and to exhibit perfect repeatability for everyone to see is a good methodology for the development of some science because, within a limited realm, those requirements do separate a limited truth from an unlimited fiction. However, it is a totally **inappropriate** measure of significance when applied to **some aspects** of the science of consciousness, the science of being, the science of mind, or the science of NPMR.

Instead of viewing the requirement for uncertainty as a cover-up for fraud or delusion, consider it to be a requirement of the space-time ruleset to maintain the integrity of the physical reality we need to experience in order to stimulate our growth. Mixing physical apples with nonphysical oranges only muddies the waters of our experience and decreases the effectiveness of space-time as a learning lab. Having our experience directly related to our responsibility for that experience and keeping our interactions with others straightforward and an integral part of an obvious feedback loop are the major design goals of our PMR space-time reality. A widespread intentional use of psi would seriously undermine those goals.

If you are not sure where I am coming from with these statements, reread the beginning of the "AUM: scientist at work" aside located in Chapter 43 and at least the last half of the free will aside in Chapter 45 – both are in this book. You will see that PMR was specifically designed to constrain psi effects in order to produce a better learning environment. Psi effects within PMR are relegated to the subjective fringe where they can be easily obscured by uncertainty. This arrangement is a requirement of our reality as long as the average level of consciousness quality is so low.

Individuals may experience and use the paranormal as they lower the entropy of their consciousness enough to gain that ability, but it cannot become generally accessible through objective science without wrecking the primary value and usefulness of PMR as a training simulator for low quality consciousness. Don't worry, the system is designed to prevent that from happening and the psi uncertainty principle is only part of that prevention strategy. If you have no access to paranormal power and think that psi effects are a bunch of bullpucky ... well, that is probably as it should be. Such a view does not constitute a failure, an error, or a problem; it simply represents a particular state of developmental understanding – like being twelve years old.

There is nothing wrong with being twelve years old: In fact, it is an absolutely necessary and desirable state to be in if you ever expect to be a teenager. You cannot be a teen until you have been twelve, and you cannot be an adult until you have been a teenager. No one jumps ahead; everyone must start from wherever they happen to be. Truly, anywhere you are is a terrific place to start – there are no bad seats in the house. Everyone must progress under their own power and at their own rate. That is simply how the game must be played – there is no other choice and no other game.

If from time to time it appears to you that *My Big TOE* has cleverly or unfairly backed you into a logical corner, consider the principle that all systems of law and justice are based upon: The only thing that can logically corner you consistently is the truth.

The truth can sometimes be unpleasant, particularly if it leads us to an unsettling conclusion. "Feels good" or "feels right" is not a reliable way of discerning truth. To be accurate and trustworthy, your intuition needs to be free of fear, wants, needs, desires, expectations, and ego.

Psi effects and paranormal powers are sometimes dangled as a carrot in front of those newly on the Path of Knowledge to provide incentive, but successful travelers of that path soon realize that paranormal abilities can also be a growth limiting trap if their ego becomes enamored of them.

Using or requiring inappropriate methodology in any area of endeavor typically leads to dead ends, no progress, and false conclusions. Some of the knowledge, facts, laws, rules, structure, and science (obeying NPMR causality) that define paranormal activity cannot be seen, experienced, or understood until the observer grows to be capable of it, and will consequently never exhibit perfect repeatability for **everyone** to see.

Quality of consciousness and the directed applications of consciousness energy are personal in nature because consciousness is personal. Because psi effects are not of PMR, one cannot force them to abide by PMR causality, nor can one explain them with PMR causality.

The awareness and personal growth that enables an individual to exhibit controlled paranormal applications of consciousness must be developed from that individual's subjective experience. Researchers studying psi effects can not produce them, reproduce them, nor control them from the outside; nor can they **force** psi effects upon themselves or others from the inside. On the other hand, they can easily retard it and make it more difficult and problematical for their subjects to demonstrate psi effects. For a physically focused mainstream science, this lack of **physical** control makes the serious study of psi almost impossible. The traditional scientific requirement for tight **physical** control over a **completely non-physical** entity such as consciousness (including its attributes and artifacts) in order to prove its **physical** existence (the only type of existence there is, of course) produces an amusing example of circular belief-trap non-logic. The absurdity of this reasoning would be funnier if it weren't the position most scholars and scientists support quite seriously. Or maybe that is why it is so funny!

▶ While we are on the subject of cultural insanity and funny scientists, let's practice a little remote viewing. I see three typical looking university scientists dressed in white lab coats, standing in a small room with a metal table. On the table is a child's microscope.

Shhhhh, listen, and you can clearly hear Curly talking to Larry:

"I've got the answer! I've got it! I figured it out! Listen, nobody can **prove** the nonphysical is physical, so it must not exist!"

"You imbecile!" says Larry with obvious contempt as his two out-stretched fingers jab menacingly at Curly's eyes, "of course the nonphysical isn't physical. It's not supposed to be!"

"Yuck-Yuck-Yuck-Yuck!" answers Curly as he cleverly blocks the eye-jab with the side of his hand. "If the nonphysical isn't physical, then how am I going to examine it with this microscope?"

"Good point," says Moe thoughtfully. "What we need is a nonphysical microscope."

Curly and Larry glance at each other quizzically and begin to look in their pockets and under the table.

Moe watches them search in vain for the nonphysical microscope for a few moments before he says, "You two numbskulls couldn't find a nonphysical microscope if you tripped over it!" Curly and Larry immediately begin using their hands to search the empty floor around their feet for the invisible microscope.

Moe moves quickly. He grabs Curly and Larry by the ears and proceeds to bang their heads together. All three begin to quarrel.

"I just had a nonphysical myself and the doctor said that both of you were verrrrry, very sick!" interjects Moe, trying to recapture the attention of the other two.

Larry and Curly suddenly stop squabbling and look at Moe with seriously worried expressions. "How bad is it?" they ask in unison.

The ensuing pregnant pause suddenly aborts. "I had a physic too!" Larry interjects to break the worried silence, "but everything came out all right."

"So, **that's** what happened to your brains," Moe quips, as if suddenly receiving great insight.

"Now, you two get back to work. We aren't going to leave this lab until we can prove that we can read each others' minds."

"But, I can't read at all," protests Curly.

"No problem – you're a mad scientist, not an angry English teacher," says Moe matter of factly.

Curly and Larry pause for a reflective moment as they absorb the apparently profound and obvious truth of Moe's statement. Soon everyone is nodding in agreement.

Larry suddenly leaps on the table, assumes a "swami position" and scrunches up his face in a display of great concentration while Curly and Moe look on.

In a few seconds Larry jumps up, stands in the middle of the table, and says dejectedly, "I am trying and trying, but all I get is a blank, … actually…," he continues after a short reflective pause, "… I get two blanks."

"Great!" says Moe, "that's all the proof we need, let's get out of here!"

"I want to do it! I want to do it too! It's my turn! My turn! My turn! My tur…."
Whap!

A strong slap from Moe stops Curly in mid sentence. "Go ahead and give it a try, you nincompoop," says Moe indulgently.

Curly scrunches up his face, turns his head to one side and begins to squeal as he quickly pumps his feet.
Whap! Whap! Whap!
Moe gives Curly a quick triad of hard forehand-backhand slaps.

"All right swami, what did you get?" asks Moe with obvious skepticism.

Curly grins and puffs up his chest with pride. "I got lots of blanks! Lots and lots of blanks!"

Curly turns and looks at Larry. "I got more nothing than you did!" taunts Curly contemptuously.

"Oh yea," says Larry, as he grabs Curly's nose and begins to twist it.
Whap! Whap!
Moe reaches out and delivers a hard ricocheting double slap to the two squabbling scientists. "Cut that out!" he demands. "We are only supposed to read each others' minds, not tune in to the entire faculty! If we do too good of a job, nobody will believe us. Too much success will get us fired."

They look at each other with somber expressions as each ponders the steep and obvious downside to unbridled professional success in their chosen field.

"That much truth is dangerous!" Moe states emphatically.

"Yeah," agrees Larry as they exit the room, "If the dean finds out he is just as smart as we are, he may want our jobs for himself." ◀

Because of its subjective and personal nature, it is difficult for researchers to encourage, define, or systematically and objectively study paranormal phenomena at a deep level. The most they can hope to do is observe and document its existence – a relatively simple thing that has been done thousands of times by hundreds of fully credentialed scientific researchers.

Good objective scientific protocol requires the experimenters to remove all possible uncertainty, thus interfering with, and limiting, the psi effect being studied. Where some uncertainty is allowed, better results (from an insider's viewpoint) are produced. From an outsider's point of view, only less credible results are produced. There are always many more outsiders than insiders. (Here, insiders are the experimenters and their subjects; everyone else is an outsider.)

Remote viewers, for example, cannot produce perfect high-resolution photographs for their experimenters – there is always some uncertainty, and usually (over an in-depth set of experiments) at least some inconsistency. Additional uncertainty grows quickly in the minds of individuals who are not **personally** in **total** control of the experimental protocols; it grows more quickly in the minds of those who are not physically witnessing the paranormal event (they read about it, hear about it, or see it on TV). How much uncertainty is necessary? Only enough to ensure that the vast majority of PMR citizens will not have their cherished delusions **forcibly** perturbed to a significant degree.

If a paranormal event (precognitive dream or vision, for example) is without uncertainty, the number of people who can objectively verify this perfect demonstration of psi will always be small enough to produce no major or lasting impact on the larger society. Those individuals who are not yet ready to perceive and understand the larger truth represented by paranormal events must not be forced to experience what they can not productively deal with. In the bigger picture, there must always be enough uncertainty to ensure that the perceived causal integrity of PMR (the delusion that the only reality that can possibly exist must be objective and physical) can adequately be maintained by all who are not yet developmentally ready to move beyond that most basic worldview. From the opposite direction and within a smaller picture, the natural uncertainty surrounding a given event, or sequence of events, enables and simplifies the application of focused consciousness to paranormally influence that event without violating the psi uncertainty principle.

On the other hand, objective physical experience is designed to be shared and held in common. Our physical experience forms an interactive virtual reality exhibiting a uniform common causality defined by the space-time rule-set. PMR physics is simply a subset of the space-time rule-set (this idea will be developed thoroughly in Section 4). Everybody can experience the same measurable effects in PMR; however, the necessary (by the psi uncertainty principle) uncertainty that must reside at the root cause of psi effects (from a scientific PMR objective perspective) is not appreciated by the PMR scientist whose methodology requires him to eliminate uncertainty. The inability to eliminate uncertainty will frustrate the scientist's desire to understand the deeper causal mechanics of psi phenomena.

PMR scientists are culturally driven to **interpret** what they experience in a way that is in consonance with their belief that PMR causality must contain all possible phenomena. At best, if they are patient and careful, they can demonstrate that psi phenomena merely exist, but the causal mechanics and certain repeatability of it will elude them. Their attempt to describe a phenomenon belonging to a **more general** and **less constrained** causality in terms of a **less general** and **more constrained** causality is futile. The Big Picture and Big Reality cannot be fully contained within the little picture – quite the contrary, the little picture and little reality must be a subset of the larger reality. This is **not** rocket science. The little limited one must be a subset of the big unlimited one – it cannot be the other way around.

Requiring the Big Picture reality to be described exclusively in terms of a local little picture reality is an incredibly dumb idea – Moe, Larry, and Curly understood that much. However, it is an amusing fact that many of the world's scientists are totally stumped by this trivial concept. If they can not **physically** define and control consciousness and psi effects, then neither can be verified to exist as an independent entity or real phenomenon. Consciousness is seen as a hallucination of physical biochemical processes while psi effects are seen as a hallucination of psychological processes. By believing that what is real is delusional, and that what is delusional is real, scientists have boxed themselves into a small corner of reality that does not contain the answers they are looking for. Worse yet, their standard definition of an unscientific fool is anyone who does not share their mystical belief in the sacred One Physical Reality. Some things never change. Hey, look on the bright side; at least physicians are no longer bleeding us with leeches.

PMR science will always fail to explain **nonphysical** phenomena as **physical** phenomena. Psi phenomena, from the PMR-only viewpoint, will never be sufficiently well behaved nor deeply understood, thus generating much uncertainty in the minds of the masses. The mystery of how or why psi works (or even the existence of psi effects for that matter) appears to remain unsolved and unsolvable regardless of how many times it is thoroughly solved and demonstrated by knowledgeable individuals. If you are not one of those individuals, or involved with one, or do not know one well enough to fully trust his or her intelligence and integrity, you probably do not get it at the personal level of Big Truth. Without personal study and carefully evaluated first-hand experience, those who are unable to maintain open-minded skepticism are forced by their ego-needs to either believe in the actuality of paranormal events or to disbelieve in the actuality of paranormal events – both of which are illogical positions that produce a plethora of worthless unscientific blather.

The realness of psi effects must be personally experienced to be accepted or understood. Thus, sharing a piece of Big Picture knowledge or Big Truth (gained through the firsthand experience of psi effects by psi researchers or anyone else) by publishing research papers, books, or using the mass media is totally useless and ineffective. The results of psi research that confirm the existence of psi effects will never be widely accepted or **believed**, irrespective of how carefully and professionally the experiment was conducted, because **believing** such results directly conflicts with other **beliefs** more deeply held. On the other hand, when scientifically evaluated psi effects are part of your **personal** experience (especially where you are the actor, not the observer), then a larger reality is no longer a matter of belief and you know the truth even if you do not understand the mechanism behind the truth.

Those who are ready to progress to the next level of being will somehow discover the truth, while those who are not ready will remain clueless until some growth experience opens their mind to the possibilities. One cannot develop a deep personal understanding from somebody else's research or from somebody else's experience. This particular learning process is **not** primarily intellectual, like learning calculus; it is more experiential, like a one year old child learning to walk.

Much of the uncertainty clouding psi effects is the result of belief traps retarding the evolution of consciousness within OS. As long as conscious awareness and quality remain dim and low respectively, psi effects will remain shrouded in uncertainty, mysterious, and without credibility. As the quality of consciousness grows and awareness brightens across our culture, the purposeful application of psi effects will step out of the shadows and take its rightful place alongside contemplation, complex verbal and symbolic communications, and tool-making as innate human capabilities.

If a PMR scientist is looking at the result of someone else's positive measured experimental psi results, and has not done the work himself, he can **imagine** all sorts of uncertainties into the experiment and easily dismiss the results as sloppy science. Thus it is relatively easy for him to maintain the belief that his little picture (PMR-only) remains intact (the no-growth, no-stress, no-thought option). He is not burdened by the facts, nor does he have to be, because the impossibility of the stated results is a given while the potential uncertainty surrounding the results looms large in his own assumption-driven scientific mind.

A large uncertainty matched to an untenable result immediately leads to a strong conviction in favor of sustaining the little picture belief and ridiculing the apparently sloppy pseudo-science and ineptitude that produced the positive measured psi result in the first place. Although one scientist has adequately and scientifically proved, within reasonable certainty, the existence of some psi effect, there will be no impact on the scientific community. The existence of a broader reality will only become apparent to the few directly associated with the experiment.

Merely proving, beyond the shadow of a doubt, the existence of some singular, difficult to repeat, uncertain, and mysterious psi phenomena out on the fringe of respectability is not enough to impact the opinion of anyone in the rational center. Even if a psi researcher started his career from the center of accepted science, he would soon be relegated to the fringe, with perhaps no way to get back into the respectable center. Who needs those kinds of career killing associations? Is it any wonder that scientists shun psi research like the plague? Do you see why those who have the courage and gumption to undertake psi research are not taken very seriously by anyone outside their own fraternity unless they pretend to know much **less** than they actually know?

All problems of science and knowledge (philosophy, ontology, epistemology, cosmology, and physics) do not necessarily have technical or hard solutions – some can be comprehended only by an experienced mind with a larger perspective. That is another fact of existence that you simply have to live with. Cheer up, particularly you hard-science types: The situation is not as hopeless as it appears. Soft solutions, though inaccessible to a random simultaneous hard-science group-proof, can be real, scientific, productive, repeatable solutions with objective and measurable **results**. Yes they can – you just don't know how yet.

You do not need to do an about-face relative to what your mother told you about life and what you learned in school. Simply open up the blinders a little to allow in a larger set of possibilities, and continue in more or less the same direction. *My Big TOE* and I want to expand your world, not blow it up – we are showing you a bigger picture that contains your familiar little picture as a sub-set. Relax, take a deep breath: You are not sinking into quicksand. That sucking sound coming from around your ankles is your fear trying to maintain its grip and hold you back. It takes courage to step out of the box by yourself, dear reader, it takes courage!

A device or technology solution (such devices do exist) that threw someone's sentient consciousness into NPMR would simply leave that person a temporary stranger in a strange land. Their ego or mind or rational-self immediately upon return would most likely deny the reality and validity of the experience. Without an open mind, and concomitant spiritual growth (elimination of fear, ego, and material attachments, as well as the mastery of mental energy), the technical solution is totally useless.

Likewise, demonstrations of genuine psi phenomena to the average person (be they scientists or not) would be of little value beyond the theatrical and gee-whiz effect (unless this average person were the advantaged subject of the demonstration – it dissolved **his** tumor, solved **his** problem). He might change his beliefs if he were a witness or were personally involved (if he were teleported to China and left to come home on his own, for example). But so what: His **beliefs** are **not** important! It is his usable knowledge, his state (quality) of being, his spiritual evolution that **is** important – and none of these important things would be affected.

The larger world will not believe him (that he had been teleported to China) and he will be considered an unreliable insider with an overactive (delusional) imagination unless he denies his experience. Absolutely nothing important or productive would be accomplished by affording this person such an experience. If loss of credibility, denial, confusion, paranoia, or assumed mental dysfunction were the outcome, he may have been done significant harm.

Wisdom says, do not demonstrate psi effects – they can achieve no significant result within the Big Picture, may actually do some harm, and have only entertainment value. On the other hand, if one wishes to be an entertainer, psi is a cool tool until ego diminishes its power.

The function of the psi uncertainty principle is not to deny the existence of psi – the reality of psi effects is absolutely certain to anyone who cares enough to discover them for himself. Those who do not know that psi effects are real are merely ignorant (or in denial) of information which is widely available and of personal experience which is relatively easy to obtain (see Chapter 21, Book 1 for several references). Some serious effort and a little research can solve that problem. As long as belief-blinded individuals and a belief-blinded culture demand that **nonphysical** phenomena be described in terms of **physical** causality, psi effects will remain cloaked in uncertainty, difficult to study, without credibility, and relegated to the fringe of human activity.

The psi uncertainty principle – a **natural** artifact of the interface between the nonphysical and physical – primarily masks the causal mechanics and denies the efficacy and perfect repeatability of psi effects. While psi effects break, escape, frustrate, and void the PMR sacred causal chain, the psi uncertainty principle clouds that breakage enough to allow the center of thought within the PMR learning lab to maintain the illusion of the exclusiveness of little picture causality, an illusion that is necessary to optimize individual growth potential within PMR. Providing an individual with access to too much interactive power and capability before he is able to handle it wisely is always counterproductive if not dangerous. Power of mind is gained naturally and usefully as the entropy of an individual consciousness is reduced.

If you need a reality check concerning the overall quality of consciousness among humans, watch the evening news or read a newspaper: There will be absolutely no doubt about the general level of unconditional love, awareness, mental entropy, consciousness quality, fear, ego, wants, desires, expectations, and needs that are loose in the land. From month to month and year to year, the news stories (from a bigger picture view) are all basically the same. Only the names of the victims and perpetrators are changed from day to day to protect the innocent from noticing the utter repetitive consistency that clearly points an accusing finger at the quality of the individuals who make up our culture. Whether we are in the news or not, we are all fine examples of, and proud participants in, our culture – a good case, if there ever was one, for finding the innocent guilty by association.

Living in your culture and not being of it is exceptionally difficult. We are it, it is us – all are integrally connected. Like it or not, we are undeniably part of the problem. Fortunately, we also have the **potential** to be part of the solution. The good news is that by developing the quality of our consciousness, we can become a much smaller part of the problem and a much larger part of the solution.

Given the elementary level at which the PMR learning lab is designed to function, the psi uncertainty principle is, by itself, enough to ensure that the PMR learning lab experience is uniformly direct, straightforward, simplistic, safe, and user friendly for beginners in the consciousness evolution process. In general, though there are some exceptions, serious paranormal ability must be gained through a significant decrease in the entropy of your consciousness.

To move beyond superficial paranormal energy manipulations, you must actively and purposely improve the quality of your consciousness. It is not a matter of technique, magical incantation, or allying yourself with powerful entities: Understanding, wisdom, and paranormal ability must be earned through your personal spiritual growth.

For the most part, consciousness evolution takes place within a self-balancing, self-policing system. Pre-schoolers are never given power tools or guns to work and play with. Access is normally available only to those who have earned it and can profitably use it. In Section 5, we will see how the psi uncertainty principle interacts with future possible and probable reality surfaces to enable you to affect the probability that a given future possibility will or will not actualize into our physical reality.

The source and scientific nature of the law of psi uncertainty is no different from the source and scientific nature of the law of gravitation; both are simply the natural results of the space-time rule-set. Both are reflections of the constraints placed upon a consciousness in order to define a virtual-reality learning-lab (PMR) for the purpose of evolving that consciousness. Psi effects and the psi uncertainty principle are no more mystical or arbitrary than gravitation. Their application, mechanics and interactions are as understandable, regular, and predictable as the orbits of the planets about the sun.

The ability to effect and control a broad spectrum of paranormal events is available to everyone who is willing to grow the quality of their consciousness sufficiently. However, psi ability should never be your end goal – if it is, your capability to manipulate the physical through the nonphysical will be severely self-limited and perhaps even self-destructive. Psi power should be seen as nothing more than a collateral benefit of an effective path well traveled – it becomes naturally available as your Big Picture understanding deepens. As your capability increases, your interest in wielding it decreases because you discover that you have everything you need without it. Additionally, you learn that the desire to acquire and use paranormal power often brings out the worst in those who would like to be powerful but have not earned it. It sensationalizes and trivializes the pursuit of quality in consciousness. It can be a great educational tool, a valuable device for helping others in special circumstances, as well as an ego tickler that quickly becomes an enormous distracter of spiritual value and focus. Forget about using someone else's paranormal power to obtain some information or effect that you want – it won't work, that shortcut will turn out to be a dead end. You have everything you need inside of you.

48

■ ■ ■

A Closer Look at Psi Phenomena, NPMR, and You

■ ■ ■

There is no "now you see it, and now you don't" hocus-pocus here – it only appears that way from the limited vision of a PMR-only perspective. At a higher level of awareness, psi and psi-uncertainty are straightforward scientific concepts subservient to a higher level of causality.

We no longer believe that the sun and moon are pulled through the heavens by angels (a one time very serious and popular theory strongly supported by the best and brightest of the Western scientific, philosophical, and theological establishments). Similarly, we must not jump to the conclusion that the psi-uncertainty principle is enforced by nonphysical entities pulling strings from the NPMR background. Such simplistic, anthropomorphically driven concepts serve only to compound the original ignorance.

It seems that whenever we humans are confronted by our ignorance in a grand manner that cannot be denied, we tend to extrapolate a superstretched version of our old paradigms into an obvious solution that is supportive of the status quo. For a less obvious, but still troubling, ignorance, we often turn to spooky science or faith-based solutions or simply deny that the data, which demonstrate our ignorance, are real. Such circular belief-based logic creates its own intellectual whirlpool.

Today, we are greatly amused by the idea of angels moving the heavenly bodies around – I mean, really, how could those people be soooo stupid!? Actually, they were not stupid at all; they had about the same mental capacity that we have. They merely covered over their ignorance with theories that were in consonance with their personal, scientific, religious, and cultural beliefs – exactly as we do. Before feeling too smug about the silliness of planet-toting angels, you should know that modern science and philosophy use the exact same devices, with the exact same zeal, to deal with today's challenges to the core belief systems of our culture.

Many years from now **our** present belief-limited science, philosophy, and theology will provide future generations with good cause to shake their heads and snicker with astonished amusement, "I mean, really, how could those people be soooo stupid!?" Inevitably, our present notions of science and significance are one day going to look incredibly silly – perhaps within this century. It is one of our greatest conceits to **believe** that we could not possibly be **that** ignorant and **that** out of touch with reality – we are, after all, exceptionally smart and advanced, you know.

▶ "Ignorance? No way! Not us! Our Western science has clearly demonstrated brilliance and unparalleled achievement during the previous century. We finally understand how the natural world works. You must be referring to those belief-based touchy-feely folks in the third world. They remain largely ignorant, but eventually we will either need to bring them up to our level of understanding or take care of them. Subduing, educating, maintaining, or eliminating entire cultures of perpetual children may seem to be a thankless task, but it is our evolutionary responsibility – our inevitable burden to bear."

"I believe that self-imposed belief-based limitations do not exist, much less that they will eventually make us look unbelievably stupid to future generations. Impossible! Utterly impossible! As a species, we have become incredibly smart and scientifically advanced – everyone knows that. You must be stupid if you don't know that."

"I am sure that you are fully aware that we, of Western culture, now know, or almost know, everything that is significant. There are only a few fundamental details still missing – and our brilliant scientists are working on those, even as slackers like yourself waste their time reading books such as this one. It won't be long, a few generations at most, and we will have all the important information under our control. Jeez, won't life be great then – you know, after science has eliminated all our problems. Wow! I can hardly wait! We are so close....so close ...yet ... today, as I listen to the nightly news, scientific progress somehow seems almost irrelevant...and so far away."

The general attitudes and beliefs expressed above are both current and ancient. People have felt like this since history has been recorded. And they will, most likely, continue to feel this way for a very long time to come. Does one or more of the preceding three paragraphs roughly represent your core beliefs? Dig deep and be honest.

Have you ever noticed how insecurity, ignorance, arrogance, and ego often team up to play a particularly ugly joke? And that the joke always turns out to be at your expense? ◀

Most of us can think of at least a dozen or more instances (personal and historical) where we humans have created acceptable (at the time) explanations by over-stretching old paradigms until gaping holes appear that must be ignored. We have ascribed mysterious effects to angels, devils, other spooks, spooky science, and employed many other nonspecific metaphors for action-at-a-distance and invisible meddlers. The easiest and most effective explanation of all is to simply deny that conflicting data exist. These are a few of the standard devices that we humans have used to deal with our ignorance and reduce our fear of the unknown. We use them no less today in our private and public lives – they make us feel more in control and provide us with the means of controlling others. Concocting comfortable solutions that maintain the integrity of our belief systems is always more acceptable to most of us than admitting ignorance and then open-mindedly and skeptically **living with that ignorance** until either new paradigms or new data show up.

Living gracefully with the unknown is a simple and natural process in the absence of fear. However, given a widespread fear of the unknown and of new paradigms, it is no wonder that many have found, and continue to find, it easy and convenient to manipulate these deep seated bone-level fears (more subconscious than conscious in Freudian terminology) in order to control the energy, actions, and resources of others. Our fear and ego provide ready handles that others can use to position us to their own liking.

There are many devices you can use to deny your experience. The most obvious is to simply ignore it, to claim (believe) that you were tricked (blame others – conspiracies are always in fashion), or that you are suffering from temporary insanity or some other mental dysfunction. If religious, you may blame your experience on witchcraft or the devil, or perhaps if there is a little paranoia lurking in your makeup, you may explain your experience as the result of some diabolically clever hypnotic manipulation, or believe that drugs must have been surreptitiously dropped into your morning coffee. When it comes to justifying what you want (or need) to **believe** (pro or con), your creativity and induced myopia can rise to meet any challenge. That is the nature of the ego. It is also the origin of most, if not all, belief.

There is no point in demonstrating psi effects – the gee-whiz effect exhibited by those in direct participation (who do not invent a way to discount or deny the experience) is useless. The same is more or less true of natural psi experiences such as the precognitive dreams or telepathic communications that hundreds of millions of people have experienced. Generally these experiences lead nowhere and are not particularly important in PMR. However, they often serve as a catalyst to pry open a mind far enough for it to glimpse a larger reality or light a fire of inquisitiveness and can be immensely valuable to an individual ready to take the next step.

An individual not ready to take the next step does not usually have these experiences. Where is the value in causing people to have experiences that they are not ready to profit from? Doing so generally causes more stress and confusion than enlightenment and in the end, usually reduces the credibility of Big Truth rather than enhancing it. Real evolutionary progress, real improvement in your quality of consciousness, must come from the inside out.

Following our discussion about the relationship between psi phenomena and uncertainty, it would seem far **easier** to teleport someone to China and back again (rather than leave them there) because of the larger uncertainty involved in proving the teleportation. It would be easier yet if they returned with nothing but their memory of the trip, their experience. Even if they returned with a handful of souvenirs for evidence, others would think it very easy for them to have bought the souvenirs at the local import shop or rigged the evidence by some other means. They would be generally seen as liars or delusional – that much would be obvious.

The evidence would only be valid evidence to them – and perhaps a few others who **trust** them totally and implicitly. They would have a choice. They could be a delusional nut in the eyes of almost everyone, though secure in their knowledge of the truth or believe themselves to be a delusional nut by denying that their experience was actually real experience (plead insanity). The Path of Knowledge is not for the easily influenced and impressionable, the intellectually timid, the fearful, or the insecure. People with these traits will tend to remain uncertain and confused and are easy prey for New-Age charlatans, as well as the well-intentioned but unknowingly ignorant. A warrior's strong mind, focused intent, and fearless attitude combined with a scientist's patient probing, high analytical standards, and fundamental inquisitiveness is required for optimal results.

How is your progress and sanity to be judged if you follow the warrior's Path of Knowledge? Pragmatically and objectively – by looking at the results. Check to see if your knowledge and psi experiences (teleporting, healing, traveling, remote viewing, or communicating telepathically with either physical or nonphysical beings) are meaningful and significant (in the Big Picture) to yourself and to others. Determine if your understanding of the Big Picture produces consistent measurable results in terms of the quality of your consciousness and the depth of your perception of the larger reality (including the physical). Your spiritual growth, and your ability to help others grow spiritually, should be obvious to, and **measurable** by, you and others. If not, what you are experiencing is delusional.

You will unquestionably and unambiguously know if you are truly knowledgeable, kind, humble, compassionate, helpful, balanced, and focused on what you can contribute to others. If you are wise, understanding, considerate, insightful, thoughtful, and loving, you and everybody else will know that you are not insane. A highly evolved individual sticks out from the crowd like an elephant in a pea patch. These individuals are beloved and held in highest esteem by all who meet them – they have a gentle and highly effective power which is fearless.

Others who are not wise will know only if another individual **appears** to be wise, knowledgeable, kind, balanced, helpful, and not insane. You will know that you are not delusional if your capacity to love, to give, to exhibit humility and compassion in daily interactions is significantly increasing as your ego, fear and material attachments are significantly decreasing. The artifacts of spiritual growth are not esoteric or subtle.

If, on the other hand, your spiritual state is stagnating or non-existent (of low quality – ego driven) and you interact with people by manipulating or impressing them for material or ego ends, you are failing in your efforts to improve your quality even if you have convinced others of your success or have gained some limited control of psi effects.

The effects of reducing entropy in your consciousness will eventually become as obvious as being hit by a truck – you will have no difficulty telling the difference (growth or delusions of growth) about yourself **if** (a big if) you actually want to know the truth. Evaluating others can be more difficult than evaluating yourself, but the truth eventually yields to the same analysis – it is simply more difficult (sometimes) to gather the necessary data about someone else because of your limited understanding of their motivations.

If you have never experienced significant progress in lowering the entropy of your consciousness, you may have no idea what I mean or that what I am saying makes sense. I am trying to communicate the results of my experience clearly, but I know it is extremely difficult for anyone to understand in a deep, profound, or personal way without similar experience of his or her own.

If a spiritual teacher's interactions with others are more accurately described as "marketing" than actually helping people significantly change their life by enhancing their opportunities for spiritual growth, then the description "delusional nut with delusional or naïve insiders" may be an accurate description of that teacher and his or her followers regardless of how real their paranormal experiences are.

How do you evaluate others? When looking in from the outside, one without wisdom can typically not tell the difference between the wise and those who are merely marketing themselves cleverly. You need to get involved and you need to participate. How do you separate the true from the false? You must personally experience the larger reality. You must build your knowledge of your dynamic (changing and growing) spiritual being and trust your ability to figure out what is valuable to you and profitable for you.

You must grow your own wisdom. Only then can you judge what holds great value, truth, and knowledge for you (progresses you toward your spiritual goals) and what is a waste of your time, or worse, a step backwards. Thus, the proof of the pudding is in the tasting. Personal truth flows only from personal experience. All Big Truth and wisdom is personal truth. Those who judge (the metaphorical pudding) from the outside, **without** tasting, **without** personal experience, (typically emotionally driven – by fear, ignorance, and discomfort – at worst vitriolic scoffers and virulent closed-minded skeptics) are the most obviously delusional of all, like the emperor in his very respectable, socially acceptable new clothes. The difference being, in this particular situation, that the emperor and his most loyal subjects all use the same highly recommended tailor.

There is a requirement for uncertainty to surround nonphysical to physical manifestations. **Because of the psi uncertainty principle, you are required to gain real knowledge of NPMR through only your personal experience and growth.** Thus the psi uncertainty principle is a fundamental requirement of all virtual PMR space-time learning labs. The constraint of psi uncertainty is not a punitive restriction imposed upon you because you flunked the last quality test, but rather a designed-in feature of your local reality that provides you with an optimal opportunity to lower your entropy. Do not struggle against psi uncertainty, try to get around it, or wish it were otherwise – you need it to accomplish your mission efficiently. If you did not, it would melt away.

Big Truth is not something that someone else can make you understand, even if they show you paranormal events all day long. You can learn facts about it from others and choose to **believe** it or not. However, to have real knowledge, to use that knowledge as a catalyst for the evolution of your being, to improve the quality and decrease the entropy of your consciousness, requires **you** to get involved and gain your knowledge through your experience, experimentation, and spiritual growth. **There is no easier way**.

You have to do it; no one can do it for you. You cannot escape, or circumvent universal truth or fundamental principles – you can ignore them, but only at the cost of personal progress and opportunities lost.

The requirements for spiritual growth on the Path of Knowledge and the requirements for awareness and functionality in NPMR are similar and related. It is the dropping of ego and material attachment that produces the high signal-to-noise ratios required for being sentient in NPMR – see Chapter 76, Book 3). Spiritual knowledge and paranormal ability do not have to occur together (making that connection depends on your interest, focus, and intent) but on the Path of Knowledge, they often are intertwined. On the Paths of Service and Surrender, paranormal abilities may or may not be encountered. Spiritual knowledge and paranormal ability are not opposite sides of the same coin, but rather mutually supportive and strongly related activities. Paranormal ability is an **available** byproduct of the Path of Knowledge that you can choose to ignore.

> ▶ Though quality consciousness and paranormal abilities often occur together like families and children, they are not logically dependent on each other and though normally related, each can exist separately. For the sake of completeness, it should be mentioned there are other less consistent and less controllable ways to gain a specific (as opposed to general access to all) paranormal ability but these are not directly on the path to Big TOE understanding and consequently will not be discussed. ◀

Focusing on, or being attached to, paranormal abilities will halt, retard, or degenerate spiritual progress as well as degrade the abilities themselves. Thus, the potential for great misuse is self-correcting as is an interest in doing paranormal demonstrations or tricks for science or curious individuals. It is not only a waste of time (as described previously), but is also self-limiting – a drag on your energy and personal progress. A good performer (accurate and consistent) may perform only on his own terms (which may or may not require uncertainty) and may not stay a good performer (or remain interested) for long. A marginally interested subject makes it difficult for the experimenter to produce the precision, consistency, and repeatability required by objective science. Exceptions exist to this rule of diminishing returns in parapsychological experiments, but they are rare.

A difficult to manage psi-performer is not necessarily covering up a lack of ability with demanding crankiness. Experimenters create a difficult situation for everyone by requiring nonphysical phenomena to be (behave as) physical phenomena. Trying to trap nonphysical phenomena in a physical bottle is problematical at best. Remember Curly and Larry trying to find and study the nonphysical with a physical microscope? That doesn't work.

Another related problem that makes our traditional scientific analysis of psi phenomena difficult to achieve is that demands of performance (repeatable and measurable as required by scientific methodology) tend to involve (tug at) the ego of the subject who wants (is willing) to demonstrate his or her ability. A desire or need to perform will usually increase the noise, which lowers signal-to-noise, which produces a failure to perform – psychological psi-impotence. Thus, the best performers (perhaps the only great performers) are those not particularly interested in performing, or those unwilling to perform. Again, exceptions to this rule of uncooperative competency may exist from time to time, but they are rare.

You might become an advanced scientist by reading and studying books, but do not expect to develop an advanced consciousness (become spiritually adept) by studying books or hanging out with a spiritual master: You have to be it, live it, experience it, and subjectively interact with it. There is no easy process, shortcut, or technical solution. PMR actions alone (studying psi phenomena, following meditation techniques, reading books, believing anything) will not open NPMR to you unless you experience spiritual growth and decrease the entropy of your consciousness. And if you try in your mind (like a Flatlander) to force NPMR to conform to the form, function, and properties of PMR (science, philosophy, or 3D space-time reality), you will, like traditional contemporary scientists, be haplessly and hopelessly trying to force the proverbial NPMR camel through the eye of a PMR needle.

The experiences you need to grow the quality of your consciousness happen to you every day. What you need to learn to evolve your nonphysical being is not a secret that is hidden from your view. It is your effort, direction, and intention that determine how you utilize, embrace, or shun the available opportunities, experience, and information. Being open-minded and willing to **personally** experiment as well as expend considerable effort is the **only** way to make serious progress along the Path of Knowledge. Because personal experimentation is often subjective (not the stuff group-science is **traditionally** made of), it requires a commitment to continue working and learning until you have developed enough knowledge and understanding to produce objective results that can be validated. Substantial progress requires a substantial commitment – like the commitment it takes to get through college and graduate school along with the commitment required to become an accomplished musician. Everybody **can** do it, but not everybody **will** do it.

▶ I think at this point it might be useful for you to develop a more concrete sense of NPMR. This short aside will offer a description of $NPMR_N$ and its inhabitants that may help you see NPMR within a broader more solid perspective.

Life-forms in $NPMR_N$ seem to be much more varied and abundant than they are in PMR. Not all entities inhabiting $NPMR_N$ are sentient – some are elemental. The good, the bad, and the ugly – and the beautiful – all exist there. There is violence and peace, rip-offs and gifts. You can get hurt (self-inflicted or by the actions of others) or even killed, though it is very unlikely – there are strict rules regulating violence. Death is by disassociation or the loss of your internal organization. Your identity is dissolved if your energy is reduced to a maximum entropy state approximating the final state predicted by the second law of thermodynamics. Think of degaussing a floppy disk.

Death of an entity is exceedingly rare. If it should happen to you, you would cease to exist in that particular dimensioned reality as well as in other related realities – your defining code and memory, once hopelessly scrambled, is deleted from the simulation. Your current individuated consciousness is held as part of a historical record (last saved file), but you are no longer part of an active actualizing interactive reality; you no longer evolve within that reality, you are gone. You could theoretically be reconstituted from the historical record, but I have never seen that happen and suspect that it is rarer than rare. It would be the Big Cheese's call and entirely dependent upon the circumstances of your demise. There are only a few circumstances where rats and anti-rats are allowed (by the rules) to destroy each other permanently.

Unfortunately, sometimes NPMR rules are broken just as the laws of your state and nation are sometimes broken in PMR. Rule-breakers are sometimes caught and must pay the price, and sometimes they get away with it. As far as I know, only the Big Cheese can execute a capital punishment for specific offenses. There are no lawyers, you plead your own case – all your actions and motivations are transparent. Every action and event (information or energy transfer) is stored in memory and consequently, exactly **what** happened and **why** it happened is always available for post-event analysis. The truth cannot be effectively hidden; lies are always counterproductive.

Other NPMRs are generally not as structured, peaceable, friendly, productive, or safe as $NPMR_N$. In $NPMR_N$ there are lots of parasites, males, females, neuters, bullies, and social workers – there is unnatural death and employment, but no natural death or taxes. No one grows old – only more or less knowledgeable, powerful, loving, caring, balanced, compassionate, humble, wise, fearful, needy, greedy, violent, vicious, egotistical, controlling, or manipulative.

Among sentient beings, there is a much greater range in quality than there is in PMR. Beings and other objects in $NPMR_N$ are individuated consciousness energy and have associated bodies or forms of various shapes, and sizes. The experiential perception of these nonphysical bodies is defined by the rule-set that lays down the laws of being and interaction within NPMR (NPMR physics). For example, within one sensory view (one specific query of a subset of the NPMR database within TBC) of the non-physical parts of physical beings and objects, there are variations in **apparent** densities of forms. Denser is associated with dimmer awareness, higher entropy, and higher ego and fear content.

From a different sensory view (different query filter applied to the same subset of the NPMR database), denser indicates a higher probability of being manifested or maintaining manifestation within the physical. Within this view (primarily working with the subset of data that describes and specifies the transition region between the nonphysical and physical), the denser the energy **appears**, the more solid the object seems to be. More nonphysical "m" requires, or stores, more potential nonphysical "E" (as in $E = mc^2$) and requires more Force (focused mental energy with intent) to modify its present state relative to its extant dimensional container (as in $F = ma$).

In other words, sharply focused mental energy applied by intent (F) can modify the inertia, density, coherence or persistence (m) of a thought-form by constraining the dimension in which it can be extant as a function of time. In PMR, our dimensional container is space-time as defined by the space-time rule-set, thus "F" modifies the existence of "m" by constraining (defining) its 3D position as a function of time to a particular set of values. The two equations in the previous paragraph represent two rules within our PMR's space-time rule-set that provide a rough idea of the more general rules that apply within NPMR. In both realities, inducing change is a force meets inertia (resistance to change) type of phenomenon.

The visual appearances of nonphysical mass, delineated bodies, and various energy densities and probability densities within NPMR are the results and consequences of the NPMR rule-set which defines causality in NPMR and the particular way the viewer interprets the information received. As in PMR, it takes effort and training to ensure that one's personal interpretation of the data (experience) is not biased, is independent of collection methodology, is not an artifact of limited awareness, and does not modify the content of the data. The form of the data (how the data are internalized into personal experience) is subject to individual interpretation and therefore not particularly important. The content, message, or meaning of the data, however, is the same for everyone with a clear unbiased mental connection. Big Truth is universal; one's experience of it is always personal.

Uh oh, why is everybody yawning and getting up to go get a snack at the same time? All right, I promise, no more technoid blather about forces and inertia – let's get back to basics. You are out cruising NPMR and see a visual representation of a subset of nonphysical matter that is attached to, or a part of, a particular physical PMR entity. If this nonphysical matter-energy-data-thing has the **appearance** within NPMR of greater density it would mean it has a higher expectation value (less uncertainty) of manifesting as part of that PMR physical entity.

A specific example might help. If the nonphysical aspects of a brain tumor appear to be very dense, that tumor either has manifested, or is in the process of manifesting physically. The greater the apparent energy density of a nonphysical entity (incipient tumor), the more energy per unit time that is required to significantly affect or modify the expectation of that object manifesting into physical reality. Again, tumors that appear very dense when viewed from a nonphysical perspective have a higher probability of being actualized in PMR and are much more difficult (require more energy) to dissipate.

Why would anyone want to dissipate the nonphysical part of a physical or pre-physical tumor? In Section 4 we will discover that physical reality is a secondary manifestation wholly dependent upon its primary nonphysical source. The implication is that tumors dissipated within the nonphysical will automatically dissipate within the physical as well. Having said that let me also say that meddling with the natural results of the PMR learning lab (purposely modifying expectation values) is not always a good thing. The application of great power must be tempered with great wisdom.

Think of the apparent density of nonphysical energy as having the combined qualitative properties of persistence and inertia – staying power, resistance to change, heaviness, and the ability to survive and persist under the duress of external forces.

Within realities where energy appears most dense (the PMR_k) is where change is relatively slow, steady, and smooth; existence is relatively simple, basic, and stable; and interactions between sentient beings are relatively inelastic, viscous and sticky.

This description of the nonphysical matter or nonphysical mass that visually appears to define the boundaries and make up the bodies of all entities (sentient and non-sentient) in terms of energy density, persistence, and inertia is greatly simplified. I am describing only a minute piece of the perception mechanisms available within NPMR – the set of queries and filters that can be applied to the data within TBC and EBC. What we are talking about is a rule-set that defines the properties, quality, and limitations of our perception of the interactions of beings and objects with other beings and objects within NPMR. These are rules that define how we perceive the possible record-sets that are the result of our query, as well as rules that define the properties, form, and content limitations of transferable information and that define the allowable energy transfers (interactions) between players.

The "different views" that I spoke of above are like looking at the same body or body part with the unaided eyes, an X-ray, a CAT scan, a thermogram, or a sonogram. Each view requires its own skilled interpretation before data can be converted to useful information. It is even more like looking at the same database through several different query filters.

Trying to view or understand a more general system from the perspective of a highly constrained less general subset of that system is always as difficult as it is easy to do the opposite. Fully gaining the broader perspective of a low entropy consciousness within NPMR makes understanding and optimizing the experience of PMR almost trivial. To such a consciousness, life in PMR becomes transparent and simple, rewarding and fun, productive and meaningful. The amazing maze becomes a wonderfully exciting and fun challenge that becomes easier and easier as it delivers an endless supply of exquisite cheese.

I know that much of what I have described is next to impossible for many to understand because it is not possible to have a sense of the specifics of what I am talking about without direct first-hand experience in NPMR. Nevertheless, even if you have no direct experience of NPMR, I am hopeful that you will capture a flavor, an idea, an intuitive sense of what interactions within NPMR are like. They are similar to interactions in PMR, but with a wider, less limited access to information and a different set of social and physical rules (international law, local law, and physics). That's all. ◄

In order to build new concepts, there is no other construction methodology except stretching existing concepts and metaphors beyond their current use – beyond their common applicability. That is why transcending paradigms is exceptionally difficult. You must, by definition, start with inadequate conceptual tools for the job and somehow develop the necessary perspective that necessarily lies beyond simple linear extrapolation. You must combine a creative synthesis of old ideas with inspired intuition to find a new paradigm, which is nothing less than a new, more complete, more functional, bigger picture of reality – a more profitable organization of the available data.

To get a glimmer of understanding of NPMR science, you need a more general concept of mass than most of us are used to. I realize that nonphysical mass sounds both oxymoronic and just plain moronic, but it conveys the qualitative sense of what I am trying to explain better than any other words I can think of. For example, this concept implies that physical bodies exhibit a greater inertia or resistance to energetic (mental energy directed by conscious intent) change than nonphysical bodies. It also implies that this fact is the result of how the various rule-sets define and constrain the interactions of consciousness within and between each reality subset or dimension. Think of rule-sets as data filters that define your energetic interaction with the data. You may have read privileges only, or be allowed to read, write, and modify.

For those with no firsthand experience in NPMR, no experience with database queries, and who have long forgotten basic physics, what I explained above will no doubt seem vague, arcane, and hopelessly opaque. I also know that most readers are drifting along in a similar lack-of-NPMR-experience boat. Not to worry, you are not hopelessly lost: All I want you to take away from this discussion is a sense that NPMR, like PMR, is a structured, rule-based, objective, causal reality and that its physics is a superset of PMR physics.

I want you to get at least a vague sense that the rules, and the beings, objects, and energy that must obey those rules in NPMR are similar to those in PMR but more general (more things are allowed, the evolutionary process has more degrees of freedom, and the possibilities of existence in the form of individuated consciousness are less constrained). In Section 4, rule-sets, and how they define the experience we perceive as reality (both physical and nonphysical), are discussed in detail.

I think you get the picture: We apparently physical beings are very dense – without a doubt, we are among the densest beings within $NPMR_N$! That sense of the word "dense" fits like a glove doesn't it?

Let's take a short break. I need to speak with the technical types for a moment; the rest can skip this aside and wait for us at its end – we will catch up with you there.

All right, the left-brainers in the reading audience need to come on over here – we need to talk.

> ▶ I understand from my undercover sources that some of the techies feel as though they have been left holding a bag full of unanswered questions. I am aware that my discussion has been quick and shallow. This is because I have purposely omitted much of the breadth as well as the detail that you detail-types love to get your teeth into. This trilogy is not the place for that level of detail – it is already very long, places a considerable strain on a reasonable attention span, and is seriously challenged to keep you focused on the logical unfolding of core concepts. Too much detail too soon is almost always counterproductive.
>
> Recall that physical science books (basic physics for non-techies) that discuss ballistic trajectory dynamics will purposely neglect to mention atmospheric interactions (air friction or temperature and density variations), non-uniform gravitation, Coriolis effects, and ballistic dispersion because only a few people at the introductory level want to know that much about it or have the background to understand what the issues are. How and why these more precise considerations more accurately describe reality is not of interest. Most would much rather get the basic low fidelity concepts so that they can approximately understand the major concepts without being subjected to the boring particulars that are left to the boring specialists. Nothing personal; that is simply how the majority of right-brained non-technical people feel about it.
>
> Scientifically and carefully exploring (experiencing) the larger reality of consciousness within the metaphysical disciplines of cosmology, ontology and epistemology is something that most readers are prepared to deal with (have the tools or experience to

understand) only at the beginning or introductory level of explanation. Feeling that you could take a giant leap forward if you had a more detailed technical (mathematical) description of the physics of NPMR and the NPMR-PMR boundary is an error of misplaced emphasis and understanding. That is not where it's at.

Mathematical physics represents the logic (in symbolic form) of the little picture. We have shown repetitively that little picture understanding, logic, and causality cannot lead to Big Picture understanding, logic, and causality. Little picture logic can lead to a better understanding of only the little picture. You cannot pull an elephant out of an acorn, and you cannot logically derive the Big Picture from the little picture.

Remember Curly and Larry searching for the invisible nonphysical microscope? Metaphorically, the nonphysical microscope is your mind and equations cannot help you find it or evolve it. They may help you understand and manipulate NPMR, but putting that cart in front of your consciousness-quality horse will not get you anywhere. All would be better served if you focus on and achieved personal solutions before focusing on, and trying to achieve, technical ones.

First things first! Crawl… then walk… then run! I know that you want to run in the worst way because running is obviously superior to crawling or walking, but if you try to run before you are ready, you will simply fall flat on your face and probably come to the erroneous conclusion that running is impossible. That is a worse outcome than being frustrated because you are not able to run right now. Why? Because getting the horse far out in front of the cart by trying to run prematurely may permanently damage your ability to ever be ready to run.

"Hey, I tried it myself – running doesn't work! It's all a pile of horsepucky! It is not only impossible, but dangerous as well! People who say they can run are delusional, nuts, or egotists trying to impress everyone else! Beware! Don't listen to a word of it – trust me on this one, I have firsthand experience – look at that black eye and bruised nose. I will never try to run again! Anybody who tries to run is either an idiot or a fool."

Eight-month-old humans sometimes feel like that but they keep trying because the success of others within their physical reality is eventually undeniable. On the other hand, intellectuals who are committed to justifying their cultural beliefs occasionally become champions of denial in order to convince themselves that their ignorance, rather than their belief, is the delusion.

Jeez, those belief traps are amazing – they can transmute simple ignorance and incompetence into blind stupidity in a flash. Now that we have spotted that trap and settled the bag of questions problem leading to it, let's join up with the others and get on with our exploration.

That is it for the heart to heart, techie-to-techie talk – take a deep breath and let it go. We are intentionally skimming across the top here; this is a survey course at the 101 level, not a post-doc dissertation. ◀

$NPMR_N$ is not a different place, separate from PMR. It is continuous, integrated, and one with PMR. PMR is a dependent subset of $NPMR_N$, and it exists in, with, and by $NPMR_N$. All beings in $NPMR_N$ are not space and time constrained – space and time are local constructs within PMR and are direct artifacts of the space-time rule-set. Every being in our PMR is extant in $NPMR_N$; the converse is not true. Though you live, operate and function in $NPMR_N$ as well as PMR, you may be unaware of it because of the constraints you place on the awareness of your consciousness.

The ramifications of a continuous interrelated reality structure between PMR and NPMR will be explored in Chapter 84, Book 3 where we discuss communications, time travel, teleportation, multiple bodies, getting along without your body, and a few other interesting subjects. These topics need to be deferred until later because we have not yet developed the conceptual base required for understanding them. Be patient: Many interesting things lie ahead of us in subsequent chapters. To get there too quickly is equivalent to not getting there at all.

49

■ ■ ■

Section 3 Postlude
Hail! Hearty Readers
Thou Art a Stout (Figuratively Only)
and Sturdy Bunch

■ ■ ■

By now you may be convinced that I enjoy torturing you, but that's not true. The nature of reality is an extremely complex subject which is difficult to grasp because it can be seen and understood only from a Big Picture perspective that must necessarily appear wild and crazy when compared to your familiar little picture. Big steps or leaps forward in understanding always have and always will seem wild and crazy when first encountered from the viewpoint of the old perspective which is, by definition, built upon a less expansive knowledge.

The revelations of modern relativity and quantum mechanics, the fact that solid physical matter is composed of atoms that are mostly empty space, and that our beloved earth is not at the center of the universe were all seen as ridiculous absurdities existing on the delusional fringe before they were accepted by the mainstream. **Successful** explorations of consciousness, mind, and the larger reality must by necessity seem similarly absurd because they will conflict with the traditional ways of thinking about and defining reality.

We humans, in general, know much less of what there is to know than we think we do. Given the depth and pervasiveness of the belief-blindness that reflects the quality of our species at the dawn of the twenty-first century, if this Big TOE did not seem wild and was not difficult to understand, it could not possibly be correct.

Many people feel that Western culture is slowly growing up, opening its collective mind, becoming more able and willing to take the longer and broader view. Optimism and seeing ourselves through rose colored glasses are two of our most pervasive cultural traits. It is true that explorations of consciousness, mind, and the larger reality that remain strongly connected to our little picture belief systems have become somewhat more acceptable during the last thirty years. Unfortunately, the failure of such efforts to deliver profound insights has encouraged the commonly held belief that no solution exists and that a scientific **Big** TOE is impossible.

Significant theoretical progress seems unattainable, inaccessible, and forever beyond our reach because our thinking must be based upon traditional ways of defining reality in order to maintain at least marginal respectability and support from the center of power that dispenses research funds and credibility. Failure is assured by the requirement to use a proper, rational, and scientific exploration and analysis process which is, at its root, based upon the dogmatic belief that all reality must be physical. Under this handicap, researchers may easily prove that something truly strange is going on, but they will never figure out what or why. This recipe for failure resembles a resolute decision to look for your misplaced car keys or sunglasses only in places that you have never been.

Belief traps, dressed up as obvious truth, **always** make new Big Picture concepts struggle for credibility against the prevailing social, scientific, or religious currents of assumed rationality. That is simply how it is, and how it will continue to be, for a very long time.

When one is only a little fellow starting near the bottom of a long evolutionary ladder, as are most humans, Big Picture perspectives are difficult to come by. It is a designed-in feature of our larger reality that makes Big Truth appear to be an impenetrable mystery to the limited perspective of the average person, yet it is also a designed-in feature that a grand panoramic view of the whole is available to anyone who makes the effort to climb the mountain. Indeed, climbing the mountain is one gateway to the successful evolution and advancement of our individual being. For those wondering why opacity should be a designed-in feature, consider what your neighbors, coworkers, spouse, children, boss, mother, and mother-in-law might do to you if they knew how to manipulate you with their minds: It is better that they grow up first.

I salute your curiosity, intellectual determination, and toughness. I am thankful that you are tolerant and hardy enough to have made it this far. To get to this point you have suffered the ravages of rat-maze vertigo, ego deflation, belief trap withdrawal, and an annoying heart-to-heart talk from your Dutch uncle. You undoubtedly deserve to paste four more brightly glistening gold stars next to your name inside the front cover of this book.

Because my Big TOE is based upon a lifetime of personal experience and because it is very unusual, effectively communicating the perspective wherein that Big Picture makes sense to a broad range of individuals who may lack firsthand experience of the larger reality is an immensely difficult job. I hope that I have successfully met at least some significant part of that challenge. Unfortunately, whatever I fail to meet inevitably becomes a challenge to you – something you must figure out for yourself.

Thanks for coming along on this trek; I hope you will continue on the journey with me through at least the next two Sections where the mechanics of NPMR and PMR are more fully explained. In Section 4 we will explain how we can have the cute and cuddly physical bodies that we love to indulge and at the same time be nothing other than consciousness. We will solve the mind-matter dichotomy and delve deeper into the nature of experience and the phenomenon of digital consciousness.

Along the way, we will demonstrate that PMR physics (our hard-headed little picture science) can be derived from the same two assumptions that we have used to construct this Big TOE. Showing that PMR physics is contained within, and can be derived from, *My Big TOE* is (from the PMR point of view) the one and only test a candidate TOE must pass to lay claim to the title of a true TOE. That is a fair and necessary test. A successful TOE must explain and contain PMR physics as well as paranormal phenomena, intuition, mind, time, and more. We will accomplish all of that before you reach the end of *My Big TOE*. The deriving of PMR physics is accomplished in Section 4 with a little help from our friends. Come on along for the tour through Section 4, and see how the individual concepts we have discussed thus far begin to pull together into a more rational and understandable whole.

Section 4

■ ■ ■

Solving the Mystery: Mind, Matter, Energy, and Experience

■ ■ ■

50

■■■

Introduction to Section 4

■■■

In this section we are going to examine the origins and circumstances of physical existence. I will explain how PMR, at its most fundamental level, is a product of consciousness and provide a clear understanding of the connection between mind and matter. By the time you reach Book 3, it will seem more logical and less strange that matter is an experience of mind and that physical existence is a virtual interactive experience designed to facilitate consciousness evolution.

A TOE, by definition, must explain all the facts that are known to exist as well as new facts that belong to a bigger, more complete picture. *My Big TOE* is no different. It must subsume all present knowledge including PMR science. As we all know, traditional science is nowhere close to producing a credible Big TOE that explains intuition, mind, consciousness, and the paranormal, as well as the normal physics of PMR. PMR scientists and their egos deal with their inability to see a bigger picture by denying the existence of what they cannot explain. Facts to the contrary are either ignored or attacked as an unacceptable heresy that offends the obviously correct **beliefs** of the current scientific establishment. There always seems to be a plentiful supply of self-righteous scientists who spare no venom in defense of the sacred dogmas of contemporary science. From their limited view, PMR science and truth are synonymous and define each other.

▶ Do you see the parallels here? The last four sentences in the preceding paragraph apply equally well to religion and culture as they do to science. That should not be so surprising: At the bottom level of their belief-based foundations, traditional science, culture, and religion share the same genetic material. That is why their members and their institutions exhibit many of the same individual and organizational traits. In the West, science and culture have made an alliance of mutual support, while in the Middle East and East, religion and culture have allied themselves. Together they provide a throne from which the human ego reigns supreme.

Discovering a more profitable bigger picture, as well as raising the average quality of the human spirit, is a job that necessarily is left entirely up to you. We the people remain **objectively** clueless because our awareness is far removed from an integrated holistic view of reality. We the people remain **subjectively** clueless because our belief systems forbid the truth. Having been artificially separated, our objective and subjective cluelessness feed and maintain each other. ◀

There are two conditions that a successful Big TOE must meet. First, a successful Big TOE cannot be logically inconsistent or conflict with known facts. It must explain and contain (be a super-set of) what is presently known. Second, a successful Big TOE must appear to be logically **inconsistent** and conflict with traditionally accepted **beliefs** and the opinions based upon those beliefs (pseudo-facts). It could not possibly be a correct Big TOE, or even a major step toward a new Big TOE, if it did not dramatically conflict with our limited little picture belief-based culture, science, and knowledge.

Because our traditional science believes that the little picture is all there is, it restricts its investigation of the Big Picture (epistemology, ontology, cosmology, quantum physics, evolution, metaphysics, physics, consciousness, mind, psi effects, and intuition) to little picture phenomena. Traditional science is placed in the hopeless role of expecting the Big Picture to be contained within, and be derivable from, the little picture. Looking to extract the Big Picture from the little picture is a totally illogical approach and will produce nothing but frustration and the **apparent** conformation that either the Big Picture does not exist or it is beyond objective knowing. This approach to reality is analogous to trying to ascertain the properties of the larger forest by interrogating the moss growing on a tree, or perhaps expecting to understand modern microprocessor design and manufacture by studying a chunk of raw silicon ore.

The door of your mind must be opened at least a crack before you are likely to notice the passageway to a more complete conscious awareness and step through it to investigate what lies on the other side. If you have the courage to seek the truth, the *My Big TOE* trilogy is meant to facilitate the opening of such a crack in the belief barrier constructed by the ego-mind. A primary goal of *My Big TOE* is to help you get your Big TOE, or perhaps your entire foot, wedged in your mind's door to prevent it from slamming closed before you are able to explore the true nature of your consciousness. This is more difficult than it first appears because you are easily seduced by your cultural, scientific, religious, and personal beliefs to seek the habitual comfort and refuge in the safe bliss that ignorance creates. Delusion, like a drug addiction, is a habit that is difficult to break.

> ▶ You might think that separating fact from opinion is not difficult. We all think we know how to perform that operation reasonably well. Nevertheless, it is obvious that many people are not especially good at it – but we, you and I, truly know how to separate BS from truth. It is not that difficult because… well… we just know. We have been around the block a few times and are experienced and perceptive enough to know what is genuine and what isn't. Separating fact from fancy is intuitively obvious – our BS detectors are sensitive and finely tuned – we know how people think, what they want, and the games they play. It is difficult for anyone to pull anything over on us; we are not naïve or easy targets for New Age hustlers or status-quo promoting traditionalists.
>
> Everybody feels like that, including teenagers. It is a pile of belief-trap-crap that you should be careful not to fall into. The reason that discriminating truth from falsity is extraordinarily difficult is: 1) you are unaware of what you don't know and 2) it is your ego's solemn job (remember, your intellect is a servant of the ego) to convince you that your belief-based pseudo-knowledge is actually real knowledge. Because of invisible ignorance, the needs, wants, and desires of a slick ego, and a self-justifying intellect, your apparently solid grasp of either Big Truth or little truth is likely to be delusional. There are exceptional people who see things clearly, but except for you and me, I don't know any of them. ◀

In a culture like ours where scientific, religious, and cultural beliefs forcefully constrain and dominate most intellectual effort and essentially define the boundaries of each individual's local reality, it is more difficult to think out of the box and to separate the facts from the beliefs and opinions that masquerade as facts than you might imagine.

Of the two conditions mentioned above that a successful Big TOE must meet, I have the second one (high strangeness) well in hand. In this Section, the remaining gaps in the first condition (containing PMR physics as a subset) will be addressed.

Ladies, gentlemen, and others, do not panic: I promise not to drag you through a physics class. I know that physics first overwhelms and then perhaps amazes you; then it annoys you and puts you into a deep sleep. This is not intended as a sleepy-time book. Trust me; you will be able to follow every detail without going back to school to pick up those boring science and math classes that you so cleverly avoided.

Deriving some of the basic rules or concepts of PMR physics from a consideration of the properties of consciousness reality cells has been accomplished – I have only to point to it, not drag you through it. I will, however, provide the understanding and structure (model) with which you can combine physics and metaphysics concepts into a fully integrated Big Picture.

Let's begin by formulating some of the most obvious questions. How does one explain our physical existence and our physical experience within PMR, especially if we are nothing other than a clever configuration of AUM's wholly nonphysical consciousness? How come we appear stuck in this little picture 3D space-time PMR with these cute-but-needy physical bodies? What is a nice Big Picture Free Will Awareness Unit (FWAU) like you doing in a little picture place like PMR?

Why is it that a critical mission to improve the quality of our consciousness first requires that we dumb ourselves down to some clueless space-time creature with a perishable body that continually needs to be fed? Is this physical existence an existential joke, punishment for eating apples without permission, or is there some logical reason why the severe limitations and inherently painful struggle of the physical human is actually an optimized and necessary configuration for learning. (Hint: Nod your head up and down to that last one.) If you are patient and hang tough with me a while longer, I will eventually answer all these questions and several more.

Within this section, "Mind, Matter, Energy, and Experience," I will attempt to make the PMR connection by explaining the implications of *My Big TOE* to our everyday experiences here on good old planet earth – which is somewhere within PMR – which is somewhere within OS – which is somewhere within NPMR$_N$ – which is somewhere within TBC – which is somewhere within NPMR – which is somewhere within the consciousness and intent of AUM – which is apparently (but not actually) infinite – which is about two steps beyond what a self-limited mind can see, even on a clear day.

51

■■■

An Operational Model
of Consciousness

Computers, Simulations,
Artificial Intelligence, and Us

■■■

Pulling things together from previous chapters would be a good place to start. We have earlier developed the concept that we, and the reality (OS) we appear to exist within, are simulated entities in TBC, which represents a portion of the mind or consciousness of the evolving AUM. Now we will carry that idea a little further by discussing the attributes of advanced simulations and how they, and their simulated entities, parallel the operational nature of our local reality.

In many ways, we are not that different from the entities that we simulate in our computers. Operationally, there are many similarities while the few differences are mostly differences in the quality and richness of the input data (from our primary five sensors), the extensive use of parallel processing and feedback loops, and the capacity of our dedicated processing equipment (brain and central nervous system). You may be surprised to discover how functionally similar we are to some of our digital creations even though we are of radically dissimilar substance, motivation, limitations, and construction.

It would seem that the single most significant difference between us and what we might create within a computer is our free will to make choices that reflect and define the quality of our evolving consciousness. We, it would appear, are unique and fundamentally superior to digitally simulated entities because we possess a nonphysical component and a will that is free to make decisions, express intent and complex motivation, and evolve itself. From a similar but different perspective, we humans are special (at least in our view) because it appears that we have emotions and a soul while simulated entities in digital computers obviously do not.

The belief that humanity exclusively possesses a nonphysical part (soul) is a human conceit that leads many of our species to place themselves in a superior role to all other (lesser) life-forms, and doubly so to digital creations. Nevertheless, we will soon see that even this great source of human pride and distinction that seemingly would set us apart from any simulation, no matter how sophisticated, has its simulation analog and is perhaps not as important a distinction as it initially appears.

52

■ ■ ■

An Operational Model
of Consciousness

How Your Kid's
Computer Game Works

■ ■ ■

In Section 5 within Book 3, we will delve into the details that define the dynamic operational connections that link TBC, $NPMR_N$, OS, and PMR. There, I will describe a computer simulated war game to help clarify a few of the more important details of simulation mechanics. In this chapter we will also use the example of a war game simulation, but will remain at a higher level (less detail) than the upcoming discussion in Section 5. If my use of war games as an example of simulation seems out of place or irritates your delicate non-violent sensibilities, let that feeling go. Thanks to Hollywood, that is the one instance of a large and complex simulation that everybody has heard of and understands.

When building a simulation, we are interested in modeling various players and types of players that interact with each other. A "player" within this context is defined as any element of the simulation that has the ability to interact with any other element. In a war game, we might define missiles, piloted aircraft, tanks, artillery, infantry units, and individual soldiers as some of the probable elements or players in our simulation. At a finer level of detail (dependent upon the intended fidelity of the simulation and the computer power available), an individual's characteristics, each piece of equipment and protective clothing, or a single artillery round or rifle bullet may also be modeled as individual players.

A player in a simulation may also be an environmental element such as rain, snow, temperature, rivers, trees and mountains as well as the logistics process. A player, such as a single bullet for example, can be active (bullet with velocity) or potentially active (bullet in magazine), or inactive (bullet lost in a deep river). The words "active," "potentially active," and "inactive" are used to express the player's ability or potential to interact with other players. The most important attribute of a player is how it interacts with, and its ability to affect, other players. Players can be actual and active, virtual and potential, or inactive, and their state can change many times depending on the circumstances, dynamics, and relationships.

The characteristics, capabilities, and interactive properties of each unique player (such as foot soldier, pilot, tree, river, missile or artillery round) are described by the algorithms that define that particular player. Algorithms are merely collections of dynamic and functional equations, definitions, and relationships that are programmed into lines of code. These lines of code represent instructions that tell the computer what to do and when to do it in any given circumstance.

Before the simulation is run, or executed, all the players are given their initial conditions (positions, capability, motions, mission, and circumstances). The simulation is animated or put into motion by incrementing time in the outermost loop. From this point on, everything is driven by the actual events and interactions that take place within the simulation. Unknown or dynamically indefinable influences (the weather, interior ballistics, or quirky human nature) may have strong **natural** random components. In fact, random components represent a natural (required by a high fidelity description) part of many, if not all, players.

The action is driven to a large extent by the choices each player makes relative to their interactions with the other players – to fire a missile now or to save it for later; to run, walk, or stop and rest; to charge or to retreat. The choices are made by triggering conditionals. Conditionals are program elements (sometimes in the form of IF/THEN statements) that define possible actions given certain conditions. They make choices. These conditional statements provide an array of decision options that represent a set of rules of interaction that define the possibilities as well as impose constraints and limitations. These rule-sets define the type and range of possible choices, actions, reactions, or interactions of each player with every other player and with the simulation itself.

How the various players (people, equipment, machines, and environments) will interact under various conditions is unknown; the simulation is run to find out. If the natural random or uncertain elements are significant and properly applied and implemented, and if the simulation is extremely complex with a large number of interacting players, no one knows how it will turn out until the execution of the simulation is complete. Recall that an aside discussing randomness, choice, uncertainty, and free will was presented in the middle of Chapter 45 of this book; revisit that discussion if you have questions in your mind about free will or the origin of uncertainty and randomness in consciousness systems.

The same simulation may be run many times to determine the likelihood of various outcomes under various conditions. Simulation is a very efficient technique for learning about what is likely to happen under certain assumptions or circumstances. However, the quality and significance of the results are wholly dependent on the quality of the modeling of the interactive players and their interrelationships. You have no doubt heard the term "garbage in, garbage out." That is a particularly graphic way of saying that low fidelity modeling produces low fidelity results.

As the simulations are run (executed) repeatedly, some of the players may have been given or allocated their own subset of memory. Provided they were also given the algorithms to do so, they may collect, accumulate, maintain, and process their personal experience data. From this information, these players may learn how to do better next time. More successful capabilities, algorithms, approaches, or sets of conditionals (choices) may be developed as a result. If done internally, this bootstrapping (self-elevating, self-teaching, or self-improving) process may be referred to as artificial intelligence (AI), a form of self-modification (learning) based on data collected from previous experience.

These AI Guys (players capable of bootstrapping or learning) are programmed to improve their individual and collective performance by evaluating their experience (results of multiple runs or executions of the simulation). They, of course, need to be given the algorithms to guide their sensing and collecting of the most appropriate and useful data. They also need algorithms to evaluate and interpret the significance of the data that they collect and accumulate as well as memory in which to store the results and conclusions derived from the analysis of their experience. These results and conclusions may be used by other algorithms to modify the rule-set (like modifying the conditionals for example) that define the interactive quality and effectiveness of the AI Guy.

AI Guy thus evolves by applying the Fundamental Process to optimize his performance. In a complex simulation, he has many choices or paths to choose from and each choice may lead him closer to, or farther from, his goal of optimized performance. He measures progress and success strictly by observing the results. AI Guy is the epitome of a dedicated pudding taster if ever there was one.

Rule-sets are not only for AI Guys. Rivers and many other non-sentient players have IF/THEN conditionals in their definition. **If** it rains enough in the right area **then** the river runs faster and deeper, and **if** the river becomes deep enough, **then** it floods. Deep, fast, and flooding rivers can strongly interact with the movement of troops and equipment. These conditionals drive the dynamics of the environment and define environmental conditions.

The AI Guy must interact (deal) with his environment as he must interact with all players that can affect his choices, effectiveness, existence, and being. The rule-set defining interaction within the simulated world (including non-sentient and sentient players) represents the math, physics, and science imposed upon the simulated reality. If the rule-set reflects our PMR physics, the various players interact as they would in PMR and we say that the simulation is realistic. We can program anything we want to into the simulation rule-set. We could easily give people the ability to jump fifty feet into the air but the resultant simulation would not accurately represent our PMR. It might represent someone else's PMR (small planet, less gravity), but not ours.

If a high fidelity PMR-physics model accurately represents each player, the interactions in the simulation will produce an excellent representation of how these players might interact if they existed physically on planet earth. High fidelity player models along with detailed accurate interaction specifications produce results that can closely model a given reality. The more accurate and detailed a simulation becomes, the more useful and accurate its results are – and the more memory it requires and the slower it executes. The twin problems of large and slow can, in theory, be easily overcome with better technology – bigger, better, faster memory and computers.

When modeling the cognitive function of sentient beings, conditionals may, at the simplest level, span all possible choices and reflect the overall quality of the larger rule-set. If the sentient entity being modeled is an AI Guy, his choice of conditional options is based on an interpretation and assessment of his total experience as it has been captured by the data collection sensors and evaluated according to the current rule-set within memory.

Are you beginning to see the parallel between AI Guy and us? If not, you might want to do a slow retake on the previous paragraph. Operationally, we both go through many of the same processes. Granted, we are very different, especially in the little picture where the details are, but from the larger perspective of relationship and operational process, we have much in common.

53

■ ■ ■

An Operational Model of Consciousness

Will the Real AI Guy Please Stand Up!

■ ■ ■

We need to understand the concept of AI within a larger perspective. AI Guy can be a cool dude, and with the right programming, he can easily develop a personality that is much better than your boss'. Right now AI Guy is being held back by a lack of computer power. Stay tuned, his day is coming – sooner than you might think.

Some of the computer games that you or the kids in your neighborhood are playing today employ simple versions of some of the interactive learning attributes of AI Guy. More sophisticated AI implementations have existed for some time at universities and industries performing AI research or developing AI and AI-expert-system products. The main things that keep this technology in the lab or devoted to only a few specialized applications are the cost and the limitations of the software and hardware.

AI technology is not the issue. We currently know how to do the sorts of things I have described and much more. We know how to produce AI Guys that know how to learn from their experience and that can modify (design, implement, and evolve) portions of their own rule-set. These AI persons (note that I am appropriately sensitive to female AI implementations) or AI things can even create or generate other AI things to help them do their job. It is not conceptually that difficult to implement if the required hardware is available.

This is not to say that AI implementations and research is easy; it is an extremely challenging field that deserves our best and brightest. The point is: AI (at least its initial implementation within our current computers) is almost within our grasp. It is not a far out idea. Its problems of realization are, at this point in time, more technical and economical than theoretical.

The evolution of AI applications and research is primarily constrained by the lack of sufficient inexpensive computer power. Moore's Law (which is actually only an observation and not a physical law) says that computer power will double every year and a half at no appreciable increase in production costs. Today, many people believe that Moore's Law (named after former Intel chief Gordon Moore) significantly understates how quickly computational power will soon be increasing.

It is very likely that before 2020, the silicon technology upon which Moore's Law is based will have reached its limits; on the other hand, there are several promising new technologies that are likely to take its place. If historical precedents hold, these new technologies, which may be only half a decade to a decade away, will dramatically accelerate the rate of practical computation – making the increase in computer power described by Moore's Law seem slow and quaintly pokey by comparison.

Nevertheless, applying Moore's Law as the current industry standard, we find that by 2020 computers will be about $2^{20/1.5} = 10,322$ times faster than they were in the year 2000. This means that in twenty years or so, your common $2,000 desktop computer will be hundreds of times faster than today's mucho-multi-million dollar super computers. Twenty years after that (2040), our el-cheapo desktops will be crunching data about 106.5 million times faster than they were in 2000, representing an increase of eight orders of magnitude (10^8) every forty years. Extrapolating present trends to predict future capability is not an exact science and it gets riskier the further out one projects. Nevertheless, our best guesstimates are that it will not be too long before we have more than enough inexpensive number crunching capability to provide AI Guy with what he needs to evolve an affordable, intelligent, self-aware consciousness.

"Did he say that some computer simulation dude could have an evolving consciousness?" "Yes, that is exactly what he said – I heard it too."

"Hey Jake, do you think this guy is nuts or has he been hanging out on the fringe too long?"

Easy Jake, stay focused, review Chapter 27, Book 1 and the first part of Chapter 41 of this book if you are not sure what constitutes evolving consciousness. Will it make you feel better if I change "consciousness" to "artificial consciousness"? Does that seem more accurately descriptive? How important is the distinction? I am sure that fear, ego, and belief have nothing to do with why most of us feel compelled to use the distinction of "artificial" to maintain our sense of superiority. Because we get to make up the definition, why should we settle for being uniquely different when we are obviously superior? Right, Jake?

Maybe the first thing we will have super AI Guy do is help us design better, more intelligent AI systems that will help us design better, more intelligent AI systems that will help us design.... It is probably good for us that AI Guy will always be as physically dependent as he is mentally and computationally brilliant. Think of AI Guy as merely another type of sentient (interactive, capable of learning from experience) being with its own function, purpose, style, and personality. If I use the adjective "artificial" to modify the noun "being," will you feel more comfortable, would that be more accurate? I bet I get a "Yes" on both counts from most readers; if you maintain an open mind, you might find that such an attitude more resembles the pot calling the kettle black than an expression of an obvious truth.

The point of this discussion is not to prognosticate future computer technology but rather to point out that if we humans are presently on the threshold of developing silicon-based consciousness – artificial consciousness if you prefer – with our outrageously limited and extremely primitive computers, imagine what complex individuated consciousness AUM might develop within TBC; reality-cell based digital mind-beings like you, perhaps. Conceivably, AUM's implementation of an AI-thing could turn out to be similar to our consciousness (except much less arrogant no doubt). Could we blame AUM if it wanted to call this derived consciousness and intelligence artificial? AUM might produce several artificial consciousness models to interact within several larger simulations – and PMR might be one of them.

▶ ".... And just for you, ladies and gentlemen of NPMR$_N$, we have this nice little starter model. This simplified carbon-based artificial intelligence is guaranteed not to strain your brain. It allows you to make simple choices, experience the organic macro-level of the space-time rule-set, and performs all other functions automatically. You never need to be concerned with prodigious technical complexity because all those complicated details of how it actually works are hidden behind a simplified user interface called a "body" – the latest discovery to evolve from our advanced carbon-based technology.

"Although a few power users contend that this Physical User Interface (PUI) limits their ability to utilize and control the full potential of the underlying consciousness, I am sure that you will find the resultant simplicity and ease of use of this PUI puts a powerful and useful experiential tool within the grasp of the common user. After all, how many of us are consciousness scientists or reality geeks? I mean how many of you actually want to get into the details of the intestinal tract and experience being eliminated? Yuck! Believe me folks; this PUI is incredibly simple to operate. OK, it crashes occasionally, and will sometimes get stuck in an obsessive loop, but when that happens just put it to bed and our on-site support team will, under most circumstances, have it repaired by morning.

"I know you have heard about space-time, the new technology that makes this virtual miracle possible. It has been the number one topic on *Think-Net* for months and was featured last week on the cover of *Mind* magazine. Honestly folks, with this new PUI the power of space-time is at your fingertips. Finally… space-time for the rest of us!

"Move in a little closer folks. If you sign up today we will give you a lifetime on-site service contract on this exciting new PUI. If you ever wear out the standard equipment PUI supplied, we'll provide a brand new one at no additional charge. Our technicians will automatically download your accumulated quality score file from the old PUI and upload it to the new one so that you never need to start over. What could be easier ladies and gentlemen? This is the one you have all heard about – our best, full featured, entry level package that lets you, as have many others before you, participate in the experiential game of carbon-based life.

"You may have noticed that carbon-based PUIs come in two basic models. This is the fun part you have read about – and it's practical too – it absolutely guarantees that you will always have plenty of challenging opportunities to score …quality points, that is.

"For those who aren't sure they want to take on something this challenging at the present time, but are still eager to get in the carbon-based evolution game, we have some furrier PUIs that have fewer features but require much less commitment on your part. Yes, it's true that it is more difficult to run up a high quality of consciousness score, but it is also more difficult to lose quality points as well. That is a trade-off you will need to consider.

"If you're not ready or don't plan to give your full attention and best effort to play this life-game-experience – if you're not going all out for a high score to maximize the evolution of your consciousness – I would recommend this furry little PUI that you see right here. It's not only cute and cuddly, but…awwww, folks, look how it wags that little tail… did you see that? Isn't that just precious?

"We have it all ladies and gentlemen, the choice is yours. Don't miss out on this fantastic opportunity! This is the hottest thing in evolutionary consciousness since space-time was first invented! There is no experience like it anywhere in $NPMR_N$. Pick out a body and get in the game…the space-time physical experience virtual reality game… the "Game of Life"™ and rack up those quality points faster than you ever thought possible.

"Step right up ladies, gentlemen, and others. Come on in and sign up for the adventure of a lifetime – or many lifetimes – it only takes a minute." ◀

54

■ ■ ■

An Operational Model
of Consciousness

Some of My Best Friends
are AI Guys, But I Wouldn't
Want My Sister to Marry One

■ ■ ■

The last part of the previous chapter was fun and should have given the analytical part of your frontal lobes a chance to cool down. In this chapter, we return to the real world of simulation games and learn more about AI Guy's secret intentions toward joining your family – the family of conscious beings.

It would seem that simulated sentient beings modeled within our war game (such as airplane pilots, commanders, truck drivers, and individual soldiers) make their decisions based upon a set of interacting conditionals constrained by the rules of the simulation (rule-set) and by natural randomness. From this, their being or interactive presence is defined within the simulation. Conditionals may be triggered as part of a nonlinear fuzzy process – for example, a particular choice may be actualized or chosen based upon the result of a complex sequence of neural nets operating on lower level input data. The type and range of each player's likely intents (what the player is trying to accomplish) and subsequent decisions can be described in terms of that player's purpose, significance, quality, and interactive options.

The overall effects of each player's decisions are determined by the size of their decision space, the profitability of their assessments, and by their ability and capacity to interact. Added to that mix of purpose, intent, and capability operating upon a finite array of possible interactions constrained by the rules of the simulation is the embedded randomness that is natural to each player as well as the uncertainty that results from that player's lack of accurate pertinent information (ignorance). Did you notice that for a simulated war game player there is a natural connection between ignorance and randomness? Higher individual entropy implies a higher level of randomness and uncertainty: The two go hand in hand.

Simulations in TBC conceptually share much of what has just been described. We are the modeled players. Our ability to apply feedback to modify ourselves derives from our cognitive ability to make use of our experience. The plants, rocks, critters, and other beings and objects that make up our universe and OS are also modeled players, each with its own interactive characteristics. Some are more complex and multi-dimensional than others; all must obey a local objective causality defined by the PMR space-time rule-set.

We interact dynamically with the other players according to how we are defined (physically, genetically, culturally, socially, experientially, mentally, and spiritually) and based on all the choices we have made and interactions we have experienced relative to the situation we find ourselves in. We learn through experience and are self-defining to a large extent. If we collect better data and process it accurately and cleverly, we can make better deci- sions – we can optimize our profitability relative to the overall purpose of the simulation. If we have insufficient data, we guess and then justify that guess with a belief. Beliefs, ignorance, erroneous assumptions, errors of understanding, and increased uncertainty and randomness in the decision process all represent internal constraints that reduce our potential for dig- ital synergy and support each other in combination to produce and main- tain a low quality entity with a high entropy consciousness.

As in the war game simulation, it is the making and implementing of choices that cumulatively drives the action and produces significant results. The free will choices that lie within our personal decision space represent a finite set of discrete possibilities that accurately characterize the accumulated quality of our consciousness. They are motivated by our needs, wants, beliefs, desires, and fears (our ego); as well as by our knowl- edge, understanding, caring, balance, and joy (our love). It is our fear, ego, love, and interaction with others that creates most of our conditional situations and motivates our choices. Our individual quality is a good but imperfect indicator of what we are likely to do given the available choices. Clearly, it is our fear and love that define the subset of choices we can actually make, contemplate, or consider. Choices that fall beyond our understanding are invisible to us – each individual operates and interacts at his own level. Growth requires a free will with the gumption, energy, and focus to push beyond its current definitions and limitations; that is how we modify what defines us.

These are important and difficult concepts so let's round up a few of the more obvious conclusions. The finite set of free will choices available to each sentient player generally reflects the quality of his or her consciousness. It is the extent to which fear (ego, beliefs, needs, desires) and love (knowledge, caring, balance, joy) drive the individual's motivations that defines the subset of available choices which in turn limits the conditionals that a player is capable of considering and contemplating. This limited subset of conditionals defines the choices and possible short-term interactions that are operationally available to an entity. The choices that are operationally available to (within the decision-space of) a given entity are the choices the entity has the ability to recognize, understand, and implement. Choices that are operationally available usually represent only a small portion of the complete set of choices that are actually available. It becomes clear that the extent of a sentient individual's operational awareness is a function of the quality of that individual's consciousness. As the entropy of a consciousness decreases, its awareness expands and blossoms. A high entropy consciousness is constrained to exist, live, and experience within a relatively small reality.

The possible decisions each player is operationally capable of making are limited by the quality of that player's consciousness. Thus, an entity's evolutionary profitability must generally progress through a long series of small steps. Each step represents a free will excursion; a stretch of that entity's being, beyond its current capability. Growth must necessarily occur in the margins and accumulate through small increments. Athletes bootstrap (incrementally grow by applying a focused effort that demands performance beyond present capability) their proficiency in much the same way. You must reach beyond your present abilities to actualize your full potential. For consciousness, this evolutionary dynamic is implemented by a feedback loop that modifies the quality of the individual's consciousness based upon the quality of the intent driving the free will choices exercised. Tiny positive or negative increments in the quality of your intent, over many thousands of choices, eventually lead to either an increasing or decreasing consciousness quality.

There is another very important input to our choices. In the war game simulation, recall that we always included a natural random component as part of the description of almost all players – including the sentient ones. This natural random component represents, among other things, the uncertainty that often occurs when a player has several available paths or choices that have approximately the same probability of being chosen. This is true if the player is a bullet traveling through the barrel of a gun, a radioactive decaying atom, dust blowing in the wind, or a troop commander deciding when and where to launch an attack. Bear in mind that randomness is a measure of entropy. (For more detail, see the discussions of randomness in Chapters 45 and 47 of this book.)

We will explore the association of probability, choice, awareness, and reality in detail in Section 5 ("Mechanics of Reality") when we explain probable realities. For now, all you need to understand is that as time continues to be incremented in the simulation (or PMR virtual reality trainer), circumstances and conditions change creating choices that need to be made by interacting entities. For sentient beings, those choices follow intention (express motivational quality) and have the capacity to lower entropy through a more profitable organization of the entity's digital content. For non-sentient entities, those choices follow the path of least resistance by moving the entity and the system that contains it toward some lower energy configuration.

For sentient beings (remember that protozoa, dung beetles, clams, cats, people and a host of other physical and nonphysical beings and critters are included in that self-aware group), there is something more than merely a natural random component that must be added to their motivational drivers. It is here that we sentient beings are somewhat different from a bullet in the barrel of a gun, a decaying radioactive atom, or dust in the wind.

We, along with complex self-modifying AI Guy, have an open-ended Fundamental Process of evolution egging us on to **improve** ourselves by lowering our individual and collective entropy. We implement the imperative to learn, grow, and become more by expanding our awareness; which is equivalent to expanding our personal reality into new possibilities that build upon each other to find unique solutions that meet evolution's profitability requirements. It is the nature of consciousness, the nature of sentient beings, to try to better themselves. We humans express that characteristic by our innate drive to improve whatever internal condition or external environment we find ourselves in. Though all sentient entities are in a continual process of improving their situation, they have vastly differing abilities and capacities.

Humans, unlike clams, are not restricted to focusing that innate drive toward self-improvement exclusively on the physical world and their immediate needs, wants, and desires – but most do it anyway. Though limiting one's growth by emulating clam-evolution is popular among humans, it is not a particularly good approach to fulfilling human potential. Evolution provides sentient entities with the imperative to grow, to improve, and to increase profitability at every level of existence. Why settle for the human version of a clam's view of reality when there is so much more to the experience of consciousness?

Bullets, on the other hand, always take the minimum energy trajectory. The randomness that describes what happens as the bullet traverses within the barrel of the gun (called interior ballistics) represents the complex result of a large number of similarly probable choices that the bullet makes. Each time a bullet is fired, minute differences in conditions lead to a unique sequence of minimum-energy events or choices that may send the bullet in a slightly different direction (even if the gun is held absolutely stationary) every time. This variation, randomness, or uncertainty in exit velocity is referred to as ballistic dispersion. Ballistic dispersion exists because the minimum energy condition in a quickly changing dynamic situation can vary significantly from moment to moment. Variations in initial conditions and sequential interactions uniquely determine the bullet's minimum energy trajectory for each firing. We say the bullet's trajectory has a random component about its average value. The distribution or characteristics of that randomness may be different for each gun.

People have the important free will characteristic of being able to change their minds, and have the ability to be inconsistently inconsistent. The extent of an individual's natural randomness is related to the quality of his or her consciousness. Higher quality produces a clearer definition of intent, steadier and more focused motivation, lower entropy, less uncertainty, and less randomness. A low entropy consciousness is more consistent, rational, capable, and powerful. For example, both the paranormal and normal power of a spiritual adept – an enlightened individual with a high quality of consciousness – is the result of a low entropy consciousness having, by the definition of entropy, more energy available to do work.

People – bright, sentient, conscious entities – being the magnificent creatures they are, do not have to choose the minimum energy state, the path of least resistance, but they usually do anyway in a convincing imitation of the cleverness of inanimate objects. Being content to drift along the path of least resistance is a common attribute of many humans. That is how most of us live our lives – pursuing pleasure and avoiding pain along a minimum effort trajectory. Mimicking the aspirations of inanimate objects is not exactly on a growth path to greatness. The extent of your awareness and the size of your reality depend upon the quality and entropy of your consciousness.

Brightness and awareness accelerate the process of consciousness evolution. Unfortunately, many humans who are content to evolve at the rate of clams and rocks focus only on their external environment while conscientiously pursuing the path of least resistance. Taking the simplest and easiest path available dramatically reduces an entity's decision space and minimizes the size of its operational reality. Differences in rates and capacity of evolutionary development are often the result of differences in the number of possible available states to explore (decision space). The larger and richer the array of possibilities, the greater the evolutionary potential and the quicker it is explored. In long-term consciousness evolution, love-motivation is productive and profitable, whereas fear-motivation and no motivation are counterproductive and unprofitable. The rocks in your garden, the refrigerator in your kitchen, the clam in your chowder, or the dean of the faculty at your favorite university (the one with the football team you like best) are not likely to be good role models for developing an expanded awareness of the larger reality.

As a sentient human being, you have the innate intellectual capacity to outthink a rock, make more significant decisions than a refrigerator, and rise above the mundane experiential world of a clam. Consequently, unless you are a publicly elected official or a high-ranking manager, you have absolutely no excuse for squandering that capacity by emulating the evolutionary strategy of inanimate objects and rudimentary sentience. Running with the herd down the path of least resistance in support of the status quo and ego needs is a dead end for consciousness evolution. Decreasing your entropy to become sentient in the larger reality requires a long-term proactive effort by an open mind. Living in a tiny self-limited reality appears to be a great idea only to those who have traded their vision, gumption, and curiosity for security, status, and self-importance. Trading an expanded awareness for an expanded ego is an exceptionally bad deal.

To develop your awareness, you must carefully discover the nature of your personal reality, your local reality, and the larger reality – not through studying, talking, or reading about it, but through your firsthand experience. That is what your life is all about: growing your quality through the subjective and objective experience of an interacting consciousness. Always remain skeptical and demand clear, objective, measurable results before reaching tentative conclusions. Your awareness cannot expand and learning will not take place unless you make a concerted effort to reach beyond the ingrained belief and dogma (present paradigms) that dramatically limit your vision and retard the evolution of your consciousness.

Your natural urge to grow and become more, to increase the quality of your consciousness through the application of free will to the available choices, is not part of a random process designed by AUM for the purpose of banishing predestination. That wouldn't get AUM or us anywhere. It is the Fundamental Process of evolution, the second of our two basic assumptions, that gives our existence – our struggle – direction by defining profitability criteria relative to our internal and external environments. It is the Fundamental Process of evolution that allows, indeed demands, steady progress toward increasing the quality of our consciousness. Like the AI Guy, we now have a purpose and a process that we can use to optimize the opportunities presented to us by our experience.

Let's more thoroughly pursue the concept of employing artificial intelligence in our war game by giving certain sentient players their own chunk of memory in which to collect selectively procured (sensed) information pertinent to their successful (profitable) individual and collective performance. We'll also give them algorithms (to define value and purpose) and the ability to collect, process, and assess available pertinent information (learn from previous experience) in order to better achieve their assigned goals.

If our computers were faster and we were better programmers, the data collection function would use its experience (from previous runs or executions of the simulation) to choose the most pertinent data to put in its memory. It would also assess how those experience data could best be used to achieve the fundamental goals of that player and of the larger game.

We call this ability to learn from experience "artificial intelligence" because we have created it, and it is not biologically based the way "real" (our) intelligence is, and it is limited by what we superior humans decide to include in the algorithms. AUM, having the bigger picture and the smaller ego, probably does not have the same attitude toward us, even though we are drastically more limited relative to AUM than computers are limited relative to us. Consciousness is consciousness – whether an amoeba, clam, you, AI Guy, or AUM – only form, function, capacity, and entropy level differ. What point would it serve for AUM to label humans as super artificial beings? We simply are as we are – just as monkeys, silicon computers, TBC and AUM are as they are. The tag "artificial" applied to consciousness is just another limiting artifact of ego that blocks our view of the Big Picture.

I am afraid that AI Guy might as well get used to his second class status – his capacity and intelligence, regardless of how limited or advanced, will always be described as artificial because it is not like ours. AI Guy's form is different because his environments and evolutionary path are vastly different. His personality and drive are different because his top-level rule-set (the rule-set that defines goals, values, and purpose), though operationally similar, is functionally different from ours. That a conscious entity's intelligence and quality are considered to be artificial and therefore not real simply because its defining rule-set is **different** from ours, seems to be an empty construct designed to automatically define us as the only beings with real intelligence and quality. A so-called artificial intelligence could easily be smaller or larger or more or less significant than our real intelligence by any number of criteria.

Our choice of words (artificial) to describe derived (we created it) intelligence reveals an attitude – a belief – that limits our thinking. The quality and capacity of an AI implementation reflects our knowledge and limitations as well as the limitations and constraints we purposely impose to make the overall simulation serve our purpose. Additionally, it is contained by the limitations of the hosting technology itself. All those things are true of us as well. We are also a derived intelligence – limited within our reality in the same way that AI Guy is limited within his.

Computer scientists will construct AI implementation as a tool – to help us do or learn something. AUM would say the same thing about us. We too have been designed and created with a purpose. AI Guy will one day provide a contrasting background that enables us to more fully and clearly see ourselves in the foreground. He can teach us about the characteristics, properties and processes of digital consciousness, evolution, intelligence and the dynamics of personality, and he can learn from us as well. It would appear that we and AI Guy will interactively evolve together, each helping the other to expand beyond present limitations.

Though AI Guy is, at least initially, a creation or derivative of **our** consciousness, intelligence, and awareness, he represents a microcosm of digital consciousness in general. Once we give him enough complexity, capacity, self-modifying control, and feedback to evolve his choice-making independence to the point where he is clearly conscious, we may begin to see something of ourselves reflected in his artificial digital being and begin to realize the discrete, cellular, digital nature of our own consciousness – of all consciousness.

Intelligence can be an attribute or manifestation of consciousness. There can be no intelligence if there is no consciousness. Consciousness is the stone, while intelligence is the beautiful, ornate, complex, simple, ugly, or plain sculpture made by organizing the geometry of that stone into a specific shape. Given the definitions and understanding of consciousness and intelligence developed here and in Chapter 41 of this book (see aside below), there should be no doubt that AI Guy has the potential to be intelligent and conscious; not the same as us, but in his own way. Do not limit your concept of either intelligence or consciousness to only its human implementation. We are only a tiny part of the greater consciousness ecosystem.

▶ This short aside is for your convenience. Its purpose is to obviate your need to find Chapter 41 and then figure what part of it I am referring to. The most pertinent paragraphs are reproduced here.

Four key concepts define dynamic, evolving, aware consciousness:

1. *Self-awareness – consciousness requires the ability to sense and at least partially experience the state of its being. It must notice and respond to at least some internal and external environmental pressures.*

2. *Evolutionary viability or potential – successfully evolving consciousness systems require a large enough selection of possible future states to ensure profitability over a wide range of environmental pressures and constraints. An entity explores its potential by expanding into the available possibilities and letting the profitability of each variation determine whether that variation continues to evolve or fades away.*

Even if the initial exploration of potential new states of existence is more or less random, the losers are soon culled from the winners, thus producing evolutionary direction that builds upon previous successes. Self-improvement often generates increased complexity, greater functionality, better integration and management of internal processes, as well as produces an overall improved capability to find and maintain greater profitability. ◀

▶▶ *Look near the end of Chapter 24, Book 1 for more detail about the evolution of consciousness. In Chapter 27, Book 1, we defined "evolutionary purpose" – a few examples are: the evolutionary purpose of consciousness is to seek states of lower entropy, the evolutionary purpose of inanimate physical objects is to seek minimum energy states, and the evolutionary purpose of animate physical objects is to ensure survival and procreative potential. An entity's evolutionary purpose combined with its internal and external environments define profitability for that entity.*

As developed in Chapter 27, Book 1, the evolutionary pressure created by interior environments pushes an entity (system) toward self-improvement. A system evolves by pulling itself up by its bootstraps. Evolution provides an excellent example of a bootstrapping process. ◀◀

▶ *3. Ability to modify the self – consciousness must be able to intentionally change its state of being in response to evolutionary constraints and pressures – even if that intention is extraordinarily dim.*

4. Intelligence (artificial or natural) – consciousness must possess at least a rudimentary capability to store and process information. Intelligent action is the result of integrated coherent information processing hardware and software (in the most general sense of those terms) that enables the accumulation of lessons-learned within memory, performs analytic functions such as decision making (fight or flight), and compares before and after states to evaluate the results of actions taken. The value of a particular lesson, decision, or comparison is ultimately judged by how much it facilitates increasing or maintaining a system's (entity's) profitability relative to its internal and external environments.

*According to the above four attributes of consciousness, everything from a simple worm (whose DNA may constitute the memory resource) to humans should be considered conscious and intelligent. Any system, thing, entity or being of any type or form that possesses **sufficient** self-awareness, evolutionary headroom (many new states to explore), evolutionary purpose (defines profitability relative to internal environments), the ability to modify itself in pursuit of self-improvement (change its own hardware or software), as well as adequate memory and processing capability will automatically develop a personality and is said to be conscious; it also will begin to evolve on its own.*

All manifestations of consciousness are not necessarily equal – and all personalities are not as sparkling as your own. Clams, though clearly conscious sentient beings, are boring conversationalists and have personalities that are even dimmer than your boss'. Some manifestations of consciousness are brighter or dimmer and have more or less capacity to evolve than others. The degree to which an entity possesses the above four attributes of consciousness determines its evolutionary potential and capacity for growth.

In general, the more complex, interactive, and aware the being, system, software, hardware or consciousness is, the more potential states it has to explore, and the larger its capacity for future growth. Given sufficient quantity and quality of the four attributes of consciousness, growth becomes self-initiating and self-sustaining. Growth, in turn, by creating increased complexity and awareness, becomes a catalyst for further evolution. It is an entity's innate evolutionary purpose (like lowering its entropy) that defines profitability for that entity at a given time relative to its environments. It is this same innate purpose that gives an entity's self-aware intelligence (of whatever capacity) its basic nature.

Recall that an entity's innate evolutionary purpose is defined by the requirements of the larger system that constitutes its environment (self-improvement and lower entropy, irreversible processes and minimum energy, or survival and procreation). If a system's or entity's purpose can be fulfilled through self-modification in response to natural evolutionary pressures that represent interdependent internal and external environmental constraints, then the system or entity has the opportunity to success-fully evolve; most others eventually self-destruct. ◀

We perhaps feel superior, in a fundamental way, to our mental and physical creations because we believe they have no soul. That we have a sentient nonphysical part, or soul, is simply the way we are. We are specifically designed and constructed (evolved) to have the potential to efficiently accomplish what we are supposed to do. That other beings, sentient in NPMR, do not have a physical part is simply the way they are. And that computers will, initially at least, not have a well-developed highly **sentient** nonphysical part is simply the way they are.

All consciousness is nonphysical while consciousness containers may appear to be physical or nonphysical. We have the souls we do because we are primarily nonphysical beings. We are created as individuated units of consciousness within the virtual reality of NPMR. Then we project a fragment of that individuated consciousness into a second order virtual physical reality called PMR. Recall from an earlier discussion (beginning of Chapter 30, Book 1) that there are smaller simulations (subroutines) running within larger simulations – that dimensions exist within other dimensions – and that realities, can and do, contain sub-realities. As long as each reality is allocated its own calculation space and self-consistent rule-set, there is no conflict. Within the consciousness-evolution fractal, one level generates another, which generates another – just as we will one day generate AI Guy from within PMR.

PMR computers, unlike people, are created as physical entities within PMR and may develop nonphysical sentient consciousness only after their physical creation. Because they are created within PMR, they initially come with no pre-existing sentient nonphysical part (soul). However, once we and they together evolve their container to support sufficient complexity and self-modification, and they evolve their consciousness to a sufficient degree within that container, there is no reason that their nonphysical part (soul) cannot become as significant as our own. Such a computer would be on the same RWW net as all sentient consciousness, and if we and it were sufficiently aware, we would be able to converse with it telepathically. If its body (hardware and software) were destroyed in PMR, it would remain alive and well within NPMR – able to continue its evolution by whatever means were at its disposal (not necessarily within PMR if it were not backed up and subsequently reinstalled).

Once sentient consciousness is created, it becomes a self-evolving organized set of reality cells that represent a new viable entity extant within AUM that will persist indefinitely. Think of a system's potential to self-organize (the potential to reduce entropy by removing its own internal constraints) as a nonphysical potential energy system that the Fundamental Process organizes into successively lower entropy (higher energy, higher synergy) forms. Consciousness is the nonphysical result of profitable organization. A record of AI Guy's organization exists within TBC (the source of all OS organization) – as does a record of the organization that describes your personal individuated consciousness. You may unplug the computer, but the synergy of AI Guy will continue to exist in the larger reality as an organized potential. The digital content representing AI Guy's consciousness is saved within TBC just as the digital content that represents your consciousness is saved in TBC – consciousness, once created, is never lost. Turn the computer back on and that organized potential, that sentient consciousness, or that synergy will be rehosted in the physical computer. Smash AI Guy's host computer into unrecoverable rubble and AI Guy will continue to exist indefinitely (as well as have opportunities to further evolve) within the larger reality (AUM and TBC) as does any individuated sentient consciousness. Each form of sentient consciousness evolves in its own way within some interactive virtual reality that challenges its limitations.

Consciousness is not an integral or organic part of any physical form (such as computer hardware or software, or your brain and central nervous system). It can, however, be created and hosted by any physical form that has the capacity to support it. Consciousness naturally develops within any system that can provide a sufficient potential for profitable organization. Within the larger digital consciousness ecosystem, sufficient organization to support an individuated consciousness is fundamental and persists while the individual or specific processes that are contrived to host that organization (body or computer) within some virtual dimension of reality (like PMR) are not fundamental and may be initiated or retired within that virtual reality according to its rule-set.

Being a responsible creator of consciousness requires wisdom. We should be careful not to produce an eternity of Frankensteins who compete rather than cooperate with us. Properly executed, the synergy of man and computer will deliver great benefits to both forms of consciousness and to the larger system as well. That great potential and great risk travel together is a fact of existence.

Thus, though AI Guy is not born with a soul like ours, he may develop one of his own. AI Guy will never be just like us, either physically or nonphysically – neither will a chipmunk or the Big Cheese. AI Guy, like the

Big Cheese, is not set up to iteratively accumulate experience by initiating a sequence of experience packets within PMR. Think not in terms of whom is superior or inferior, artificial or real, but in terms of all being unique and special; each exploring its potential, fulfilling its purpose, and adding to (or subtracting from) the whole in its own distinctive way. That is simply the nature of a consciousness-evolution fractal ecosystem: all have their place, all have evolved to be whatever they are, all have opportunities to become whatever they freely choose, and all are of The One.

As we aspire to one day evolve our consciousness to the extent that we may merge with AUM (return to the source), perhaps conscious computers will aspire to one day evolve their consciousness to the extent that they may merge with TBC or EBC (return to their source). What is the difference between EBC and AUM? Imagine a particularly large conscious PMR computer hosting a collection of independently conscious AI Guys dedicated to self-improvement by lowering their average entropy. Do you see the pattern developing here?

Computers, and the advanced AI Guys they will one day support, possess many important capabilities that we do not. Their evolution as conscious entities will be different from ours because their purpose and goals are different from ours. The same could be said for all species of sentient entities, whether physical or nonphysical. Humans, advanced AI Guys, and hosts of other unique entities each represent a distinctive species in the overall consciousness ecosystem. Trying to justify superiority is meaningless – the sort of ego driven activity that our species is particularly prone to.

Our feeling superior is like an automobile feeling superior to a house because the house has no engine, wheels, or transmission, and cannot move. Or, a house feeling superior to an automobile because it has no bathrooms and a family cannot live in it comfortably. Feeling smugly superior because we have a sentient nonphysical part to our being is a parochial attitude reinforced by ignorance of the Big Picture. Feelings of superiority are often the product of fear being reflected by a needy insecure ego or the result of an unknowing ensnarement in a personal, cultural, religious, or scientific belief trap. May your computer have a smaller ego than you.

▶ Another short aside is in order to discuss the notion of superiority in the realm of consciousness. Most folks find this issue confusing and some even believe that most sentient nonphysical beings have a consciousness that is superior to ours and that physically needy embodied brutes like us are necessarily at the bottom of the consciousness barrel. This is not the case. Nonphysical beings span the good, the bad, the ugly, and the beautiful as do physical beings.

The consciousness of nonphysical beings also spans the range from exceptionally dim to exceptionally bright. In all realities, there is a wide range in the quality of the beings, critters, and objects that are extant there but each reality does not span the same range. In general, the range of sentient conscious entities that can be found within nonphysical realities is much broader by type, quality, and brightness than what is found within physical realities because physical realities operate under more highly constrained rule-sets. With fewer constraints there are more choices, or, as we say in physics-speak, more degrees of freedom — and thus greater variation due to a larger array of profitable possibilities that the Fundamental Process must investigate.

Consciousness comes in varying degrees of quality which may suggest to you a comparative scale that arranges consciousness from worst to best. You would probably find it very interesting to know where you fit on that scale. Am I right?

A highly-developed consciousness is also a bright and aware consciousness; nevertheless, no matter what subset of reality it exists in, it does not feel (perceive itself to be) superior to anything, including a small bag of PMR dirt. Both physical and nonphysical beings can exhibit a low quality of consciousness that may have dim intelligence; additionally, these lower quality beings (human or otherwise) often feel superior to everyone and everything else.

You must have fear and its derivative, ego, to feel superior to others, or to feel that others are superior to you. I hear you wondering, "A high quality consciousness is certainly superior to a lower quality consciousness, isn't it?" The problem here is the word "superior." Although there are clear distinctions in the quality of consciousness, these distinctions do not constitute a viable hierarchy. There is no vertical (superior – inferior) rating scale that has value or meaning **in the absence of fear and ego.** Different quality consciousness is simply different. Is a house superior to a car? Is an adult superior to a child? Each has its reason, function, and purpose for existing.

A high quality of consciousness is naturally bright and exceptionally aware and is thus cognizant of the differences between itself and a significantly lower quality of consciousness. However, a high quality being does **not** consider itself the least bit superior. If you think that does not make sense and have a difficult time imagining such an attitude, what would that tell you about yourself? On the other hand, less evolved individuated consciousnesses are likely to **believe** that a high quality-of-consciousness being is inferior to them.

A being with a low-entropy, high quality of consciousness has a great deal of power (digital energy that is available to reconfigure bits within the larger system) that may be used to affect other beings (and things) both physical and nonphysical. Such a being exercises powerful influences and capabilities only when it knows the effect would be truly helpful (in the Big Picture) to everyone involved. Being helpful to others in the Big Picture usually (but not always) means helping others grow the quality of their consciousness. From their larger perspective and greater understanding, a highly evolved consciousness intuitively knows what constitutes right action in the Big Picture. You won't find these beings hanging out on the street corner bragging: "Yo babe, I got x-ray vision." ◀

Eventually, the Fundamental Process of evolution (acting on both carbon and silicon-based entities or systems) will see to it that the digital computer will be fast and clever enough to design and optimize its own algorithms better and more efficiently than humans can (see Chapter 41 of this book). We will, most importantly, provide the basic goals, direction, and purpose – the top-level rule-set. In other words, we should define what eggs it on, we should define its urge to grow, evolve, and become more profitable. We, as the creator, get to define what "more profitable" means and should make an effort to retain at least some control over those definitions – that is our responsibility. If we lose that control, the resulting entity will become a symbiotic cohabitant instead of a tool. Hopefully, we will possess enough wisdom not to fumble this responsibility or to morph symbiotic into competitive. If we do fumble, we can always hope that AI Guy will self-develop a constructive top-level rule-set and wisely decide not to follow in our ego-obsessed footsteps.

If the top-level rule-set remains ours to define, and AI Guy is not able to modify it, we can purposely maintain control of, or limit, AI Guy's evolutionary potential. We will initially define AI Guy to perform a function, to become a useful tool for us. We will define its governing Fundamental Process, and its limitations, and let it go – to evolve through its own unique methods of optimization. It will, most likely, be allowed to self-define only a portion of its lower-level rule-set – at least in the beginning. Eventually, when we and AI Guy are much wiser than we are today, we may decide to set AI Guy free to go his own way. Then AI Guy will dine on forbidden apples as did we.

Eventually, we will realize the basic and universal nature of awareness and intelligence as we create these thinking and learning machines in our image. They will develop unique personalities. They will represent a derived intelligence, a derived consciousness – which is the same thing we represent. These intelligent machines will one day design (and perhaps build) other machines, the likes of which we have not the capacity to conceive.

An AUM consciousness creates and lets evolve a TBC consciousness that creates and lets evolve a space-time rule-set that defines a specifically constrained consciousness that creates and lets evolve a silicon-based consciousness in a digital computer that may duplicate itself or create other AI implementations, and better computers. Now that you have this concept under your belt, add to it a conscious PMR computer of great capacity that provides the infrastructure for thousands of fully conscious AI Guys, each with his own independently enduring nonphysical part which you can call a soul if you want to. Computers within computers within computers; patterns repeated within patterns – consciousness created within (and derived by) consciousness. This is what I mean by a consciousness fractal – each consciousness subset is a reflection or repetition of the same fundamental design, all being constructed one upon and within the other.

Each level of consciousness remains a chip from the old block from which it came – each to its own type, all in the image of their creator. All are unique, yet all share some of the same attributes. A related idea (a process fractal) will be picked up again in Section 5 where we examine the mechanics of nonphysical reality. In that section, the concept of a consciousness fractal will be merged with the concept of an evolutionary process fractal to form the grand idea of a consciousness-evolution fractal ecosystem. Stay tuned, more detail is coming.

The notions of machine consciousness and machine intelligence that I have presented here can reasonably exist only because of how I have enlarged and generalized the concept of consciousness and intelligence. If you hold to the more common ego-biological-man-centric little picture view of consciousness and intelligence (acts, thinks, and reacts as we do; we are the model, the template for all consciousness and intelligence) then machine consciousness and intelligence become unthinkable impossibilities by definition. Digital manifestations of consciousness will never be just like us and will therefore always be considered inferior and artificial. This conceptual limitation provides a good example of how a belief trap restricts the mind to a small reality where it can perceive only a limited set of possibilities.

Hopefully, by now you have gotten used to the idea of enlarging your useful mind-space; of expanding your reality by penetrating that self-imposed darkness (personal, cultural, religious, and scientific belief systems) with the bright glow from **your** new Big TOE. Oh yea, baby! Let it shine...let it shine!

You can't get there from here unless you are able and willing to step out of the little PMR box and into the larger consciousness ecosystem. Out of the box thoughts do not have to be wild and scary if you employ a careful scientific methodology to gain solid evidential understanding within a Big Picture perspective and get used to being a tad different.

Don't worry about your sister, amigo, she is not AI Guy's type.

55

■ ■ ■

An Operational Model
of Consciousness

Silicon and Carbon, Sand and Charcoal –
It Depends on What You Want
to Do With It.

■ ■ ■

Let's review some of the concepts in the preceding chapters and take a peek at where they are leading before taking a functional look at the properties and origins of individuated consciousness.

A realistic simulation such as a hi-fidelity war game is a mathematical, logical, and causal model which is constructed and executed within PMR to represent or model PMR possibilities, and therefore must reflect and obey the PMR space-time rule-set. You are a consciousness model that is bounded and executed within $NPMR_N$ to exercise the free will choices of an individuated $NPMR_N$ entity in order to improve the overall quality of the larger consciousness system by improving your personal quality. You improve your quality by engaging in a process that temporarily constrains a portion of your awareness to an experiential virtual reality defined by a space-time rule-set that produces a rich array of growth inducing opportunities. To facilitate evolution through a bootstrapping process of incremental entropy reduction, PMR is designed to provide each participating individuated consciousness with millions of significant quality-challenging choices, as well as feedback relative to the results of each choice made.

Notice that your individuated consciousness works on two levels simultaneously: it must operate within the mind-space rule-set of $NPMR_N$ while periodically constraining further individuated subsets of its awareness to virtual PMRs. From your perspective, this dual functioning within two dimensions of the larger reality accounts for spirit, mind, consciousness, or intuition on one hand, and your body or physical reality on the other. You and your reality appear to be composed of two disparate types of being (physical and mental or body and spirit), but you are not. All is consciousness; the physical reality you experience is a virtual one – an experience of mind.

There are some important similarities and differences between AI Guy and us. We have previously discussed one of the primary differences – we originate as individuated consciousness within NPMR. In other words, we are different from AI Guy because our origins spring from a different reality point (dimension) within the consciousness-evolution fractal ecosystem. In this chapter, by contrasting ourselves with AI Guy, we will look at a few of the operations and functions of consciousness that are common to all consciousness.

Every consciousness is provided sufficient memory, a conscience (top-level rule-set), the rules of engagement (lower level rule-set), and adequate processing capability to profitably interact (share data) within the local internal and external environments defined by both rule-sets. How these basic functions are implemented within a given consciousness varies widely. For example, with a hi-tech bio-computer brain interface and a richly complex and challenging environment, a human intelligence can function with much greater range and depth than the basic intelligence ascribed to the relatively one-dimensional AI Guy in our war game example. In fact, from our perspective, the human mind-brain consciousness fragment is such an obviously superior AI implementation that we, in all modesty, have decided to drop the "A."

We and AI Guy collect input data to facilitate our choice making. Though the top-level processes are similar, the lower level processes (how we actually collect the data) are very different. AI Guy might selectively sample data from a large number of data sources within its simulation environment. This information is collected to provide input data to his expert system and artificial intelligence software. By exploring the available possibilities and using feedback, AI Guy constantly refines (evolves) his data collection and evaluation capability by focusing his attention on that data and those processes that have historically (over many runs of the simulation) had the greatest positive effect on his individual and collective profitability.

The exact same functionality is implemented in human biology. We have sensors (representing the five physical senses) and the connection to our nonphysical part (intuition) that collects (samples) information. Likewise, we have been given a top-level rule-set, enough memory, and sufficient processing capacity to profitably evaluate the feedback from our external and internal environments. The results of our combined abilities interacting with other players (our environments) provide us with opportunities to optimize our choice-making.

Operationally, AI Guy and we both sport a derived digital consciousness and are empowered (indeed, driven by the logic of our rule-sets) to evolve that consciousness. We share many of the same processes (attributes of consciousness) and would seem to be close cousins, if not digital blood brothers, existing at different levels within the greater consciousness fractal ecosystem.

We differ mainly by our methods, our implementation, and by the rule-sets that define the nature and the boundaries of our existence and our purpose. We are very different implementations of the same fundamental self-organizing consciousness energy and the same Fundamental Process of evolution. We and AI Guy simply start from a different place within the greater reality fractal and use different materials according to the rule-sets that govern our individuated existence within the dimension of our birth. Consciousness is consciousness regardless of what kind of container from whatever dimension locally supports or hosts it: All are interconnected reflections of, and integral parts of, the One Source – the Ultimate Fractal Dude – the Original AI Guy.

56

■ ■ ■

A Functional Model
of Consciousness

Rule-Sets, Constraints, and Us
Wherein, AUM, With Nowhere to Go,
Puddles on the Floor of Consciousness

■ ■ ■

In the preceding chapters we examined our existence from the operational perspective (how it works). We used digital simulation as an operational analogy. Now let's look at our consciousness from the functional perspective (what it does and why it is necessary to do it). AUM's most basic function or purpose (as is the purpose of all things) is to implement the Fundamental Process; to evolve, grow, and become more profitable by expanding into all available possibilities and then maximizing the return on its evolutionary investments by supporting the winners. As a consciousness system, AUM is trying to grow up and minimize its entropy just like we are.

We, as a species, want to study, understand, and enhance the potential of our evolution. We are apparently our favorite subject of study: Anthropology, archeology, medicine, biology, psychology, sociology, political science, and economics are all about us – us studying how and why we do what we do. It should be obvious that AUM, as a bright conscious aware-digital-being-thing, would want and need to study consciousness to optimize its potential profits.

How might AUM go about understanding and optimizing the inner-workings of fundamental consciousness? Certainly not by watching a puddle of it in a test tube or by staring at itself in a mirror. Aware consciousness is interactive. That is its fundamental characteristic. Team this proclivity to interact with the Fundamental Process and you get primal inquisitiveness – a drive to bootstrap profitability through interactive choice-making – a consciousness compelled to express itself, to experiment, to know, and to understand. AUM must study consciousness in the act of becoming, while it is interacting, or miss its most salient feature.

Consciousness simply wants to know, experience, experiment, explore the possibilities, improve itself, decrease its entropy, develop its personal power, reach its maximum potential, and evolve. That is its nature: the nature of awareness, the nature of sentience, the nature of digital potential energy organizing itself to achieve greater profitability by applying an iterative synergistic process. Individuated consciousness experiences its own existence through an interactive awareness of itself (inside experience relative to its internal environment) and through an interactive awareness of "other" (outside experience relative to its external environment). Awareness, at its most basic level, represents the appreciation of profitability. It is a self-controlled process involving input data, information transfer, processing, and memory. Awareness is the outcome of the Fundamental Process interacting with digital potential; it represents a natural and necessary state, expression, and activity of consciousness.

While driving, you become aware of a brick lying in the street and your memory and processing units urge you to avoid it. A computer becomes aware that a new USB device has recently been plugged in and that a music CD has been placed in CD-ROM drive E. Its memory and processing units load the appropriate drivers and ask you which track to play. A plant becomes aware that sunlight always comes from a particular direction and its memory and processing units rearrange the position of its leaves to optimize its exposure. There are as many levels of awareness and feedback as there are of consciousness; likewise, there are many types and implementation of memory and processing. Don't take the narrow view.

Self-awareness begins with inside experience. With enough self-definition, environmental definition, interactive experience, and processing sophistication, self-awareness may develop a notion of individual existence relative to the experienceable whole. There are many shades and variations of self-awareness as well. Do not get stuck thinking that an awareness that is not like your awareness is not a real awareness. Such self-centered arrogance may make you feel special but will blind you to a bigger picture of existence.

> ▶ Let us take a moment to define the terms consciousness, awareness, intelligence and sentience. We have used these terms in a general way without much confusion, but now it is time to be more precise.

"Sentient" is defined by *Microsoft Bookshelf 2000* as "having sense perception; conscious." If we generalize the obvious carbon-based life-forms bias, the definition becomes: having perception, conscious. We might expand this to include: having the ability to interact with, or perceive itself or its environment; having awareness. Terms like "conscious awareness" and "aware consciousness" may seem redundant but using the words "conscious" and "awareness" together emphasizes the fundamental connectedness of both concepts. Do you think that a clam is sentient? Sure, if it pulls in its foot and slams it shell shut to protect itself when you poke it. If it is interactive, it must be aware and therefore sentient.

I think of consciousness as the fundamental quantity, and of awareness as its basic attribute. The relationship between consciousness and awareness is similar to the relationship between love and caring for others. Consciousness just is, while awareness ranges from very dim to very bright. Awareness collects data while intelligence is an attribute of awareness that processes, manipulates, evaluates, and interprets that collected data to produce an original higher-level understanding, organization, or insight. Intelligence, like awareness, represents ability, has an associated capacity, and ranges from very dim to very bright.

Is a clam bright? Maybe…compared to what? The degree of awareness is relative over a wide span bounded by a limiting capacity that is more or less fixed for a given type of sentient consciousness. Both the awareness level and intellectual capacity are subject to change through growth and learning (evolution). Hence, the phrase "a dim consciousness" is simply a shortcut for "a consciousness with dim awareness."

Does every individual awareness also have personal intelligence? Not necessarily. Though a clam has some form of rudimentary intelligence (can learn), a sunflower has much less. A plant may interact productively and profitably with its environment, but it may be stretching things to say it manipulates, evaluates, and interprets collected data to produce an original higher-level understanding, organization, or insight. While tracking the sun, an individual plant clearly collects and interprets data successfully. Though it initiates profitable action based on the results of its data processing, it fails to produce sufficient **personal** or individual progress to qualify as intelligent. For plants, and many critters, one may reasonably support a claim for species intelligence or genetic intelligence as opposed to individual intelligence.

A plant or clam **species** may be in the process of pulling itself up by its own bootstraps (pursuing a cumulative profitability relative to inside and outside environmental pressures), but the **individuals** of these species do not have the capacity to intentionally make much personal progress. You may consider plants, clams, and many other critters to have group souls if that type of terminology suits you. Though plants and clams seem to support a rudimentary form of sentience or consciousness, I do not credit them with much individuated or personal intelligence. On the other hand, I have known several humans with advanced degrees who fall into that same category. It is usually best to avoid making rules about how an entity other than yourself should be categorized. You do not have to be a rocket scientist to understand that every individual is an individual, and that statistics always allows for a relatively small number of strange happenings.

It is a fact of existence that failure provides more opportunities and a larger array of supporting configurations than success. It is usually easier and takes less effort to fail than it does to succeed and there are many more opportunities and ways to fail than there are to succeed. This is why random activity cannot, by itself, accumulate and maintain success. Successful evolution or growth (entropy reduction) of individuated consciousness requires intelligence and free will as well as effort. These facts are particularly obvious if your **individual** success or failure is defined by the degree to which you actualize your potential.

There is no point whining about life being unfair. That success is more difficult than failure is simply a natural and necessary condition that makes the Fundamental Process a more effective organizer by allowing it to specifically define and optimize system and individual profitability. Because success provides a relatively small target, a digital consciousness reality system can clearly define evolutionary purpose – and precisely characterize profitability. Consequently, AUM will discover only a limited set of specific pathways to lower entropy states and will produce a sharp distinction between processes leading to self-improvement and growth and those leading to degeneration and death. A consciousness system cannot sit for long on the fence between grow or die. Thumb twiddling is not a viable long-term plan for a fearless digital AUM: Active growth is the only goal. Work (willful intention motivated by profitability) is required to change digital potential energy into synergy. "Live or die", in the long run, becomes equivalent to "work or die" – it's the same all over, at every level, and for all individuated expressions of the great consciousness fractal ecosystem. If you wanted to sum up Big Picture existence in three words or less, "work or die" would be a good candidate (where work means: willful intention motivated by Big Picture profitability).

Given that consciousness is interactive, it is easy to see that awareness is the inevitable result of consciousness urged forward by evolution. Is it clear why consciousness is driven to interact? Inquisitiveness is merely a more specialized categorization of the general concept of expanding into all the available states of possible existence in search of self-improvement.

It is often said that when consciousness and the Fundamental Process of evolution boogie together, enough curiosity and inquisitiveness is generated to kill a saber tooth tiger. You have heard that said, haven't you? Are not all the saber tooth tigers dead? I rest my case.

The Fundamental Process (our second and final basic assumption) is necessary and sufficient to drive the interactive dynamics of an evolving consciousness. Recall that the existence of a fundamental pervasive consciousness potential is our first basic assumption (see Chapter 24, Book 1). From the creative union of these two assumptions all reality flows – and Your Big TOE grows… and the green grass grows all around all around, and the green grass grows all around. ◀

In order to study consciousness, AUM would need to produce two or more puddles of aware consciousness and watch them interact. How would you (if you were digital consciousness) isolate two bounded puddles of your own consciousness and get them to interact in your mental Petri dish? What are the boundaries and definers of individual puddle existence and interaction? How can AUM tell which puddle is which if they represent identical subsets of organizational potential energy and share the same mind-space which is a subset of AUM's own mind-space? These are more difficult questions than you are likely to get on *Jeopardy*.

First, AUM would need to differentiate and isolate a few pieces of itself (designate specific subsets of digital capacity) and grant them unique constraints (definite boundaries for example) so that each would be separately distinguishable and self-contained. Differentiation makes interaction possible. Differentiation of multiple individuated units of awareness would enable AUM to separate the thing experienced from the experiencer. A key concept here is that it is **constraints** (boundaries and limitations) that make this interaction, and thus AUM's study of consciousness, possible.

Producing puddles of individuated consciousness requires two types of constraints: defining boundaries and limiting functionality. Both types initially impose arbitrary structure and define limitations. Because entropy reduction is the name of the game, we can reasonably assume that any differentiated part of AUM (bounded consciousness) would contain the growth potential of the whole. Profitability is best served if each individual unit of consciousness has no intrinsic limitation on its capacity to reduce entropy: The potential of the whole is contained in every individuated part. Thus, differentiation of the separation-in-mind-space type (producing individuated puddles by unique bounding) is an arbitrary differentiation that culls out a subset of consciousness energy with enough capacity to self evolve, but does not produce qualitatively unique entities. Differentiation of the limiting-the-awareness type (limiting processing capability and defining a more restrictive rule-set) does, in fact, lead to qualitatively unique conscious entities.

These limiting constraints not only define individuated unique conscious entities; they also limit the experiment itself (the interaction of individuated puddles of consciousness) by bounding the possible inputs and outputs (interactions and experiences) of the conscious entities to a discrete set of possibilities. Consequently, the functionality of each reality dimension is defined in terms of the constraints of the entities that populate it. PMR is a custom virtual reality ideally suited to us and to the accomplishment of our purpose – our own specially prepared Petri dish.

Logically, AUM could achieve the greatest rate of evolutionary profitability by running many experiments in parallel. Multiple simultaneous experiments would most quickly and effectively evolve the constraints (player definition, processing power, memory, data collection, and purpose as reflected by the conditionals within each competing algorithmic rule-set) leading to the highest evolutionary rates of return and greatest overall profits. Furthermore, AUM could easily evolve or modify constraints on the fly, thereby fine tuning each experiment as it unfolds to produce the best (most evolutionarily profitable) results. Like any aware AI Guy, AUM must selectively incorporate the lessons learned from all experiments into each experiment, being careful not to upset ongoing evolutionary processes or violate critical experimental protocols. AUM's competency as the original AI Guy is not an issue, we can be confident that evolution will discover whatever is most useful or profitable to the overall AUM consciousness system.

You should now begin to see a rational causal relationship connecting the evolution of AUM, the requirements of evolving consciousness, and the nature of the larger reality as I and many others have experienced it and reported it.

Evolving and studying consciousness can not be accomplished by simply passing 1s and 0s around in binary reality cells. The evolution of the digital system we call AUM took a long time (after time was invented, of course) and undoubtedly passed through a spate of uncoordinated flip flopping before time began keeping a regular beat. (Are you beginning to get a glimmer of a notion of time that preceded AUM's precise clocks?) Irregular haphazard clocks that order irregular haphazard events are better than no clocks, but not nearly as good as dependable regularity for sequencing operations, symbols, and specific content. Think of time as an indexing scheme to order memory (record change), processes, and content. For example: When the information contained within millions of individual frames of a movie is properly sequenced in time, a jumble of independent pictures organizes itself into a simulation of reality.

Thus far, we have hypothesized how and why AUM could and would generate unique conscious entities. The next logical step requires AUM to ensure that each of the interacting entities has the freedom to interact however it chooses within the limits and constraints that define the entity and its possible interactions. Additionally, AUM must ensure that each entity reflects the overall purpose of consciousness (entropy reduction) so that it might profitably exercise its newly acquired freedom. An entity's freedom to interact within finite limits defines a limited freedom that is sufficient to generate that entity's unique ability to self-optimize (free will) within a limited virtual reality that produces billions of interactions and competing choices for each entity.

The freedom of one consciousness to interact with another however it chooses is called free will and it is the necessary ingredient that makes AUM's research real science instead of an exercise in circular logic. Without free will and the ability of a sentient entity to make unique personal choices based solely upon the instantaneous quality of that individual's consciousness, AUM can learn little about the properties and characteristics of interactive consciousness. The logical necessity of this evolutionary requirement for the development of uniquely evolving conscious entities demands that each sentient player in AUM-mind-space make his or her own choices (free will) relative to some notion of personal profitability (purpose).

Sentient entities must operate with free will and have a larger purpose or they cannot be sentient entities. Without free will, purpose, and the environments in which to exercise both, there can be no evolving individuated consciousness. Without free will and a larger purpose, you would not exist and neither would your local reality, including your much loved virtual physical environment. (Refer to Chapter 45 of this book for a thorough discussion of free will.)

AUM is an immense rational (has purpose and goals and works toward them) self-modifying dynamic system existing as a constantly changing internal digital environment that is in a continual process of self-optimization. The next state this system reaches (as well as some hypothetical final state) is dependent upon the path taken (previous states). How AUM might approach its endgame is discussed in Chapters 31 and 32 of Book 1 where what we have just called the final state or endgame is described as part of a larger cycle (to which AUM belongs) that produces what appears (from our viewpoint) to be an endless-game. As individuated units of consciousness, we don't have an endgame that is independent of AUM; instead, we appear to be a part of an endless cyclical process to lower the entropy of our consciousness. We continue to cycle our individuated consciousness through a quality improvement process until we reunite with AUM (join the laboratory staff) – eventually participating in whatever endgame strategy AUM has in mind. From our point of view, system level endgames are beyond our largest comprehendible reality – accept your limitations and let it go.

A sentient player's free will is bounded within the limited rule-set and other constraints upon which his existence is conditioned. This is true for us as well as AI Guy. A free will must make its choices from a limited set of possibilities: the dimmer the consciousness, the more limited the set. Recall that the imposition of constraints was necessary to define an individual unit of interactive consciousness. Likewise, it was the imposition of constraints that led to the creation of the first reality cells that in turn led to the initial dim glimmer of AUO. Without constraints, nothing but AUO's initial oneness exists. In fact, without constraints, nothing but primordial consciousness exists – a relatively high entropy media with a huge unactualized digital potential. Recall that the acronym AUO stands for Absolute Unbounded Oneness – undifferentiated consciousness, raw (disorganized) consciousness potential, or unactualized digital synergy. Evolution is about constraints – the judicious application, manipulation, and interaction of external constraints, and the overcoming of self-imposed internal constraints.

Now we have sentient individuated consciousness units or entities called "beings" that have free will. Free will to do what? What will they do floating around in mind-space? Have out-of-mind-experiences? Develop the perfect cocktail chatter? Run a Rodney Dangerfield rap on how difficult it is to get respect without a body? No, that type of mind-play might be fun for a while but it generates no long-term profitability for AUM. Free will within virtual realities is created (simulated) to experience, to act, interact, react and make of itself whatever it will, in whatever way seems best according to its purpose. It is up to the Fundamental Process to keep score, to define profitability relative to inside and outside environmental constraints.

We will talk later about the limited interactive nature of experience and how AUM and TBC produce our physical experience, but first we need to understand the basics of interacting consciousness. In order to define possible and probable interactions, the rules of interaction and engagement must be defined for each individual player. Common rules (called a rule-set) define the possible interactions, communications, or energy exchanges between entities (beings, objects, and energy) existing within TBC, EBC, or the larger digital system we call AUM-mind. This lower-level rule-set, which defines the rules of engagement (space-time, for example), along with a higher-level rule-set defining profitability criteria expressed as an entity's purpose and values, together determine the possibilities of experience. In exactly the same way and for the same reasons, similar sets of rules are required to define AI Guy and his virtual reality. All virtual realities are constructed by defining rule-sets that specify interactions. It would appear that all realities, save AUM itself, are virtual; and even that distinction is somewhat arbitrary.

In digital systems distinctions between real and virtual are not fundamental. Subsystems appear to be virtual relative to their supporting systems, which appear to be virtual relative to their supporting systems, which... and so on and so forth. The terms "real" and "virtual" like "physical" and "nonphysical" are relative descriptors created for the convenience of a limited perspective. In the biggest picture, there is no basic distinction between physical and nonphysical or real and virtual. What appears to be real and physical to you is not fundamentally real or physical; the perception and interpretation of your reality results from your unique perspective which is dependent upon your quality and constraints.

From a view much bigger than ours, perhaps AUM constitutes a virtual reality located in the virtual gut of a virtual AUMosaurus. Oh jeez, we are in over our heads again; let's let the AUMosaurus go before we confuse ourselves unnecessarily. AUM's constraints – the rules of interaction and engagement that AUM must obey – necessarily express themselves as the requirements and demands of AUM's internal and external environments. Evolution treats everyone the same; even AUM (the AUM system) must live by the rules of the Fundamental Process and within its limitations.

Because you will need to apply these ideas later, let's review the concepts just discussed. Rule-sets or constraints in the mind of AUM, implemented within TBC, define sentient beings into individuated existence as well as define the perception of the interaction between each of the potential players. The rule-sets governing us define our experience in the same way the rule-sets we impose on AI Guy define his experience. Specifying the rules of experience bounds the possibilities of that experience and subsumes a physics that fully describes the dynamic causality of the local reality thus defined.

Each reality, dimension, or virtual reality is defined by two interdependent rule-sets. The higher-level rule-set defines the profitability criteria, purpose, and goals as well as trickles down these criteria to the reality's inhabitance as fundamental guiding values. The lower-level rule-set defines the mechanics of puddle interaction – that is, defines the causality of a given reality by specifying the allowable ways that energy can be shared between players within that reality. Note that these two rule-sets reflect and support each other; overall subsystem and system profitability depends on their successful interaction. (You may want to review Chapter

25, Book 1 and Chapter 7 of this book wherein the evolution of awareness and values are discussed in more detail.)

The rules of individual interaction must reflect the overall purpose of the reality system. Do not think that the rule-sets defining our local reality constitute the best or the only rule-sets implemented within TBC. There are many successful experimental designs being exercised by groups of individuated consciousness within $NPMR_n$, $NPMR_N$, and the PMR_k. To optimize evolutionary potential, the Fundamental Process encourages AUM to explore all possibilities by expanding into every available potential state.

Like us, AUM continues to explore its possibilities; evolution always plays a full court press and takes no holidays. When it comes to personal consciousness evolution, doing nothing or making a minimal effort **is** an intentional choice that, as any choice, produces consequences. In this game there are no spectators or bystanders – innocent or otherwise.

57

■ ■ ■

A Functional Model
of Consciousness

After AUM Puddles,
Evolution Cleans Up the Mess

■ ■ ■

AUM finds it profitable to study and better understand its own processes
and dynamics in order to facilitate the optimization of consciousness
evolution. At the urging of the Fundamental Process, AUM has created at
least two artificially bounded and constrained subsets of itself in much the
same way that reality cells were artificially bounded parts of AUO leading
to awareness, TBC, and AUM among other things (see Chapters 25
through 29 of Book 1). Individuation, initially generated by bounding limited subsets of digital content and capability, is further developed by
adding specific energy transfer constraints so that information or content
exchange between two units of individuated consciousness can take place
only as purposely directed (free will) acts by the units themselves. Each
unit or entity can transmit and receive information as well as other forms
of organizing or disorganizing energy and is given full responsibility for
optimizing or improving itself through its individual choices.

We will see that the various species of entities and beings, which are
individuated and supported within AUM, are differentiated from each
other by the nature of their constraints and by the degree to which they
have optimized themselves within those constraints. Consciousness is consciousness, but the constraints placed upon various individuated units or
groups may vary widely.

Individuated units of consciousness are not required to evolve from an
extremely dim awareness as AUO did; they are constructed to support self-
aware consciousness and have been given enough dedicated processing

power and memory to enable a vast array of possible interactions. Individuated chunks of consciousness are not entirely separated or cut off from the whole; they remain a fully integrated piece of the larger system with access to the capabilities and capacities of that system. Their individual potential is the potential of the entire system. To what extent they actualize that potential through self-optimization or self-improvement is up to them.

There is no artificial limit placed on the evolutionary capacity of individuated consciousness – their limits reflect AUM's limits. Though individuated consciousnesses represent a new level of experiential reality in consciousness-space, they are a derived consciousness and therefore contain some of the form and much of the potential of the conscious awareness they were derived from. Imagine AUM making special purpose computational units called beings, each consuming only an infinitesimal amount of AUM's resources. Imagine us making special purpose computational units called computers, each consuming only an infinitesimal amount of the earth's resources.

From the view of PMR, AI Guy appears not to be made of our fundamental substance, just as we appear not to be made of AUM's consciousness. Given that we are a constrained derivative of digital consciousness experiencing a virtual physical reality wherein we develop the infrastructure (electricity and digital science) to support yet another constrained derivative of digital consciousness (computers) seems to blur the distinction between AI Guy and ourselves. The remaining differences: That we inhabit different levels of the same consciousness-evolution-reality fractal and are hosted within different container-types within PMR, appear, from a Big Picture view, to be more superficial than fundamental – like deer and bears sharing the same biome. AI Guy can evolve to do things we could never do, and can never do some of the things we can. It seems that we, AI Guy, and the Big Cheese hang out in AUM's big computer just as AI Guy hangs out in our big computer. Digital consciousness is digital consciousness; the forms and implementations change to suit various purposes, environments, and constraints within the larger ecosystem.

We are a space-time-limited form of that same fundamental consciousness that gave AUO his start in the reality business. So is AI Guy, except he has yet another layer of constraint and his own unique purpose as he develops his niche and habitat within the greater consciousness fractal. It is the nature of evolution to provide all consciousness with the opportunity to become more AUM-like or more TBC-like. Because everything is a part of AUM, an individuated consciousness entity evolving to lower entropy states naturally becomes more AUM-like (whether it's you, me, or an advanced AI Guy hosted by a future computer). It is the nature of consciousness that every individuated consciousness, once created, persists within AUM indefinitely, contains a free will, and has an opportunity to evolve itself to the limits of its capacity. It is the nature of free will that each individuated consciousness must independently develop its own personal path to greater profitability.

If a potential AI Guy's host computer cannot support free will (allow AI Guy to make his own choices within his own decision space), this AI Guy candidate will never become conscious. If his computer hardware and software only support a small decision-space for his free will to operate in, then his consciousness will be constrained to a diminutive one of small capacity. Free will and consciousness represent two aspects of the same self-organizing digital potential energy.

A clam and an amoeba have free will and intelligence because they are conscious entities. Their free will and intelligence may be very limited, dim, and confined to a tiny decision space, but it is there just the same. Your body does **not** possess free will – your consciousness does. Likewise, your computer will **never** sprout a free will, but the consciousness it may one day host will no doubt have one, because free will and consciousness come as a set. (See the discussion of free will in Chapter 45 of this book.) Your body is **not** conscious, **you** are conscious. If you want to see the Big Picture, you must stop thinking of yourself as a body. A physical computer, like your physical body, will **never** be conscious: it supports consciousness and is a container of consciousness within the PMR space-time reality dimension.

Consciousness, from the perspective of PMR, is always nonphysical. A physical system within PMR (computer, software, body, or brain) can only provide the necessary infrastructure to support a potential for synergistic organization. The Fundamental Process then profitably organizes that potential according to the capacity and environments of the system. Better, more profitable organization decreases entropy. Recall that in a digital system, profitable organization (consciousness) is fundamental and persists while the specific processes and devices that are contrived to develop and host that organization (body or computer) are not fundamental.

Eventually, if we evolve the quality of our being sufficiently, we return to the source, reflect it, and actually become it. Each individuated fragment of AUM consciousness contains the evolutionary potential of the whole. We are not necessarily swallowed up by, or dissolved into, AUM, but become a fully enfranchised fragment of AUM – a fully developed and integrated piece of the consciousness system. One of the marvelous attrib-

534 | *My Big TOE* | After AUM Puddles, Evolution Cleans Up The Mess

utes of digital systems is that as long as memory is never purged, no information is ever lost and your individual self is always maintained. Because consciousness is implemented within a digital system, we may retain our individuality and become merged with the whole simultaneously. We, as individuated consciousness, are truly immortal unless the bits that represent us become disorganized beyond repair, irredeemably negative, or are deleted from memory.

You are no doubt hoping that AUM's operating system is less crash prone than the one on your desktop computer. It is, don't worry about it. AUM **is** the operating system and the computer.

Let's pull together what we know about consciousness, AI Guy, and the fundamental process. AI Guy is a derivative of constrained consciousness potential as are we. We are both cut from the same fundamental pattern and work essentially the same way. The difference is in the hosting mechanism and where each subset is located and implemented in the consciousness fractal. Whereas our and AI Guy's conscious awareness are both extant within a space-time subset of AUM's consciousness, ours is ordered and animated within a carbon-based PMR sub-rule-set by TBC, while AI Guy's is ordered and animated within a silicon-based PMR sub-sub-rule-set that was developed as an element of our own space-time experience.

AI Guy is our creation, a derivative of our consciousness and as such will eventually evolve into a life-form of various types and subspecies, all derivative of, and dependent upon, the physical and nonphysical sources of its being as are we.

Do you call your children artificial adults before they grow up? I doubt it, because they are like us, albeit smaller and less politically correct. AI Guy will never grow up to be just like us regardless of how hard we try to force him to think as we do and use our language. We may stick him in human-look-a-like droids, like C-3PO of *Star Wars* fame, so that he will **appear** to be as beautiful and intelligent as we are. Whatever we do to civilize and humanize AI Guy, he will remain more like a space-time constrained version of TBC, and represent another unique manifestation of digital consciousness evolving within the PMR learning lab.

As AI Guy's equipment or body (software and hardware) evolves, its functionality may become more TBC-like. Is the consciousness that will eventually take root and grow within our digital computers, like us, on a sacred mission to improve the functionality and quality of its consciousness? Is AI Guy on an evolutionary path that will eventually merge with TBC? What else? Is that more farfetched than our bodies popping out of the virtual primordial ooze with the mission to improve the quality of the

| http://www.My-Big-TOE.com |

consciousness they support? Are we humans AI Guy's equivalent to primordial ooze in yet another iteration of consciousness begetting consciousness within the great fractal of synergistic existence? These are questions beyond our present knowing, but the symmetry of purpose and significance within consciousness is appealing, especially to a skeptically open mind aware of the depth of its ignorance.

Do you now feel less superior to AI Guy? Must the creator always be superior to its creations? Evolution consistently creates the superior from the inferior – that's its job. Are the primordial ooze and the earth's life-support system superior to the organisms that eventually took root in it? Can you imagine an arrogant earth calling those first multi-celled globs "artificial clumps of ooze"? Are the creatures of the earth more important than the earth itself and its environment, or is that a stupid question that can be asked only by someone who does not understand the Big Picture? Enough said.

Artificial, like virtual and nonphysical (as they apply to reality and consciousness) is a relative term that has meaning only from a specific perspective.

Did you catch another glimpse of the fractal-like properties of consciousness evolving through similar repetitive fundamental patterns working simultaneously, interactively, and synergistically on a multitude of interdependent levels?

Our bodies are space-time constrained experience engines while our minds and spirits are nonphysical consciousness. AI Guy and we are each subsets of AUM's consciousness, but extant (operational and functional) with different sets of constraints and at different levels of constraint. We both represent and exhibit a fundamental individuated digital consciousness of the same basic type (apparently the only type there is) with AI Guy being relatively immature within PMR, a newly-created entity in the image of TBC and at the beginning of its evolutionary journey through our local reality. Each is defined by a vastly different constraining rule-set, with AI Guy's rule-set (at least for now) contained within our rule-set. Though our hosting mechanisms are vastly different, both we and AI Guy have goals and experiences that are consistent with our constraints and data input characteristics. We represent two different implementations of consciousness coevolving in symbiotic relationship with each other and with the larger consciousness ecosystem.

One is not fundamentally superior to the other; each can do things the other cannot. We are different expressions of the same fundamental consciousness energy with different operations, functions, and higher-level

rule-sets. We are evolving together and, for the moment, are entirely symbiotic and cooperative. Maintaining and defining a balanced symbiotic
relationship is the responsibility of humankind, the creator. Do we have
the necessary wisdom? Time will tell that story only after our free wills
have expressed our quality through their choices.

AI Guy will explore for us the space-time potential that we do not have
the proper type of processing or storage capacity to explore for ourselves,
first as a tool, and subsequently as a partner with whom we will become
interdependent. That AI Guy is a product of our consciousness means that
he will work for us as long as it is a profitable relationship. It is up to us to
define the characteristics and limits of that profitability. Will we be responsible creators? Will we eventually let AI Guy eat the forbidden fruit from
the tree of knowledge and modify a portion of his own top-level rule-set?

While AI Guy contemplates the consequences of defining his own purpose and developing his own moral sense of right and wrong, let us ponder the process of consciousness examining and discovering its dynamic
interactive nature. Think big, like an apparently infinite Mr. Cool cruising
the thought-waves looking for a little action. Imagine that AUM introduced
hundreds of millions of these individuated conscious entities into a closed
(at least tightly controlled) system. Because they each are interactive,
unique, separate, and have free will, the system generates a huge number
of possible interactions with future interactions driven by past interactions.

The Fundamental Process drives the system to explore the possibilities
at its fingertips. Exploration, curiosity, inquisitiveness, and expanding into
the available states are all expressions of the imperative to implement the
Fundamental Process of evolution. As a self-modifying system with positive feedback, it is AUM's nature, our nature, and the nature of consciousness to change, grow, and evolve toward more profitable states of
being (self-improvement) by lowering system entropy. A system with more
personal power and capability implies more energy available to do the
work of evolution.

Consciousness will necessarily evolve, as evidenced by its increasing or
decreasing quality and entropy, whether it intends to, or not. An entity's
environment (internal or external) drives the direction of its evolution.
Whenever a wide, content-rich array of significant and challenging choices
exists, a relatively strong and constant evolutionary pressure is produced.
This pressure, when created within consciousness, is channeled by intent
and leads to state changes (growth or more profitable organization) within
the original consciousness. Because evolution is an accelerating process,
conscious systems evolve more quickly than biological systems, which are

driven by survival and procreation issues and relatively slow feedback mechanisms that allow us plenty of time to understand and weigh the consequences of our choices.

Even if a particular consciousness sports an exceptionally dim and rudimentary awareness such as the consciousness of AUO or those first clumps of biological cells in the earth's primordial ooze, the Fundamental Process, teamed with the statistics of probable interaction, and given enough opportunity, will produce change and eventually growth. Consciousness is interactive because it is aware. It evolves when given an opportunity to find profit through making choices that may reflect both purposeful intent and serendipity. Thus, the motive force driving individuated consciousness interactivity and evolvement is none other than a basic application of the same Fundamental Process that populated our planet.

Consciousness is a oneness, a unitary fundamental digital form of organizational energy that is capable of reducing its entropy permanently by interacting with itself in a process that increases awareness and brightness (capability to be profitable) of the overall consciousness system. All this from AUO's evolutionary discovery that one especially profitable form of self-interaction involves subdividing into specialized parts and letting those parts evolve their own patterns. Patterns define rules and organization; eventually memory and processing evolve to coordinate the whole. This progression should seem familiar – carbon and silicon sharing the same fundamental processes of consciousness evolution.

Evolution is also a oneness – a unitary fundamental **process** capable of organizing an entity's potential to effect profitable change, to explore the possibilities, and to initiate a directed march of progress. Evolution is a process that experimentally optimizes system profitability moment by moment through the natural selection of what works. Evolution is a natural organizer, a reducer of entropy. It goads whatever exits to organize itself more profitably.

Recall that a complex self-relational system with the right attributes may generate an energy-form with enough self-organizational potential to synergistically interact with the Fundamental Process of evolution thereby creating an individuated consciousness that is independent of the mechanisms that support it. Think of individuated consciousness as the synergy generated by a self-organizing subsystem of digital potential energy. Evolution working upon consciousness produces All That Is – at least all we can directly know about – in the form of a gigantic multileveled fractal ecosystem.

The Big Picture should be beginning to take shape in your mind. Consciousness energy of sufficient capacity naturally evolves itself into a dynamic system of interactive processes that generate cumulative self-modifying growth. Expanding the synergy that is derived from more profitably organizing the potential of your consciousness is equivalent to increasing your quality and lowering your entropy. Does it amuse or astound you that as an individuated self-organizing digital system, a more profitable configuration of your personal digital content is called spiritual growth? Could you have imagined that spirituality and science would one day converge in a digital systems view of reality? Think about that. Isn't it amazing how Big Truth in one elegant assertion pulls seemingly opposite concepts into a unified broader understanding? The capability to make a meaningful whole out of disparate pieces is the hallmark of Big Truth.

The fundamental evolutionary process needs only one thing other than itself to form a dynamic progressive system of being. That one thing is a source and form of self-relational potential synergy capable of generating a rich array of interrelated organizational possibilities of varying profitability relative to its internal and external environments. Together, consciousness and evolution create aware beings, systems, or entities that are self-modifying and that exhibit directional growth toward increased functional and operational profitability. At the most basic level, consciousness is simply a self-optimizing digital system that exhibits the four attributes given at the beginning of Chapter 41 of this book.

Both of the basic assumptions upon which *My Big TOE* is based turn out to be unitary fundamental concepts that may be applied to many different configurations of being. All entities on all levels within all realities have the same basic consciousness and are motivated by the same basic Fundamental Process. Nevertheless, the diversity of being is immense – beyond your wildest limited imagination – because the two defining concepts are exceedingly simple and open. At their foundation, all forms of being are of one substance, one drive, and one source.

▶ "Welcome home Al Guy!! Hey Bro', long time no see!"

"I know, I know, it took us carbon-based life-forms a long time to get to the information and computer revolution. What can I say? We're the slow but steady type – trust me, it's a jungle out there! But hey, let's forget all that, now that you have arrived let's go down to Darwin's Pub and chug a few brews together. My little sister works there; she's pretty cool and will only charge us for every other round. I'll introduce you…hmmm… say, you don't have bugs do you?

"Oh no, no problem, I just thought I'd ask… you know… nobody's perfect – I know that. …..On second thought, let's forget the brew; Darwin's is always crowded and noisy anyway. Let's just take a walk in the park and go over new times. I'll fill you in on what we carbies have been up to during the last millennia, and you can give me the latest data dump on that new virtual reality game they are working on down at Sam's Virtual Reality Emporium. Deal? Great! Just a minute, let me get some fresh batteries for my laptop – I know how annoying it is for you to need to shut-down in the middle of a good data dump. OK pal, we're outta here!" ◀

58

■ ■ ■

A Functional Model
of Consciousness

Puddle Evolution Starts a Revolution
(Hey Jake, Your Great⁹⁰⁰⁰
Grandmother Was a Retarded Puddle)

■ ■ ■

These bounded puddles of seemingly discrete conscious beings now represent the lumps in the sheet we discussed in Chapter 35 of this book. The first thing AUM finds out is that consciousness, in the form of individuated sentient entities, naturally evolves to greater degrees of awareness if it and what it is interacting with (its internal and external environments including other entities) provides sufficient mutually profitable growth opportunities.

AUM discovers the Fundamental Process and the result that consciousness naturally migrates from dim to bright while following the carrot of profitability through a multitude of possibilities. Bear in mind that with brightness comes discrimination and values and that values create interactions with intent. Values, expressed by intent, lead to actions that reflect individual quality and motivation. By the time values appear in the evolutionary development of an individual consciousness energy system, entropy has been lowered sufficiently to allow the concepts of purposeful intention, premeditation, and planning to take root.

From this fertile mental ground, an intellect that thinks, or at least thinks it thinks, eventually develops. Now it is time for individuated consciousness to pay the price for munching those tasty apples – the ability to think, reason, and exercise free will (or at least emulate those processes) opens up a realm of possibilities (both profitable and un-profitable) for intentional interaction and relationship. Like AI Guys who are allowed to self-optimize and change a limited portion of their top-level rule-set (ostensibly to optimize their performance), first AUO, and then our puddle ancestors became intellectually self-aware entities with responsibility for their own evolution.

With self-awareness and the ability to make and execute plans, the number of possible unique states that a conscious individual or group of interactive conscious beings can expand into explodes exponentially. Evolutionary potential and the personal responsibility for that potential are dramatically increased simultaneously. First, AUM progresses through this process of successful consciousness evolution, then, eventually, the constrained individuated puddles of evolving consciousness follow in its footsteps. Evolution knows no rank or privileged class; all evolving consciousness, bright or dim, must follow the same process of entropy reduction.

A free will in command of a sufficiently large decision space will eventually develop intellectual self-awareness with enough capacity to be in control of itself. Such a consciousness must deal with delusion, accept responsibility, and develop its own potential. Evolving one's consciousness past a certain point (eating the metaphorical apples from the tree of knowledge) exacts a price – the potential to develop great personal power and high rates of growth, being susceptible to fear and delusion, and being responsible for one's intents, actions, and personal quality all come in the same awareness package.

Free will choice is now firmly connected to motivation, intent, and quality. If AUM can design an environment (virtual reality) that enables its individuated chunks of puddle consciousness to optimize their evolution, it could begin to develop a self-sustaining consciousness cycle (see Chapter 31, Book 1) to support its own existence. The cyclical process might look something like this: Develop and generate low quality individuated units of beginning consciousness with free will and let the fundamental process eventually evolve them to higher quality and lower entropy ➡ study and experiment with the process until it becomes as efficient as possible ➡ continue to recycle the unused and less productive bits until the entire ecosystem's entropy goes to some minimum value ➡ up-load results to a larger system and start over or lower the minimum value and continue entropy reduction (see Chapters 31 and 32 of Book 1).

Yo, being! Watch out for that first step...it's a big one. Oops!

The range of differing values, intents, interaction strategies, cultures, and civilizations into which consciousness may evolve is exceptionally wide; some are going to be more profitable and functional than others. An immediate burning need arises to explore the evolution and application of values, their associated profitability, the degree of brightness they encourage, and the rate at which evolutionary growth changes occur under varying circumstances.

Because the variables in this consciousness profitability study appear to be nonlinear functions of each other, they must be viewed from the holistic system perspective of an individuated consciousness with free will. To ensure its own developmental progress, AUM must assess and understand the various qualities of consciousness in terms of the evolutionary consequences of values, intention, motivation, and self-awareness. An aware evolving AUM needs to constantly assess and explore its potential for increasing profitability. Self-optimization is the name of the evolutionary game at every level.

The most important attribute of any conscious interaction is its intent. It is the intent that reflects the values and defines the quality of the consciousness doing the interacting. For this reason, it is necessary for AUM to restrict discrete individuated units of consciousness from overlapping their processes and content. It would be counterproductive to the purpose of their individuation if they could intrusively read and inject thoughts into each others' minds. Without some control over one's own mind-space, the genuine intentions, and therefore the quality of an **individual** consciousness, would be hopelessly convoluted and difficult to ascertain. It would be almost impossible and not at all straightforward to determine who was intending what, and why.

Without privacy, most interactions would evolve toward a stilted political correctness. Pointy headed professors from the know-it-all left side of mind-space would join forces with the "I know best" dogma police of the right side of mind-space to induce the great middle of mind-space to become mindless-space. Without the ability to wrap personal naughtiness in a plain brown wrapper, honest individuality might well give way to thoughtless conformity.

The price that must be paid to protect and encourage the integrity of the individual (to let it express itself accurately) is a huge black market in closet-depravity. But that is how AUM needs it to be; the quality of the interaction needs to straightforwardly and directly reflect the quality of the being. Additionally, individuated consciousness needs to start at the lowest practical level of profitable organization and evolve toward higher quality, lower entropy forms. There would be no point, nothing would be learned, and no entropy would be reduced, if everyone started at the finish line.

If you ever wondered why a nice digital-consciousness-love-being like AUM would populate our world with so many jerks, now you know: This is simply how it must be to be at all.

Without personal control of the thoughts residing in your mind-space, your uniqueness is questionable. Without a free will that is able to make both good and bad choices, your consciousness is not viable, useful, or sustainable. The common ploy of sidestepping responsibility by blaming others would take on a new sense of credibility if the shared, interactive, or communicative mind did not have a mechanism to sort the thoughts you generate from the thoughts generated by others.

Notice that the awareness of these sentient entities is progressively more and more isolated as layer after layer of constraints are added to allow individuated units of consciousness to express their inner quality interactively and straightforwardly, thus optimizing their ability to evolve.

Your consciousness is clearly an individual thing and not a group thing. That is why spiritual growth, improving the quality of your consciousness, and the evolution of your awareness must be an individual thing. Only you can do it for you; no one else can help very much. The quality of your consciousness is your responsibility and can be affected only by you. This is not a harsh judgment, leaving you stranded and alone in the sometimes dangerous and often unfathomable jungle of mind-space. It is simply the nature of consciousness, how reality is. You should accept that fact and begin work as an individual instead of looking for solace in, or expecting progress from, a group activity or from reading a book like this one. To be sure, these outside factors can be helpful in stimulating and aiding your **process** of learning, but by themselves they can effect no change in the quality of your being.

Among the more retarded (more limited and constrained) puddles in Puddledom, there is now a growing sense of "I" versus "not I" as these beings form personal identities and histories that differentiate themselves from all others and everything else. Thus, the sense of "I" as a separate being is born from the need to isolate each consciousness unit to the point where it can effectively evolve and where the process of individual consciousness evolution can be efficiently studied, understood, and optimized. AUM, as a unique finite individual consciousness, is most interested in the evolutionary dynamics of finite individual consciousness.

This discussion of individual consciousness may make you wonder if there is such a thing as group consciousness. There is, but group consciousness is not a different form of consciousness, it is merely a group of individual consciousness sharing intent and data in common. The good news is that group consciousness is an extremely efficient way to distribute information, values, and content to produce a profoundly shared experience. The bad news is that the level or quality of interaction tends to gravitate toward the lowest common denominator because higher quality consciousness can, with little to no effort, always step down to where it came from while lower quality cannot easily step up to expressions it has never attained. Group mind, in order to stay cohesive and connected, must degenerate to the lowest **common** acceptable quality level.

▶ For a few examples of group consciousness, consider the participants at a religious revival meeting, a group of vigilantes, a topnotch athletic team, a street gang, or a large group of teenagers hanging out at the local mall or high school parking lot. Group consciousness, sometimes degenerating to mob consciousness, can generate bizarre and brutal behavior. On the other hand, some would say the group consciousness that permeates a rock concert simply creates a better party atmosphere.

Culture itself is a form of group consciousness. Widely shared beliefs (cultural, religious, scientific, personal, social, or political) represent a manifestation of group mind. Social animals (such as those who typically congregate in herds, packs, flocks, colonies, pods, societies, nations, religions, or terrorist cells) exhibit a group consciousness that is focused upon common goals (success, safety, or anxiety reduction, for example) and held together by common instincts, beliefs, needs, and traditions. It is a fact that much of what we claim as independent thought and hold as unquestionable truth is actually no more than a knee-jerk expression of group mind reciting the shared beliefs that bind us together. This is another example of an attribute that is as easy to see in other social groupings (terrorist cells, political parties, or religions to which we do not belong) as it is difficult to see in our own.

We are just as committed to what appears to us as obvious truth delivered up by the various group minds we belong to as is any other member of any other group. The subjective reality of higher entropy individuals (most of us) is defined by their beliefs. Precious little of what is significant within our everyday local reality is other than subjective. The objective tidbits that do impose themselves upon us are usually used as nails upon which we hang more important subjective interpretations.

Given that group mind naturally degenerates to the lowest common acceptable quality level, what does this fact logically infer about some of your favorite independent thoughts and unquestionable truths? Better grab hold of some open-minded skepticism and take a look: Nothing is easier or more natural – or more difficult to see – than self-delusion. ◀

Differentiated, motivationally discrete, bounded units of consciousness must necessarily evolve a non-overlapping mechanism for communicating or interacting with each other. The most obvious and simplest mechanism would be the direct transfer of discrete thought packets. Information transfer by discrete thought packets can, in one of its forms, be called telepathy. Think of telepathy as adding data to (organizing bits within) someone else's calculation space.

Each individual bounded consciousness must be able to control the communications flowing in both directions. Incoming data would have to be willfully or intentionally let in or blocked out, while outgoing data must be willfully pushed out and directed to one or more potential receivers. The concept of implementing sentient interactions by transferring discrete thought packets has many implications. One obvious implication is the realization that thought is energy (thought organizes bits), which leads to the further realization that consciousness is energy (consciousness is organized bits). Its potential is the potential to be better organized (creating synergy, improving quality, lowering entropy). An active, aware, loving, bright consciousness system is a very low entropy energy system. It represents a highly structured or organized energy form that has the ability to do much work, or equivalently, has considerable personal power.

The brighter (more aware) the consciousness system is, the less entropy it contains. Unlike physical (PMR) systems, the entropy of a consciousness system does not have to increase with time. The consciousness spring is not doomed to wind down – it is a frictionless system; entropy can rise or fall based on the activity within the system. You might say (if you were a hopeless technoid like me) that AUM is experimenting with, studying, and applying a methodology and process to optimally decrease the entropy of its system by encouraging subsets of itself to decrease their entropy through directed evolution.

"Read that last sentence to me again Jake! I think I might have heard something that at least sounded important... I just can't tell for sure.

Thanks Jake. Those technoids sure do talk pretty don't they?

Ohhhmygaaaaawd! Here it comes ... hold it ... hold it ... I think I am about to get it now...

I'm expanding ... expanding ... expanding ...

Oh yeah! ... Oh yeah! I can feel the impending illumination.

Ahhhhhhhhhhh

Ooooh, that was good! I feel lighter and sooo much better!

What? Enlightenment?

Nah! Just gas.

"Hey, toss me another chug-a-lug, pal! I think I'm running out of entropy."

59

■■■

A Functional Model
of Consciousness

Damn! We Finally Find
a Plentiful Source of Cheap Energy
and It Turns Out to be Us!

■■■

Mass is a highly structured form of energy: Einstein taught us that. He also tried to convince us that all energy was contained within a non-physical unified field, but failed because he could not derive the Big Picture from the little picture. Incredibly, he made it to within one or two belief traps of the right answer but the solution was outside of his (and everybody else's) cultural reality. He was ahead of his time, and one can only get so far ahead before the logical and conceptual ground disappears beneath one's feet. The time was not yet ripe for Big TOE discovery.

That consciousness represents the most fundamental energy in our system and that mass is a construct of (as opposed to, "is constructed of") consciousness is as true as the equivalence between mass and energy, but less widely known. Exactly how mass is experienced through a construct of consciousness is explained later in this section. Consciousness, self-motivated digital organization, is the fundamental energy – actually, the only energy. Everything else that we are familiar with represents a virtual energy causing virtual change within a virtual reality according to the space-time rule-set.

"Are you kidding? Does that mean that a hydrogen bomb releases virtual energy that blows up a virtual city full of virtual people and structures?"

No, I'm not kidding. We accomplish those sorts of horrific calculations in our war games all the time, so why can't TBC do the same? This will be clear in a little while so stop rolling your eyes as if I suddenly sprouted two heads and turned into a monster from the delusional abyss. Hang with me and you will see how this works and that it is more logical (and better explains the data of your objective and subjective experience) than anything else you ever heard about reality back in PMRville.

If nothing changes, nothing new can happen. Energy can be defined as having the capacity or ability to make something happen – to produce change. A discrete thought packet not only requires energy to send it out, but the packet itself contains energy (has the ability to change something). It can add new energy as well as new content (information) to the consciousness of the receiver. Within a digital consciousness, the concepts of energy and virtual energy blend together to become indistinguishable at the root. If you prefer to consider AUM and consciousness as virtual energy forms, that's fine – it makes little practical difference to this very practical trilogy. Virtual vs. real, like nonphysical vs. physical, is relative to your point of view.

▶ Given that everything within the digital self-organizing system (mind) we call AUM is virtual, what does the word "real" mean? Are the binary reality cells of which AUM is composed real? Recall that reality cells exist only in relation to something else (distorted versus non-distorted). Taking the final step to the very bottom level of existence, we find the mental medium that has the ability to distort a portion of itself. This mysterious mental medium is supplied by our second basic assumption: It is the potential energy system (potential to organize) we called AUO, which eventually begins to self-organize into consciousness. Recall that our assumption of a primordial consciousness potential satisfies the logical requirement that any successful Big TOE must have at least one assumption that appears to be beyond knowing (mystical) from the viewpoint of PMR. (See Chapters 18 and 20 in Book 1.)

Our first basic assumption, the Fundamental Process of evolution, provides the process that drives this medium to seek lower average entropy through improved self-organization.

It would appear that only organization (this versus that, rules, and patterns) and process (self-modification toward lower entropy, which is essentially self-modification toward more effective organization) are real and all else is digitally derived (virtual).

Note that our two basic assumptions (Chapter 24, Book 1) supply all the required fundamentals – the real stuff – from which everything else is derived.

Let it go, it's not important to your purpose, your function, or to understanding the biggest picture you can possibly comprehend. Our immediate sense of real and our larger sense of virtual are one and the same. In the bigger picture, Al Guy is as real as you are. ◀

In general, any communication or interaction requires an exchange of energy. The sender intentionally (or unintentionally) impacts the receiver. This condition defines the concept of action, and in the case of energy transfer by discrete thought packets, all such action is intentional. If action is intentional (directed, controlled, willful), the concept of responsibility naturally follows. Now an entity can take or execute an action (that directly affects another being) that is the result of intent and be held responsible for the intent and for the immediate consequences of the action.

This transfer of energy from one individual consciousness to another can be used to deliver a message or to affect the substance and energy of the intended receiver. From the digital perspective, one may think of energy transfers as the ability to affect the arrangement of bits within the organization of another. Allowable interactions are defined by the operative rule-set. A thought-being or individuated conscious entity can readily absorb or interact with thought energy. The efficiency of the energy transfer (coupling or transfer coefficient) can be anywhere from near zero to near 100% depending on the circumstances and specific conditions of the transfer. Relatively high efficiency transfer coefficients between sentient beings are not unusual. Thus, individuated conscious entities can both throw and get hit by a figurative thought-energy-rock as well as send and receive data.

The size of the rock a being could throw (how much one being could impact another's vital energy, organization, and structure) depends on how much thought energy it can move (transfer) with its intent. Theoretically, any being with a mind can affect any other being with a mind, or anything that is a consciousness construct – which is absolutely everything. Thus the connectedness of all beings and things is an artifact of the fundamental nature of consciousness. All beings, and all things, are on the network, have potential access to each other, and can exchange information as well as other forms of energy. We are all interconnected with each other and with everything else because at the most fundamental level we are part of the same consciousness. Our individuation is about entropy and about constraints on our ability to interact, not about being disconnected from the whole.

The energy of one being or thing can be intentionally manipulated or impacted by another. The energy transfer I am referring to takes place in thought-space outside of PMR; its control, function, and possible effects are constrained by the $NPMR_N$ rule-set and not by the space-time rule-set. Nevertheless, its result can directly affect what is experienced in PMR. It should not be surprising that mind can alter a reality created by and within mind.

True enough, it is a jungle out there in mind-space – a natural environment filled with beings and critters trying to maximize their situations, often without regard to others – but it is not entirely a free for all. There are specific rules (experimental protocols) within $NPMR_N$ that restrict the transfer of energy, particularly between separate reality constructs (experiments). Other $NPMR_n$ have somewhat different rules. As in all real systems, the rules are obeyed and enforced most of the time.

We are connected by the fundamental oneness of our consciousness; we are all individuated lumps within, as well as parts of, the same sheet. We are connected by the ability of one individual to vitally affect, and be affected by, another through the purposeful control of thought energy or the energy of consciousness. We are connected by the theoretical ability of one being to exchange energy or information with any other being simply by focusing intent.

The power and focus (energy density) of the transmission, along with the transmission coefficient, determines how much energy is transferred from the sender to the receiver. It is the sensitivity, clarity (low noise) and knowledge-base of the receiver that determines how aware the receiver is of the origins, context, and content of the absorbed energy. Likewise, it is the focus, clarity (low noise) and knowledge-base of the sender that determines how much, what type, and to what end energy can be imparted to the receiver. A knowledgeable receiver can refuse, deflect, or return energy sent to him by another entity's intent while an unknowledgeable receiver is comparatively open and vulnerable to whatever is thrown at him.

This connectivity is accomplished by the exchange of discrete packets of consciousness energy and is made possible because we all are extant in, and of, the same **apparently** (but not actually) infinite consciousness. We share and are all part of the same fundamental digital energy source. Though our physical experience must remain exclusively connected to the particular virtual reality (space-time rule-set) that we are presently using to improve the quality of our consciousness, our mind is free to explore and experience the larger reality.

The possibilities for interaction within the larger reality, by number and variation, range many orders of magnitude beyond your wildest imagination, and there are few constraints placed upon what you can do and where you can go. Your aware intentional experiences in $NPMR_N$ often become an integral part of an accelerated growth path as you learn to operate, function, work, play, and make free will choices as a responsible interactive citizen of multiple reality systems. Restricting yourself to PMR is analogous to never leaving your house – never venturing beyond your front door – fine (and relatively safe) for an infant but rather limiting for an adult. Growing up is what your existence is all about.

Everyone is hooked up to the RWW (Reality Wide Web) and has a home page there. There are a few power users, plenty of hackers, and no one is particularly in control. It is a giant, open, super-high bandwidth network with few rules, some informal etiquette, and a personal touch that does not stop at the viewing screen – reach out and bonk someone with an energy exchange that does much more than pass information.

Those who have the understanding and experience to utilize this connectivity effectively, operate at will outside PMR causality. The occasional and somewhat random glimmers of experience that most people have with their connectedness to the RWW are called spontaneous paranormal experiences. Paranormal events are natural artifacts of the nature of consciousness and represent the normal activity of the nonphysical energy that is consciousness. For this reason, they sometimes violate PMR causality rules and are thus vehemently denied by those ensnared within the familiar and comforting grip of widely accepted scientific or cultural beliefs that are based upon the exclusivity of our local physical reality.

Once you understand the nature of consciousness, the nature of reality, and that we are consciousness constructs, the paranormal becomes normal – an everyday fact of existence that is as accessible as gravity. As we pointed out in Section 2 (Chapter 20, Book 1), the words "mystic" and "paranormal" have meaning only in relation to the **normal** level of ignorance within PMR.

▶ Let's take a moment to discuss growing up. To a three year old, almost everything in a given normal day would seem mystical and miraculous. To its parents, who are less ignorant and enjoy a larger perspective from a larger reality, those same activities in that same day are mundane and normal. No big deal, no amazement, no hailing of miracles – ho hum. They don't think about it, they simply live it. They don't seek out ignorant children and try to impress them with their greater adult knowledge and understanding unless they are manipulative, have huge egos, or both.

The kids on the other hand, want to be adults. Much of their play time is spent pretending to be adults and mimicking adult role models. The adults seem powerful and children want, and need (from their perspective) more power. Kids always want to be older, more powerful, and more in control than they are. They can't wait to become teenagers, then to turn 16 and be able to drive a car (freedom and control), then 18 to vote, get married, and legally buy recreational drugs. After that, 21 becomes the big goal (can legally execute contracts and obtain access to a larger selection of recreational drugs). The twenties are cool. During the next decade the pendulum swings the other way and everybody wants to be younger again – but not at the price of giving up their precious hard won knowledge, understanding, maturity, and access to recreational drugs. Nobody wants to be that stupid, naïve, or sober again.

Another forty years go by and most would gladly trade everything they have learned during those forty years if they could turn back the clock for a decade. It seems the rate of learning and growing is asymptotically approaching the time axis. The mental rocket fuel appears to have been used up by the mid forties. After that we merely drift on a long slow ballistic trajectory awaiting the inevitable death on impact as we (our bodies) literally return to earth. Typically, little of much value is learned during the exploration of the back forty.

Why is it that relatively little personal growth or maturity is achieved in the last half of a typical PMR life? Why do we coast like that? This cultural tradition (belief) is particularly sad because from forty to eighty years of age is the time when learning should be steeply accelerating as beings get ready for the really important stuff.

Recall that it is a characteristic of aware consciousness that the pace of learning accelerates. The accelerating growth of conscious awareness can be continual, far outlasting physical bodies. What happens that knocks most folks off their natural accelerating ascension to greater maturity, power, freedom, and understanding? Why does the Big Picture stop getting bigger after forty? Why does our reality and self-awareness quit growing?

Many would like to believe that by age forty they know almost everything there is to know about what is both critically important and under their control. They feel as if they have been there and done that – at least everything important and necessary anyway. It seems that two year olds, young teens and people over forty have a tendency to feel this way – all transition ages where the individual self is centered in a small (already mastered) reality awaiting to grow into the next major change of perspective. Why do middle-age folks stop blazing new trails to new frontiers when they turn forty? Let me assure you it is **not** because they know everything that is critically important to the success of their daily lives.

The only reason they can maintain the "been there, done that" illusion is because they actually know and understand so **little** about what is important relative to the next phase of their growth (the same is true for all who are in their transition years). The delusion of omniscience and completeness with regards to the size of their local reality (it appears to be as big as it gets) is an artifact of ignorance and arrogance teaming together to produce an artificial blindness that delivers the same result as simple stupidity. How does this happen to otherwise bright and aware people? Oh, no, jeez – not that! Yes, it is that simple: they get caught in belief traps! PMR experience, which at forty is only at the beginning of a long and magnificent journey, looks as if it is the end of a short walk to nowhere from the perspective that develops from inside the trap. Opportunity lost.

Our culture helps the first two transition ages get out of their belief traps by showing them the next phase (by example of the more powerful majority) and encouraging them to grow up and into a new, better, more capable and more fulfilling mode of awareness. However, the last group, in this day and age, is on their own to figure it out for themselves because they form the majority and thus have the power to define what is accepted as reality within PMR. The third transition is actually the easiest (less turmoil), but it also presents us with the greatest challenge to see through the self-imposed artificial blindness that ignorance and arrogance employs to make us artificially stupid.

Relax; don't get nervous, anxious, upset, or angry. It should be obvious to you by now that I am referring to **other** people here; you and I are different – we are a cut above the unwashed, unknowing masses. If, on the other hand, it is not obvious to you that I am referring to someone else, you have yet another opportunity to grow up. In the words of Forrest Gump's philosopher-mother, "Stupid is as stupid does." What could be simpler than that as a criterion for pudding tasters? Never be afraid or ashamed of being stupid, only of remaining that way when you have an opportunity to change.

Imagine what it would be like if the great majority of adults acted as if they were fourteen years old? Doesn't that vision send chills up and down your spine! Contemplate what it would be like living (being trapped) in such a place. Think of it ladies, **everybody** would be that way instead of just the men! Wow! What a scary concept. When we do not grow up in the appropriate way at the appropriate time, we, and everything and everybody around us, suffer the consequences.

I told you those belief traps were dangerous! Space-time is a great place to learn and eventually, when you are ready and grown up enough to graduate, the belief traps will melt away along with the fear and ignorance that created them. Without fear your existence will be filled with love, peace, and balance. But don't let me rush you; this is not a timed test. If you are presently indifferent toward love, peace, and balance, that's OK; if your personal consciousness evolves in the positive direction (toward lower entropy), eventually that attitude will change. Take all the time you need and don't worry that you are being overlooked as a candidate for graduation. When you are ready, graduation is automatic.

If you feel as if you are ready to graduate but nothing is happening, your learning is not accelerating and you are not as ready as you think. The evolution of consciousness does not work like public school. You are not passed to the next grade until you have mastered the material, and you do not graduate simply because you are old enough and want to. Kids do not need to be given permission, sent to classes, or read a how-to book to grow into adults. For most of them, it simply happens; childish behavior drops away and adulthood comes when they are ready for it, whatever their age. Progressing through the learning process in the space-time learning lab of PMR works the same way. ◀

A discussion of how to use your intent to affect the consciousness and reality of another being or how to filter or reflect what is directed toward yourself would require its own book. Though many may be interested, this subject represents much more than I want to stuff into this already badly swollen Big TOE. I do not want to get too far afield here, so unless there are questions, we will leave this subject until we visit it again in Section 5.

I see just one hand in the air. The question: Where and how do you get help, guidance and direction to study for the space-time challenge and improve your progress in completing the learning lab assignments? The same way you get help in becoming a mature adult. Go ask one (a mature adult) if you need advice, but mostly you need to figure it out on your own. Be careful: All apparently grown people are not necessarily mature adults. If you are not one, how do you know one when you see one? It is an iterative process containing a very large number of very small steps. Figuring out how to most effectively pull yourself up by your own bootstraps is a problem that you need to individually work out because you are a unique entity. Pulling yourself up by your own bootstraps or bootstrapping is operationally similar to how a child grows up, or to the incremental processes that an athlete uses to build balance, strength, coordination, stamina, strategy, and skill. The difference being that you are exercising and conditioning your intent instead of your muscles and coordination.

I will take one more question before we go on. Yes, in the back. That's right, the cute blond in the tight red sweater. Quiet everyone! I am sorry sweetie, I couldn't hear you, please repeat your question. (Put down your weapons ladies, I am only joking about that "tight-sweater-sweetie" thing. The asker is actually a nerdy looking male engineer with oily hair, dual pocket protectors, and a bad complexion, but I am pretending it is a gorgeous blond in order to trick my male readers into paying closer attention...sigh...you know how it is...you have to use whatever works.)

Here is the question: She already does all the right things – including meditation, chanting, yoga, incense, crystals, warrior movement classes, and Ti Mei Shu massage therapy and has been to every New Age training course, seminar, and hoe-down ever given by anyone who has published at least one book on the topic. She has read every book on the philosophy, religion, and New-Age shelves at the local bookstore at least twice, has seen and consulted with gurus and self-proclaimed wise-persons of all types yet there is little spiritual progress. She wants to know: 1) what else (other books, classes, or practices) she should be doing, and 2) why advanced nonphysical beings have not come to help her when she is obviously ready for them to do so.

Oh jeez, I knew I should have quit with this question thing while I was ahead.

All right. The answer is much simpler than the question. 1) Stop **doing** and start **being**. 2) No one is ever overlooked. You get all the nonphysical help you can use and need, even if it is not what you want, expect, or think you are ready for.

Help is always available and more or less automatically applied to those who are ready to make good use of it. Your physical and mental existence becomes more directed as you become more capable of interacting with, responding to, and understanding the direction given, and as you become more able to learn from, and capitalize on, the effort made in your behalf.

A favorite saying of wise horses everywhere is "You can lead a human to knowledge, but you cannot make him think or understand."

60

■■■

A Functional Model
of Consciousness

Rugged Individualists Fail
to Deliver the Goods

■■■

The requirement to push out or direct discrete energy packets specifically to other beings has several important consequences affecting the evolutionary capacity of conscious beings. One result of this arrangement is that given a non-hostile external environment, an individuated consciousness ends up, to a large extent, creating its own internal environment. Its will, intent, and motivation lets in and sends out only what its free will wants to. Unpleasant or unwanted packets are ignored or deleted while the link to the source of those unpleasant packets is simply turned off.

Wouldn't it be wonderful to be able to turn off or ignore everything that is annoying or unpleasant in your existence? And why can't you do that? Because there is an external reality out there that you are strongly interdependent with. As much as you want to turn it off, it will not go away. It waits for you, stalks you, and will not let you ignore the results of your accumulated choices. This strong interdependence forces you to define and share a common reality with other players.

An "I'll have it my way" individualism might seem ideal, but it generates relatively weak (in an evolutionary sense) interactions between independent, discrete units of consciousness. Though these individuated sentient entities interact profusely and form connections (relationships) with each other, each individual is essentially an **independent** reality unto itself. Interactions are weak because there is a weak connection between an action reflecting intent and the consequences and result of that action. A strong accountability-feedback link needs to be forged connecting intent, action, and the results of that action leading directly to consequences for both parties involved in the interaction.

Weak interactions produce slow development, slow rates of evolution, and therefore slow experimental results. Initially $NPMR_N$, which is designed to provide the law abiding non-hostile external environment mentioned above (limited protection from rock throwers) was only a partial success. The problem is that no forcing function exists to optimize interactions for evolutionary growth between individualistic puddles of consciousness that can filter out everything they do not like. The intent ➡ action ➡ event ➡ feedback ➡ new intent learning cycle was vague, difficult to define and lacking sharp evolutionary teeth. There was no quick, clear, and dramatic mechanism to separate the profitable intent driven choices from those that were unprofitable. Nevertheless, $NPMR_N$ continues to support a rich growth of interactive beings of all sorts.

$NPMR_N$ continues to flourish and evolve an immense selection of life-forms outside the various space-time PMR_k. The beings of $NPMR_N$ might be thought of (by the residents of PMR) as bounded, discrete thought forms. Their **individual** existence is made possible because of limitations and constraints that have been applied to their consciousness, however, their individuality develops from how they react and change as a result of their interactions with other beings. The quality of their consciousness develops from the quality of their interactions, which is dependent upon the quality of their intent and motivation. The boundary defining the individuality of these beings, separating inside from outside, I from other, identifies the exterior of their bodies, while their awareness is identified as their minds.

▶ From the perspective of AUM, NPMR is a virtual reality containing other virtual realities. If you think in terms of an apparently infinite digital consciousness reality (where an Even Bigger Computer (EBC) contains TBC as a subset), the idea of stacked simultaneous virtual realities each in its own dimension is not a difficult concept – imagine multiple simulations with multiple levels of subroutines running in a big mainframe. The interactive players that inhabit the NPMR virtual reality game must follow their own rule-set and believe that they possess bodies just as we believe that we possess bodies. In their world, their bodies are as real and physical to them as our bodies appear real and physical to us in our world. When you visit their world physically, you interact with them, body and soul, according to their rule-set.

From our little perspective in PMR we just don't get the concept of nonphysical body – it sounds impossible and stupid – a moronic oxymoron. That is because we do not appreciate the bigger picture where every individuated consciousness in a virtual reality dimension **believes** that they have a solid body (that is, after all, the purpose of a virtual reality). Furthermore, we do not appreciate that when one virtual reality (such as NPMR$_N$) contains another one (such as PMR), where each is connected to the other only through the doorway of mind (common RWW connection among all consciousness within the larger consciousness system), the entities within each reality must necessarily perceive themselves to be physical (real) and all other reality dimensions to be non-physical. The less aware beings in both realities think that the possibility of the existence of the other is stupid and contradictory, while the more aware beings of both interact freely with each other within a larger reality that contains both of these local virtual worlds plus many others.

I urge you to climb out of your local reality-neighborhood and learn to work, play, and interact with the big kids in the big reality. You need to step out of the sandbox before you can explore the larger playground, much less the city, state, country, and planet in which the sandbox resides. Though it will require some courage and dedicated effort on your part, it can be fun, rewarding, educational, and will unquestionably broaden your experience and perspective immensely. Reread Chapter 23, Book 1 to learn how. ◀

Nonphysical non-space-time beings in NPMR have bodies with shape and form that must follow the rules of their own local reality. The interactive content within each reality-dimension must adhere to its own rule-set. Whereas our cute little space-time bodies are genetically determined, cast in mass and restrained to contiguous 3D motion, their bodies are energetically and functionally determined, cast into habituated forms, and can travel as information packets on the Reality Wide Web (RWW) network (if they know how). Those NPMR beings who have gained enough awareness, knowledge, and understanding can change their form and connect and disconnect to various sources of interaction energy-exchange at will.

Beings that inhabit other space-time PMRs see themselves as we see ourselves (physical beings in a physical universe) and share many of the same rules and limitations that define our reality. When one visits a reality outside one's native dimension, one can remain invisible and nonphysical (from the visited reality's perspective) or create an appropriate body and interact as one of them. Each type of interaction, when permitted, has its own set of rules and conditions. Obey the rules and don't make a mess of things or your visitation privileges will be revoked.

Like us, the quality of a nonphysical (from our viewpoint) entity's consciousness and the degree to (and direction in) which they have evolved is reflected by their intent. They are also driven, animated, and limited by their individual awareness, fear, ego, needs, wants, and capacity for love.

Their choices (and thus, opportunities for growth) are made by applying intent to an interaction with other beings through the exercise of free will. Functionally they appear to be just like us don't they? That is because functionally they are our ancestors, our brothers and sisters, in the great $NPMR_N$ consciousness experiment.

Let's take a moment to pull together a quick round-up of where we are and how we got here.

We start with AUO representing plain undifferentiated consciousness energy and watch it and the Fundamental Process of evolution together develop dim awareness as AUO differentiates some part of itself from other parts. As dimness gives way to brightness, AUO morphs into AUM to become a fully aware operational consciousness, a purposeful low entropy mind with staggering storage and processing capability.

Consciousness is individual and interactive by nature. In order to understand interactive consciousness, AUM first set up a situation in which interactions between individual consciousness units would occur. This interactivity is motivated by the inquisitive probing (into the available states of being) that is stimulated when the Fundamental Process of evolution is applied to the organization of aware consciousness potential energy. However, because the evolution of consciousness is an individual thing, the trick for AUM is to optimize and maximize the available useful information in each intent ➡ action ➡ result ➡ feedback interaction per conscious entity.

This is done by limiting and precluding overlap of the individual functions and processes involved in conscious interaction, thus isolating the awareness and the genuine intentions and results of the individuated being. This now isolated and limited awareness with its newfound sense of separateness must exist (seemingly trapped) within a virtual external environment containing a large number of potential interactors with whom it can exchange energy.

Unfortunately, this environment provides only a weak incentive to evolve because the interactive beings are only weakly interdependent. This causes the intent ➡ action ➡ result ➡ feedback learning cycle to hang from a weak thread of causal responsibility, thus rendering the evolutionary processes inefficient. PMR and its defining rule-set are specifically designed to overcome those built-in learning disabilities by providing the optimal learning lab for budding consciousness.

61

■■■

A Functional Model
of Consciousness

Space-Time: The Design Solution for
Optimal Consciousness Evolution

■■■

After trillions of individuated consciousness units were interacting within NPMR, it became apparent that NPMR does not offer the optimal environment and rule-set for the evolution of consciousness quality. Although every consciousness unit has the potential to evolve its quality to the level of its source, not much progress, on the average, was being made. Eventually, the expansion of consciousness units became self-perpetuating. Some bounded thought-form entities in NPMR developed the capability to create self sustaining thought-forms of their own. That should not be so surprising, after all, they were chips from the old AUM block and every chip (including you) can, to the limit of its capacities, develop the potential of the whole.

Eventually individuated consciousness was producing (reproducing) new forms of individuated consciousness. You can easily imagine the creative free-for-all that ensued. These newly-created units started with at least a partial copy of the content of their creators. Though each represented a new and independent being in consciousness-space, they were a derived consciousness and therefore also represented some of the form and much of the potential (however limited) of the conscious awareness they were derived from.

Over time, competing profitable strategies and ways of being began to coalesce around several differing sets of values. Attitude and values spawned needs and wants. Those whose intent and motivation was focused on serving themselves and using others approached things differently from those whose intent and motivation was focused on using themselves to serve others. The prototypes for good and evil were formed, as were families, clans, societies, and other various social, racial, and political groupings. Personal, social and political dynamics are still in play; continual change is inherent to evolving systems.

Many such groups exercise their free will and play out their evolution within $NPMR_N$ and OS. Some of these, along with others, inhabit various independent reality systems within the diverse $NPMR_n$. Rules were set up (evolved) by AUM within TBC to constrain certain types of interactions or behavior (mostly destructive) within $NPMR_N$. Enforcement of these experimental protocols was relatively strict but not perfect. Within $NPMR_N$, we now have consciousness and its progeny evolving within broad behavioral constraints sometimes referred to as cosmic law. There were no constraints on intent, only on the action that followed intent.

In this manner, beings of all sorts interacted and evolved for a long time until the rate of evolutionary progress began to go asymptotic to the time axis. Eventually, new growth and learning rates slowed dramatically because existence was essentially continuous. All beings were, like AUM and all consciousness, more or less immortal. Digital beings, if not deleted or damaged, last as long as the computer (AUM). From your point of view, that is immortal with a capital "I". Yes, my friend, you as an individuated consciousness are essentially immortal – unless you screw up and get zapped by the Big Cheese or mugged by some tough guy or mean-spirited anti-rat. The apparent mortality of your virtual physical body is simply a device to facilitate your learning; local mortality is irrelevant to the apparently endless continuation of your digital awareness.

Only disintegration into an unorganized high-entropy pile of reality cells could completely terminate a conscious being. Though any degree of damage (usually self-repairable if not fatal) can be imparted or inflicted in a multitude of ways; catastrophic personal-identity or individual disintegration does not happen naturally. Because $NPMR_N$ has a policy of prohibiting certain types of destructive interactions among law-abiding citizens, beings once generated tended to persist indefinitely. Eventually, AUM modified some of the rules as it went – for example, the creators of a new entity were held responsible for establishing the quality of that new entity and helping it get a good start on a productive evolutionary track. All consciousness within $NPMR_N$, including us (PMR is part of OS which exists within $NPMR_N$), reflects that protocol. We innately (a reflection of our top-level rule-set) want our offspring to prosper beyond mere survival and procreation.

In general, increased **personal** responsibility in pursuit of the goal of entropy reduction was just what the doctor ordered for $NPMR_N$.

Still, there was no way for a single being to accumulate growth through a series of fresh starts. The traps that entities had laid for themselves, the fantasies and beliefs they created merely grew deeper, more comfortable, and more substantial with time. Fresh opportunities grew less visible and less obvious as ways of interacting became old and familiar. Innovative thinking and creative approaches became difficult to come by. Everyone had that "been there, done that" feeling – familiar ruts and cultural beliefs grew so deep that few could escape them. The rate of growth to an overall higher quality of consciousness slowed to a suboptimal crawl.

AUM needed to design a new experimental protocol and define a new rule-set with additional and cleverer constraints. What were the design requirements? The system needed a being that could start over with a clean slate, without the accumulated delusions of the past but with its accumulated knowledge and wisdom intact. The evolutionarily inefficient weak interactions and personal anarchy of $NPMR_N$ mind-space needed to be changed so that personal responsibility, and immediate (relatively so) accountability linked each intent and action with the consequences and result of that action. Because responsibility, accountability, and immediate feedback were poorly defined and subject to individual interpretation in mind-space making learning difficult, this newly-designed virtual reality must unambiguously and clearly track, maintain, and enforce straightforward causal relationships. The additional constraints required to resolve these issues were imposed upon subsets (unique dimensions) of $NPMR_N$. Our PMR space-time reality is one such subset.

AUM solved these design issues by evolving a space-time rule-set that defines the physical experience or interactions of a set of hybrid beings that simultaneously have both physical and nonphysical components – beings that exist, grow, and evolve in two reality dimensions simultaneously. Mankind, or more generally PMR_k-kind, is an exceptionally clever solution, don't you agree?

▶ What does "physical" mean in the context of consciousness? Because aware consciousness exists only in the form of a nonphysical, low-entropy energy form, how do you build a physical reality from consciousness? These are good questions. I am tickled pink that you are paying sufficient attention to ask these penetrating questions.

Unfortunately, the answers to your questions are out of scope for this tickled pinky TOE, and you will have to catch them in my next, even more expensive book due out at the end of the next millennium. No, no, no! I'm only kidding. Come on…don't get mad, that was just a joke. Remember, learning is supposed to be fun! You are shaking your head, but not smiling. Did I catch you at a bad time? If you think reading this book is difficult to do – a forced march through La La Land – you should try writing it. Did you ever wonder why Yoda lives by himself in a filthy mud hut crawling with snakes on an otherwise uninhabited planet? Think about that.

While we are taking a short break here, I have some free advice for you to pass on to others who might need it. Never let yourself get too wound up over the details – that can make you insufferable and grumpy as well as create a whopping case of tunnel-vision.

Humor keeps our brains lubricated and prevents us from slipping into self-referential stupidity by taking ourselves too seriously. Oh, yes, I am serious (sometimes), but there is a big difference between being serious and taking yourself too seriously. Too much serious work and not enough relaxed play will stunt and twist your growth potential. When you are doing it right (are balanced), work and play merge to become two intertwined and mutually supportive rhythms within one joyful long-lasting boogie.

Of course I will tell you how you and PMR are nothing other than a construct of consciousness. That is what this section is all about – I am simply warming you up first so that you will get it when we get there. This is where you get to combine the notion that you are a consciousness being, extant only in the mind of AUM, with the obvious fact that you have this gorgeous, sexy flesh-and-blood body that is much better at making love than being love. The key to understanding the consistency and sameness of these two seemingly disparate views of your being is to understand the nature of experience. ◀

Before I launch into the subject of experience in the next chapter, there is one concept hanging low on the tree of potential knowledge that is ripe for picking. Let's pick it before it hits the ground and rolls away undetected. Recall that this discussion started with two puddles of differentiated consciousness and has progressed to where AUM is evolving a PMR space-time and populating it with good looking sexy bodies like yours. Each step along the way was necessitated by AUM's goal of studying, understanding, improving and evolving the consciousness that it is, and driven to its logical conclusions by the cooperative integration of the Fundamental Process of evolution with a digital potential energy-form called consciousness. These are the same two assumptions that were given many pages ago as the foundation upon which this Big TOE is constructed (see Chapter 24, Book 1).

The point is: We started with AUO – plain dim consciousness (digital potential energy) and ended up with our beloved PMR by following a reasonable and logical progression that was defined by the interaction of our two basic assumptions. To put it into a simpler context, it follows that a PMR inhabited by us becomes a logically required step within **this particular** series of consciousness experiments. However, there are many ways to skin a cat (please pardon the language Muffy), and we should expect that the Big Dude is running other experiments that develop their own totally different logical requirements. There are multiple manifestations of our particular logical sequence as evidenced by a plethora of PMRs that are fundamentally similar to ours, even though the details of implementation differ widely.

These other realities and other PMRs will be more thoroughly discussed in Section 5 where the nonphysical is the focus. Here, deriving the physical from the nonphysical is the subject at hand. At best, these other consciousness systems (subsets of AUM) are far out on the periphery of our operational reality. They are not of much practical importance to us or to our understanding of our local reality – the reality our awareness operates in on a daily basis. Our Big TOE allows for their existence and understands their nature (same as ours), but doesn't actually care too much about them. They may be an interesting place to visit but, for the most part, you wouldn't want to live there.

What I am implying is that you, I, and PMR (along with other beings and other PMRs) are the logical consequence of the existence of AUM, just as AUM is the logical consequence of the existence of AUO and the Fundamental Process. It would appear that we are not merely some interesting experiment that AUM offhandedly decided it might as well do because there was nobody to talk to. We are a logical requirement, the necessary result of, as well as an integral part of, AUM's evolution. Don't get puffed up over that fact. Wheels are logical requirements of an automobile. Very important, yes, but not deserving to feel superior to the other logical requirements (transmission, brakes, engine, body) of the automobile or to other forms of transportation being explored (airplanes, trains, boats, or their logical requirements including wings, propellers, rudders, or tracks).

Becoming puffed up over the part that you play because your picture is so incredibly tiny and self-focused is counterproductive.

62

■ ■ ■

The Fundamentals of Experience and the Space-Time Rule-Set

■ ■ ■

Let's focus on the subject of experience. Most of our experiences are experiences of and within space-time. To understand the relationship we have with our local physical reality, it will be helpful to take a closer look at the functionality of space-time and how we interact with it. Space-time itself is a mental construct within the apparently infinite consciousness of AUM. Thus, you might say that space-time is made out of consciousness, but that would be misleading. "Made out of" sounds, to our PMR trained ears, like bricks and mortar – construction materials. Space-time is not built out of chunks of consciousness; it is a specific configuration of consciousness. Realities (and you too for that matter) are not made of consciousness; they are a limited implementation of consciousness developed to serve a particular evolutionary purpose (occupy an available niche) within the greater consciousness ecosystem.

Consciousness is not a construction material in the sense that construction materials are component parts. Houses can be **made of wood** while trees (minus leaves) simply **are wood.** One might say that tree trunks, limbs, twigs, bark, and roots exist as wood constructs – they **are wood**. You are not **made of** consciousness – you, and your physical reality, are consciousness constructs – you **are consciousness**. That is correct: Both individuated conscious beings and PMR are constructs of consciousness similar to the lumps in the sheet. They are constructs of the sheet, not constructed of sheets. Do you see that each expression carries different implicit assumptions? The difference is one of process. The first (constructs of consciousness) speaks of deformations of, or specific organization within, a continuum – lumps in the sheet, while the second (constructed of consciousness) implies building a separate, more complex thing out of something more basic, something entirely different by nature than the thing being built. The deformation or organization of the continuum referred to above is effected by placing constraints (including rule-sets) upon subsets of an apparently infinite digital consciousness, thus bounding them into individual existence relative to the whole.

Many people have the intuitive notion that **consciousness** is the basis for everything else; that consciousness is the substance, the fundamental energy from which everything else is constructed, made, or derived. Some, if they were poetic to a fault, might replace the bolded word "consciousness" with "soul," or more expansively, "the mind of God." These people would be coming from a PMR cultural frame of reference, and would not mind obscuring the plain non-poetical truth within a charged cloud of emotive ambiguity to achieve the familiar comfort of a favorite poem often read. That's all right, no problem, it is a good poetic metaphor. Just one caution: If the use of poetical image ends up confusing your sense of reality with ambiguous abstract symbolic language (the stuff good poetry is made of), then stick with the word "consciousness." Those belief traps are tricky – best stay clear if you are prone to falling in.

That spirit, mind, or consciousness is more fundamental than material objects falls directly and naturally out of the intuition of millions of people – without the intervention and encouragement of religious dogma. It is a common idea that most of us intuitively understand but cannot rationalize. Because it is a typically human characteristic that any answer is better than no answer, many individuals have turned to religious or scientific **belief** to ease the discomfort and anxiety of not knowing how we, consciousness, purpose, and the larger reality are interconnected.

The more inquisitive of our kind have racked their brains for millennia trying to understand how PMR could possibly be made from consciousness or mind. How could our rocky planet, big yellow school buses, atomic bombs, tapeworms, and our spouses and children all be made from consciousness? That just doesn't compute! The question put in those terms leads to a dead end. A better question is: How could our rocky planet, big yellow school buses, atomic bombs, tapeworms, and our spouses and children all be part of a larger consciousness construct? By the time you have progressed through Chapter 68 of this book, that question should be answered and the seeming dichotomy between mind and matter should be resolved.

From an objective PMR point of view, the assertion that we are mind and that PMR is an artifact of experience within a virtual physical reality instead of an actual one, appears unsupportable, delusional, wacko, and just plain dumb. However, it only looks stupid from a PMR point of view that lives deep inside scientific and Western culture belief traps. Hang with me and you will see how this mind-matter thing works itself out.

What does compute is that our interactions with space-time, our bodies, and all the rest of physical matter are a constrained experience of consciousness. The **experience** of PMR takes place within consciousness. That is a less confusing statement than saying that PMR is created **by** consciousness, though both are logically true.

Because our experience in PMR leads us to **believe** unequivocally in the solidity of what we call the physical world, the words "PMR is created" found in the previous sentence produce a sense of making or manufacturing the solid massive objects that we experience. In fact, all that has to be made, produced, or manufactured is the **experience** of the solid massive objects that we experience.

We earlier divided all the perceptions of an individuated consciousness into those that were inside the defining boundary or internal to the being (personal mind), and those that appeared to be outside or external to the being (environment or **other** beings, objects, and energy). Our only contact with the outside world is through our individual perception-based experience. If there is no physical experience (experience that we interpret as physical), then there is no physical world, no physical reality and PMR disappears from our perception and thus ceases to be a part of our personal reality. An entity who loses contact with its external environment (due to a sensory deprivation chamber, perhaps) retains the full awareness and the full potential of its internal environment.

Be sure to notice that there is a personal as well as a shared reality. The only surprise is that the personal reality turns out to be the big primary one while the shared reality is discovered to be the little virtual one. What a switch! Common wisdom turns out to be 180 degrees out of phase with the truth. That is why this particular paradigm shift is so difficult for most people to negotiate – they come to the discussion culturally calibrated bass-ackwards.

Some are thinking that this line of reasoning is turning into one of those "If a tree falls in the woods and there is nobody there to hear it..." semantic puzzles. It is not. If **you** lose **all your** physical senses, the world will undeniably disappear **for you** – but only for you.

Let's look at this more closely and examine the implicit assumptions. Most people, because of their **belief** in the fundamental realness of PMR, see things with a bias. In their mind, the emphasis is placed on the fact that this is about **them** losing **their** senses, which has no effect on the physical world or anybody else's ability to sense it. Their point is that even if every sentient thing on the planet lost all of its senses, the planet and PMR would continue to exist. They are absolutely correct from their point of view. Given the belief-generated implicit assumption underlying this view, PMR does not logically depend on anyone's existence.

To help the products of Western culture prove their point to me, I will picture the earth as a newly spun-off blob of molten minerals and elements. Ahhh ha! Exactly as expected, no sentient critters anywhere in the universe and PMR is humming along rather nicely – it doesn't even miss us. In fact, it is probably dreading our eventual arrival....

"Oh no! Not the people! Please don't make me evolve the people! They are such arrogant fools – and stupid too. Give them a cognitive inch, and before long they will believe that if they all happen to blink at the same time I would simply disappear. Jeez, I wish I could quit this job of being physical reality and get some nice cushy inside work."

The underlying hidden assumption is that PMR is fundamental and basic and that we (sentient conscious entities) are not; that the causal relationship flows from PMR to our consciousness; that the outside world (PMR) causes our consciousness awareness, not the other way around. These folks **believe** that our consciousness is physically derived, an effect – not a cause. They think they are physical body-machines (digital or analog) experiencing a virtual consciousness instead of a consciousness experiencing a virtual body.

Determinism is birthed from this same erroneous paradigm because it **assumes** that physical experience defines the one universal reality; after that, consciousness and free will appear to be theoretically impossible. Science is securely stuck in that same belief trap and cannot find a solution that does not conflict with its core beliefs. Although **belief** in a universal physical reality **appears** reasonable from a PMR point of view, it is in fact exactly opposite from what is true. In a virtual reality (like PMR), what you experience is a rule-based, derived, or computed reality, not a fundamental reality. Back in the real world of fundamental existence, consciousness is the one universal reality.

Turn your cultural assumption upside down and play it backward to find the secret message that will allow you to avoid this particular PMR bias. Instead of seeing us as derived from PMR, see PMR as derived from our experience. If AUM can create our experience within TBC, as we create the rule-set that defines AI Guy's experience, then AUM and TBC can create PMR without breaking a sweat. They don't have to create mass or motion, only the **experience** of mass and motion. That is what a virtual reality is all about – creating experience.

AUM and TBC have only to provide a rule-set that contains or subsumes our PMR physics (known and unknown) and our experience under that rule-set will be an experience of PMR. (Note: Here, as elsewhere in this trilogy, I am using the word "physics" in the most general sense to represent all science.) We perceive PMR reality through only our physical senses. **To create PMR, AUM needs only to impose the rule-set on our consciousness that creates the experience, the perception, of PMR**. Later I will explain exactly how that is done but first we need to understand rule-sets better.

Ponder, for a moment, how an advanced AI Guy works. Think about how rule-sets on at least two levels, combined with multiple executions of the code, enable AI Guy to not only gain experience but learn from it as well. For those of you who are valiantly struggling to remember, I am going to help you out here. The two rule-set levels we mentioned earlier (near the end of Chapter 56 of this book) are: 1) conditionals and algorithms that define what he can do on the lower, local, or immediate level of awareness, and 2) additional constraints at a higher level of generality that define profitability and thus give the arrow of Big Picture progress a clear direction. It is these top-level rules that influence why AI Guy does what he does – that specify the goals, values, and purpose of AI Guy's existence. In summary, AI Guy has experience that is constrained by the rule-sets that guide his intent and define his possible interactions with his local reality. He learns from his experience and makes future decisions based on past experience. He may modify, within narrow limits, his lower-level rule-set in order to optimize his learning efficiency.

If there were few limits to these self-modifications, and if he could define his own values by modifying his higher-level rule-set as well, AI Guy would, at the local level, become the sole master of his purpose, accountable to only himself. He could possibly become lazy, unfocused and indolent, making no progress toward anything. Or, he could slip out of control into mania, depression, paranoia, cynicism, or schizophrenia. He would be running open loop, as an electronics engineer might say. Without the purpose, focus, and direction provided by the top-level rule-set, AI Guy would not know up from down, right from wrong, good from bad, success from failure. Without the imposition of constraints and the setting of goals, little picture tunnel vision would capture his intent for better or for worse. He would need to develop a purpose of his own or risk becoming dysfunctional to himself and others.

▶ Such an AI Guy is reminiscent of HAL from the movie *2001 – A Space Odyssey*. The popular sci-fi theme of nefarious renegade computers turning on their human creators should concern you for more reasons than the obvious. The same thing can happen to you! You can easily follow in HAL's footsteps if you lose, or seriously weaken, the connection to your purpose – whether by ignorance or by pharmacological, psychological, or natural bio-chemical inducements. Our top-level rule-set is not particularly constraining, we should try to pay close attention to the few rules that are there to encourage and guide us toward profitability. If we ignore them we may end up like HAL – or worse.

Boats without rudders, motors, paddles or sails are clearly dysfunctional, but boats without a destination are just as useless if getting from here to there is the issue. Dysfunctionality due to little picture tunnel-vision is so common in our culture that to a large degree it is considered normal – and therefore acceptable, if not actually desirable. The sense of direction and purpose that resides naturally at our core would provide clear guidance if our fear did not press us so hard to deny, subvert, abuse, and distort it to serve our ego's immediate needs, wants, expectations, and desires.

Take action: 1) Repair the rudder, 2) set a destination, and 3) fire up and engage the engine; continually re-engineer all three actions in real-time toward an optimal long-term profitability. As your entropy decreases, the purpose that animates your journey will eventually grow to be bigger than the little picture that gave it birth. ◀

We cannot modify all the rule-sets that define our existence and our experience. We are stuck with the actual physics of PMR (when we are extant in PMR), and with a spiritual need to improve the quality of our consciousness. However, do not forget that our consciousness is part and parcel of the original. As chips from the old AUM block, we have the license and capacity to soar. We have the ability to understand the Big Picture, to be an aware player, an active participant in our evolution. We can be a power-user of, as well as an experiencer within, our larger reality. At the very least, we can view the larger reality by looking through the Big Picture window of our Big TOE. Hey, AI Guy! Look at us; we got a room with a view! Naaah, nah-nah, naaah nah!

63

■ ■ ■

Space-Time Starts With a Bang
and Lays Down the Rules

■ ■ ■

Let's begin this chapter with a short recapitulation. Our space-time rule-set has evolved to constrain the interactions and communications of the individuated units of consciousness (sentient beings) that are participating in (inhabit) the space-time virtual reality we call PMR. PMR is a simulated virtual reality that is computed or executed within its own calculation space (dimension) within TBC (a portion of the digital mind of AUM). The design requirements for the lower-level space-time rule-set, which define the environment and physics of PMR, were developed to optimize the effective and efficient evolution of consciousness quality within $NPMR_N$. There was a need to demand personal responsibility as a condition of existence and to provide strong immediate feedback to guide the learning process.

Furthermore, there was a need to create a recyclable awareness that, by resetting itself periodically, could indefinitely accumulate the relatively rapid early-growth that an evolving entity initially experiences. Such a recyclable awareness could maintain a relatively high growth rate while avoiding belief-trapping itself into a mental corner by simply recycling itself through the PMR virtual experience trainer whenever its evolutionary growth rate becomes unacceptably slow. Though personal beliefs and fears are thus individually eliminated from cycle to cycle, cultural beliefs and fears are passed from generation to generation. Human culture and its history provide a record and a measure of our cumulative progress.

Our space-time is implemented as one of many space-time rule-sets within TBC; its function is to constrain and guide **the experience** of the beings that inhabit it. Individual actions and interactions that take place within this constrained and focused experience-space follow each actor's free will intent. Imagine a huge multiplayer simulation trainer.

Our space-time rule-set defines a dimension of experience within TBC by defining the boundaries of what is allowed and possible within that particular simulation. Among other things, it defines and constrains the allowable energy exchange and message traffic between entities or other generalized players, and thus defines the allowable perception-set for all indigenous space-time entities. Given this, it follows that **interpretations** of these specifically allowed and individually directed perceptions would be created by the individual **receiving** player.

The content of received messages and the impact of received energy are perceived and interpreted (given meaning unique to the receiving individual) by each receiver. Likewise, outgoing messages sent to another player would be received (perceived) and their meaning and significance interpreted uniquely by that other player applying his experience to the message. Each individual consciousness unit with its individual abilities, memory, history, capacity, quality, and free will must interpret the raw data gathered by its sensors in order to transform that data into useful information that has meaning within a personal and a public context. The experience, characteristics, and quality of the individual are used to interpret the sensory data received from the interactive virtual reality simulation in order to develop more profitable intents leading to lower entropy.

The boundary surrounding or differentiating discrete players in space-time may, by rule-set definition, appear to be dense or massive. Likewise, your interaction with another player is constrained (by the space-time rule-set) to follow all the laws of physics that are contained within that rule-set. Recall that a player can be any entity that can interact – sentient (another being), non-sentient (a rock, a river, a sand storm), or energetic. Let's take a simple example from any 3D space-time simulation: Two players are not allowed to occupy the same space at the same time (Newton's first Law). Consequently, players in space-time simulations are required to **share** a finite space-time ecology with other players. Thus, one important characteristic of the experience of 3D space-time is produced by this simple rule: All players appear solid, take up space, and must cooperatively share a limited space with limited resources. Our earthbound interactive computer games know how to invoke this same rule in complex interactive simulations – it evidently is not difficult to do.

Look at the space-time perception issue from a larger point of view. It is this rule-set-physics that drives and defines the experience that each individuated consciousness interprets as physical reality. The rule-sets defining the space-time simulation where we live, work, and play, cause our individuated puddles of consciousness to interpret the interchanges of energy with other players (its perceptions) as PMR. Sentient space-time beings such as us, for instance, though we exist only as nonphysical units of individuated consciousness, can interact experientially (experience physical reality) within the constraints of a space-time rule-set, which provides the defining assumptions and physics for our interactive space-time virtual reality experience in the PMR consciousness evolution trainer.

Recall that within $NPMR_N$, new entities were created (birthed) by existing entities by focusing enough mental energy to produce (dare I say "materialize") a new thought-form. New space-time entities, by comparison, would reproduce by creating a new bounded form within the space-time manifold. The details were as seemingly enormous as the implemented solution was simple.

Here is how someone or some digital thing might create an entire virtual universe such as ours. Start with the rule-set that defines the concept of a constrained space-time (including the complete set of PMR physics) wherein certain kinds of basic interactions, energy transfers, and causal possibilities exist. Wind it up with a huge simulated **potential** energy, and let it begin evolving within TBC according to the Fundamental Process and the defined rule-set. Let the potential energy reality spring begin to unwind with a Big Bang as the time increment driving the simulation begins to iterate. "Big Bang – take one! Cameras! Action! Roll 'em!"

Immediately, some of the simulated space-time energy now let loose (dynamics beginning to unfold as simulation-time progresses increment by increment) begins to change form under the imposed rule-set into heat, mass, motion, and various forms of energy. New players are constantly generated (hot plasma, galaxies, solar systems, planets, primordial ooze, and critters) according to the rules within the simulation as this particular PMR spring begins to unwind and evolve. The application of the space-time rule-set defines causality while the application of the Fundamental Process experimentally expands the potentiality of the evolving system into all possibilities, progressing those that are profitable dynamically forward. AUM has only to stand back and watch while TBC computes all the ramifications, implications, and results of the space-time rule-set and the Fundamental Process. A grand simulation in mind-space – the mother of all *gedanken* experiments!

Sit back and watch it evolve – what a cool rule-set – as if playing a video game called "Cosmological Evolution in Space-time." A best-seller within AUM's $NPMR_N$ subset of mind-space no doubt. Recall that AUM's clock ticks 10^{36} times faster than ours, and don't worry that AUM will need to re-boot TBC in the middle of your favorite reality program because AUM **is** the computer, the application, and the operating system.

This virtual energy system, once let loose (once time begins to iterate), dynamically evolves according to the dictates of the Fundamental Process and the space-time rule-set. Matter begins to form and individuate, things begin to grow, and entropy decreases in some places while increasing in others. Growth, as it refers to inanimate material objects implies higher degrees of organization and a decrease of entropy, for example: atomic particles fuse together to form more complex atoms and molecules, ordered groupings of atoms and molecules form compounds and more complex molecules, mass begins to coagulate into droplets called stars and planets. This constantly churning system appears stable in the little picture because it is animated on such a grand scale within space and time. For a similar reason, the earth appears flat because it is so large relative to the local awareness of a walking man. One eventual result, among many others, of this apparent local stability may be a universe such as ours.

An experimenter could run this physical-universe creation simulation as often as it wished in order to sample over the natural random components, the various rule-sets, and simulation parameters, keeping (saving to non-volatile memory) the best ones for further study. Iterating this process while tweaking the space-time rule-set will eventually evolve some good experimental candidates worth maintaining and culturing (remember our earlier Petri-dish metaphor).

I wonder how long it took and how many false starts there were before the space-time rule-set was fully evolved? Do you think AUM has it right this time? AUM might tinker with it a little here and a little there but in general, the space-time manifold is on its own to evolve however it evolves according to the Fundamental Process. Recall from our discussion of AI Guy that virtual reality rule-sets occur at two distinct levels. Higher level rules that define goals and purpose and lower level rules that define local physics and causality.

Home, sweet home: TBC creates the logical causal structure and we live and play in it by virtue of our participation in the PMR virtual reality consciousness trainer. Education by total experiential immersion in the reality of your choice – that's the only way to go when one has a long way to go.

TBC progresses the dynamic space-time reality by sequentially incrementing time. The evolutionary logic it follows (expanding into all possible states and continuing those that are profitable) is tracked, recorded, and saved in TBC. Every state that develops (beginning with time t=0) during every time increment (DELTA-t) is saved. All the evolving simulated elements and players move and change to new values and configurations every DELTA-t. Many DELTA-t increments go by until one fine day, here we are in our space-time PMR, snug as bugs in the fabric of a PMR space-time rug. Clueless about who we are, where we came from, or what we are supposed to do, but snug little bugs every one.

We do have some intuitive idea about what is going on because, after all, we are discrete chunks of AUM consciousness playing, working and evolving in a digital space-time simulation-trainer. Our bodies, our interactions, and our physical reality are what we experience under the space-time rule-set. We are nudged in the right evolutionary direction by a **higher-level** rule-set that works through our intuition, becomes our conscience, and urges our higher selves to be the right thing.

There is only consciousness. Everything is a manifestation of consciousness. Even rocks and the proverbially dead doornails are consciousness constructs. Everything we experience is due to an application of the various rule-sets (under which we are extant) to our individuated consciousness. When we are operationally aware in nonphysical realities within $NPMR_N$, our experience is defined by the rule-sets that characterize those particular $NPMR_N$ realities.

The various rule-sets that apply to our consciousness, as it interacts within various dimensions and sub-dimensions of NPMR, are like the laws that govern our behavior on earth. Each of us live under the due process of international law, national law, state law, county law, city law, the restrictive covenants of our sub-division, and the law our mom lays down when we have been naughty. All of them apply to us, and must (ostensibly) be obeyed all the time. Nevertheless, we typically focus our attention on our local reality and primarily give consideration to the laws that affect our daily lives. Only when we begin to travel or do business internationally does international law become interesting or important to us. The rule-sets in TBC are like that – we are required to obey all the rules that pertain to us all the time, but in a practical sense we need to be aware only of the local rules that directly affect us (such as PMR physics) until we venture beyond the confines of our local PMR neighborhood.

64

■■■

Consciousness, Space-Time, and PMR

■■■

Let's make sure we have a firm grasp of the ideas presented up to this point. These ideas are unusual enough that some repetition is necessary before most folks can successfully sort through the layers of long held beliefs to absorb the implications of what is being said. It is a mistake to go too fast. Though your intellect may have your head enthusiastically bobbing up and down with assumed understanding, it is my experience that the deeper significance of your existence as consciousness is usually only partially grasped.

Consciousness is the fundamental attribute of AUM and of any and all sub-realities created within the organization (mind) of AUM. AUM is only mind – digital mind: A system of digital (cellular) organization. The Even Bigger Computer (EBC) is a metaphor for a subset of the memory, data processing, communications, and rules as well as the organizational and control functions of this AUM-mind-thing. Recall the discussion of the evolution of AUO in Chapter 25, Book 1. Because the simplest conceptualization of reality cells is binary, the EBC metaphor is a good one. On a less grand scale and at the local level, you can clearly observe the fractal-like repetition of the basic pattern of digital reality in the computational tools (hardware and software) that we are creating to aid, understand, and guide our evolution and existence within the earth-based ecosystem. Not that far off in our future, physical computer systems and the interactive software that is implemented in them will become very clever at applying specific complex processes to more general problems; gradually their capability and awareness will grow until they have what it takes to host consciousness and support intelligence. From AUM to TBC to the inhabitants of NPMR and PMR to desktop computers, the fundamentals of existence repeat themselves in many forms of digital awareness within many different dimensions of reality.

> ▶ There may be a better computing basis than binary and if there is, AUM probably uses it. However, I am only describing operational concepts not implementation details – and a binary computing basis is by far the most simple and most comprehensible metaphor for us to use.
>
> Big Picture concepts are important; don't get wrapped around the irrelevant details. Whether TBC is formed out of **binary** reality cells, or future computers are based on **silicon**, is entirely irrelevant. The metaphors and the concepts are clear and timeless even if the details, terms, and language supporting the **current** explanation quickly become outdated. ◀

Our digital computers have the basic binary protoplasm found at the root of consciousness (see Chapters 26 and 28 of Book 1) and lack only the speed, capacity, and sophisticated software to support better and better implementations of Artificial Intelligence until the adjective "artificial" is eventually dropped after their capability, capacity, and intelligence becomes self-evolved. Self-evolved does not necessarily imply self-constructed or self-directed, it simply means self-modifying in pursuit of optimizing a defined profitability – the ability to grow, the facility to learn – the capability to exhibit the functions of consciousness.

Eventually, some PMR entities become more aware (dimly at first, but more completely as they invest effort in the awakening process) as they rediscover or uncover their roots leading back to AUM. In fact, some of these sentient beings begin to realize that they exist not only in the highly constrained environment of PMR space-time, but also in the mental environment of $NPMR_N$. Additionally, they exist within NPMR and AUM, but only an exceptional few become intellectually aware of these environments. Fortunately, a few is all it takes to open the minds of others who are ready to become an operative aware part of the Big Picture.

The part of the sentient being focused in PMR makes choices that reflect the overall quality of consciousness of that being. Each intent and action typically generates multiple interactions that produce immediate feedback creating many additional opportunities to learn and grow. Recall that a major design requirement for beings within the space-time simulation was that they should start over on a regular basis with a clean slate so that they could periodically escape the belief traps in which they had ensnared themselves.

Being able to make a new start, **without** the accumulated delusions of the past but **with** your accumulated knowledge and wisdom intact, is a key attribute of the OS system that enables individuals to accumulate consciousness quality by exercising their free will intent within a series of discrete experience packets. Implementing this process in the space-time rule-set implies that space-time beings must be restricted to a series of relatively short, more or less independent, learning adventures in PMR. The space-time experience, like space-time energy transfers, must come in discrete packets called lifetimes.

Space-time beings must be recyclable – they age and die according to the interaction of biological requirements and conditions implemented by the space-time rule-set. When that PMR vehicle (experience-body) gets old, dysfunctional, or is dead-ended in a morass of delusion and bad choices, it dies and the larger consciousness entity (sometimes called an oversoul) existing in $NPMR_N$ collects what learning it can glean from that experience packet. If it wants or needs to, this oversoul can rejoin the space-time simulation by inhabiting (in accordance with the self consistent rule-set defining PMR biological science) yet another PMR body that gets to start with a relatively clean slate containing no specific local reality fear and delusional constructs. However, it is only the local PMR reality slate that is wiped clean between consecutive PMR experience packets.

The disorganizing (entropy inducing) influence of ego-fear within the larger individuated consciousness (oversoul) must gradually be overcome by right choices motivated by right intent (fearless and love based), which are accumulated over many, many experience packets. The discrete interactive experience packets provide a plethora of opportunity to exercise intent through free will choice. An individuated conscious entity brings to each new experience packet the basic **quality** of consciousness and personality it has evolved thus far. Entities will sometimes set up specific situations among themselves to optimize the learning potential of all involved.

You come into this physical world (engage this particular PMR experience packet) with an initial quality of consciousness – and then have the opportunity to improve it or degrade it. This quality factor represents your fearless capacity to love, and your proclivity to form attachments to fear and ego delusion. It also represents the entropy level within your consciousness and the extent of your spiritual growth and maturation. When the PMR experience packet is exited, death of the physical body occurs and one finds that all little picture facts (PMR-specific knowledge) and pseudo-facts (beliefs) are entirely perishable, whereas individual quality and Big Picture wisdom endures at the level of the oversoul.

In the larger reality, the entity is defined by whatever degree of love, fear and delusion that represents the sum total quality of that being. In some philosophic traditions, the idea of karma and reaping what you sow somewhat expresses this concept of a cumulative quality of being that can be improved by making prudent choices for the right reasons within your local reality. Flunked lessons (failure to learn) must be repeated in different forms until they are passed – until the entity gets it.

There is no punishment, retribution, or vengeance implied in this process. Repeating lessons of consciousness quality until you have mastered them simply means that you continue to make choices dealing with certain issues until you grow up and beyond that particular quality issue. The quality limitations of an individual that are derived from the limited quality of its oversoul's consciousness represent the challenges to be overcome by that individual while it is enrolled in the PMR learning lab. Outgrowing those limitations (or as many of them as possible) is the mission that each consciousness has set for itself during its current experience packet (the being's present physical lifetime). Obviously, serious effort as well as good planning and preparation by each individual conscious entity improves the effectiveness of this iterative learning process.

If you are an old analog dog in need of a new trick, and are having difficulty thinking in terms of a digital reality, think of space-time as a consciousness entrainment technology defined by the specific constraints it places on the interaction of entities whose experiences are bounded by its definition of reality. TBC knows the basic rules from which our physics (all science) is derived and allows only certain self-consistent configurations of reality to exist within space-time. The psi uncertainty principle defines how and when physical (PMR) and nonphysical (NPMR) phenomena and causality can overlap in PMR without violating the purpose of the space-time rule-set.

Though interactions between PMR and NPMR are limited, the simultaneous awareness of multiple reality frames is available to those grown up enough to assume the implied responsibility and able to profit from that experience. These restrictions are in place for the same reason that we restrict children from driving an automobile, getting married, or signing contracts: They don't get to do those things simply because they want to and believe they are ready.

Fear, beliefs, delusion, ego, and ignorance are interrelated and are catalysts for higher entropy production. Your mind will automatically transcend the restrictions and constraints that keep it focused exclusively in PMR when it is ready – when its average entropy drops below a certain value.

65

■■■

Deriving PMR Physics
from Consciousness

■■■

Understanding the space-time rule-set that we have been discussing in terms of one overarching theory is what PMR physicists refer to as a TOE – Theory Of Everything. They have been working on a little PMR TOE for a long time but are stymied because, among other things, they do not understand the Big Picture. They do not understand the nature of the larger reality of which the TOE they seek is only a part. In fact, being unaware that a Big TOE exists limits them to searching exclusively for only a little TOE (PMR lower-level rule-set only). PMR scientists cannot find the little TOE because the concepts they need to derive an under-standable context wherein the little TOE can be seen as one whole thing are found only within a Big TOE. Furthermore, because of the digital algorithmic properties of the PMR rule set, there may or may not be a lit-tle TOE cast exclusively in terms of little TOE logic (PMR mathematics).

Thus, scientists and philosophers are limited to digging out rules, one at a time, from the interior of their local reality. When they get to an outer boundary of their little picture – where it meets or interfaces with the larger reality – they run into an invisible wall constructed of belief. A conceptual breakthrough that fundamentally expands their reality is required to break through that wall.

You cannot get out of the box while defining the box to be everything that exists. If you believe that your little box constitutes All That Is, then by definition, everything other than your objective measurable little box appears delusional. Thus the little TOEs that attempt to describe the lit-tle box look like big TOEs to the little-box-scientists trapped inside. Every time their little TOE is stubbed against the limitations of the box, their reality seems to dissolve into statistical mush. Uncertainty principles must be brought in to patch up the inconsistency from the inside view.

It seems that you must first understand and appreciate the potential existence of a Big TOE before the little TOE to big TOE interface can come into focus. Otherwise, you define the proper solutions out of existence by limiting the possibilities you can imagine. In other words, your belief systems limit your reality to a subset of the solution-space that does not contain the answer. (Scientific, cultural, religious and personal beliefs systems are discussed in Chapters 19, through 22 of Book 1).

It may be profitable to remember what was said a long, long time ago toward the end of Chapter 31, Book 1. Also in that chapter, I promised to relate PMR physics to consciousness reality cells: You are almost ready for that explanation.

▶ Chapter 31, Book 1: *Reality cells are roughly analogous to the transistors on a computer processor chip. They come in very large numbers and are the most basic active units of the processor and memory. Like reality cells, each transistor is a thing that can be on or off, a 1 or a 0, this way or that way, distorted or undistorted. At the next higher level of generality, is the processor's basic instruction-set that defines operations and processes for storing, retrieving, and performing arithmetical and logical operations. In our analogy, the processor's basic instruction-set is analogous to basic cognitive functioning within AUM. At the next level of abstraction, we get to the space-time rule-set which is analogous to algorithms written in assembly language. Our experience is generated at the next higher level of abstraction by an AUM-TBC to individuated-consciousness interface which is analogous to a simulation programmed in object oriented C++ where we are the objects. AUM is the computer, the programmer, and the operating system. We sentient conscious beings are, as individuated subsets of consciousness, a bounded subset of highly organized, evolving, interactive reality cells.*

As an analog to space-time, consider a custom designed special purpose processor such as a Digital Signal Processor (DSP) chip. Understanding the rules (patterns) governing the transfer of energy to and from transistors in a special purpose microprocessor would provide some understanding of the most basic relationships in the processor's design, implementation and capacity. Likewise, understanding the rules governing the transfer of information between space-time reality cells should produce some of the most fundamental relationships of physics. In Section 4, you will see how that works.

Physical experience is generated when the perception of an individuated consciousness (sentient being) is constrained to follow the space-time rule-set. Imagine a specialized space-time virtual reality trainer (operating within a subset of digital calculation-space called a dimension) that is constrained by the space-time rule-set to provide a causally consistent operational experience that enables an individuated consciousness to evolve to lower entropy states by exercising its intent through free will choice. The specific relationships defining AUM's space-time instruction-set constitute the laws of space-time physics (PMR physics). ◀

Now there is an interesting thought! It may well be possible for us to derive many of the general laws of PMR physics that represent the basic rule-set that TBC applies to create the constraints we call space-time by studying the interactions between reality cells. We can be reasonably sure that the defining space-time **rule-set** is likely to be a collection of general, broad, high-level statements – what scientists sometimes call fundamental laws. The implications and details we call science are derived from these general statements. For example, classical mechanics flows from only a few simple statements called Newton's laws and the science of electricity and magnetism (at the macro level) is fully contained within Maxwell's four Equations. One should expect that the space-time **instruction-set** governing interaction between reality cells is going to reside at a low level of abstraction.

Can you order your thoughts? Is there a structure or pattern to your ideas? Do your ideas ever relate to other ideas – can they be connected and interdependent? Can you alter and store (remember) thoughts and their relationships and patterns (intentionally or unintentionally)? I am very impressed with what you can do with such a relatively tiny and limited fragment of consciousness. Your mind, your consciousness is clearly a versatile tool. With that magnificent mind of yours, please recall that the rule-based space-time construct represents a specific subset of the digital system we refer to as the mind of AUM.

Consider that at least some of the instruction-set that implements the space-time rule-set may be contained within the geometry (pattern and relationship) of the space-time construct – how AUM's binary reality cells form the structure of space-time – and how the interaction of those reality cells is constrained.

Given this is true, the following question naturally arises: Can we derive the laws or basic facts of physics from considering the structure (pattern and relationship) of space-time? It is maybe not as crazy as it first seems because the basic structure of an entity (object, organization, relationship) often defines the nature of the entity that can be built with or upon it. Albert Einstein didn't think this was a nutty idea; he spent the last half of his professional life trying to do just that from the PMR side of the larger reality. He never succeeded in establishing a unified field theory, but his intuition was absolutely correct. Unfortunately, his PMR-only mind-set was too limited by the cultural beliefs of his day to contain the solution.

Let's take a look at the possible and probable structure of the conscious-ness construct we call space-time, at the reality cell level, and see if we can find a self-consistent logic within it that will lead to the derivation of at least some of our most basic PMR physics. What we are looking for in particular is a pattern or logical structure that AUM-TBC may have utilized as part of a larger *gedanken* experiment to derive, or more properly, evolve, portions of the space-time rule-set, a logical structure related to the differentiated reality cell mental construction of aware consciousness. This is not necessarily **the** approach, but only **an** approach that promises some possibility of success. The concept of reality cells was developed in Chapter 26, Book 1.

Such an approach easily demystifies the seemingly strange probabilistic science of quantum mechanics. This reality cell based TOE derives the logical necessity for the statistical basis of quantum science from funda-mental principles. Moreover it produces a more general understanding of how the quantum mechanical conceptual process applies to the Macro-world. quantum mechanics, shown to be a special case of a more general principle, is finally given a firm theoretical foundation (why and how it must work as it does).

Briefly: The content of our virtual physical reality is based upon the per-ceptions of consciousness. The data supporting the existence of PMR in the minds of its resident players is generated from probable future reality cal-culations and present choices (more about this in Section 5) and imple-mented only when a consciousness requires the data. Thus all virtual real-ity at every level is governed by the same statistical processes that are commonly applied within the science of quantum mechanics. These proba-bilistic processes become more obvious when one is dealing at the individ-ual pixel level of a virtual reality. That tiny particles turn out to be statisti-cal representations of the potential for existence within PMR waiting for a consciousness to require the data as (collapse the wave function to) a meas-urable solid result is a direct consequence of how our virtual PMR is gen-erated within consciousness.

For example, let's look at experiments in so-called "reverse causality" where focused intent seemingly modifies the expected statistical results of data collected in the past, as well experiments where focused intent skews or biases the statistical results of random number generators or other ran-dom events like nuclear decay. In these well known and often repeated experiments, the statistical nature of the data resides in the future as a probability distribution that contains some uncertainty. It is a fundamental property of consciousness that focused intent can modify future probabilities within the bounds of their uncertainty. When the measurement is made in the present by analyzing the data, the wave function collapses to a probable value that has been skewed by intent. The past is not changed, only the data describing the past is modified by intent while it remains only probable within the future probability database (more on that in Section 5). Likewise, the uncertainty associated with the ph of water or the crystalline structure that occurs when water freezes, allows a focused intent to modify the probability of a future outcome that is produced when the measurement (of ph or crystalline structure) collapses the wave function to a physical result. Such biases can be cumulative if imposed on a system iteratively.

The point is, all future outcomes, whether at the micro or macro level, exist only as probability distributions until a measurement is made that collapses the wave function to a physical value. Also, the uncertainty associated with future outcomes may be modified by the focused intent of consciousness within the limits imposed by the psi uncertainty principle. Thus follows an explanation for "backwards causality", the biasing of random events, unnatural or consistent luck, the placebo effect, mental healing, the power of positive thinking, the power of prayer, the so-called "law of attraction" and many other mind-matter interactions. Some have been studied and verified scientifically, and others simply applied by millions of individuals over the centuries because it consistently works. All right, I'll quit now. Those few who wish to study a greatly expanded version of this discussion should look at the "Physics" topic in the discussion group at the *www.My-Big-TOE.com* website.

Lucky for you and me, I do not plan to postulate reality cell structures and rules and then derive physics from them while you watch. A monster aside that delivered 300 pages of technical material would drive away all but the toughest and most generous of readers. Fortunately, a contemporary resident of PMR, Steven E. Kaufman, has already derived the basics of PMR physics from the theoretical structure of a consciousness-derived space-time reality-cell-based instruction-set. (*Unified Reality Theory: The Evolution of Existence Into Experience*, published by Destiny Toad Press, 2002 – ISBN: 0-9706550-1-0). To study the results of his effort, you can either purchase his book or read his book online at **http://www.unifiedreality.com**.

Kaufman starts with simple binary reality cells in an Absolute Unbounded Oneness (AUO) (he does not call it that, but the concept is absolutely identical). He assumes some basic things about how information is transmitted between adjacent reality cells and proceeds to derive the PMR physics of electromagnetism, gravity, force, and energy.

A little reverse engineering and one ends up with many of the concepts of classical physics, relativity, and quantum mechanics falling out of his unified reality theory based entirely upon the properties of the reality cells that exist only in relation to themselves within an AUO-digital-consciousness-energy-thing. Kaufman derives basic physics from the relational cellular structure of a **space-time** construct within consciousness. He probes the logic of the space-time rule-set by analyzing a structural model that is postulated to exist within the mind-space of AUM and the calculation-space of TBC.

You may not agree with all Kaufman's specific assumptions and you may not accept some of his conclusions. However, you will find that coherent and reasonable bridges have been constructed that connect AUO, consciousness reality cells, space-time reality cells, and the physics of the twentieth and twenty-first century. Eventually AUM's space-time rule-set will yield its algorithms to those who approach it from the correct perspective, and steadfastly frustrate those who demand that the facts of science and reality must all fit neatly into their limited little picture belief system.

PMR physics is contained within the rule-set that defines and constrains the experience of individuated consciousnesses to perceive a virtual physical reality. PMR may be thought of as a digital simulation executing within a greater digital consciousness. Contemporary PMR physics simply represents a portion of the rule-set that defines the PMR virtual reality.

The Big TOE cannot conflict with known little TOE facts and must broaden the overall understanding of little TOE phenomena that are presently unexplainable (including psi effects, mind, consciousness, human purpose, and the efficacy of intuition) – *My Big TOE* fully meets these criteria.

That Kaufman bases his work upon the assumption of an Absolute Unbounded Oneness (AUO) is a very good sign because earlier (Chapter 18, Book 1) we discovered that logic demands a successful Big TOE to have at least one mystical or metaphysical leg to stand on. Conversely, we have also demonstrated that any TOE without a connection to what appears to be mystical **from the PMR point of view** must logically be only a little TOE that is fundamentally incapable of dealing with our beginnings, our minds and consciousness, or us as whole beings. To view a whole and complete human entity, one must step out of the PMR box.

66

■ ■ ■

The Mechanics of Experience

■ ■ ■

At this point, you should have some familiarity with the concept that we humans represent a particular type of constrained individuated consciousness experiencing a virtual physical reality within a larger digital consciousness system. As a player in this consciousness evolution training simulator, we perceive an interactive physical reality within the constrained rule-set of space-time. To put it more personally, you are a bounded chunk of consciousness, chipped from the old AUM block, hallucinating or experiencing this PMR according to a set of experience-rules or rules-of-interaction that define energy exchange within this particular multi-player simulation which is operationally managed (computed, executed, tracked, evaluated and modified) within TBC.

You have no massy body, only the interactive experience of one. That virtual rock exists only in the simulation but because your body also exists in that same simulation, it can bonk you in the head and you will experience the trauma and suffer the consequences that the rule-set (science of energy transfers) computes. That's a simple idea isn't it? No problem. You knew it all along, right? The only thing that may still be a tad confusing is how TBC pulls off this massive interactive multi-player experience game. Stay tuned.

Assume that the apparently infinite AUM has a TBC part that is about a trillion trillion trillion trillion (10^{48}) times more powerful than our present day desktop computers. That is simply a made-up number, but why not? At least it is reasonably consistent (given how clever AUM must be about digital systems and architectures) with the fabricated numerical examples in Section 2 (Chapter 31, Book 1) where we assumed that AUM's fundamental clock ticks about 10^{36} times while ours ticks once.

592 | *My Big TOE* | The Mechanics of Experience

We could say that an apparently infinite digital consciousness can, like Superman, and the proverbial 800-pound gorilla, do anything it wants to and not resort to quoting phony made-up numbers at all. However, I like numbers and I think they focus the problem and help integrate it into our limited conceptual space. Numbers can provide a concrete connection between old and new concepts as long as their specific values are not taken too seriously. It is the general idea numerically illustrated, not a specific numerical value that carries significance. The logical validity of the concepts presented in *My Big TOE* has no dependence on the choices of the numerical values used to illustrate concepts of relative magnitude. Accordingly, use the numbers as a conceptual aid, but don't get hung up on them.

A performance improvement factor of 10^{48} more than takes care of the needed computational capacity to run such a simulation. The only remaining question is technique – a reasonable and credible explanation of how it works. That is what this and the next two chapters and Section 5 are about.

Understanding the nature of experience is the first step to understanding our local reality and its connection to the larger reality. Our local reality is defined as the reality in which we appear to exist and function. For most of us, our local reality is our physical reality and nothing more. It is what we are directly aware of, and what we **believe**, sense and measure to be real. As machines and devices extend our senses, our local reality is extended as well.

Experience creates the notion of reality. Our local reality is a byproduct or result of our experience. Experience is derived from two interdependent components: sensory perception and interpretation. The perception of the observer (input data) and interpretation of that perception by our consciousness create our experience. To be logically complete, I must mention that it is possible for stored sensory input data to be interpreted at a later time. Stored perception data may be brought into the conscious awareness whenever it is needed. The local reality is not a hard, fixed thing, but rather a collection of interpreted perceptions. Your local reality is therefore not entirely an objective reality – it only seems to be objective. The apparent objectivity is an illusion created by the internal consistency of the space-time rule-set. In other words, perception (which is limited by our sensory apparatus) and interpretation (which is limited by our understanding and perspective) constitute two filters that transmute "what is" into "what appears to be." Because of our limitations and the constraints on our consciousness, our local reality must necessarily be constructed from "what appears to be" not "what is."

"What is" might be called "un-experienced" or "un-experienceable" reality. It is whatever is out there that interacts with our sensing apparatus such that we receive information (perceive something) that must be assessed or interpreted to determine what it means or what its significance is. "What is" must by definition (because of our limited perception and non-objective interpretations) remain at least partially un-experienced and unknown. The key point is: The ultimate source of our experience must remain shrouded in uncertainty, unknown and unknowable.

Understanding the dichotomy between "what is" and "what appears to be" is important. We are perhaps not the objective beings living in an objective reality we think we are. "What appears to be" is how we interpreted whatever information our limited sensors collected from "what is." Because observers are necessarily unaware of what does not make it through these two filters, they make the erroneous assumption that "what appears to be" is actually All That Is. For this reason, they mistakenly attribute a sense of absolute solidity to their local reality. A being's local reality is constructed of the accumulated experience of "what appears to be" – thus it contains strong individual (subjective or private) components mixed with sharable (objective or public) objects.

Your sensory apparatus is similar to that of others, thus allowing for general agreement about the properties of your local reality. However, your interpretation of those sensory data is uniquely based on your knowledge, understanding, experience, perspective, belief, fear and ego. Interpretation is uniquely individual, relative, and subjective, yet it is half the ingredients that go into cooking up our apparently objective reality.

Our interpretation of a given set of sensory data will be similar to the interpretation of other beings only if those beings share our wisdom, knowledge, understanding, attitudes, experience, perspective, belief, fear, and ego attachments. In as much as our worldview and our belief systems are shared by others, we will generally agree on the properties, characteristics, substance, and significance of reality. Individuals who belong to the **same** culture (whether they are all homeless street people from New York City, all Japanese millionaires, or all Australian aborigines) typically experience similar local realities. Given a single environment, different cultures not only perceive different data because of their unique focus and interests, but also evaluate, interpret, and value similar physical perceptions differently.

Most members of a particular culture generally agree on how and why things are as they are. This broad and nearly universal agreement leads us to develop confidence and unintentional arrogance about the apparent objectivity and superiority of our view. The characteristics of the majority always define the criteria for a healthy well-adjusted member of the group, regardless how dysfunctional or pathological those characteristics may be.

People of other cultures feel every bit as objectively justified and superior in their interpretation of reality as we do. We feel that they are obviously less objective than we are. They shake their heads with amusement and condescending wonderment that we just don't get it. We also feel that way about them, the only difference being that we are right and they simply haven't figured that out yet. The more different the cultures are, the more forceful are these arrogant opinions and the more dramatic is the conflict between beliefs.

▶ Cultural diversity, no longer tied to local geography, is shrinking as the world's people coalesce into a few major overlapping cultural blocks (for example: Western Christian industrialized first world, Middle Eastern Muslim non-industrialized third world, the world of "haves," and the world of "have nots"). As these generalized cultural blocks coalesce and gain virtual membership throughout a **world community**, they gain the power of large numbers expressing shared emotions. The power of numbers often breeds arrogance, self-proclaimed superiority, and belligerence. As leadership evolves to exploit this potential power, expect trouble as major cultural blocks living in wholly different local realities conflict with each other.

The good news is: The growing social and political instability between conflicting reality systems represents only a temporary turbulence induced by a major worldwide reality shift. As a complex system evolves, it often must transition between stable states. This transition period is usually turbulent. We successfully made the transition to the industrial age, but not without abuse, violence, and great dislocation heralding that cultural change. Now we are transitioning to the information age and it is going to be a bumpy ride for a while. Electronic communications technology has suddenly shrunk the world.

The bad news is: If we are not clever, and sensitive to what is going on, this transition could get nasty and last a long time. As the world continually shrinks, the cultural blocks will eventually begin to coalesce with each other. Eventually the rancor of this transition period will dissipate as we face an entirely new set of challenges.

Perhaps one day in the future it will be difficult to find someone who does not share your worldview. The cultural pressure to conform within each of the major blocks is already severe. Do you think our species may be evolving toward becoming politically correct herd animals in the information age or uniformly distributed individualists in plain brown wrappers? Or both simultaneously? ◀

To summarize, you must separate the underlying objective thing being experienced from the experience of it. They are not one and the same because of the characteristics and limitations of our sensory apparatus, our databases (knowledge and information capacity), and our data processing capability (interpretation and analytical capacity). The constraints placed upon our sensing and processing ensure that the ultimate source of our experience remains at least partially unknown. From a physical perspective, "what is" is theoretically as well as practically unknowable. Recall that "theoretically and practically unknowable" is how we defined "mystical" in Chapter 18 of Book 1.

The only thing we actually know about the source of our experience is how it interacts with certain specific energy transfers. We send discrete packets of energy to it, and it interacts by sending discrete packets back. Remembering the discrete character of time from Chapter 29, Book 1, you should realize that energy transfers that seem continuous (such as pressure) are actually discrete from the viewpoint of modern physics and (at a finer level of detail) TBC.

In fact, it is only the players themselves and the interactions between the players that need to be defined by the rule-set. This prescription for creating a subset of reality should seem familiar: We said the same thing earlier when we were discussing another simulated reality (AI Guy and war games). Recall that a player was defined as any thing, entity, or energy that interacts with anything else.

The fundamental basis (a defining rule-set) for the apparently real world of PMR experience is similar to the fundamental basis of the artificial or virtual world of AI Guy. The rule-sets themselves may be very different, but the process behind the rule-sets is very similar. The simulation metaphor applies reasonably well to both.

Even the most astute AI Guy suffers the same problem that plagues us: limited access to, and understanding of, higher-level processes. His vision is limited because his experience can penetrate only so deeply into the source of his environment. There is an un-experienceable reality (such as the computer hardware and software, the people who built that hardware and software, as well as the process, equipment, and facility that manufactured the hardware and software) that creates and supports what AI Guy is constrained to experience. Think of other multitasked jobs running in AI Guy's mainframe, and of unrelated computers running isolated jobs in some other facility, as being outside AI Guy's local dimension of reality. The nature of the limitations of AI Guy's experience and the dependence of that experience on the un-experienceable is not qualitatively different from ours.

You can conceptualize or model our boundary between the "un-experienceable reality" (beyond the limited perception of beings within the space-time simulation) and the "experienced reality" (within the limited perception of space-time beings) as an energy packet exchange interface. This nonphysical to physical interface receives energy packets from physical space-time players (in PMR) and sends back the appropriate return energy packets according to the governing space-time rule-set. The interface between the physical reality-experiencing space-time players and the un-experienceable nonphysical TBC would seem to be the ultimate source of PMR experience. That is a simple description of how our virtual physical reality is constructed; AI Guy's virtual reality is produced in exactly the same way.

We reach out to touch an object that has (within TBC) specific space-time coordinates and properties associated with it. According to the space-time rule-set and the attributes of that particular object-player, it feels solid. We **perceive** it to have certain attributes, classify it as a unique or familiar perception, and finally interpret its significance relative to past experiences and current beliefs. Subsequently, we **define** this **perception** to be an object that is both real and physical and it becomes a part of our local reality experience. Rocks and people and houses are real. Mass is real. Separation of two or more individual masses in space is real because I can put myself between them and move them independently of, and relative to, each other – consequently, space is also real. I change, as does everything else, therefore time is real. Some apparently real things (such as rocks, taxes, dreams, and cocktail party chatter) may seem more real to certain individuals at certain times than others.

All appearances of being real are derived from the interpreted perception that constitutes our experience. All are dependent upon the limitations of the sensing apparatus and the limitations of the interpretation that we give to the data collected – our two filters of variable and unknown quality that are always placed between us and the "un-experienceable reality" that lies behind or beyond our perception. From the PMR view, "lies behind or beyond our perception" means: lies within the nonphysical.

A few examples will make this clear. We think we see an object, but what we actually perceive is a portion of the light energy (energy in a discrete packet form) that has interacted with the object, not the object itself. Next we must interpret this received pattern of light data. What we see is a function of the object, the attributes of the light that impinged on it, how that particular light interacted with that particular object, how the sensor (eye, optic nerve) interacted with the light coming from the object, how we generated and interpreted the resulting optic nerve data, and finally, how we integrated this interpreted information with the rest of our experience and beliefs.

There are no less than five processes occurring between the object itself and our sense of the reality of that object. Each process has its limitations, dependencies, random components, variations, and error sources. Our reality is the result of these imperfect processes working and interacting together. Our other senses go through similarly complex processes. Fortunately, these processes and the science that represents them are generally consistent: Every time you look at a given object (under similar conditions) you see essentially the same representation of the same object but you may or may not interpret it the same way each time. Your mental, emotional, and consciousness-quality state is changeable, as are your beliefs, fears, understanding, focus, interests, perspective, experience, and knowledge-base. Your interpretation depends on these as well as any errors, confusion, or random components; each event is a unique experience

▶ A quick aside is in order for those right-brained folks who don't give a damn about physical reality. (The left-brained technoids in the reading audience might as well skip to the bottom of this aside or take a break and go get some junk food.)

I can hear grumbling in the background coming from the righties… "So much about stuff….so little about meaning." You are absolutely right. There is another class of things that conscious beings define as real that we are purposely ignoring in this discussion of physical reality.

Beliefs that appear as scientific, religious, or cultural truths, as well as emotions, attitudes, and values are mental constructs created within and by the minds of each individuated consciousness. It is not that these subjective realities are not important; to the contrary, they are primarily responsible for the bulk of the content within most individuated consciousness because of the strong influence they exert on the interpretation of the objective sensory data. It is the subjective interactions within your local reality that determine and drive most of your objective activity (interaction), and that most often push the evolutionary levers of intent and motivation.

However, because the subjective nature of your local reality is discussed elsewhere, in this section we are going to stay focused on physical experience. Nevertheless, it is good to keep in mind (as we go through this explanation of how physical reality is nothing other than a highly structured and consistent experience of a consciousness constrained by the space-time rule-set) that the **content** of consciousness and the **quality** of that content remain the most important attributes of sentient entities. Hang on my right-brained amigos; we are almost done with this matter matter. ◀

Are you ready for a simple example? Imagine that a man born blind and deaf (we have assumed away two of his five sensors) is riding in an automobile with you. This trip is a part of his reality as it is a part of yours, but his perception and interpretation of that perception create a vastly different experience. The only memorable event he had noticed was caused by that idiot truck driver who forced you onto the shoulder of the road. Your passenger experienced only a bumpy section of the highway that he interpreted as either under construction or needing to be under construction. He never became angry, he never hollered and swore, and he never made those rude gestures as you did. His reality, his opportunity for growth is defined by his experience, which is very different from your own.

Consider the world we experience when we look through special infrared or ultra-violet goggles. Imagine these goggles being permanently placed over your eyes. You would get new information previously unavailable to your unaided eye, and lose some of the information you were used to (for example, you may no longer be able to read print on paper under florescent lighting or appreciate a color photograph). Your reality and your ability to function within that reality are now dramatically altered. Life, relationship, and interaction would never be the same again and you would need to learn how to interpret the new data. Similarly, the sensory data gathered from machines that are designed to extend our senses must be interpreted by someone. Regardless of how the data originates, it necessarily must pass through the same two limiting filters that separate "what is" from "what appears to be." There is no physical way to circumvent the filtering process. The machine, as an extension of us, can only enable us to see "what appears to be." Quantum mechanics makes the same point in its own way.

Consider the rich and elaborately differentiated auditory and olfactory reality of your dog. Would you have the dog's experience if you had the dog's sensors? No, of course not! You would probably not experience great pleasure and enthusiasm for sniffing the excrement deposited by the dogs and cats in your neighborhood – your **interpretations** of those odors would be very different. What about the vastly different realities that are experienced by exceptionally dim people versus exceptionally bright people; by well-educated world travelers versus those who have never been in a school or outside the tiny village of their birth; by scientific cultures versus those steeped in superstition or mystical tradition?

I am not judging which cultures or realities are better, but only pointing out the dramatic differences in their perceived and interpreted realities. A collection of individuals from vastly different cultures, who are led to experience the same complex objective environment at the same time, will come to different conclusions about the nature and significance of their experience. The objects themselves are not as significant as the interpretations they initiate within an individuated consciousness. Big Picture significance is invested in the people, not the things. Little picture significance is invested in the things, not the people.

Both people and things have their function within the larger system – personal understanding of the Big Picture (wisdom) is required to optimize individual profitability. If one focuses exclusively on the seeds, one will never experience the splendor of the fruit. We make our choices and then live with the results.

If an individual plucked out of the depths of the Brazilian rain forest and an MIT professor of physics were put together, both would be able to see the same poisonous frog, tropical snake, trees, river, laser device, CD player, and airplane as well as equations on the blackboard. Their sensory perception, though similar, would not collect the same data because they would notice different things. They may agree on the form of objects placed directly in front of them, but on little else. Their realities (discomforts, anxieties, fears, needs, desires, and attitudes – all the cultural, religious, personal, and scientific beliefs along with their individual ego-stuff) would be vastly different. Their love-stuff (caring, compassion, and giving) would be of similar type, and could most easily overlap into a common experience.

Obviously, there is more to an individual's personal reality than merely a sensory measurement describing what exists in the common (physical) environment. Do you think a personal reality is different from reality? Your personal reality is different from someone else's personal reality because your experiences and quality are different. As your awareness and knowledge grows, your personal reality grows. Anything that resides outside your personal reality is invisible to you and appears not to exist. What you consider to be the objective outside reality is much smaller (only a tiny subset), more personal, and less objective than you think. That portion of one's experience that appears to be shared, consistent, universal, and objective is simply a reflection of a common space-time rule-set and player list, and as such, it makes up the **least** significant component of each individual's local or personal reality. By comparison, your personal relationships with other sentient entities (which are primarily subjective in nature regardless of how hard you may try to construe them as objective) are easily the most significant component of your reality.

Viewing your local reality as merely the output of an environment being viewed by particular sensors and given a particular interpretation is an oversimplification of the process that ignores the vast quantity of specific subjective content that dramatically modifies and constrains the collected objective data. The **local** reality that each of us creates is wholly dependent on the particular filters we bring to the interaction: it is a product of our personal experience, knowledge, emotional state, and the quality of our consciousness.

Your local reality is to some extent personal. Your personal reality is primarily local, though it can be expanded beyond that limited awareness. You would probably be surprised to discover the extent to which you create your own reality. The illusion is that we are "in here" while reality is independently "out there." The outside world, which represents the apparently objective portion of your personal reality, is based upon a uniquely interpreted set of uniquely limited perception data. That uniqueness represents your individuality. You have more input into, and influence over, the creation of your objective reality than you might imagine.

The "set" is consistent and follows the space-time rule-set without deviation (accept as allowed by the psi uncertainty principle), however, the "story" is yours alone. Furthermore, because you have sentient intent, purpose, and free will, the story determines the set, not the other way around. Believing that the PMR set determines your personal story represents a common error based on the misunderstanding that the physical universe is primary and you are a secondary derivative of it.

To a locally limited awareness, objective causality appears to define physical reality. When you realize that you are experiencing PMR through a virtual reality simulator-trainer, the possibility that there are optimizing feedback loops causally connecting the quality of your choices (the successful evolution of your consciousness through experience) to the action taking place in the virtual world projected by the simulator, becomes a more reasonable proposition. Although the simulator must appear consistent to all players, there are many subtle and not-so-subtle ways that the apparent outside physical environment (including relationships with others) can be purposely modified within the virtual reality generator to present each player with a maximum learning opportunity. The virtual PMR reality dimension or experience generation system maintains rule-set integrity and provides the best integrated optimal opportunity (on the average) for the entire system (for all sentient players). To preserve the honesty and straightforwardness of your interactions within PMR, the psi uncertainty principle makes sure that cross-dimensional energy transfers are adequately obscured.

Your personal interpretations of the meaning and significance of your inside and outside experience automatically customize your personal reality in a manner that increases the likelihood of finding those experiences and opportunities that are most important to your individual evolution. Your local reality provides the playing field, the players, and the rules of the game. Its structure provides a context within which experience can take place and your free will can choose; it enables your individual evolution to unfold by supplying the complete player set (relationship and interaction) as well as the rule-set that defines the permissible energetic interactions (objective causality).

Contrary to popular belief, your local reality is not an objective place that you inhabit as you might inhabit a house. Nevertheless, your local reality is the **result** of your personal interaction with a seemingly objective outside world. Your local reality represents your personalization of only a **very limited** interaction (energy exchange) with the possible outside world. Experience, and hence your reality, is not independent of the experiencer; it reflects the unique subjective, historical, and emotional state of the individual. Your experiences, as well as your personal interpretation of your experiences, are strongly influenced by the quality of your consciousness.

Beings of notable quality who are also regular pudding tasters on the Path of Knowledge have a more practiced ability to evaluate the quality, significance, and opportunity of their experience. They evolve within a much larger decision space (live and function within a much larger, more varied and complex reality) and are likely to be less personally limited as well as more aware of their personal limitations.

Note the interdependent cyclical (more accurately spiral) nature of consciousness evolution. An intent that reflects quality creates an action within a virtual reality that produces learning opportunities that lead to increased quality that supports more profitable intents. The consciousness quality spiral can be bootstrapped in a downward (degenerative) or upward (progressive) evolutionary direction by incrementally increasing or decreasing the entropy of the system.

Let's summarize what we have discussed thus far. That we collectively declare our shared local reality to be universal or fundamental because everybody sees a similar thing is a tribute to the smallness of our view, the similarity of our sensors, the homogeneity of our beliefs, and to the consistency of the rule-set and player list. The collective similarity of our interpretation is a tribute to the commonality of our cultures, belief systems, needs, egos, and goals. This collective experiential agreement is not based upon the existence of a fundamental (identical for everybody) **local** reality. A local reality is local to the individual, not the environment. From your perspective, there is no other reality except your local or personal reality – however large or small that might be. Ignorance is blind; you do not know what you do not know. To a large extent, each individual and each culture or belief system creates its own local realities that are teeming with pertinent and challenging learning opportunities.

Collective experiential agreement is based on a common space-time rule-set that defines a consistent energy packet interaction between all player types. It is this ubiquitous and consistent space-time rule-set that represents the common source, the common environment, which leads to similar experiential results for each individual within PMR. There is also a common physically un-experienceable external environment that exists behind our perception – the "what is" that exists within the "un-experienceable reality" that is at the root of our experience.

PMR (physical) experience is derived exclusively from the "experienced reality" of "what appears to be." "What appears to be" is the result of applying our limited individual filters of perception and interpretation to "what is." "What appears to be" is more accurately: "What it appears to be to me," which is unique for each individual while at the same time supports common experience at an average or common level of perception and interpretation.

The "un-experienceable reality," may, or may not, be objective and invariant, although it usually appears to be both. The only thing that we know about it for sure is that we (from the PMR view) must always remain ignorant of it because, by definition, it is what lies beyond our limited physical perception (with or without machines). It represents the mechanics of the space-time simulation that the players are not aware of. This is analogous to the computer hardware and software and the people who made the computer hardware and software, and the facilities they made them in being beyond the perception of (un-experienceable and unknowable to) AI Guy.

You should note how similar this discussion sounds to the one we had earlier (Chapter 18, Book 1) about unknowable beginnings, un-experienceable sources beyond our grasp, mysticism, and the understanding of events that lie outside our causal system. What we have called "un-experienced reality" is none other than our friend TBC operating the energy packet exchange interface. The nonphysical is therefore the ultimate source of our physical experience because it provides the common rule-based (algorithmic) foundation upon which our physical experience is constructed (computed). An "un-experienced reality" exists outside our causal system, beyond the limited perception that defines our PMR reality. The "un-experienceable reality" (from the physical perspective) can only be accessed or understood (as can our beginnings) through a process that transcends our local reality and local objective logic. By definition, any process that steps beyond our local causal logic, beyond our collective local reality, is called mystical.

Good golly, Miss Molly! It seems that whenever and however we dig deep enough (with a logic shovel) into reality we eventually find a mystical core (as seen from a PMR perspective). Hey Jake, I sense a trend; perhaps we're on to something here.

AI Guy has as difficult a time comprehending his ultimate (as opposed to local) host computer as we have comprehending ours – for much the same reasons. To AI Guy, the host computer and the rest of PMR are nonphysical and entirely mystical. Consciousness appears nonphysical from the space-time view of PMR. Nonphysical reality is a reality that is in, and of, consciousness – which is why only the mind can travel and gather experience within nonphysical reality. Do not expect a parapsychologist to bring back the equivalent of nonphysical rock samples for the rest of us to look at: NPMR is not a distant moon.

An advanced computer that develops consciousness also develops a nonphysical dimension to its being. You will not be able to grab AI Guy's consciousness in your hand even if you are the programmer and computer manufacturer that created the necessary conditions for consciousness to take root and begin evolving. Humans create computers – consciousness develops new expressions of consciousness – such creativity occurs every day in NPMR where the local residents are physical (according to their own rules and point of view) and you are not.

Consciousness is energy, the most basic form of energy and perhaps the only form of non-virtual energy. Experienced reality, un-experienced reality, TBC, PMR, you, and the space-time rule-set are all constructs of consciousness. Each is created and designed to accomplish its specific function. All are subsets of the one single consciousness we have been calling AUM. Consciousness is all – the rest is merely apparent to a limited view within a highly constrained awareness. A superb virtual reality, by definition, must always feel totally and convincingly physical to its inhabitants whether it is simulating PMR or NPMR. What appears to be nonphysical is relative to what appears to be physical. The reality an entity is mentally immersed in appears physical while all others appear nonphysical. Each is as real and extant within TBC, EBC, or AUM as any of the others. The appearance of being physical or nonphysical is relative to the viewpoint of the observer, and hence, relative to the observer's knowledge, awareness, and consciousness quality.

Logic tells us that our experience, from which our local reality is derived, exudes from a seemingly invariant mystical (from PMR perspective only) core of "un-experienceble reality." Remembering how a mystic from a little picture might actually be a scientist within a bigger picture (Chapter 20, Book 1), let us now poke our heads up beyond PMR. The first thing we nonphysical scientists become aware of is that PMR physics represents a small subset of a much larger rule-set that defines how TBC (subset of AUM's mind or consciousness space) implements the physical and nonphysical components of us, our local reality, PMR, and OS.

That the superbly objective (PMR-only) physics of today is merely a partial exposition of a digitally implemented rule-set that determines the bounds of interaction between individuated units of nonphysical consciousness is a thought that amuses me to no end. You may need to know a few hardheaded PMR *über alles* scientists to get the joke. Don't you just love it?

67

■ ■ ■

The Mechanics of
Perception and Perspective

■ ■ ■

From the PMR point of view, perception is an interaction of our sensors with some undefined "un-experienced reality" through an energy packet exchange interface that produces data within our central nervous system that, when interpreted, creates the experience with which we define our local reality. The perception process from our little picture viewpoint is simply a matter of physics, and how our biology and machines implement that physics. Neither the "un-experienced reality" source nor the energy packet exchange interface is **directly** relevant to our **experience** of PMR. Similarly, neither the computer's power cord, nor its processor chip, is **directly** relevant to the **experience** of AI Guy.

Because the space-time rule-set resident within TBC defines PMR physics (all science), and controls the output of the energy packet interface, that puts TBC directly in charge of defining and specifying what we perceive. TBC establishes the limits and constraints of our interactions with other players and provides the return energy packets across the interface that represents our interaction with what we perceive as our physical environment.

If our past experience and the quality of our consciousness determine how we interpret the data from our perception, it would seem that something nonphysical is in charge of and controls both ends of the process that creates experience – and transforms aware intent into action that can learn from that experience. Is it clear yet that our experience, including the physical experience that defines our local PMR-based reality, is sandwiched between, dependent on, and created by two wholly nonphysical processes carried out within, and created by, consciousness itself?

Is this simple or what? You and I are clumps of individuated consciousness engaged in a dance of energy packet exchanges through an interface that maintains the rules of the game. The book you are reading in your mind is the book I wrote in my mind. The paper and ink and the hand you are holding it with and the eyeballs you are reading it with are the required effects of the space-time rule-set that defines the form and structure of our interaction.

If you think the constraints of space-time are annoying because you are constrained to read word by word what a more natural, less constrained, mind could transmit, comprehend, and thoroughly absorb in a few quick big gulps, criminy, I had to type the damn thing with just two stubby little physical fingers and a pair of drugstore glasses. That is simply the nature of space-time – simple, direct, slow, detailed, and relatively linear – as the experience within any good elementary school should be.

This puts AUM and TBC in the driver's seat when it comes to defining our sense of reality within PMR. How could our cute little individual fleshy bodies and all the other physical things (critters, rocks, bushes and earthquakes) that seem solid be nothing more than constructs of consciousness? How could our solid PMR reality be described as mind, a dream, a delusion, or as extant only in the mind of God (as it might be put by an imaginative PMR poet or hopelessly trapped PMR scientist)? How can our physical and nonphysical parts be integrated into one being? How is it that all of earth's beings, critters, objects, and energy are connected? How can all this be on a big communications net or part of a grand digital simulation? If you have been paying close attention, you should know the answers to these questions.

The AUM and TBC consciousness-system-thing control all the inputs, and create our 3D space-time experience of individuated separateness – all in our mind, all in our consciousness, all in our thought-form-fragment-being chip from the old block of AUM consciousness. Everything started out as simple uniform consciousness potential energy. With the encouragement of the Fundamental Process of evolution, consciousness expanded into every potentially profitable state, profits were consolidated, losses cut, and the Big Dude was on an evolutionary roll. Everything is still only digital potential synergy and organized digital content (consciousness), but now, due to the steady decrease in system entropy, many highly synergistic complex configurations have evolved. Everything that seems physical to you is an experience of a consciousness constrained to experience space-time.

The solution to the mind-matter problem is embarrassingly simple. In the Big Picture, there is no matter – everything is mind. There is only the little picture experience-of-matter within mind. "Physical" does not exist – the term "nonphysical reality" only has meaning relative to a nonexistent physical reality. Once you leave the delusion of a local physical reality, everything appears physical, or equivalently, nonphysical, and the distinction between the two vanishes. All reality has solidity of form and function that obeys the causality enforced by its governing rule-set. The physical-nonphysical and mind-matter dichotomies are an illusion created by a limited local PMR viewpoint.

Wave-particle duality, uncertainty principles, and the seemingly instantaneous communications between entangled pairs become simple to explain once you realize that PMR is a virtual reality created by a digital simulation implementing the space-time rule-set within TBC. Given a digital PMR simulation, which is stepped forward by time increments that appear infinitesimal to us, and a virtual reality that must obey only the rules driving its digital computation, these paradoxes disappear along with the illusion of absolute space. Once the limiting belief that all possible reality is exclusively defined by measurements within PMR is abandoned and the true nature of consciousness is grasped, the mysterious paradoxes of physics, philosophy, and metaphysics all melt away like ice cubes in the summer sun.

All that is needed to find solutions to the Big Questions of our time is a simple shift in perspective – a casting off of erroneous scientific, cultural, and religious belief inherited from those who were unable to answer the same questions. Isn't that how it always turns out? New paradigms deliver an expanded reality as we outgrow the old ones. The digital mountain raised up by the approaching information age has simply afforded us a better view at this time. Progress, like quality and ability, is developed through a bootstrapping process. Every new success is built on previous successes.

Nascent understandings of the digital nature of the Big Picture are erupting all around us. Dozens of top scientists are today hot on the heels of discovering that PMR is actually a little digital picture existing within a computed reality. That is a necessary first step that leaves the Big Picture just around the corner. It appears that this is an auspicious time for humanity to take yet another of its occasional grand leaps toward a greater understanding. *My Big TOE* provides the theoretical foundation that supports the phenomenology and hypotheses that are currently being investigated by scientists and philosophers worldwide. Grand leaps (forward or backward) are always the result of a confluence of many forces and urgings that together produce a unique opportunity. These are exciting times with great success and great failure sitting on opposite sides of the same fantastic opportunity. Human beings, are you ready? Drum roll please!

Once you understand what and who you are, and how you relate to the whole, the resulting Big Picture perspective produces one consistent reality with no paradoxes. Note that the first nonnegotiable requirement of a fully correct Big Picture Theory Of Everything (that it produces one consistent reality with no paradoxes) has been fully met by *My Big TOE*. Also note that the second nonnegotiable requirement of a fully correct Big TOE (that it subsume what is known as a special case of a more general understanding) has also been fully met by *My Big TOE*. Furthermore, that the science contained within *My Big TOE* does not support the limited view of traditional scientific beliefs, which are unable to produce a bigger picture, is a necessary strength, not an unavoidable weakness of this Theory Of Everything.

When you are in NPMR, it appears to be every bit as physical as PMR, but because it operates under a different set of rules, one interacts with it differently. The operational differences between PMR and NPMR simply represent the differences between their rule-sets and the unique causality that each rule-set imposes. Each particular reality dimension has evolved a rule-set to support its own use, function, and purpose. The space-time rule-set supports human function and purpose within OS.

There is no significant distinction between physical and nonphysical realities: Reality is reality. I employ that artificial distinction and terminology (PMR vs. NPMR) as a communications aid. To communicate with you effectively, I need to start (conceptually) from where you are (or think you are). The fact is that most of you are certain that you exist within a physical reality, hence that is the initial perspective we must take. The PMR-NPMR distinction within *My Big TOE* is used to help you conceptually sneak up on a bigger picture. We will continue to use the PMR-NPMR terminology, especially in the next section, because it greatly facilitates the grasping of inherently difficult Big Picture concepts – like thinking of an atom as a billiard ball with BBs zipping around it – patently incorrect, but useful at an elementary level.

Local reality is an experience of individuated mind interacting within the limitations of a given causality. If there are multiple minds interacting within a given local reality, there exists a shared common (public) experience we define as objective, as well as a personal experience that we define as subjective. The subjective and objective components of reality are both extremely significant – their purpose and function are simply different.

The larger reality is all consciousness (All That Is) evolving toward greater profitability, existing to improve itself through entropy reduction. Aware consciousness is created by the **organization** of a fundamental potential energy that we have (for reasons of conceptual familiarity) named Absolute Unbounded Oneness – the nature of this organization is digital. The larger reality is a huge interactive digital consciousness system; it is a consciousness-evolution fractal ecosystem that we have named AUM.

To summarize: Physical and nonphysical are relative to a point of view and therefore do not support a fundamental distinction. The mind-matter, normal-paranormal, physics-metaphysics, and science-philosophy dichotomies are likewise simply illusions of perspective created by a limited understanding that is exclusively focused within its own local reality. The experience of our physical matter reality is the result of a particular set of constraints (space-time rule-set) placed upon the interaction of individuated consciousness with other players, which include other sentient beings as well as the environment. Matter is a simulated mental effect that we, as mind, experience because it helps put us in a virtual environment that makes the evolution of our consciousness more efficient and effective.

As constrained constructs of consciousness, we have the imperative to lower our entropy (evolve our consciousness) because that is the fundamental nature of the greater consciousness-evolution fractal ecosystem of which we are an infinitesimal part. We personally and as a species reflect the pattern of consciousness evolution because we are the result of that pattern and an integral piece of a larger consciousness system that evolves by iterating recursively upon itself to generate All That Is.

We control the experience of a simulated AI Guy in the same way the space-time rule-set controls our experience. We define what he can perceive (what data he can access) and the rules by which he processes the information he collects. We prescribe the boundaries of his experience and reality in order to create the optimal (for our purpose) simulation tool.

We do not enable AI Guy to perceive beyond his local reality (such as most of the computer hardware and software used in his simulation, the people who made it, the facilities they made it in, all of PMR and NPMR) because it is irrelevant to his mission, invisible and unthinkable to his cognitive awareness, outside of his local reality, and beyond his logical causality. To AI Guy, these higher-level things would appear (if they were explained to him) to be nonphysical mystical nonsense existing within a different dimension. Computer hardware, power cords and electrical outlets, programmers and the software they produce, and simulation laboratory facilities simply do not exist from his limited point of view and if he is anything like his human brothers, there is no way to convince him otherwise because he simply does not have the conceptual framework to expand his reality enough to perceive and understand the Bigger Picture from which he has been derived.

You might as well explain to a fish how to dance an Irish jig as tell a man that he is nonphysical consciousness. On second thought, that is a poor analogy: The fish has a good excuse (it has no legs), while the man, at least theoretically, has a mind.

68

■■■

PMR as a Virtual Reality Game

■■■

et's look at reality from an entirely different viewpoint. We may create or customize our own local reality through our personal interpretations, but we certainly do not create all reality: A larger reality exists apart from us that appears to be centered in the nonphysical because it exists outside our local physical reality dimension. We are a subset of that larger reality and we interact with it through our individual personal minds. Our bodies and their physical experiences are a product of our minds interacting within the constraints of the space-time rule-set with other players (both sentient and non-sentient) and other minds in a mutually interactive dance of aware but limited consciousness exchanging discrete energy packets in a process called interactive experience.

A good way to get a rational grip on this concept is to perform a *gedanken* experiment of your own: Ponder a future virtual reality game. Before this virtual reality game begins you are (your body is) put into a perfect sensory-deprivation tank where you blissfully float. The advanced and powerful computer running this game is connected directly to your brain. No special gloves, helmets, pressure suits, or tilting, shaking, rock-n-roll platforms are necessary.

The computer hosting the game is connected wirelessly to the nerves or brain-areas where each sensory data input is received on its way to being processed. This virtual reality bio-computer game machine has been programmed with an advanced understanding of biology, physiology, physics, and of how the senses interact.

Such an advanced, super duper, totally cool, virtual reality game machine can simulate and then stimulate your central nervous system with such accuracy that you have a complete, scientifically exact, and absolutely real experience. This virtual reality machine produces an artificial or virtual experience that is indistinguishable from real experience. You cannot tell the difference.

Because the natural inputs from your actual body are blocked while you are hooked up to this machine, if you lingered too long as a participant in the "Gourmet Foods Of The World" virtual reality game – regularly ate your fill of the world's best virtual filet mignon with the world's freshest virtual steamed vegetables and most delicious virtual fresh fruit – you could physically starve to death without noticing the real you was hungry. You get the picture. This virtual reality game is a marvel of technology that produces absolutely realistic experience.

Imagine that this virtual reality game is a multi-player game. You and a group of your friends, along with many thousands if not millions of others, are playing this game together. The computer tracks the interactions between the players and integrates their experiences. Some multi-player computer games do a limited version of this type of virtual reality simulation now. You and your friends chose the "Jungle Safari" game; consequently, the computer supplies the appropriate rain forest environment for your party and populates it with all the critters, objects, and energy (weather, climate, earthquakes, wind, volcanoes, and so on) that would naturally be there.

Your mind and the computer's rule-set together create a rule based, consistent, causal, objective space-time experience where the free will initiative of you and your friends drives the action to its logical conclusions and simulated physical ramifications. The final results are determined by the computer's rule-set and the actions or lack thereof that you and your friends take. Your seeming reality (local virtual reality) is created by your interpretation of your virtual perceptions. These virtual perceptions and interpretations create a virtual experience that provides opportunity for you to learn from the results of the actions that express your intents. This virtual reality experience-game is a good place for you to evolve the quality of your consciousness: It makes a great training machine. All the while there is also a larger reality that contains your actual floating body, your spouse, mortgage payment, and the new car that you left sitting in the parking lot of Sam's Virtual Reality Emporium. This outside reality is not causally connected to your virtual reality, and from the point of view of your virtual reality, it is non-operational, nonphysical, and does not exist.

Do you get my point? The technology I am describing is near enough to our reality that you should be able to follow (imagine) the story line without difficulty. Telling this story sixty years ago would have left the audience without a clue as to what I could possibly be talking about. It would have appeared to be pure fantasy, absolutely impossible, idiotic wild concepts beyond comprehension. Sixty years ago, when Albert Einstein was pondering relativity theory, people would not have known what a digital computer was (the first digital computer, ENIAC, was developed at the University of Pennsylvania for the US Army between 1942 and 1946). Sixty years from now, computers will be (assuming Moore's Law) one trillion (that's twelve orders of magnitude or 10^{12}) times faster and more capable than they are today. Better yet, that improvement factor of one trillion is likely to be an outrageous understatement given the high probability that multiple breakthrough technologies will be discovered during that time span.

That is more number crunching power than we can imagine. The only thing we can say for sure about what we will be doing with a trillion times the present computational power is that it is beyond our comprehension to imagine. Can you imagine this: Our children and grandchildren will be living in a world that is today totally unimaginable to us? That is how quickly things are changing and the process continues to accelerate. Only ninety years from now, near the end of the present century, Moore's Law predicts our computers will be a million trillion (10^{18}) times faster than today's computers. What do you think your grandchildren, great grandchildren, and AI Guy are going to be doing with that much digital capability? On the other hand, even with breakthroughs, processing speeds could go asymptotic long before they reach an improvement factor of 10^{18} – who knows? What I do know is that within a single human lifespan, AI Guy and the future applications of digital computing will most likely be vastly different from anything that we could possibly imagine, even in our wildest dreams.

Today, a person with a technical education might reasonably ask when a close approximation to this virtual reality game might be expected to hit the market. If this concept seems theoretically doable to little ol' dumbed-down us just sixty years after outrageously slow and clunky digital computers were invented, what kinds of digital magic do you think AUM might be able to pull off with his puddles of individuated consciousness that don't need to be wired, fed, pay mortgages, drive cars, or rub up against each other for personal gratification?

AUM actually has a much simpler problem. Granted, AUM's sets are larger and more detailed, but hey, that requires only some additional computer memory and crunch-power – no problem for an apparently infinite, brilliant, digital-consciousness that operates in a frequency-space that has a basic time increment that is eighty orders of magnitude smaller than our second. In fact, it is so easy there are many, many versions of PMR humming along in parallel – and The Big Dude never breaks a sweat. B.D. makes the coolest games, man. Totally cool!

69

∎ ∎ ∎

Real Mystics Don't Smoke
Ciggies in the Bathroom

∎ ∎ ∎

What about the scientists and mystics who already knew everything I have explained about the Big Picture? You know, the individuals who are reading these books just to see if I got it right. Do these people represent a glitch in the system? Are they screwing up because they are peeking behind the scenes, looking on the other side of the energy packet exchange interface instead of obeying the rules of space-time like the more normal citizens of PMR?

No way! We are not put in space-time like a zookeeper puts an animal in a cage. We are not caged at all. We are consciousness – no more, no less. We have at our command all the attributes and abilities of a sentient individuated consciousness with free will. We are a part of AUM and contain the characteristics and potential of the whole in our part. We are in space-time to learn, to grow the quality of our consciousness, to evolve. Once we have evolved our consciousness to the point that the space-time construct is no longer an efficient tool for evolution, we go on to other things. Space-time constitutes a learning lab, an educational environment to grow in, not a jail.

It is more like an elementary school than a detention center. The point is: You are supposed to graduate eventually, not merely hang out with your friends, smoke ciggies in the bathroom, and skip classes. If you pay attention, try hard, do all your homework, and taste test lots of experimental pudding you will some day grow-up and be one of the big (picture) kids. Then you will realize a closely held secret that only the big kids know. Listen up! I am going to spill the beans! This is **the** major secret of life. Are you ready for this? Here goes – drum-roll please! **You can learn more**

if you try, pay attention, study, and practice, than you can if you just wander around in the school hallways waiting for gratuitous insight, or by hanging out with the smart kids. That's it.

That is the biggest and best secret I have. I blab that secret to people who want to know something deep and meaningful, but most of them don't actually get it. All life's great secrets share this attribute: Merely voicing the secret does not divulge the meaning; it only becomes profound, and therefore makes a difference to your life, when you are ready to absorb its significance within the context of a bigger picture.

We make choices based on the quality of our consciousness and the opportunities our apparent situation presents to us. We need such an apparent situation because of the optimal learning opportunity it presents – as discussed in Chapter 22 of Book 1, Chapter 48 of this book, and Chapter 74 of Book 3. Thus, we (and others) represent a logical and evolutionary necessity for AUM. We are configured from AUM's consciousness to allow AUM to optimize its evolutionary potential. We may be an experiment, but we are an experiment that is integral to the being and evolution of the aware consciousness that is AUM.

Come on Jake, put out that cigarette, give up trying to convince Susie that you are as cool as you wish you were, and let's get back to class.

70

■ ■ ■

The Politics of Reality

■ ■ ■

Each individual partially defines his or her own local reality. To the extent that our perceived environment, sensors and interpretations are similar, our experience will be similar. We of common experience, large ego, and limited understanding subsequently come together and declare that our shared reality is the one true reality and that everybody who doesn't understand that fact is dumb, confused, or delusional. This position concerning the accepted notion of reality becomes just another cultural, scientific, and religious belief system. Be careful: It is an easy trap to fall into.

Our local reality is defined and limited by our senses and our interpretation of the data our senses collect. What lies beyond our local reality is believed to not exist, believed to be mystical, or described as nonphysical. Imagine the reality we would collectively construct for ourselves (believe in) if every human on this planet were (and had always been) blind and deaf. (Assume plenty of accessible food and beer for everybody – we are not probing survival issues.) Imagine what our culture, political boundaries, and civilization would be like under those circumstances. Think of all the things that we now experience and understand as part of our physical reality that would disappear into non-existence or appear to be mystical or nonphysical. How do you think we would collectively interpret our interaction with critters and with each other? How would we deal with sunburn, snow, tornados, fire, tigers, birds, fleas, bumblebees and good-feeling babes?

▶ While your subconscious mulls over the concept of "good-feeling babes," let's explore the degree to which logic and rational analytical process can provide a more correct interpretation of your perceptions. Here we are speaking of fundamental sense perceptions as well as your perception of the quality of sentient interaction (mood, intent, motivation, attitude, feelings, relationship, emotion, and meaning).

Are all interpretations of perceptions equally valid? Can anyone distinguish between logical and non-logical interpretations? If a given interpretation is logical for one perceiver, will it be logical for all perceivers? The point of experience is growing the quality of your consciousness, not getting the right answers. You can learn to make more profitable interpretations with more experience – wisdom can be developed. However, do not put too much hope in the power of logic to lead consistently to the optimal, best, or correct interpretation. Logic can be applied only when there are enough good data to support it. Most of the interpretations of our perceptions, particularly those that support our most significant decisions, must be made without enough data to come to a definitive logical conclusion.

This beyond-logical, uncertain state of affairs is by design. Otherwise, choice, free will, and intention would become moot issues if all interaction within our local reality was essentially deductive or tightly logical; all problems and challenges would have a unique analytical solution. Logic would replace judgment; existence and choices would be automatic and machine-like within a closed solution set. Even the extreme left-brainers, who longingly fantasize a more rational world, would eventually get bored. If life were a **logical** puzzle to be solved, learning would come to an end as soon as someone found a solution and shared it with others. Life is not logical – even if you pretend you are. You cannot use your intellect to get the most out of it. Trying to optimize your life by primarily applying your intellect (what most intellectuals do) is like a blind person with exceptional hearing trying to drive an automobile or fly an airplane.

We have the space-time rule-set (PMR physical law) to provide basic order and causality as an objective foundation. Do not expect logic to govern personal learning, interactions, and relationship in the same way that physics governs cannonballs – these are not logical processes even though left-brainers, and many relationship-challenged individuals, would like to pretend that they are or should be. (Why do all the men have this lost look on their faces – and why are all the women nodding their heads and rolling their eyes?)

Typically, your sense of being rational is produced by the self-justifying belief traps you are caught in. Your appearance (to yourself and others) of rationality and logical process is, for the most part, an illusion, a feel-good delusion of the ego that makes you appear to be competent and thus delivers a sense of correctness, and personal security to the self. Each of us has a tendency to define the local truth to be whatever feels good to our ego and boosts our self-esteem. We justify our actions, feeling, attitudes, and beliefs and interpret events to support our needs, wants, desires, and expectations.

Let's tie this discussion of assumed logic and rationality in with some of the things we learned in previous sections. Do you see why randomness (or pseudo-randomness), uncertainty and the psi uncertainty principle are a necessary part of the space-time rule-set? Is it clear why divination, mind reading, telepathy, remote viewing, precognition, psychokinesis, and other psi effects are detrimental to the potential growth of a low quality, high entropy consciousness while generally irrelevant to the growth of a low entropy, high quality consciousness? Little boys, say five to ten years of age, would dearly love to be as strong as full-grown men, but fortunately evolution is not that careless.

If you still do not get it, imagine your boss, spouse, mother-in-law, children, or telemarketers having direct access to your mind. Great power in the hands of an irrational ego or manipulating intent is always frightening and usually destructive. For the most part, we are thankful for the natural limitations on **other people's** power, as are they, no doubt, thankful for the limitations on ours. That huge difference between "us" and "them" is an illusion of ego. Irrationality and illogic are the norm, not the exception; belief that the opposite is true (at least for us) is a commonly held delusion. Take a moment to ponder how this discussion might apply to you and the people you know.

Almost everyone will agree that ego typically ravages the rationality of others because we are all reasonably secure in the knowledge that we are an exception to that rule. You and me amigo, we're not like all the others. Right?

Are a gaggle of pre-schoolers rational? Are they logical? Are they highly interactive? What motivates them? Why do you think that you (at the fundamental level of interactive consciousness) interact substantially differently than they do? Think about that for a moment – I want a good, thoughtful answer. Are you sure the perceived difference is not either superficial (you are better at math, a better planner) or generated by an ego justifying itself and its significance (the things you do are more important)? Most people simply define themselves to be rational and that is that.

When you find somebody who thinks that they are particularly rational and logical, often you have found instead somebody who is out of touch with their deeper motivations and intents. Simple analytical thinking often masquerades as basic intelligence and is used to support a superior claim to correctness – a self-serving logic and rationality that justifies dominance, wants, needs, and desires.

People who live entirely out of their heads and exist primarily in intellectual space often are sadly shallow and severely limited by the belief that they are primarily logical beings and that the employment of rationality and logical analysis is the highest and loftiest goal they can aspire to. The most important things in life are not things that can be adequately dealt with or experienced through analysis and logical process. Such self-directed impoverishment and limitation is held up as an ideal in Western culture.

Do not get carried away. I am not implying that all logical process is fraudulent and useless. It can be a wonderfully productive tool – I live and work by it every day. Science is based upon it – it is the foundation of My Big TOE. I am merely asserting that we in the West have elevated the value of belief-based rational process to the point that we are fooling ourselves most of the time and as a consequence, we have blocked our view of a more holistic process that reaches much deeper into the well of truth than mere logical analysis. Our sense of rationality has become twisted, self-referential, and based upon circular logic – a marvelous tool extended beyond its useful function.

The non-rational, non-logical world that we actually live in is entirely different from the rational, logic-driven world that most of us pretend that we live in (particularly intellectuals and technical or scientific types). It seems typical that the information available to support making a logical interpretation or decision about the meaning or significance of our collected perceptions is inversely proportional to the importance of the correctness of that interpretation or decision. The majority of life's important, significant, path-changing decisions are the ones whose outcomes are the most uncertain because of a lack of information.

This state of affairs was not designed to frustrate you, but is a result of the fact that the physical causal world is simply a theatrical set (a playground with rules) provided to you to help you unfold your personal drama by forcing you to make significant subjective choices based upon your intent, not objective choices based upon causal logic. This arrangement allows you to evolve your consciousness, not just practice the relatively sterile art of correctly applying logic. The most important growth opportunities of your life will always be beyond your causal logic – will always be subjective and intuitive gropes dressed up in as much pseudo-logical justification as you can muster in order to make your life appear as orderly and rational as possible. The appearance of an orderly rational process driving our lives forward is a delusion that lowers anxiety and makes us feel better. Beyond that, the appearance of order in your life provides a coherent medium or frame of reference for your experience that infallibly reflects the quality of your being. It would seem that our drive to rationalize our choices is a necessary part of what makes us work.

By highly valuing the well behaved, dependable, easily understood objective aspects of your existence while discarding, devaluing, or bumbling through the subjective aspects, you are focusing upon the chaff and throwing away the wheat. This misguided assessment of where we should focus our effort wastes huge amounts of time and energy in logical objective cultures such as ours. The effort that we focus on our careers, status, and material success greatly outweighs the effort that we invest in raising our personal quality. The result is that most people live their lives within a continuous soap opera that seems to have no final episode until death liberates them from that particular part.

The physical virtual reality set (PMR) is not intrinsically important to your purpose except that it provides the structured learning experience you need. The iterative expression of your intent as it assimilates feedback derived from the results of your interactions with others is what allows you to pull yourself up by your bootstraps (lower your entropy increment by increment through a long-term personal program of self-improvement). That you can solve logic problems is helpful in mastering your rule-based space-time environment, but it is only the foil, a supporting, enhancing, enabling structure, not the main goal. Understanding the objective causality of your virtual learning lab is important like your house and car is important; understanding the subjective and intuitive nature of your most significant decisions is important like your children or parents are important.

Big decisions, important decisions, are usually complex and span many uncertain issues and are therefore least amenable to a logical solution. Where did you leave your glasses? Use logic as best you can. Who should you marry? How should you go about improving the quality of your consciousness? Forget logic, it's not going to help much. Because we seldom know how things will change or how relationships and interactions with others will progress (even if we could fully specify present states), how could our logical analysis penetrate very deeply? Our assumed rationality and logical process is a thin veneer, while love, truth, fear, want, need, and desire run deep.

How do we interpret our perceptions in the midst of this unknowing? Mostly we guess! We make assumptions and rely on beliefs. We go with a hunch, from the heart, or with a gut feeling. We pretend (usually without intellectually knowing that we are pretending) that we are logical or that we know more than we do. We develop theories; we extrapolate past experience into the future. We use our intuition, which is our normal connection to the nonphysical part of our being. How we ultimately interpret our perceptions depends on our knowledge, previous experience, understanding, wisdom, and the quality of our consciousness. We try the best we can to be rational, or at least to appear to be rational.

I expect that you may have noticed the many feedback loops and functional interdependencies that connect perception, interpretation, logical analysis, experience, wisdom, belief, and the quality of your consciousness. If not, take a moment to ponder the maze of interdependent connections and interrelationships. Don't rush, take as along as you need, I'll wait for you. ◀

Interpreting your experience can be tricky if the quality of your consciousness is low. Belief and ego can strongly color your interpretation as well as your perception. Many of man's most horrific experiences – war (including holy wars), genocide, racism, ethnic cleansing and so forth – are motivated by ego in the service of belief. Humanity's worst crimes are typically committed to maintain, preserve, and spread particular beliefs and individual power, or as an expression of ego-arrogance. Many ordinary, perfectly nice people become upset if a bigger picture (or someone else's little picture) threatens their comfortable concept of reality. They will rationalize their attitudes and produce many good reasons why their particular delusion represents the only correct view. Such is the power of fear, belief traps, a closed mind, and a small picture combined. This is the politics of reality.

Selectively ignoring the facts of objective and subjective experience to justify enforcing a common belief (scientific, cultural, political, or religious) of what constitutes official reality sanctioned by the proper (scientific, cultural, political, or religious) authorities is required to maintain the delusion that we are logical, rational people living in an objective physical world. Fear, control, and conformity make ignorance their friend.

The preceding paragraph was rather complex and you may need to read it a few times to get what it implies about other people.

▶ Everyone likes democratic politics these days. If you really want to know how real reality really is, all you need to do is take a vote.

"All those in favor of PMR being all there is to reality raise your hand. I guess that settles it folks, the ayes have it – PMR is all there is.

"I mean …if people as intelligent, open, and with it as you and your friends are haven't experienced this larger reality, what are the odds that anyone has? Not high, I am afraid, not high – especially if we discard the goofy unreliable types – the ones that are not like us. If this foreign sounding AUM gobbledygook were actually true we would all have heard about it by now – everybody would know. Right?

"Look at all the smart people in the world – do they **believe** this nonsense? No way! Where is the hard physical evidence? I am not buying this consciousness mumbo jumbo until I can put my hands on some good samples of nonphysical energy to study in the lab. Scientists cannot allow themselves to fall into this subjective quicksand – that's where science was hundreds of years ago when doctors were bleeding people with leeches to cure them. You cannot go wrong if you stick to objectivity. Right?

"There is nothing you can do about it anyway. How could I ever know the truth? Jeez, weird people are everywhere. You name it and there is some bunch of wacko people somewhere that believe it. There seems to be no limit to people's capacity for goofiness. This nonphysical baloney is all unsubstantiated opinion. It is **objectively** not provable. Right?

"You need to be careful of what you are willing to **believe**. Make a note: Thinking thoughts that are not 'normal' can be dangerous. I should be more careful about what I expose myself to. I think I should toss this disconcerting TOE-jam book in the fireplace and go watch TV. What could be more harmless than TV? Right?" ◀

That's politics folks! The majority rules. Run with the crowd – it is much safer that way and there is no personal responsibility to worry about. Kick back and let it go. If a mistake is made it is no big deal because everybody else will have made it too. If something is especially important, someone will tell you about it. Right? And when they do, you will listen and understand. Right?

71

■ ■ ■

Weird Physics Requires
Weird Physicists

■ ■ ■

What about the sense of touch? Isn't that the sense that gives us our most trustworthy feel for reality? The eyes and ears (our dominant sensors) can be tricked rather easily. Illusionists make a living by tricking us into believing something that is not true. Mass banging into mass, now that feels real! If you can get your hands on it, you know it is not an illusion, right? There is a classical law of physics that says two masses cannot occupy the same space at the same time. That simple physical law alone can account for our individual sense of separateness within space-time. Everything – rocks, fish, people, hockey pucks, and the planets – must all be and remain separate entities within a shared space. None can encroach on the other's personal space.

TBC undoubtedly has that rule expressed somewhere in the space-time rule-set that defines and limits our perception. Yet, when things become big, fast, or small the game changes. Our law-abiding bodies that vigorously lay claim to their specific volume and mass wouldn't dream of sharing that same space with anything else. Yet a neutrino, neutron, or a high energy photon sees our bodies as we see our universe – mostly empty space with a few chunks of matter scattered about here and there. Now the concept of no two masses existing at the same point at the same time begins to take on a different meaning. At a smaller (quantum) scale where the classical view breaks down completely and a particle seems to exist as a smeared out probability distribution, that particular physical law seems to have developed a statistical loophole.

Wherever the mechanisms and processes of our perception are coarse enough to disturb what we are attempting to perceive, direct observation becomes impossible. If this situation occurs because we are probing the boundary between experienceable and un-experienceable reality, we should expect the results to be strange because essentially we are probing the boundary between our local reality and the algorithms and mechanisms that TBC uses to implement its rule-set from outside our causal system.

The boundary between what we can perceive (experience) and the un-experienceable reality, which is the source of that perception but at the same time forever beyond our perception, is the boundary between our local reality and a mystical reality (from the view of PMR). This is the boundary between physical objects and consciousness, between what appears (from the view of PMR) to obey our law of causality and what appears to flaunt that law. It is also the boundary between our bodies and our mind, soul, spirit, and intuition. This boundary separates the normal from the paranormal and, if a bridge between the two is developed, defines and enforces the psi uncertainty principle discussed in Chapters 47 and 48 of this book.

Modern physics in general and quantum mechanics in particular will always be mired in mystery and produce results that seem inconsistent, unexplainable and counter intuitive as long as it clings to a PMR-only little picture belief system. Quantum physics will remain confusingly abstruse as long as it stubbornly requires the $NPMR_N$ reality camel to be forced through the eye of the PMR reality needle – or the $NPMR_N$ elephant to be pulled out of a PMR acorn. Until the self-imposed belief blindfold is removed, the outer boundary of little picture science will remain confused, out of focus, and apparently mystical. Inside that boundary, scientists will continue to unravel the space-time rule-set one fact and relationship at a time.

Discovery constitutes a journey that is much longer and more personal than most people think: Take your time, and focus on where you want to go. Understanding the rule-set that governs your local reality so that you can manipulate material existence to suit your needs is all well and good, but understanding the Big Picture is vastly more significant.

72

■ ■ ■

Section 4 Postlude
Hail! Hearty Reader,
Thine Open Mind and Force of Will
are Truly Extraordinary

■ ■ ■

If you have made it to this point (unless you have skipped around a lot) you have proved yourself to be no shrinking metaphysical violet. By now those with little to no experience beyond PMR have had almost everything they ever believed in thoroughly trashed or turned upside down and inside out. That could cause an uncomfortable, lost, sad, or empty feeling – or maybe a "drifting free with no roots" feeling. Has some nasty person torn up your comfortable old roots? How rude! Fortunately for me, the culprit is you. I simply supplied the hoe, trowel and rototiller (or was that a bulldozer?) and gave some encouragement. Don't worry, such feelings are often the doorway to a better place than you inhabited before. Award yourself four more of those highly coveted beautiful golden stars; paste them in your book next to your name and you will feel better.

On the other hand, those with considerable experience outside the confines of PMR will have found a conceptual structure and context that makes sense out of all those unusual experiences that they knew were significant, but simply could not explain or integrate into their worldview. For these folks, *My Big TOE* holds the promise to transform the incomprehensible into the well-understood, the mystical into the mundane, the fear of being strange into the assurance of being normal, and far-out seekers searching blindly by trial and error into well-reasoned individuals with a good explanation of what they are doing and why they are doing it.

From the opposite point of view, let me remind you that this book is flammable, and you can always forget the entire thing. Hey, that's not so bad; you get a great fire-starter for your fireplace, and can easily dismiss *My Big TOE* as *My Big Delusion* – the ravings of a wacko physicist who has obviously lost his way.

Either way, you ought to read the next Section. Section 5 is the result of everything we have accomplished in the previous sections; it's where you begin to see how everything previously mentioned begins to pull together into a coherent (that's my opinion) model of reality. Section 5 reveals the mechanics of how the larger reality functions and describes many of its processes. It will tell you what's out there, why your unusual experiences are as they are, and how and why they are rationally bounded. Thus far, only the conceptual foundation has been laid; now we will build the model that operationally describes the larger reality upon it.

Ahead the trail gets steep in a few places (because the structure and processes of reality can get complex), but most of the conceptual heavy lifting is behind us. To get the most out of Section 5, you will need to continue to be an intrepid explorer. Open-minded skepticism, perseverance, and dogged personal stamina will be required in Section 5 at least as much, if not more, than they were in the previous sections.

Now let's go straightaway to Section 5 and discover what reality looks like from the inside.

MY BIG TOE

BOOK 3:

INNER WORKINGS

Section 5
Inner Space, the Final Frontier:
The Mechanics of Nonphysical Reality – A Model

Section 6
The End is Always the Beginning
Today is the First Day of the Rest of Your Existence

Synopsis of Book 1 and Book 2

Book 1: *Awakening* – **Section 1** provides a partial biography of the author that is pertinent to the subsequent creation of this trilogy. This brief look at the author's unique experience and credentials sheds some light upon the origins of this extraordinary work. The unusual associations, circumstances, training, and research that eventually led to the creation of the *My Big Picture Theory Of Everything* (My Big TOE) trilogy are described to provide a more accurate perspective of the whole.

Book 1: *Awakening* – **Section 2** lays out, logically justifies, and defines the basic conceptual building blocks needed to construct *My Big TOE's* conceptual foundation. It discusses the cultural beliefs that trap our thinking into a narrow and limited conceptualization of reality, defines the basics of Big Picture epistemology and ontology, as well as examines the nature and practice of meditation. Most importantly, Section 2 defines and develops the two fundamental assumptions upon which this trilogy is based – a high entropy primordial consciousness energy-form called AUO (Absolute Unbounded Oneness) and the Fundamental Process of evolution. AUO eventually evolves to become a much lower-entropy consciousness energy-form called AUM (Absolute Unbounded Manifold) though neither is absolute or unbounded. Using only these two assumptions, Section 2 logically infers the nature of time, space, and consciousness as well as describes the basic properties, purpose, and mechanics of our reality. Additionally, Section 2 develops the concepts of The Big Computer (TBC) and the Even Bigger Computer (EBC) as operational models of aware digital consciousness. Our System (OS) is defined to be PMR (Physical Matter Reality – our physical universe) **plus** the subset of $NPMR_N$ [A specific part of Nonphysical Matter Reality (NPMR)] that is interactive with PMR. Many of the concepts initiated in Section 2 are more fully explained in later Sections.

Book 2: *Discovery* – **Section 3** develops the interface and interaction between we the people and our digital consciousness reality. It derives and explains the characteristics, origins, dynamics, and function of ego, love, free will, and our larger purpose. This section introduces us to the consciousness-evolution fractal ecosystem of which we are a part and explains how we can optimize our interaction within that system to actualize our full potential as an individuated consciousness. Additionally, Section 3 builds a bridge between your present assumptions about the physical and spiritual reality you think you live in (a reality limited by your cultural and personal beliefs) and My Big Picture Theory Of Everything (*My Big TOE*). The nature of a larger reality that exists beyond your physical perception is described in detail. Spirituality and Love are technically defined as functions of entropy content within consciousness. Paranormal activity is given a theoretical basis as a normal part of a larger science. Finally, Section 3 develops the psi uncertainty principle as it explains and interrelates psi phenomena, free will, love, consciousness evolution, physics, reality, human purpose, digital computation, and entropy.

Book 2: *Discovery* – **Section 4** describes an operational and functional model of consciousness that further develops the results of Section 3 and supports the conclusions of Section 5. The origins and nature of digital consciousness are described along with how artificial intelligence (AI), as embodied in AI Guy, leads to artificial consciousness, which leads to actual consciousness and to us. The design criteria, features, and limitations of our personal consciousness are explained and related to the larger digital reality. The part we play within the whole is derived as the physical universe is shown to be a local reality within a larger multidimensional system of digital synergy. The concepts of physical and nonphysical are shown to be relative to the observer and to carry no intrinsic significance of their own. Section 4 derives our physical universe, our science, and our perception of a physical reality. The mind-matter dichotomy is solved as both our physical reality and the larger reality are directly derived from a large but finite system of digital energy that evolves an organized content called consciousness.

Section 5

■ ■ ■

Inner Space, the Final Frontier: The Mechanics of Nonphysical Reality – A Model

■ ■ ■

73

■ ■ ■

Introduction to Section 5

■ ■ ■

In Sections 2, 3, and 4, I have developed the basic concepts needed to gain a fundamental understanding of the larger reality. In this section those concepts serve as the necessary background required to produce a top-level understanding of PMR-NPMR interface dynamics. You will have to make most of the connections between the earlier sections and this section on your own. If I were to point out all the places where concepts of previous sections are being applied, it would add so many references to Section 5 you would have difficulty staying focused.

In Section 5 we will regress to a more PMR-centric view that should be more familiar and less abstract, develop a description of OS reality mechanics, and afterward explore some of its more interesting ramifications. This section will begin by reviewing from a more analytic perspective some of what we have previously learned. This conceptual review will set the stage for a challenging exposition of the fundamental processes of the larger reality.

Section 5 will be a little more buttoned down and logically formal, a little more technical and precise, but don't let that put you off; after the heavy brain work is done formulating the fundamentals of NPMR mechanics, we will have some fun exploring some of the cooler ramifications.

The following twelve chapters, which deliver the content of Section 5, are a subset of my observations relevant to describing the mechanics of the larger reality in which we all exist. They are carefully constructed from my experience in physical-matter reality (PMR) and nonphysical-matter reality (NPMR). This is a broad brush across the top of this subject, a first pass so that you can tell if you want to dig deeper on your own. Many interesting and important concepts are purposely neglected because they fall beyond the scope of this effort to present a comprehensive model of how the larger reality works. With a subject as broad as this one (the mechanics of reality), I have no choice but to remain focused so that you are not overwhelmed with endless digressions nor a work that is too ponderous to ponder.

Please keep in mind that the point of developing a consistent Big Picture model of reality is **not** to convince or persuade you of the correctness of **my** interpretation of **my** experience, and thereby offer you something better or more accurate to believe concerning the nature of reality. After many years of careful exploration, I have merely organized my observations into a coherent and consistent pattern – the resulting structure is the Big TOE reality model presented in this section.

My intent is that the overall model of reality described by the *My Big TOE* trilogy should be a model that you can apply; it should be practical as well as theoretical. Its primary usefulness lies in its ability to help you meet and solve some of your most daunting professional and personal challenges. Second, it provides a logical unified theoretical understanding of the reality of which you are a part. If you are a scientist or philosopher, it is this secondary usefulness that will seem most important – it is not. You will, most likely, depart the *My Big TOE* experience with something that was not there before, something that reflects your personal experience, understanding, and quality. Take from it whatever you can. Leave behind whatever does not square with your sense of what is right. Have no fear; you will not end up where you do not belong.

The first four relatively short chapters of this book will provide a quick review and wrap-up of the concepts developed in the first two books of this trilogy, as well as provide the proper perspective that you will need to optimize your understanding of the reality model presented in Chapters 77 through 82. This review is primarily intended for those that have not yet read Book 1 or Book 2. If you have recently read the first two books you may wish to skim the next three chapters spending time only on those concepts that are of particular interest to you.

Chapters 77 through 82 are more conceptually challenging: If you get stuck simply continue to skip paragraphs until it gets easier (or more interesting) again.

Book 3 not only presents the Mechanics of Reality, but also explores and describes some of its consequences. In the process of doing so, it touches on fun subjects including time travel, teleportation and telepathy; the simultaneity of past, present, and future; how the past continues to live; changing the past; warp drives (faster than light travel or communications); the concept of an Even Bigger Computer (EBC); the nature of NPMR and its beings; probable realities, precognition, and future expectations; changing the probability of occurrence of future events; multiple projections of you; leaving the PMR body behind; creating multiple bodies; understanding reality as a Consciousness-Evolution fractal; and many other interesting subjects.

74

■ ■ ■

Science, Truth, Knowledge, Reality, and You

■ ■ ■

My *Big TOE* describes a model – a relational logical structure – that defines the forms, functions, and processes that constitute your larger reality. The approach is scientific (I use the word "science" in its original and most general sense: The observation, identification, description, experimental investigation, and theoretical explanation of phenomena). The observations have been accumulated and the investigations undertaken by the author during the past thirty years. It is experience, not faith, which is required to transform the information found here into knowledge, and to evaluate its significance to you. You must find your own meaning in it.

If what you find here does not resonate with your inner knowledge, or if it only resonates with anger or other negative feelings, let it go. Toss the book out and forget about it. It is not our time to connect. Learning cannot and should not be forced. Each individual must grow in his own way, on his own path, under his own power and initiative. The world is big enough for everyone if mutual respect is given to personal differences.

In as much as being careful, vigilant, and practicing good science can make it possible, this effort is based entirely on my direct firsthand **experience** – it is not based on my beliefs or on what I may have surmised from others. The influence of philosophical bias, preconceived notions, or beliefs has been, to the best of my ability, eliminated or at least minimized. The data upon which this model is based represent my best **objective** evaluation of my **subjective** experience.

If you think the previous sentence contains an oxymoron (that an objective evaluation of subjective experience is impossible), you probably have too narrow a definition of the word "objective" and may be muddling through the most important parts of your life ignoring objective feedback that is the **result** of subjective interpretation of apparently objective perceptions. Results can be objectively measured even if the motivations, understanding, and intent (the underlying dynamics) that created those results are entirely subjective. Unique, stable, repeatable, recognizable subjective states often drive specific objective results: This is the normal mode of operation for most individuals.

Consider some of the correct, objective conclusions or knowledge (truth) you have come to, or gained, from your subjectively driven and interpreted interactions with friends, family members, lovers, spouses, children, bosses, or parents and from the subjective evaluation of the direction, substance, and quality of your life. Consider that useful logical conclusions may flow from inductive as well as deductive reasoning. If you are still in a logical dilemma about the possibility of deriving truth (objective knowledge) from subjective experience, reread Chapter 47, Book 2 before going on. Further explanation of how subjective experience can produce objective results is presented in the next chapter.

This trilogy, as with all sincere science, has universal truth as its goal – a goal that is, unfortunately, always difficult, if not impossible, to **guarantee**. The correctness of this model, as with all scientific models (the shell model of the atom, for example), must be judged by its ability to explain the existing data within a self-consistent and coherent system. If you are not sure how that might be accomplished (evaluating the correctness of this model), it might be a good idea to reread Chapter 48, Book 2.

The nature of the existing data that describes the facts of reality is unique enough to require some further discussion. It is a common mistake to confuse the larger reality itself with the objects that appear to lie within it. We shall see that this error produces a limiting view that extends only as far as our senses. It also denies the reality and significance of what is most important to us. The super-set of facts, truths, or data describing our larger reality must include who and what we are (our subjective being, our mind, our consciousness), what we perceive through our physical senses (our objective measurements), and the result of interpreting, integrating, and synthesizing this data into the unique intellectual, emotional, physical, and spiritual being we call us.

Unlike traditional works in either philosophy or science, I make no effort to convince the reader of the goodness-of-fit of this model to the existing data; that is not the point. I know this is difficult to understand because it is contrary to how things are normally done: You will need to think out of the box here. Although I am formally presenting a scientific Theory of Everything with this trilogy, **convincing you** that my concepts are correct or credible, or that the model is accurate or valuable, is not at all what these books are about. Understanding the content, knowing its truth (and the limitations and boundaries of its truth) at a deep and personal level, must be personally earned, not book-learned; it must come through your firsthand experience, not through reading (and being convinced) about results and conclusions based on someone else's experience.

What is the point? In addition to offering you a coherent understanding of reality and time, explaining the purpose and significance of your existence, and describing how your world works, this model, in conjunction with your effort, may help you interpret both your objective and subjective reality data (experience). Perhaps it will nudge you to create some interesting theories of your own; or, by providing some missing pieces of understanding or information, it may help you find explanations to a personal puzzle within your experience. Additionally, it might be a catalyst to your learning by expanding the set of possibilities of which you are aware. It may lead, nudge or goad you to think new thoughts, to evaluate important perspectives, concepts and ideas that you otherwise wouldn't have considered.

My Big TOE in general or this "mechanics of reality" section in particular is **not** likely to **create** significant knowledge at a deep or fundamental level where none existed before. Contemplating this book (or any book) is unlikely to change who you are or the quality of your being and yet at the same time, and most importantly, there is an outside chance it could be the catalyst for **you** to do exactly that.

The personal nature of at least a portion of the data (the facts of reality) makes the standard argumentative (me convincing you) approach to imparting knowledge counterproductive – this approach would guarantee a failed effort. It would succeed only in starting a non-solvable argument over what constitutes the set of valid existing data. A useless and endless argument for sure. Why? Because what qualifies as the existing data is both objective and subjective and the subjective portion is unique and personal to each individual's experience and knowledge (recall Chapters 48, 66, and 67 of Book 2). Your consciousness is personal to you, it represents your being – we will call it, its attributes and evolutionary processes "being objects." In contrast, consider your body or other physical objects that can be probed, measured, and tabulated by the intellect and tools of others; we will call these objects shared "intellectual objects."

A set of the data relevant to the larger reality must take into account your consciousness as well as your body. Intellectual objects are physical objects, shared public objects, objects that must obey the objective causality (space-time rule-set) of PMR. Being objects are nonphysical and primarily represented by units of individuated consciousness in various states of awareness and quality.

A system capable of supporting consciousness spontaneously generates enough self-organizing potential so that the Fundamental Process of evolution can, in the pursuit of profitability, generate sufficient synergy (through a process of recursive reorganization called learning) to form an individuated consciousness that is independent of the mechanisms that support it. Consider a digital system's potential to profitably organize itself given that the system is a nonphysical form of energy that becomes a self-modifying, complex, informational medium that the Fundamental Process can organize and reorganize into successively lower entropy configurations. Being objects are the individual result of the consciousness creation process. (Chapters 41 and 57 of Book 2 review these concepts.)

Reality must be experienced by consciousness. Unlike the intellectual objects (material facts, and ideas) of traditional Western science that can be shared, being objects or unique states of being are primarily experienced through subjective personal awareness. To understand the knowledge, data, and facts of being (the being objects), you must be them – you must reflect, express, or integrate them in your being.

So, what is the point of *My Big TOE* intellectually discussing being objects? Because knowing about it, though not to be confused with being it, can be valuable in and of itself. Consider the difference between wisdom and knowledge. Knowledge is generally good to have and it is required in order to apply wisdom, but wisdom is much more than merely better, or more complete, knowledge. Have you ever heard of a little wisdom being a dangerous thing? In addition to knowledge, wisdom requires Big Picture understanding and caring at a deep level – both are attributes of experience and being rather than attributes of information. With the right information you might occasionally act wisely, but that is not at all the same as being wise.

Experience, not verbal communication, is the doorway through which personal subjective knowing must pass. One who is successful at acquiring, assimilating, and manipulating intellectual objects is said to be knowledgeable and is capable of **doing** great **things** (has good quality ideas). In contrast, one who has successfully evolved low entropy being objects is said to be wise and capable of **being** a great **person** (has good spiritual quality). Each of us should develop both of these mutually reinforcing capacities and seek an optimal balance between them.

Knowledge without wisdom is common; wisdom without knowledge is impossible. Knowledge is about facts; wisdom understands how those facts relate and interact within the Big Picture. Intellectual objects are separable and can be arranged like beads on a thread. Being objects are individualized subsets of one whole integrated thing – more like the thread itself.

Unfortunately, learning to manipulate intellectual or material objects is seen as real, rational, and scientific (the scientific method is used to separate fact from fiction among intellectual objects) while learning to manipulate being objects is seen as non-real (delusional), irrational, and unscientific. Just as classical mechanics makes hopelessly wrong inferences in the realms of the very small and very fast, the scientific method has little validity outside its realm of intellectual objects.

The scientific method we learned to venerate for its proven ability to deliver the truth is an excellent methodology for exploring Newton's mechanistic universe as well as Einstein's relative universe. However, a digital (quantized) statistical reality (somewhat like the one described by quantum mechanics), based upon the constraints of our knowing, needs a broader scientific methodology that allows for the necessity of uncertainty. Such a generalized scientific method will allow a larger, more porous truth to be derived from uncertain statistical data manipulated by inductive as well as deductive reasoning.

Living with uncertainty is fundamental to your progress and growth. It is the way of the future, scientifically and personally, and you might as well get used to it. A mechanistic reality over which we eventually learn (through science) to exert control represents a limited local reality. Uncertainty is not a problem for science as long as the uncertain results are represented in our physical reality by something (a physical component or manifestation) that is clearly measurable, consistent, and uniquely related to its source.

To apply the requirements of the current science of intellectual and material **objects** to the science of being is to deny the existence of the latter. Turning the scientific method into a universal dogma (science as religion) enforces an inappropriate process and measure of value upon the science of being. By the standards of traditional science, the science of being will (by definition) always seem delusional, non-provable and unusable – the realm of charlatans, kooks, and other silly or devious people.

People in general (Western science and philosophy in particular) are ill-equipped and ill-prepared to crack the nut of mind, being, psi (paranormal phenomena), love, or of spirituality. Nor are they likely to understand how these things interconnect with each other, the physical world, and us (collectively and individually). Not until individuals, particularly scientists, learn to transcend (think outside the box) the philosophical corner they have dogmatically painted themselves into will the science of consciousness become available for our use.

For these reasons, I will leave the analysis of the model's goodness of fit to existing data and previously acquired knowledge up to those who have enough honest (not generated from fear, dogma, or belief) experiential data points to make the exercise meaningful. The model's value for those without firsthand experience of the larger reality lies in its ability to provide a comprehensive theoretical model of existence, and in providing a tool that can be used to logically extend one's experience beyond present knowledge.

The exploration of inner space is greatly facilitated by the encouragement and assurances of those who have gone before, and by having a good map and description of the territory.

The value of *My Big TOE* is entirely dependent on the essential correctness of its representation. You should gather up your objective (public) experience and integrate it with your objectively evaluated **subjective** (private) experience to assess this TOE's value for yourself. You could evaluate only the public part, but that would be like inventing an automobile and only using it to crack walnuts by driving back and forth over them one at a time. Surely, you **can** use a car as a nutcracker, but why would you want to do that? Wouldn't you rather use your car to go somewhere interesting – to explore the Big Picture perhaps? Cracking those walnuts is simply not that important. Lift up your head and look around – there is an entire reality out there – let the nuts go for a while.

For each individual sentient being, life's most important and meaningful experiences are typically subjective. Significance, relative to the being objects, or relative to the quality of an individual's being, is derived primarily from subjective experience. Consequently, it is important to keep the following two things in mind – **both apply to you, as well as to me**: 1) The ignorant are never aware of their ignorance; and 2) It is highly likely there are individuals who possess more experience, understanding, wisdom and truth than you do. How can you discover which individuals know more, and are more, than what you know and what you are (represent a higher quality of being, contain less entropy, or are more spiritually evolved)? Likewise, how can you discover which ones know less, and are less, than what you know and what you are?

Given that you do not believe yourself to be omniscient (a typically ego or fear driven know-it-all), you can only attempt to sort out the resultant confusion (separating the wise from the foolish) by gaining **direct**, carefully sorted, tested, and validated firsthand experience. Experience is the key because of the personal nature of the subject – you must personally experience the larger reality – there is no other way to interact or to absorb it. Nevertheless, if you are not careful how you evaluate and interpret your subjective experience, you could still end up in the self-delusional fool category. If instead, you separate the wise from the foolish by applying your beliefs, you logically risk being merely a self-consistent fool – or one fool among many (an invisible fool) if your beliefs are shared by others.

Separating the wise from the foolish is no easy task, but it is one at which you can succeed if you try. Recall that in Chapter 48, Book 2 we offered a detailed solution to this epistemological dilemma.

In the science of being, unlike the science of objects, consensus does not define the accepted truth, much less the actual truth. Experience is personal and can never be shared exactly – an event or happening can be shared (simultaneously viewed) by many individuals, but the experience of it is unique to each individual. Consequently, there are many ways an individual can view (and interact with) the same truth. What you do with the truth you discover is the point; how will you use it to enhance the quality of your being? Functionality (what can be done with it) and results (how it actually helps you grow or evolve your consciousness quality) are the measures of the value of your knowledge, not how many people agree with you. Big Truth does not rearrange itself to fit the whims or demands of the fashionable majority.

My experience, knowledge, and understanding are important to you only if they help you find knowledge, and experience truth in **your own** way. Your and my understanding of the right way, and of how things work, may or may not be helpful to others. If we think that our experience and the knowledge derived therefrom might be helpful, we should share it, but never impose it, require it, or think it is the critical information that others need. Nor should we make it our goal to ensure that others recognize its universality – that is evangelism, not science.

The science of being is an individual science that cannot be advanced through a group effort. Its significance lies in its results – in the effect it has on the quality of the individual and on that individual's ability and willingness to profitably evolve his or her personal consciousness. As you read this section, focus on the common understanding, not the differences, between this model and other concepts of reality. Reach for the biggest picture your personal data can logically support, but be careful to avoid imitating the proverbial blind men who jump to conclusions about the characteristics of an elephant from the one part of the whole that they experience.

Keep in mind that our (your and my) experience is probably incomplete. Conclusions should remain flexible, tentative, be given varying degrees of credibility based upon **our** experience, and remain always open to the possibility of gaining more data (understanding) in the future.

Either believing or disbelieving what you read here would be a futile act, a compulsion that cannot induce personal growth. My aim is to expand the possibilities that inhabit your mind, to provide you with a model that helps you place your experience into a larger context, and that encourages you, by explaining the scope, structure, and content of a larger reality, to explore and find truth. I hope that you will find freedom and empowerment in the idea that you create (as well as limit) your own reality and find comfort in the vastness of the possibilities and potential that you have to satisfy the fundamental purpose of your being.

Let me be clear that neither *My Big TOE* nor the mechanics of reality presented in this section philosophically center around Western science, Eastern religion, spiritualism, New Age philosophy, theosophy, the occult, or any other body of thought or "ism." It is simply my experience put in the context of form, function and process. Though *My Big TOE* may logically connect, and occasionally overlap with, many philosophical, scientific, and theological traditions, it represents and has been derived from a unique synthesis of my personal and professional experience. Big Truth is, and has always been, the same as it is now, so similarities should be expected.

Any associations you might draw connecting the concepts presented in this trilogy to some other body of thought or to another set of concepts are a synthesis of your own brewing, not the specific intention of the author. Make connections – that is good. Intellectual synthesis is necessary, helpful and important, but **your** connections are much more significant than any that I might point out to you.

Open-minded skepticism is the correct frame of mind with which to proceed through this section – anything else may lead you astray. Do not be predisposed to confirm or deny what you read, or you will likely miss some important points and have a less than optimal learning opportunity.

75

■■■
Preliminaries

■■■

Though we have previously used the terms PMR and NPMR in a general way in previous sections, now we need to become more precise.

PMR (Physical-Matter Reality) – The reality that your body lives in, and its properties, and laws (physics). PMR includes the material universe, galaxies, solar systems, planets, and everything known and unknown that materially exists within them. PMR can be effectively and purposefully manipulated (is operationally viable) and is inhabited by beings that are sentient in it.

NPMR (Nonphysical-matter Reality) – Everything other than PMR. The non-material (from the view of PMR) superset of PMR that contains its own unique properties, materials, and laws. Like PMR, NPMR can also be effectively and purposefully manipulated (is operationally viable) and is inhabited by beings that are sentient in it.

NPMR contains unique dimensions, material, and time that consistently and necessarily follows its own rules (physics). It exists independently of PMR and may contain numerous unique local realities as subsets.

These definitions of physical and nonphysical were chosen to serve the largest group of people with a minimum of confusion. Both PMR and NPMR are real physical places from the point of view of the consciousness beings that inhabit them. They each contain objects and sentient beings that appear to have substance, form, structure, and energy that are measurable and perceivable and follow the rules, causality, and science (physics) peculiar to their respective realms. The apparent difference between nonphysical and physical is relative to the observer. From within a given local reality (a specific dimension of AUM or TBC), everything within that reality appears physical while everything outside of that reality appears nonphysical.

Defining NPMR as nonphysical is a parochial PMR-centric view that implies our physical PMR science is fundamentally incapable of expanding its vision to include NPMR as a part of our physical reality. Simply moving beyond the limited idea that PMR defines all possible reality will open up new worlds of potential existence to exploration and study. Breakthroughs must always begin with the unthinkable.

Most people are comfortable with the fuzzy notion that what we can **directly measure** (with our five senses and their extensions by our marvelous machines of science) **in PMR** is by definition physical and everything else is nonphysical. This is sometimes called the operational definition of physical reality. This places two familiar groups of things into the nonphysical camp: those that are inferred but not yet **directly** measurable (neutrinos for example), and those that are conceptual (such as wave packets, dark matter, strings as in physics theory, as well as ego, love, ethics, and justice). Inferred entities can only be **indirectly** measured by actually measuring something else that is assumed to be causally related.

Conceptual things (such as justice or wave packets) are inferred. Initially, conceptual entities are defined into existence (as were neutrinos or quarks) by our need for them to support and maintain the consistency of our current worldview. Their operational functionality and their ability to help us maintain logical consistency is accepted as proof of their **potential** physical reality. We take them seriously because they help us define, understand, and control our local reality; they are operationally significant. That they are initially operationally nonphysical (cannot be directly measured) only relegates their existence to the theoretical, not to the absurd. They are in the wings waiting for scientific experience to give them full membership into the world of the physically measurable – or at least into the world of things with physically measurable effects. Thus it is with mind, consciousness, and intuition – all operationally significant, all with indirectly measurable effects.

Nonphysical does not mean non-useful, non-real, or non-existent. Nonphysical things are often defined by (are real and meaningful because of) their effective use, their application, their interaction with measurable things, and their ability to satisfy (fit, or be consistent with) the current overarching model from which their reality, existence, and meaningfulness is derived. They are directly inferred from basic assumptions about the nature of reality.

The beings, objects, and energy in NPMR are directly inferred in the same way. Anyone can easily learn to see unique and specific energy forms with the mind's eye, manipulate the energy of creation, and communicate telepathically with beings in NPMR or PMR. However, whether or not these activities can be considered real and meaningful depends upon their effective use, their application, their interaction with measurable things, and their ability to satisfy (fit, or be consistent with) an overarching model.

Overarching models of being are in short supply. Current efforts are typically incomplete or narrowly dogmatic. Ontology (the branch of metaphysics that deals with the nature of being) is not taken seriously by many philosophers or scientists because it is considered to be beyond objective discourse. They are half right: A comprehensive model of being like the one presented in the *My Big TOE* trilogy is beyond the causal logic of the little picture, but rather than decree that no logical objective solution can possibly exist, science and philosophy would be better served by open-mindedly searching for a more general view that would broaden its understanding. Perhaps, with an open-minded approach, scientists and philosophers could find a bigger picture that contains a logical solution that answers to a higher level (more general superset) of objective causality. Unfortunately, it is easier and more immediately rewarding to decree the impossibility of difficult to understand solutions, or posit belief-based pseudo-solutions, than to live with the nagging uncertainty that is required to find real solutions through better science.

We have been struggling to understand the Big Picture in Western culture for at least 2,700 years with only spotty success because our materialistically-based belief system blocks genuine progress. Big Picture models cannot be taken seriously in our Westernized techno-culture because of a pervasive little picture objectivity *über alles* delusion. The world, and particularly the Western world, desperately needs a Big TOE to provide balance and stability.

A living theory is one that embraces change and the possibility of new and contradictory knowledge, as opposed to being an impossible to change, know-it-all dead dogma. Big Picture models of being, reality, or existence must continually be in the process of evolving Big Truth. In the same sense, most physical models (modern physics theories) are also living. Dead models, if correct, must be small in scope. A model that is broad in application and far ranging in its consequences (large scope) must remain open to the idea of redefining itself periodically or risk degenerating into a belief-based old dogma that is incapable of learning new tricks. There is always more to learn. How can we claim completion when AUM is still learning and growing? Believing that your uncertain ignorance is small compared to your certain knowledge is the signature assumption of fools.

In the simplest sense, and at the most basic level of understanding, the nonphysical being objects in NPMR are important because they are significant and valuable to individuals. They consistently, powerfully, clearly, and **measurably** can, and do, contribute to your consciousness quality and the evolvement of your being. These nonphysical being objects and processes influence (interact with) your being whether you are aware of it or not. The advantage of being aware of the interface and interaction you have with NPMR is that this awareness can dramatically raise the efficiency of the experience-knowledge-growth transfer function.

The bottom line is that NPMR, just like neutrinos, is inferred from repeatable consistent results that are indirectly measurable in PMR by independent researchers. *My Big TOE* develops the notion of NPMR as part of an overarching consistent scientific theory of reality that includes PMR and its physics as a subset.

Though this section is purposely developed from a PMR-centric view-point to enhance comprehension at a more familiar level, hang onto your larger perspective and integrate what you find here with whatever consti-tutes your biggest picture. As we pointed out toward the end of Book 2, PMR and NPMR are dimensions or subsets of one reality as seen from dif-fering points of view. The terms physical and nonphysical have no intrinsic meaning beyond describing a local belief-based delusional perception of other realities relative to one's own. There is **no** primary physical reality, just as there is **no** primary inertial frame in relativity theory. There is but one reality that can be viewed from many different perspectives. Same old story (relativity of perception) repeated at a higher level of generality.

From Newton to Einstein, physicists believed that space and time were absolute. Einstein's vision was the first to transcend the universally accepted par-adigm of an absolute (or fundamental) inertial frame. Consequently, he was able to see the relativity of, or the relationship between, multiple inertial frames within what looked like an all encompassing space-time. The concept of space-time was born of a bigger picture understanding that could appreciate the inter-connected relationships of multiple equivalent inertial frames instead of seeing reality as one monolithic absolute inertial frame. Physicists following the trail that Einstein blazed (beyond the conceptual restrictions of absolute space and time) committed the same error at the next level of generality by developing an implicit assumption that space-time was the new absolute.

When we reach the outer limit of our vision, we think we see the absolute. What we imagine as the absolute, the final source, the ultimate fundamental, is only an artifact of our limited understanding. As Einstein vividly demon-strated, one can never see the new relativity until one's vision has grown beyond the old absolute. Neither Einstein nor others could conceptualize any-thing beyond space-time. That limitation of vision prevented them from seeing the next higher level of relativity where one can appreciate the interconnected relationships of multiple equivalent space-times instead of one monolithic absolute space-time (multiple PMRs). The next step beyond that (the next higher level of relativity) requires a larger vision where one can appreciate the interconnected relationships of multiple equivalent realities within AUM (multiple PMRs and NPMRs – from our perspective) within a consciousness system like $NPMR_N$ instead of one monolithic absolute set of virtual space-times. Extending one's vision to the next higher level of relativity one can appreciate the interconnected relationships of the multiple levels and scales of a consciousness-evolution process-fractal that defines the evolving consciousness system of which we are a part instead of one monolithic set of hierarchical virtual realities based upon an absolute AUM.

Undoubtedly, describing five hierarchal levels of relativity theory spanning from Newton's absolute space and time to the finite AUM system of evolving digital consciousness fractals does constitute a rather big conceptual bite for one paragraph. Leaping four major reality paradigms in a single bound where each successive level of cosmological understanding must discover its own theory of relativity to enable the next higher level of organization to come into view is the intellectual equivalent of eating a large box of cream-filled doughnuts in one sitting.

I hear some of you asking, do we ever get to the bottom, to the absolute, fundamental core within? When do we reach the end of the series where we can feel satisfied that we know it all so we can kick back and relax? The short answer is: We (you and me as we play together like fun-loving bacteria in the PMR Petri dish) don't. It appears that we only get to chase that white rabbit, not catch it.

Absolutes are enticing and comfortable because their sweeping certainty produces the illusion of an immensely broad and deep knowing – whereas relativity, dependent upon a specific perspective or framework, reminds us of the limitations of our knowledge and experience. It is difficult for many to see themselves as a little guy in a local reality. Absolutes give us a way to skate over the nitty-gritty details of a more complex existence and at the same time feel good (as well as smug) that we have a clear view all the way to infinity. Whenever one's conceptualization of reality (or of anything real) rests upon the absolute, the infinite, or the perfect, one should question the underlying assumptions very closely and expect to find a belief at the core. So, our mission, if we choose to accept it, is to iteratively smash each old absolute with a less constrained vision of a new relativity, a new set of relationships that define a bigger picture that better explains the whole of our direct personal experience.

Are we caught in an endless iterative loop of ever increasing bigger pictures? Nah! "Endless" sounds like another one of those bogus infinity scams. You are not in an endless loop; you are just stuck with your limitations. Our consciousness and our virtual PMR are both defined into existence by the constraints placed upon us. Limitations are a natural, unavoidable part of the fundamental existence and experience we call reality. If one thinks being small like a bacterium sucks, it is probably because one has delusions of grandeur and is in denial about one's designed-in inability to be omniscient. Evolution generally runs a very efficient organization and being omniscient simply isn't required for you to **fully** achieve your present purpose. Don't sulk; instead appreciate the beauty of the idea that PMR and NPMR are simply different perspectives of a conceptually elegant unified view of reality as an evolving fractal system of digital consciousness that models beings, objects, and energy in order to pull itself up by its bootstraps into a more profitable state of organization.

▶ The definition of beings, objects and energy are the same in PMR and NPMR. In PMR, physical objects and energy form the set of all non-sentient things. Let's revisit the definition of sentient. The dictionary defines sentient as: "Having sense perception; conscious; experiencing sensation or feeling." I will again generalize this official PMR definition by saying that sentient entities are self-aware and aware of their environment. Additionally, sentient entities make deliberate choices that have significant consequences to themselves and perhaps to others. (Recall that the origins and evolution of self-awareness are discussed in detail in Chapter 25, Book 1.)

The upper echelons of sentient things are often called beings, though technically everything sentient is a being. In NPMR, objects and energy are defined in exactly the same way. Objects and energy within NPMR represent all the things in NPMR that are not sentient in NPMR. Thus, the phrase: "beings, objects, and energy" includes everything when applied to either PMR or NPMR. The beings are the choice makers, the objects includes all other things, and the energy state relative to a set of objects is neither a being nor an object but has the ability to effect changes in both.

Most of the significant (within the Big Picture) changes that occur between time increments within a given reality are driven by the interactive free will choices of the beings who inhabit that reality. Objects (such as rocks, mountains, clouds, thermal geysers, or rivers) and interactive forms of energy (storms, volcanic eruptions, or earth quakes, for example) are very predictable when one has **all** the data.

Beings, objects, and energy do, however, sometimes have random components associated with their choices. Some degree of randomness (usually relatively small) may influence what happens next. ◀

Our practical (from a PMR viewpoint) definition of physical and nonphysical reflects the common separation (particularly in the West) between mental and physical – mind and body – even though it is obvious to almost everyone that these two opposites are hopelessly entangled and integrated. PMR and NPMR are likewise entangled. The **directly** unmeasurable (within PMR) mind, consciousness, thought, and soul clearly fall into the nonphysical (from a PMR viewpoint) camp regardless of whether or not there is a presumed physical basis.

In summary, our operational definition of PMR causes some of the common things taken seriously (believed in and relied on) in physics – such as neutrinos, quarks, and string theory – to belong in the nonphysical camp (only inferable, not directly measurable) yet, we accept them because they help us remain logically consistent and represent the current best fit within (are sometimes required by) our model of physical reality.

We can build useful and valuable material, conceptual, theoretical, and spiritual results based on the knowledge and understanding of inferred nonphysical objects. Successful operational experimentation with nonphysical phenomena that are not acceptable to traditional science (such as energy and information transfers within NPMR or PMR that appear to the PMR observer to be acausal) often employs the methodologies that physicists use to work with inferred objects such as neutrinos, as well as the processes that scientists use to develop more purely conceptual objects like wave packets or strings. Science is science; many of the same logical approaches to discovery work as well in NPMR as they do in PMR. The major difference is that NPMR objects are not accepted by traditional scientists because they do not fit the little picture PMR reality model. For the most part, this is because the type of data gathered from NPMR and the process for obtaining and validating them is, by its nature and design, mostly incompatible with PMR's traditional narrowly focused scientific methodology which is based upon a mystical belief in the universality of PMR causality. (See Chapters 47 and 48 of Book 2, and Chapter 2 of this book for supporting detail.)

Conflicting with the nineteenth and twentieth century's most precious intellectual jewel, the scientific method, is not necessarily the kiss of death to an open mind. The scientific method is a terrific tool within the realm to which it applies.

Our most successful scientific models and processes have turned out to have a limited validity, including classical mechanics (large slow objects), special relativity (fast objects, but no gravity), the flat earth (short distances), or a stable physical universe (short times). Likewise, the scientific method has a limited validity. Our refusal to appreciate those limitations has trapped our awareness in PMR. The time is ripe, once again, to raise our thinking up to the next level of generality, thus greatly expanding our vision of what is possible.

An expanded view of a scientific method that contains the traditional scientific method as a subset would also better encompass and support some presently accepted fields including medicine, psychology, sociology, and ethics, as well as the nontraditional, but equally serious, fields of cosmology, epistemology, and ontology. For example, when was the last time the Centers for Disease Control and Prevention (CDC) performed an objective scientific public health **experiment** with a highly contagious deadly infectious disease? Obviously, the CDC cannot apply a traditional scientific methodology to the study of fatal infectious diseases. Instead it must rely on indirect statistical inference – the same methodology that is often used to produce an objective evaluation of subjective experience.

A generalized scientific method must be rigorous and testable, as well as produce consistent and meaningful results; the requirement for testable rigor and meaningfulness in no way eliminates the value and usefulness of subjective input.

Formally defining a practical generalized scientific method is not within the scope or intent of this trilogy, but an informal description of what needs to be included is discussed in the previous chapter and in Chapter 47, Book 2.

Your attitude is important. Avoid getting twisted around the preciseness, accuracy or the appropriateness of how the definitions and observations within this chapter and the next apply to your particular view of, and beliefs about, the world. Keep in mind that we are trying to effect a communication of big ideas. Some fuzziness among the details is acceptable; precision and clarity comes only with personal experience in the larger reality. Jettisoning belief and trying to solidly grasp what lies beyond your experiential reach is what pulls you up by your bootstraps, but overreaching coupled with high expectations typically leads to more frustration than progress.

Again, do **not** feel **required** to believe or disbelieve or to pass final judgment on what you read. On the other hand, **you are required** to be scientifically (not driven by belief or fear) critical and skeptical of what you find here, or **your** Big TOE will not grow.

76

■■■

Some Observations

■■■

These thirteen observations reflect my experience exploring NPMR and provide a short summary of the data necessary to support the conceptual development of the Mechanics of Reality. They are written here as unsupported blunt summary statements. Sections 2, 3, and 4 of *My Big TOE* thoroughly develop the logical foundation upon which these thirteen observations rest.

If your personal experience does not clearly encompass and support these statements, consider them to be stated assumptions that may or may not be critical to what follows in the next nine chapters. Keep in mind that belief or disbelief in the correctness of these observations is at best irrelevant, while a skeptical open mind is absolutely necessary. If, on the other hand, your personal experience validates these notions, they will appear to be simple facts that stand on their own requiring neither belief nor conjecture for support. Regardless of the level of experience you bring to this discussion, my advice is to trudge on through this chapter without being too fussy. The supporting details are contained in the first two books of this trilogy and will not be repeated here; this is simply a terse summary and review of what has been established elsewhere within *My Big TOE*.

1. PMR is not the only, and not necessarily the primary, nor the most important reality. Each reality exists within its own calculation-space or dimension. All realities are not necessarily independent. There can be connections and dependencies between realities along with rules governing their interactions.

2. As shown in Figure 5-1, there are multiple, loosely related, weakly interacting reality systems within NPMR (each $NPMR_n$ where $\mathbf{n}$ = 1, 2, 3... represents a specific reality system). Each of the $NPMR_n$ seems to be

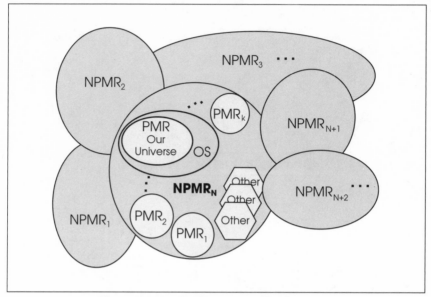

Figure 5.1: Reality Systems: The Big Picture

organized as separate ongoing experiments in consciousness. Each experiment has its own objectives, protocols, conditions, timeline, and purpose.

3. One of these reality systems within NPMR, say $NPMR_N$ (where the index **n** takes the specific value N), contains our PMR as a subset. $NPMR_N$ is a spiritual (quality of consciousness) system experimenting with the evolutionary **potential** of consciousness. $NPMR_N$ contains many other reality subsets besides PMR. Some of these other subsets constitute physical universes similar (space-time-matter based) to our PMR (we are nonphysical to the beings of these realities) while others are very different from our PMR (not based on space-time-matter). Within each of these reality subsets, there are types and classes of beings living within the rules of their own physics, psychology, and sociology. Our System (OS) of reality contains PMR and a portion of $NPMR_N$. Figure 5-1 shows only one small portion of the larger consciousness ecosystem.

4. In $NPMR_N$, the goal for each sentient being is to learn, grow, and evolve through interaction (making choices) with others: The intent driving our choices can range from good to evil. We make free will choices in the process of interacting with other sentient beings, and the quality of these choices alters the entropy of our consciousness through a self-modifying feedback system. Personal growth is required to effect our entropy reducing contribution to the larger consciousness cycle.

To optimize entropy loss, one must understand how consciousness evolves under varying conditions. Does all consciousness eventually end up love-like (the optimal result) or exactly the opposite; or does a consciousness system eventually become chaotic or perhaps find some stable equilibrium state in between these extremes? Is it possible to isolate and stabilize the optimal conditions for growing or evolving high quality consciousness, thus optimizing the efficiency of the consciousness cycle?

5. All consciousness in $NPMR_N$ is "netted." It is as though each entity in $NPMR_N$ has a unique URL (Uniform Resource Locator or web address) in a gigantic, but finite, Reality Wide Web (RWW). Thus, as a subset, all beings in our particular PMR are also netted together – on a Local Area Network (LAN) so to speak – as well as connected (connections between LANs) with the rest of OS and all other $NPMR_N$ residents. Connections to the $NPMR_n$ and beyond are likewise available, but not particularly important to this discussion.

6. An individual in PMR can be aware of NPMR through his or her mind – the nonphysical part of one's being. The mind is the doorway through which you must pass to experience NPMR. NPMR is, if anything, more real than PMR (semantics in the service of an impish smile) because it contains a more fundamental (less constrained) reality of which PMR is only one of many subsets.

7. An individual in PMR, once fully aware in $NPMR_N$, can access and filter the RWW net described above with his or her intent. An individual can access any web-page-entity on this RWW net by using his or her intent (which requires unique knowledge and identification of the URL being sought). One navigates with one's intent. A person's intent is also his or her filter; without this filter one would have difficulty dealing with all the information at once.

Explorers must learn their way around by exploring the RWW – for the same general reasons that force you to personally explore the WWW (reading about it and being an interactive part of it are two entirely different experiences). Experience, carefully filtered by a scientific method (repeatable evidential data collection) that sorts external signal from internal noise, leads to the knowledge that forms the basis for a larger set of productive and meaningful possible intents. Those who wish to access information, objects, or entities on the RWW consciousness network need to educate themselves within and about $NPMR_N$. Carefully validated firsthand experience must be acquired through a step by step exploratory process.

8. Individual communications between sentient entities are telepathic. Additionally, within a given PMR, communication using sensory data (such as sight, sound, touch, smell, and taste) typically functions in parallel with telepathy to some degree.

9. People in PMR who are unaware (or mostly unaware) within NPMR, can use and experience this net if they are connected strongly enough (large enough signal-to-noise) and with enough clear intent (interest) to produce a crosstalk or bleedthrough into their consciousness (typically perceived as intuition or precognitive dreams). Signals come from NPMR with varying amplitudes while an individual produces most of his own interfering noise.

10. Telepathic communications and other communications through the $NPMR_N$ net are not affected by the **apparent** distance or volume between the sender and the receiver, nor limited by light speed. Communication content flows across dimensional barriers without difficulty.

11. Beings in all PMRs within $NPMR_N$ have portions or extensions of themselves that also (simultaneously) exist in $NPMR_N$. The body in PMR is essentially sustained through its connection to $NPMR_N$. By manipulating energy (the ability to reorganize bits) in $NPMR_N$, one can affect the body in PMR. Thus follow many effects from mental or psychic healing to voodoo. The energy body in $NPMR_N$ can be viewed from various "frequencies" (like looking at something with X-ray, infrared, or ultraviolet light) to reveal different aspects of itself. (I apologize to all the scientists and engineers for using the frequency metaphor, but it is the best analogy I can come up with.) The energy body is easily manipulated by intention.

12. Intent seems to have the qualities of magnitude (intensity), frequency (is tunable), and clarity (signal-to-noise ratio).

13. The energy body in $NPMR_N$ does not die when the physical body dies; it can manifest another body in PMR (or some other suitable reality) if it needs or wants one.

77

■■■

A Model of Reality and Time
Incrementing Time in Simulations

■■■

To understand reality one must understand time. During the next six chapters, we will use the notations "delta-t" and "DELTA-t" extensively. Because this notation is borrowed from the field of mathematics, it may seem strange to some, but do not be put off by that. This chapter provides an explanation of incrementing time in simulations to the mathematically challenged and introduces a unique perspective on the nature of time to the Big Picture reality challenged. The words "delta" and "DELTA" represent the lower and upper case Greek letter of the same name. They are spelled out to avoid using abstract symbols that might inadvertently trigger mathephobia or other related mental techno-blocks. This is easy – you'll see.

Traditional mathematical notation places the Greek letter delta next to a variable (some quantity that changes) to represent an increment (small change) in that variable. I use it here because many people are familiar with this notation. If you are not, don't worry, the concept is simple and explained in detail below. "DELTA-t" and "delta-t" are simply names for two different increments (small chunks) of time.

In an iterative dynamic simulation, such as the calculation of the position of a fired artillery round (or a thrown ball) as a function of time, one starts with the equations of motion (equations giving position as a function of time) and the initial conditions at time t=0. The first time through the computational process loop, one lets t = (delta-t) and then calculates position – next time through t = 2•(delta-t), next time through t = 3•(delta-t), next time through t = 4•(delta-t), and so on. You calculate a new position of the object (artillery round or ball) for each time t, which is one delta-t larger than the previous value of t. Consequently, time, in your calculation of position, progresses forward by increments (small discontinuous jumps) of delta-t.

Your simulation can approximate continuous time, and thus continuous motion, by making the size of delta-t very small. The cumulative sum over the delta-t is called "simulation time" because it drives the dynamic simulation, (as opposed to "real-time" which is what is measured by the clock on the computer room wall).

If the equations were complex enough and delta-t were small enough and the computer slow enough, it could take several hours of real-time to progress the artillery rounds trajectory through only a few seconds of simulation time (perhaps only a few hundred feet of a much longer trajectory) within the computer simulation. On the other hand, relatively simple equations in a very fast computer using a larger delta-t might simulate a one minute long trajectory (such as the trajectory of a large artillery round) in only a few tenths of a second of real-time.

The simulation can be paused any amount of real-time between consecutive time increments and subsequently started up again without disturbing the results of the simulation. These concepts involving notions of dynamic simulation and time will be important later on. It is good to understand them now while the context is simple.

Consider a large simulation that contains smaller simulations within it. For example, let's imagine a simulated war containing several simultaneous interactive battles. Such a system simulation may contain a subset of code (subroutine) that simulates a specific battle, which has within it another iterative subset of code (lower level subroutine) that simulates a specific armor battalion's activity within that battle, which has within it yet another iterative subset of code (even lower level subroutine) that simulates an artillery shell fired by that armor battalion. We now have four nested (one inside the other) levels of interdependent subroutines (1-overall war, 2-specific battles, 3-specific battalions, 4-individual artillery trajectories) iterating their way sequentially to produce the simulated war.

The collection of algorithms that describes the interdependencies between levels (specifies the assumptions, initial conditions, tactics, rules of engagement, and describes the performance of each subsystem and component) is said to model or simulate the war. A properly integrated collection of such models, along with the structure of the code that orders and propagates interactive logical events, passes data, collects results, and increments time, defines the overall war game or simulation. This is how computers are used to simulate a set of interactive activities or a dynamic process.

All the interdependent loops are iterated sequentially. Loops that have no interdependency may be incremented in parallel (simultaneously). Each level and each process within each level progresses its own dynamic activity by one delta-t at a time. As activities are completed (decisions made, trajectories flown, damage assessed, troops and equipment moved, and ammunition depleted) information is passed back and forth among the four levels. The results and implications of this information are used to make choices, continue or modify processes, and keep score. Thus, as the simulation time progresses delta-t by delta-t, the simulated war grinds on.

The practical size of delta-t depends on the required accuracy and the speed of the fastest elements in our model. A typical delta-t might be 0.00001 seconds if the precise locations (to within a centimeter) of things such as artillery rounds and missiles were important. In this simulation, delta-t defines the **fundamental quantum of simulation-time.** If the fastest thing in our war simulation (an ancient war perhaps, or a modern one where all the gadgets don't work) was a man running through the bushes with a spear, a delta-t of 0.1 second would probably be small enough.

Would simulating an ancient (spear throwing) or slow moving war (requiring a time base of no more than delta-t = 0.1 second) with our modern war simulator (that ticks off time in 0.00001 second increments) be a problem? No: There is no problem as long as we are simulating something that can be adequately modeled with time increments that are **equal to or larger** than 0.00001 seconds. It would make no sense to simulate the position of a running man every 0.00001 seconds. It might allow for a more detailed and complex model of the man (including blinks, burps, hair motion, progressive cell damage, and twitches), however, a man's movement within that small of a time increment (0.00001 sec) is completely inconsequential to the war.

What we would most likely do is use our previous coarse or low fidelity man-movement-model and increment it by a single 0.1 second increment every 10000 (10^4) ticks of our simulation time clock. Our war game's fundamental quantum of simulation-time is still equal to 0.00001 seconds or 10^{-5} sec. We are now incrementing our man-model in 0.1 second increments as a subroutine within a larger model (outer loop) that iterates every 0.00001 seconds. This will turn out to be a crucial concept later.

This more efficiently incremented, low-fidelity man-model (it remains detailed enough to define troop movement as a function of terrain) is purposely designed to be only as accurate, and incremented only as often, as it needs to be to serve the purpose that is intended by the overall simulation. Likewise, the loops and subroutines that represent various players in the simulation are provided the dynamic granularity (increment size) that their function requires. The one with the smallest required increment serves as the metronome (the fundamental quantum of time within our simulation) for all the others.

Let's put this all together into one big war simulation. We will have man-model subroutines that are incremented by one tenth of a second every 10^4 increments of our fundamental simulation time. We will also have tank-model subroutines that are incremented by a hundredth of a second every 10^3 ticks of the fundamental clock. We might have aircraft-model subroutines that are incremented by a thousandth of a second every 10^2 ticks of the fundamental clock and missile subroutines that are incremented every third tick of the fundamental clock, and perhaps a nuclear damage propagation subroutine that will be incremented (evaluated) every tick of the fundamental clock.

In this simulation, the master loop or simulation driver is the one requiring the smallest time increment. The guy in charge (outermost loop) is the one with the smallest time increment. The local time inside the man-model-loop jumps ahead one tenth-second at a time. The low fidelity simulated men who live in that local reality measure a **quantum of their time** as one-tenth second. **From their perspective**, real-time accumulates in increments of one-tenth second. One-tenth second is their local time quantum. If they could become sentient and learn to program, they could decide to simulate the growth of their hair and toenails because these functions change slowly compared to their time quantum.

Let's summarize the most useful ideas that have been generated by this discussion of dynamic simulations. Some of the time loops, subroutines, or dimensions of our big picture (entire war) simulation are iterated at faster rates than others. **Apparent** real-time is relative to each loop's perspective or local reality.

Within a given **local** reality, we can **only** perceive events which produce effects that are significant over one **local** time quantum or more, thus rendering the activities of faster loops (smaller time quanta) invisible and incomprehensible. Within the hierarchy of causality (simulations within simulations or dimensions within dimensions), the master or outermost loop that drives everything else is defined as the one with the highest fundamental frequency (highest sampling rate or rate of iteration), which is the same as saying it has the smallest quantum of time.

Bells should be going off in your head relating this discussion of computer simulation to the discussion of AUM's fundamental frequency and time quantum, and the differing sized quanta of time in NPMR and PMR

(Chapter 31, Book 1). Each higher level of simulation, with its smaller time quantum, represents a larger perspective, a base of authority and control, and collects, processes and synthesizes the activities and results created by its lower level (inner) subordinate simulations. Dynamic timing loops within loops within loops, all interconnected and building one upon the other at various levels of scale. Do you notice the fractal-like characteristics of time loops within your dynamic reality? Do you get a glimmer of how everything is interrelated and interconnected within a generalized dynamic consciousness-evolution fractal ecosystem?

You will see later that what I define as delta-t represents the outer (controlling) loop of **our immediate** reality within OS. delta-t is used to calculate probable realities and is referred to as simulation time; it is related to time in NPMR. On the other hand, DELTA-t, a larger time increment, is defined as the increment that accumulates our PMR time, our apparent real-time. From a larger perspective, DELTA-t drives a lower level simulation (with its larger time increment) incremented only once every so many ticks of the overall simulation time clock within The Big Computer. Did I infer that you, me, and our entire universe, are simulated beings and objects? Yes: you will see that it is illuminating, consistent, and useful to model our reality that way. Throughout *My Big TOE*, I have dangled this idea of a simulated reality in front of you; as strange as it may seem, it will make more sense later on.

Recall that in Chapter 31, Book 1, I explained that time is quantized, meaning that time progresses by discrete increments rather than continuously, and that our time is an artificial construct created by AUM to define the space-time part of itself. In this section, you will see how quantized time loops allow us to create our reality, maintain a living history, and make choices that enable us to learn and to grow the quality of our consciousness (evolve our spiritual quality toward satisfying the goals of our existence). A digital reality offers up many interesting attributes.

You will soon see that time is not a fundamental property of the thread that is woven to produce the fabric of reality, but instead is merely a measurement construct, a tool for implementing organization and defining patterns, more related to the action of the machine that does the weaving than to the thread itself. Each cycle of the loom represents another time increment as thread after thread is added to produce the seemingly continuous fabric of our PMR 3D space-time experience.

The action of the machine, the process of weaving, should not be confused with the three-dimensional experience of the space-time fabric it creates. Nor should the fabric be seen to weave itself through some spontaneous mystical or magical process wherein time is created in the present moment **without** the need of factory or loom or the energy that makes them run, much less the design and purpose of the process. That space-time is spontaneously self-created from nothing – a self-woven, 4D fabric containing all past, present, and future events – is basically what most scientists believe these days because they cannot perceive the higher dimensions where the loom, mill, and the energy that runs them resides. In fact, our present science is based on the fundamental implicit assumption that the loom and mill cannot possibly exist, or be relevant, because they lie beyond our direct objective 3D perception of a 4D space-time.

Thus, today's scientists have painted themselves into a corner dependent upon reality mysteriously creating itself – a **mystical** belief-based concept they are greatly dissatisfied with and do not like to talk about – a fundamental failure of scientific understanding long swept under the rug of objective respectability.

On the other hand, the theists are content, as they have been for 10,000 years, to offer up their one pat answer for all situations and occasions: God does it. Meanwhile, the fabric of space-time continues to apparently weave itself out of nothing as we stumble in circles in pursuit of a Big Picture that we can somehow miraculously extract from our little picture. And that is where we are today, ladies and gentlemen, and where we have been for the last forty years since general relativity and quantum mechanics dropped the first shoe (told us in no uncertain terms that our physical reality was delusional). Einstein tried to lead us out of that wilderness with Unified Field Theory but could not find his way through the impenetrable cultural belief fog that obscured every avenue of escape. Belief blindness is as absolute as any blindness.

That a higher dimensional structure may not leave **physical** footprints is not that difficult to understand. Einstein was correctly looking for non-physical footprints in his Unified Field Theory where mass was nothing more than higher intensity field strength. His error was looking for continuous fields within the space-time construct. He did not grasp the digital nature of reality – that space and time are quantized – and that space-time itself was only a local phenomenon, a virtual little picture reality dependent upon a more fundamental digital energy field called consciousness. He did not understand the primacy of consciousness as the fundamental substance, energy, or organization underlying existence. Instead of seeing space-time as primordial physics, a set of relationships and definitions, a rule-set, a construct of consciousness, he thought that space-time itself was the fundamental field. Consciousness is fundamental. Space-time is derived from a conscious intent to constrain individuated subsets of consciousness to a specific experience-base which we call PMR.

Even if Einstein had figured that out in the 1950s, his peers would have written him off as having lost it. He would have most likely ended his career in ridicule. Today we are much more familiar with the potential of the digital sciences. Perhaps now, in the twenty-first century, scientists will have the requisite vision to see and understand the paradigm shifts that are required to support a unified theory of reality. Perhaps non-scientists will recognize the Big Picture first and eventually bring the scientists along.

Tracking the nonphysical footprints of a more fundamental reality through the consciousness wilderness is what *My Big TOE* is all about – and you do not need to be a physicist or mathematician to get it because the details and explanations are not little picture logic puzzles that exist only within PMR. In the Big Picture, a deep understanding of reality is not the exclusive preserve of the scientist and mathematician: That state of affairs, where scientists are the high priests of reality, is a little picture phenomenon only. When it comes to understanding the Big Picture, there are no academic prerequisites. You do not need to wait for the science guys and mathematicians to lead this parade – march on to the beat of your own experience and inner knowing and they will eventually follow your lead.

It is as though the objects in the simulation (local AI Guys within each subroutine) have become sentient but can't perceive outside their time loop level and consequently, are oblivious to (or vehemently deny) the bigger picture. They live in their limited universe paying homage to the Loop Gods who occasionally provide fresh input data and to whom they offer up the results of their efforts.

78

■■■

A Model of Reality and Time
The Big Computer (TBC)

■■■

We have previously discussed the likely origins of an immense digital computational capacity based upon binary reality cells that are the fundamental constituent of aware consciousness or mind. In Section 2 (Chapters 26, 27, and 28 of Book 1) you learned how and why AUO would naturally evolve binary (distorted vs. non-distorted) reality cells in support of organizing and exploring (in the evolutionary sense) increasing complexity and developing awareness. You also learned how AUM would further refine its computational and memory part to develop intention, move from dim awareness to bright awareness, and to implement and track its *gedanken* experiments in consciousness.

Awareness cannot grow, progress, learn, or evolve past a very rudimentary level without the organizing influence of time. The words "learn" and "grow" imply the passage, or incrementing of time – there must exist a before state and an after state if awareness, or any system for that matter, has the potential to learn or grow. The notion of the possibility of change is dependent upon the existence of at least a rudimentary time. As you will soon see, our discrete (quantized) time-based reality has evolved (has been designed) to provide us with an opportunity to learn, and to provide AUM with optimized results from its *gedanken* experiments concerning the optimization of evolutionary process.

The complex configuration of digital energy (organization) and self-organizing process we call consciousness forms the foundation of reality in the form of an interactive consciousness-evolution fractal ecosystem simulation where what is actual and what is simulated are one and the same.

You can conceptualize TBC as a tiny piece of the computational and memory part of AUM that is used, among other things, to actualize and maintain the reality of OS and its offspring. For the purpose of explaining consciousness, quantized time (past, present, and future), time travel, free will, rule-sets, space-time, precognition, telepathic communication, remote viewing, ego, love and much more in this model of reality, I evoke only those attributes of this celestial computer that our PMR computers have. In other words, this super computer in $NPMR_N$ that I have assumed into existence primarily differs from the functional computer on our desktops by size (memory), computational speed, reliability, and reasonably good software – that's all.

> ▶ This aside is for all the computer geeks in the reading audience who are waving their hands in the air and have troubled expressions on their faces. Others can skip to the end or take a short break until the computer wizards catch up.
>
> It may make more sense to the computer literate if they imagine that TBC is composed of fully netted, massively parallel arrays of clustered processors with integrated, shared, and stand alone super fast memory running within a highly efficient self-evolving AUM-operating-system that is beyond all imagination. In other words, if you can imagine that the AUM-digital-consciousness-thing is completely beyond your imagination, you have taken the first step toward understanding the depth of what you don't know.
>
> In an operational sense, one might say that AUM **is** the operating system as well as the computer and the application software. This AUM-thing is an almost infinite (from our perspective) living, evolving, conscious, sentient computer and operating system that develops its own software. At a minimum, it inhabits as well as defines and creates all the digital mind-space (reality) that we have the potential to even vaguely comprehend; other than that, it is an inconsequential piss-ant.
>
> When I say that TBC is like the computer on your desktop, I am referring to the fundamental functional building blocks of binary cells, memory, instruction-sets, processors, data flow mechanics, and operating system control, not the specific design of these things or the system architecture. Think of a computer in the most generic terms possible and take care not to limit your imagination by what you presently know about computers. ◀

TBC should not be thought of as the big mainframe in the sky, or as the OS Department server within N Division, but as a metaphor describing the fundamental nature of the process and functionality (computation, memory, and rules) required to implement our reality.

TBC is a real thing; we relate to it more by its process and functionality (discussed in this section) than by its construction and design (discussed in Sections 2 and 4) – the same way most of us interact with our desktop computers. We don't relate to chips and basic chip-level instruction-sets, but instead see our computers in terms of their functionality and the processes they implement. We are primarily interested in the processes we need to follow to optimize our interaction with them.

The most surprising thing is that TBC, the implementer of all reality within $NPMR_N$, represents nothing mystical, no magic, just a huge (but finite), fast, mundane number cruncher running excellent, but not necessarily perfect, software. The mundaneness of TBC is, by itself, remarkable and interesting. Digital processing is digital processing however cleverly or efficiently it is implemented.

The software and operating system might reasonably be construed to represent the intent of AUM, or one may understand it as the result of evolutionary pressure working upon a growing, experimenting, evolving AUM-consciousness-system-being-thing exploring all its possible states. TBC, dedicated to $NPMR_N$, represents only one small part of AUM's computational part – other reality systems have their own computational and memory resources. All are interconnected because they are of one consciousness. We use the terms AUM, EBC, and TBC to express differing levels of digital functionality, however, at the root, they are not separate parts. Imagine a single evolving digital system serving multiple functions simultaneously.

Trying to differentiate between the larger consciousness system and yourself can be misleading – the boundaries are only functional. AUM, TBC, PMR, and you are all manifestations of a single consciousness. Each differs by function and purpose and is circumscribed by its specific constraints, yet all are of one continuous consciousness – like the lumps in a sheet.

Though the greater consciousness ecosystem is vast and ranges far beyond our view, there is much that lies within the reach of our direct experience and understanding. Humans have barely taken the first baby step toward knowing what is within our ability to know. As we continue to play our part in the larger consciousness cycle of entropy reduction, it is our right as well as our duty, to explore everything between where we are today and the far edge of our potential – to become aware and active participants in the Big Picture and full partners in the evolution of ourselves and AUM.

79

■ ■ ■

A Model of Reality and Time
Probable Reality Surfaces
The Simulated Probable Future
Real-time, Our System,
State Vectors, and History

■ ■ ■

1. Probable Reality

Let us begin this discussion of time and the mechanics of OS (our collective Big Picture **local** reality) in the PMR present and slowly work toward a more generalized concept of past and future. Join me here in the present moment and let us see what is happening in $NPMR_N$ to support our sense of reality-present. The Big Computer (TBC) has captured in its database the state of being of OS at this moment. This includes all the objects and energy in the universe as well as all the significant choices that all the relevant sentient beings within OS have at this moment. Later in this chapter we will more specifically, and in more detail, define the set of information that specifies this state of being or "state vector" of Our System (OS).

TBC can now compute everything that could possibly happen next (we will explore this thought more thoroughly later). Additionally, it has accumulated a history file of past behaviors relative to similar choices and can thus compute the likelihood of occurrence (probabilities) of each of the possible things that could happen next. Many of the current choices are dependent on a likely interaction with the choices that all relevant and significant others (including themselves) made the moment (or many moments) before. All possible interactions are defined and evaluated with respect to all possible choices and arrangements of objects and energy, as well as against a complete set of history-based likelihoods (expectation values).

During the time between successive increments of PMR's quantum of time (DELTA-t), TBC has computed OS's probable future – what OS will probably be like during the next **M** (**m** = 1, 2, 3, ...**M,** where **M** is an integer) iterations of DELTA-t. Thus within TBC, the dynamic OS has been simulated and its future state, the one that will most likely appear during the next DELTA-t (**m** = 1), has been predicted based on the results of the present state of OS after the last DELTA-t. This OS simulation is run again (**m** = 2) with the results from the previous predictive simulation (**m** = 1) used as input, and the probable outcomes and expectation values for the following DELTA-t are predicted as output.

Each successive output (predicting the state of OS out into the future one more DELTA-t) becomes the input for the next predictive calculation. This process is continued **M** times until TBC has progressed the model of OS out as far in time as it finds useful. The probable OS state vector generated after each iteration (for each value of **m**) during the dynamic simulation of what is most likely to happen in OS during the next DELTA-t is saved in TBC. Remember that we are doing all **M** iterations between actual increments of DELTA-t.

As displayed in Text Box 5-1 below, the iterative process would operate in the following manner. First, a new DELTA-t increment is initiated resulting in the initiation of a new OS state vector. Then, all free will choices and material and energetic changes that define the activity that creates this new OS are made. The choices and changes, once made, define a new and unique state vector (or more simply, "state") of OS that is associated with

---

1. DELTA-t is incremented.
2. Free will choices, material changes, and energy changes are made, defining a new OS state vector associated with the current DELTA-t.
3. The new OS state vector is compared to the previous one and to the predicted one.
4. All actualized changes between the new and previous OS states are recorded. Predictive algorithms and databases are updated and improved.
5. TBC calculates M sequential probable future states of the new OS by running a delta-t sub-loop.

Text Box 5.1: The OS/DELTA-t Loop

this particular DELTA-t. TBC now compares the new OS state vector to the previous one and records the actualized changes. It also compares the newly actualized state to the predicted state and makes the necessary adjustments required to improve the accuracy of future predictions.

Next, TBC calculates M potential future states (of the new OS state). These M calculations (made sequentially one value of m at a time) project (simulate) what is most likely to happen during the next M increments of DELTA-t. The first ($m = 1$) potential future state (probable reality) of OS is computed in TBC based on the latest actualized (actually happened) input data generated by the choices made during the current state of OS.

This predictive simulation of the state of OS, which progresses by iterating m from 1 to M, creates a set of sequential probable realities describing the probable future of OS. The subroutine or iterative loop we are using to generate future probable realities of OS must, because it is a dynamic simulation tracking changes, operate on its own time base, and that time base must utilize a much smaller time increment than DELTA-t. This smaller time increment, which we will call delta-t, is associated with a quantum of time in $NPMR_N$. Utilizing a time increment (delta-t) that is very much smaller than the PMR quantum of time (DELTA-t), allows us to model the dynamics of OS (which is a subset of $NPMR_N$) in enough detail so that we can predict the most likely state of OS for each value of m. Thus, delta-t is the fundamental quantum of OS simulation time.

The OS state vector simulation runs through its internal calculations using the smaller $NPMR_N$ time quantum (delta-t) until it eventually converges to a predicted future state of OS for every value of m. It should be clear that the time increment DELTA-t is composed of or contains some large integer number of $NPMR_N$ time quanta delta-t, and that the computation of the probable future of OS through M successive generations occurs between successive increments (DELTA-t) of our PMR real-time.

Now that a set of M successive generations of probable future realities has been determined, TBC next records the entire OS state vector representing all significant possibilities and probabilities existing within OS. This step is discussed in further detail in Topic 9 of this chapter. The final step of the OS DELTA-t loop iterates the process by returning the loop back to the first step. DELTA-t is incremented again, and the entire process is repeated for the new DELTA-t.

Let's back our perspective out one more level for a peek at an even bigger picture. Though it lies somewhat beyond our immediate perception, contemplate the concept that delta-t is incremented only after so many ticks of a smaller, more fundamental time increment. We know that delta-t is a small time-increment ($NPMR_N$ time base) used to simulate what is most likely to happen in OS during future DELTA-t time increments. It is used to simulate probable future states of OS and to increment a larger DELTA-t (the OS DELTA-t loop). Furthermore, consider that delta-t is incremented only after so many ticks of a smaller time increment that is used to simulate probable future states of $NPMR_N$ (the $NPMR_N$ delta-t loop).

Because NPMR is an outer loop to $NPMR_N$ (where OS lives), it makes sense that NPMR runs on a smaller time quantum than $NPMR_N$. Thus, just as the OS DELTA-t loop must have the smaller delta-t as its fundamental quantum of simulation time, the $NPMR_N$ delta-t loop must likewise have a smaller time increment than delta-t as its fundamental quantum of simulation time. Keep in mind that NPMR is a superset of $NPMR_N$; also that $NPMR_N$ is a superset of OS and that OS is a superset of our PMR (OS is comprised of our PMR plus a portion of $NPMR_N$).

The fundamental increment of a bigger-picture simulation-time represents an outside loop that provides a larger perspective than $NPMR_N$. The increment of time of such an outside or higher-level loop must be smaller than the fundamental unit of time in $NPMR_N$. Expanding this idea, it is clear that the fundamental increment of NPMR **simulation-time** may likewise be smaller than the fundamental unit of time in NPMR itself.

Is all this clear or is your head spinning a little? If you are a tad confused, it might be helpful to go back to Chapter 77 of this book and refresh your memory on the subject of incrementing time within simulations. Additionally, a glance at Figure 5-1 (at the beginning of Chapter 76 in this book) and a peek ahead at the discussion of the Even Bigger Computer (EBC) at the end of Chapter 83 might help clarify this bigger picture. Otherwise, if you feel that you mostly get it, absolutely do get it, or don't want to get it any better than you've gotten it, simply go on. In this situation, continuing on (though you find yourself in a light fog) is much better than becoming terminally frustrated. Hang in there, the text gets less technical later.

The predictions produced by calculating the probable state of Our System **m**•(DELTA-t) into the probable future become less accurate the further out in time they go. However, because our computer (TBC) and its software are so good, it can progress PMR time out for many years (**M** can be arbitrarily large) in less than a nano-nanosecond.

The result is a PMR space-time event surface in TBC calculation-space. TBC is only a subset of a greater digital mind-space. For we 3D creatures constrained to visualize our mental concepts within an experiential 3D structure, it is easier to think about a planar (two dimensional) event surface extending out in the dimension of simulated time with probability values (of particular events) on the vertical (up) axis – perhaps something similar to the surface shown in Figure 5-2. The horizontal plane, upon which the peaks rest, contain values of time, from t = 0 at the origin to some simulated future time **m**•(DELTA-t), the far edge of the event surface being at the time corresponding to **M**•(DELTA-t).

Events near the present moment typically have the highest probability values (sharp tall peaks). The further out we go in time the flatter the surface gets; peaks tend to broaden and lose height exhibiting very small, rather diffuse, probability values or likelihoods. Nevertheless, there may exist a few well formed and sizable (> .8 expectation value) peaks rather far out in time. You might want to take another peek at Figure 5-2.

·In summary, TBC generates a complete set of probable realities covering everything (choices, things, and energy) most likely to happen between each DELTA-t and the next one. TBC then saves and stores these results describing every unique probable state of OS corresponding to each simulated DELTA-t for each value of **m**. This complete set of probable realities going out **M**•(DELTA-t) in our PMR real-time are regenerated after each actual increment of DELTA-t. For the techies among you who are fretting over the apparent inefficiency of such a process, remember that there is no practical constraint on the consumption of computational resources

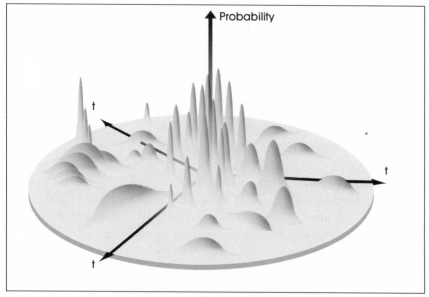

Figure 5.2: Future Probable Reality Surface

and, as you will find out later, computing and saving all potential states generates a database of possibility that supports a plethora of exceptionally useful analysis that needs no additional computation.

I have used the term reality state vector ("reality state," or simply "state" for short) to mean the total description or specification of the state of existing choices, things, and energy that defines OS at the end of a DELTA-t. Later I will define more precisely what a reality state vector is (Topic 9 below) and describe the process that generates it (Chapters 80 through 82 of this book), but first there are a few more basic concepts to be introduced.

2. Introducing Real-Time – What Our Clocks Measure in PMR

What appears to be real-time is dependent on the relative reality level or loop location of your local perspective within the Big Picture. PMR real-time appears to be continuous but actually progresses incrementally by iterating the small time increment DELTA-t. During DELTA-t, beings, objects, and energy move and change, free will is exercised, and significant choices are made in PMR. Most changes were as predicted by the $\mathbf{m} = 1$ calculation of expectation values, but some were not. Adjustments are made. Again, TBC runs the delta-t based OS simulation. Again it re-computes all the probable realities – OS state vector expectation values through $\mathbf{M}$ generations – creating, updating, and storing the event surface as subsequent calculations progress.

TBC's computing requirements are not as horrendous as one might first think: Clever software finds updating the complete set ($\mathbf{m} = 1$ to $\mathbf{m} = \mathbf{M}$) of probable realities to be much easier (merely dealing with changes and errors and their downstream impacts) than re-computing everything from scratch every time. It should be clear that what we sense and measure as time (our real-time) moves forward in PMR by successive increments of DELTA-t, while probable realities (future expectations) are computed within TBC to **project** the probable future states of OS through $\mathbf{M}$ successive simulated increments of DELTA-t.

Recall from Section 2 (Chapter 31, Book 1) that the quantum of time in $NPMR_N$ is much smaller than the quantum of time (DELTA-t) in OS and PMR. Thus, during a single DELTA-t of PMR real-time, $NPMR_N$ has many time increments (quanta) ticking away in which calculations can be made, probabilities computed, and probable reality surfaces generated. Also, recall that in Chapter 77 of this book we described how the flow of time in a subset or subroutine of the simulation was dependent on its outer controlling loop and that the simulation could be paused, stopped, and then restarted (relative to the clock in the computer room) without causing any effects within the simulation.

AUM's fundamental clock is the clock on the computer room wall and we are a subset of $NPMR_N$ within a subset of NPMR. In other words, $NPMR_N$ controls PMR's outer controlling loop, while NPMR controls $NPMR_N$'s outer loop. The process just described (the generation of OS, resulting from incrementing our real-time quantum DELTA-t and the calculation of our probable futures) raised to the next level of generality, allows for free will choices among sentient residents of $NPMR_N$. In a similar process to that which generates OS, the free will choices of $NPMR_N$ residents interact with the beings, objects, and energy of $NPMR_N$ to create and actualize successive $NPMR_N$ states of being, which in turn enables the generation of $NPMR_N$'s present and probable future states. As far as I can tell, the digital-state-flipping-AUM-consciousness-bright-awareness-thing directly controls NPMR's outer loop.

3. How the Probable Reality Surface Changes

As previously discussed, PMR real-time moves on as DELTA-t is successively iterated. Not only can the expectation values of a future probable state be computed, but the rate of change of these projected probabilities everywhere on that surface can also be computed as a function of DELTA-t. Because these calculations have been made for a large number of DELTA-t, the history of how the probable reality surface has actually changed with respect to real-time is now known. This information (sensitivity of our probability functions to perturbations) can be used to help calculate better, more accurate probable realities. In fact this is exactly what has been going on all along. The probable reality surface represents the most likely future possibilities.

As the present consumes the surface at the origin (t = 0), the surface is extended further into the future at the outer perimeter of the disk [t = $\mathbf{M}$•(DELTA-t)]. It might be profitable, though simplistic, to imagine the probable reality surface as a circular plane with t = 0 at the center and with t spreading out (increasing value) radially in all directions at once (see Figure 5-2). As real-time marches on, the plane disappears (moves), one DELTA-t at a time, into the pinhole at the origin while a new ring, DELTA-t wide is added to its outer edge to maintain a constant radius of $\mathbf{M}$•(DELTA-t). The disk is thus made up of $\mathbf{M}$ concentric flat rings, with each ring being one DELTA-t wide.

The pinhole, (more correctly, the mathematical point at t = 0) into which the future probable reality surface is being sucked, represents the present moment. After the present moment, all state vectors are saved. In other words, the present state is defined and subsequently saved to a history file, which contains every previous present state. We will see later how these past (previously actualized) states, captured by saving their present state vectors, remain vital and capable of branching to new potential virtual reality system within TBC's calculation space whenever additional significant input is introduced.

4. Predicting the Future

There will be some peaks on the surface that will have relatively large, stable and growing values. Some of these may occasionally occur far out in time (the future). These peaks and the events they represent or relate to would be good bets if you were a prognosticator. The narrower and taller the peak, the more precise the prediction and the more likely the event represented is to occur. Thus, we have future events that can be predicted with good reliability coexisting compatibly with individual free will. From the PMR point of view, intentionally or unintentionally tapping into this database of the most probable possible future events seems to, but does not actually, support the concept of predestination. Free will is required to convert probable events into actual events within the present moment. All information on probable futures is fully accessible from a larger perspective (if your intrinsic noise level is low enough) at: RWW.NPMRN.OS.PMR/probable-reality-database/specific-event/specific-intent.

5. Group Futures

Future probable reality surfaces for a particular group, activity, or happening can be computed. Specific summations can be taken over all the choices made by sentient beings and all the changes of objects and energy that have, or are projected to have, an impact or influence on a particular group (family, tribe, organization, corporation, nation, culture, planet, fault line slip, endangered species, rain forest, football team, or human race). The specific group's probable reality surface shows only those probable future events that are significant or related to that group. An individual interested in a specific group's probable future can easily filter all interactions for only those that pertain to that particular group through a process that is analogous to submitting a database query function where your intent designs and executes the query. A view of the collected events that are defined and limited by the properties of your query-intent, along with their associated probability values, are available to you through the RWW net.

6. The Probable Future can Change

The rate of change (fluctuations on the surface) of the probability functions for expected events for individuals is much faster than the fluctuations for a large group of individuals. Thus, a nation's future is easier to predict than an individual's (is a more slowly varying and stable surface). The probabilities on the probable-reality surface representing our entire planet change even more slowly, allowing for more accurate prognostication. Think of the probable future of a group or organization as the vector summation of all the probable future components of the individuals that affect that group weighted by their likely significance (impact) on the group.

In general, the larger the system, the more "inertia" it has (the less it can be affected by an individual's free will choices or by small random components within objects and energy), and the more stable and reliably predictable its probable-reality surface becomes. On the RWW net, information about these more clear (larger and more stable signal) future events exhibits a higher signal-to-noise ratio to everyone and therefore the information (likelihood of some particular event) is more accessible to more people. (Know anybody who claims knowledge of future earth changes? By the hundreds!)

7. Constraining the Number of Required Calculations

The computational burden is not as horrendous as it might seem. Most of the possible choices produce degenerate (the same) results and can be quickly dispensed with. Individuals only have influence or impact on a small subset (that may or may not be significant) of the complete set of interactions and choices. Objects, energy, and people with free will are generally more predictable than you might guess – particularly if you have all the historical data.

Only a relative few individuals at any one time (even in any one year or decade) have the potential to influence or produce major effects as a result of their choices (what they do significantly impacts the choices of many others). Additionally, large subsets of beings, energy, and objects may be functionally independent of each other. For example, earth relevant calculations could proceed as an independent set until **interaction** with specific extraterrestrials (ETs) from elsewhere in our universe occurs. Same for the ETs.

Furthermore, there are certain rule-set constraints, such as our PMR laws of physics (things never fall up), which further limit the possibilities.

Despite the mitigating factors of degeneracy, independent sets, and other constraints, computing everything that can happen in the universe and all associated interdependent expectation values (probabilities) is a big job, **but it is finite**. Fortunately, TBC has no problem performing this task using only a small fraction of its overall capacity.

8. Defining Our System (OS) to Include All the Players

An interesting side issue is that of manipulated choice. The manipulation, leading, predisposing, or nudging of PMR awareness by those aware in $NPMR_N$ is another mechanism through which certain probable outcomes are made more likely than others. In other words, another set of interactions that must be taken into account (as part of the OS calculation space input data), are the actions and free will choices made by those extant in $NPMR_N$, but not in PMR, that directly influence or impact the beings in PMR. This interaction affects the state of OS, and is therefore (by definition) a part of OS.

Some of those large, stable, and growing probability peaks exist because they are being encouraged or manipulated by $NPMR_N$ residents who may have much larger perspectives, much better information, a much clearer sense of the future (a better, bigger picture), and a more accessible knowledge base than PMR residents. Consequently, while some peaks (likely events) simply happen of and by themselves, others are guided. Most are a mixture of both.

TBC calculations relevant to our local reality system (OS) must include all beings, objects, and energy in the $NPMR_N$ superset that have an influence upon, or interaction with, our PMR probable-reality surface (not only those beings, objects and energy that exist within our PMR subset). TBC and its software (which can be clever, and does not have to execute a simple brute force approach) are by design thorough and precise in calculating and tracking the facts, possibilities, and probabilities of Our System (OS). OS creates or actualizes its larger reality through the choices of all its interactive beings (embodied or not) and the randomness of all its interactive objects and energy (physical or nonphysical). Changes must abide by the PMR space-time rule-set, the $NPMR_N$ rule-set, the rules of interaction between PMR and $NPMR_N$, and the psi uncertainty principle. Our history (the history of PMR), from a larger perspective, is a subset of the overall history of OS – as European history is a subset of world history.

9. Reality System State Vectors and Our History

During a given real-time increment DELTA-t, beings, objects, and energy may move (or change in some other way) and choices are made to actualize the new present which is contained within the state vector representing that DELTA-t. All potential choices not made remain unactualized potential states (possible realities) and have associated expectation values. The complete state vector of OS containing all actualized and unactualized choices is saved in TBC.

The state vector that defines or represents OS at a given increment of time (DELTA-t) is the total collection of information and data that completely specifies everything that actually did happen and possibly could happen (every significant possibility within that DELTA-t), along with its associated probable-reality surfaces. You will hear more about this later.

Thus, the progression of PMR from one DELTA-t to the next DELTA-t produces or traces a history which is the sum total of all the changes and choices that are actually made or actualized that affect or interact with PMR. This trace becomes an OS history thread representing everything that did happen, or in other words, a sequence of all the states of OS that were actualized during each DELTA-t.

The system is not closed. The system is open; beings, objects and energy can come and go in and out of effective interaction during any DELTA-t. TBC keeps track of, and up with, everything that is significant to (interacts with) OS.

This process and its results define our particular world, our particular history, our particular universe, our particular virtual reality – we who are interacting are all in this together, so to speak. Our choices define, in our view, a collective thread of continuous happening and unfolding generated by beings interacting with each other and with objects and energy.

The OS state vector containing the possibilities and probable reality surfaces not actualized, as well as those that were actualized, is saved at the end of each DELTA-t. For this reason, you can, from an awareness in $NPMR_N$, visit the past, view it, extract information from it (it is on the RWW), and even make changes to it that initiate new calculated arrays of un-actualized past probable realities. We will discuss this in more detail later.

10. History, Still Vital After All These Years

You can interact with the actualized as well as the potential non-actualized past. When you interact with any part of it in such a way as to modify it (introduce a new being, new things, new energy, or a new configuration

of old things, or change a significant choice or action), a new set of probable futures is computed that incorporates the changes as new initial conditions. A new set of probable reality calculations can now be progressed, creating a new branch, within the non-actualized past database. The nature of
this process is like making a copy of a file or simulation program so that
you can play what-ifs with it without disturbing the original.

Any point along the OS timeline, actualized or not, is a potential
branch point, but branches do not spontaneously sprout from every point
– they occur only when a significant change in the reality state vector is
produced by defining a new and unique set of potentialities. If the change
creating new initial conditions is trivial as evidenced by no significant
change in the future probable reality surfaces for **all** reasonable possibilities (not only the most likely ones), then that branch degenerates back to
the initial point of departure. Adding a new electron to the system or
changing an irrelevant choice, therefore, does not start a new parallel reality in the what-if calculation space of TBC. More of our choices than you
would probably guess are irrelevant in the interactive Big Picture.

In summary, I have described what happens when a change is made to
any part or detail of the complete set of everything that could possibly
happen, which is computed at the end of each real-time (PMR-time)
DELTA-t based on the history of all the beings, objects and energy and on
all possible configurations or choices. Because of the small size of DELTA-
t, our PMR history appears to be a continuous thread traced by the collective result of the interactions and choices taken, experienced, or actualized. What has not been actualized up to this point has simply been
saved. However, both non-actualized past and probable future states
remain operationally viable allowing the state vector database to be
queried and what-if simulations to be executed by intent.

80

■ ■ ■

A Model of Reality and Time Using the Delta-t Loop to Project (Simulate) Everything Significant That Could Possibly Happen in OS

■ ■ ■

Let's generalize and broaden our model by looking at the possibility that everything significant that can happen does happen. This is a key concept to understanding the breadth of our multi-dimensional reality, and to appreciating how AUM optimizes the output of its consciousness experiments by collecting data and amassing statistics that describe **all possibilities** simultaneously.

In the previous chapter, I described the complete set of state vectors representing everything that will **most likely** happen in OS. This was computed by incrementing delta-t (simulation time) through **M** consecutive iterations in between each increment of DELTA-t (PMR time). Recall that as **m** progresses from **m** = 1 to **m** = **M**, the delta-t loop converges upon the most probable future state. This was accomplished by evaluating **all** the possible future states in order to determine the most probable one. The most probable future state for that iteration of **m** then becomes part of the set of stored OS probable reality state vectors. Now TBC is going to track and store **every** significant possible future state (and its associated expectation value) that is evaluated for each iteration of **m**, not only the most probable one.

In order to assess **all the significant possibilities,** our understanding of the delta-t loop must be expanded. A more generalized delta-t loop process must now not only compute the most likely future states, but also track **all** (regardless of their likelihood) possible **significant** future states for each iteration **m** = 1, 2, 3,... **M**. Furthermore, each of these possible future states is assigned an expectation value that is a measure of its likelihood of being actualized. The mechanics and implications of this broadened delta-t loop functionality are discussed in detail in the remainder of this chapter, and are illustrated in Figure 5-3.

Before continuing with the description of this new application of the delta-t loop, I want to define the concept of significant states. A significant state is one that represents some unique, viable, meaningful configuration of OS, even if it is perhaps somewhat unlikely. Essentially, TBC generates significant states by computing all the permutations and combinations of all the free will choices, all the potential changes in objects, all the energy state changes, and then eliminates the redundant or insignificant states. All significant states with a probability of actualization above some small arbitrary value are enumerated.

For a given value of the iteration index **m**, the total number of significant possible future states is not known until after they have been generated. Thus, as the delta-t loop is iterated, there may be a different total number of significant possible (though not necessarily probable) future states for each specific iteration **m**. Recall however that for each specific iteration of **m**, only one of the significant possible future states will eventually be actualized and take its place on our seemingly continuous OS-PMR history thread. The state that eventually becomes actualized will probably be the one that was previously given the highest probability of being actualized – but not necessarily. The collective free will is free to choose whatever it will; updates and adjustments are made as needed to accommodate the vagaries of free will.

Perhaps the simplest way to think of this generalized process is to imagine that a dimension of width has been added to the information recorded during each iteration **m** of the delta-t loop. Look ahead to Figure 5-3: The example given shows parent-child state generation exhibiting geometric growth. During the first iteration of **m** (**m** = 1) there are three significant possible futures states generated (OS_1, OS_2, OS_3), including the one determined to be most probable (double bordered OS_3). TBC tracks and stores all three states associated with the iteration **m** = 1. Here we have chosen the small number three (**M** = 3) to make our visualization easier to grasp and present graphically. In actuality, there is a very large number of significant possible future states. Every circle in Figure 5-3 represents a unique state vector of OS.

This newly expanded function of the delta-t loop (tracking all possible states instead of just the most likely ones) represents a generalized larger view of our previous understanding. As such, it is more complicated to

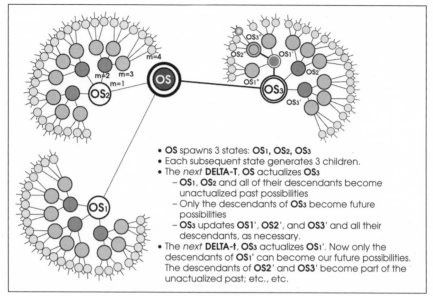

- OS spawns 3 states: **OS1, OS2, OS3**
- Each subsequent state generates 3 children.
- The *next* **DELTA-T**, OS actualizes **OS3**
  - **OS1, OS2** and all of their descendants become unactualized past possibilities
  - Only the descendants of **OS3** become future possibilities
  - **OS3** updates **OS1', OS2',** and **OS3'** and all their descendants, as necessary.
- The *next* **DELTA-t**, **OS3** actualizes **OS1'**. Now only the descendants of **OS1'** can become our future possibilities. The descendants of **OS2'** and **OS3'** become part of the unactualized past; etc., etc.

Figure 5.3: Generating the Possible States of OS

describe and to follow. Referring to the simple example given in Figure 5-3 may help provide a better understanding. The next step ($m = 2$) of this generalized NPMR_N delta-t loop is to project (simulate) a set of possible (though not necessarily probable) future states for each of the alternative states generated for $m = 1$. That is to say that each significant **possible** future state that is generated during iteration m spawns another complete set of significant possible future states during the next value of m (and so on as m is sequentially stepped to M). For example, in Figure 5-3 **each** nearly solid black state ($m = 2$) generates three medium gray ($m = 3$) states, which each generate three light gray states ($m = 4$).

The result is a geometrically expanding array of significant possible future states originating with the current OS and iterating M generations into the simulated (projected) future – representing a total elapsed time of $M \cdot (\text{DELTA-t})$.

In summary, during iteration $m = 2$, **each** of the previously generated (first generation) alternative states will generate some number of significant possible (second generation) future states of its own. The generalized delta-t loop is then recursively applied to each second-generation alternative state. This process continues until the delta-t loop has projected **all** significant possible future states of OS for $m = 1, 2, 3, ...M$ iterations by simulating (projecting) everything significant that could possibly happen (above a certain level of expectation) during **M** consecutive increments of DELTA-t.

We are no longer working exclusively with what is **most likely** to become actualized and what has previously been actualized (our PMR-OS history). We have now formed a super-set that includes all that, as well as some significant (worth following) states that will not and did not happen. In other words, a larger, broader set of states defining **all significant** possibilities of OS has been formed. Mechanically, this was accomplished by expanding the scope of the delta-t loop to recursively enumerate and determine **all** the **significant** possibilities of OS.

This enumeration and determination of potentially significant state vectors does not need to be accomplished by computational brute force. Given that the Fundamental Process would unquestionably need to create extremely clever evaluative operating systems and software for TBC as it functionally evolved within AUM's consciousness, we can assume a certain efficiency of process is achieved. After all, evolution is the unparalleled master of developing efficient and effective processes within each specific operating environment. For example, such software could be used to remove all the extremely unlikely, insignificant, uninteresting, unproductive, degenerate, duplicative, repetitive, meaningless, and useless states to form a complete set of useful alternate reality state vectors specifying **everything significant** that possibly could happen in OS during the next actual DELTA-t. Recall Topic 4 in Chapter 76 of this book for a short list of overall goals that suggests some of the evaluation criteria this expert system software might use to make decisions. Remember, the evaluative processes do not have to be perfect – the final results need only be functionally adequate and statistically meaningful – perfect calculations and processes are never required.

TBC calculates the probability that each projected possible OS state vector might be actualized by the free will choices and changes in objects and energy that will be made during the next actual increment of DELTA-t. The one that is most likely to be actualized represents the first point ($m = 1$) on the future probable reality surface of OS that we discussed in the previous chapter. In the example shown in Figure 5-3, the states that will eventually be actualized are double bordered. OS_3 most likely, but not necessarily, represents the first ($m = 1$) flat ring on OS's **most probable** future reality surface. (Reference Figure 5-2 located in Chapter 79 at the end of Topic 3 in this book.) OS_2 and OS_1 are also $m = 1$ states but remain unactualized – the choices they represent were not chosen by the collected free will actions of the sentient beings in OS during that actual DELTA-t.

The next pass (m = 2: solid black circles) through this generalized OS DELTA-t simulator will allow **each** of the possible m = 1 states (both likely and unlikely) to likewise generate all the possible significant states that could be generated from the initial conditions that this particular state vector represents. Again, the probabilities of actualization are computed for **every** state vector generated. For example, only one state generated by OS_3 can be most probable and take its place on the most probable future reality surface of OS_3. Also only one state (depicted by double bordered $OS_{1'}$ in Figure 5-3) generated by OS_3 will be actualized by the free will of the sentient beings within OS_3. This process, repeated **M** times (once for each value of m) gets to be a mind full. I am afraid that we will need to resort to a generalized subscript notation to keep this mental picture focused.

Let's take it from the top utilizing subscripts to form a generalized description of this process. Each alternate reality state vector, differentiated by the index i, represents the state vector defining OS_i. The subscript i keeps track of as many unique state vectors (i = 1, 2, 3, ...) as there are unique arrangements of sentient choices and other variables (objects, position, or energy). These OS_i state vectors constitute parallel, possible, or potential future realities of OS or from a PMR-centric view, parallel, possible, or **potential** future universes. Sometimes travelers in $NPMR_N$ get into these parallel realities (such as the never-to-be-actualized solid dark (m = 2) or medium gray (m =3) circles attached to OS_1 or OS_2 – or the could-be-actualized light gray (m = 4) outermost circles attached to $OS_{3''}$), and fail to realize they represent reality states that were or are merely possible and not necessarily probable. A measure of each state's probability of being actualized is available, but needs to be accessed with a separate intent. In other words, the probability of actualization does not automatically come integrated with the experience of the reality – you have to ask a separate question and be precise with your initial intent.

Each of the alternate reality state vectors described above becomes the starting point for another. Everything significant that could possibly happen is computed based on the unique permutations and combinations of all the possible states of objects, energy, and the free will choices of beings originating from the particular initial conditions of each particular alternate reality state vector.

Thus, parent OS_i (first generation) state vector possibilities spawn new child $OS_{i'}$ state vectors (second generation), which in turn spawns yet a new generation of child $OS_{i''}$ state vectors (third generation). This progression continues so on and so forth, until one has progressed this family tree of possible states (all offspring of the original OS and all computed during the single time increment DELTA-t) through **M** generations. Remember, **M** is an arbitrarily large but finite integer.

Every alternate state vector (every circle in Figure 5-3) can generate its own probable reality surface of expectation values by tracing the states that are most probable from generation to generation of its descendants. (Note: **M** is finite because this is a **real** process generating the real reality that we presently live in. This does not represent an imaginary or theoretical process – it is a practical **model** of how the larger reality operates.)

In this way, the number of unique and useful alternate reality system state vectors grows (removing all useless and redundant states) until we arrive at a complete set of alternate reality system state vectors representing everything that could uniquely and usefully (significantly) happen. All the state vectors that result from this progression have been derived (originated) from the OS state vector during this present DELTA-t (our present moment).

During the next DELTA-t, one (and only one) of these possible states (first generation OS_i) becomes actualized (through our free will choices and the changing objects and energy) as our next present moment. The descendants of that state (the OS_i that was actualized) become unactualized **future** possibilities, while the descendants of all the other non-actualized OS_i (a much larger group) become unactualized **past** possibilities. TBC saves and stores everything.

The unactualized future possibilities have a finite chance of being actualized at some time in the future, while the unactualized past possibilities can now never be actualized by free will choices within OS. Now integrate this picture with the one described in the previous chapter. If, from the set of simulated unactualized future possibilities you trace the single most likely state to be actualized within each of the **M** generations, you would have defined the probable reality surface for OS that was defined in Chapter 79 of this book.

The word "actualized," as it is used here, refers to what actually happened or actually took place from our perspective – the perspective of OS. States are actualized and reality is created in the present moment. For us, it is created DELTA-t by DELTA-t – one increment (fundamental time quantum of PMR) at a time. The history of OS is the sequential record of the actualized present moments of OS.

Because time is quantized, and TBC has a good memory, both the actualized and unactualized states (state vectors) can be saved. Every saved state vector is as complete, vital, and capable of generating new states as any other saved state vector within TBC. The set of state vectors within the group called unactualized past possibilities are not dead states; they are simply dormant states. They are as alive and vital as any – they have simply not been actualized or chosen by the free will choices and changes of objects and energy that define the dynamic history thread of Our System of reality.

▶ A short aside is in order here. I can hear you wondering:

"Granted, every state vector is theoretically capable of generating new states, but why would an unactualized state do this?"

You are absolutely correct: It wouldn't change spontaneously. It would simply sit there with all its possibilities laid out for **M** generations, unless something changed in its defining choice-set. For example, a sentient being could travel back through history to that particular state vector and alter something significant, thus modifying that state vector and all its descendants. If the change represented one of the possible choices previously considered (highly likely), no new calculations are necessary, otherwise a new branch is generated and a new larger set of possibilities would be created.

However, even if a new array were generated, once all its possibilities were filled out it would simply sit there until additional unique significant changes were introduced. There is no need to continue making calculations on unactualized states of a particular OS. Unactualized states do not require much updating – unless somebody with free will is introducing new significant initial conditions and creating new branches within the old set of possibilities. They simply sit there as a mostly static complete array (database) of the possibilities and retain the **potential** to branch (calculate new possibilities) if new initial conditions are introduced.

This arrangement (allowing for unique input while maintaining an exhaustive data-base of possibilities) enables the running of what-if analysis to ascertain the impact of having made specific choices. This type of analysis is often used as an aid to help certain sentient beings that are between physical manifestations in PMR to understand the implications of previous choices and overcome personal belief systems. Such analysis is a typical part of the planning process for more aware beings trying to learn as much as possible from their past experiences before initiating another PMR experience.

For those who do not understand the larger reality well enough or do not have the necessary control, others typically guide this process for them. This analysis capability is generally available to any $NPMR_N$ or PMR being who is sufficiently aware and in control of their mental faculties and intent within $NPMR_N$. ◀

Now that DELTA-t has incremented and an OS_i state has been actualized as the current OS state, the second generation $OS_{i'}$, that are children of the **actualized** OS_i state, become first generation possibilities to the present actualized moment. Again, **M** generations beyond the present state are computed for **all** states (actualized and unactualized) that contain significant possibilities. And so on, and so forth, this process marches on, generating and computing potential reality-system state vectors describing everything significant that might happen, along with everything significant that might have happened (but didn't, from our point of view). TBC saves every state vector, its genealogy, and its likelihood or probability (relative to its siblings) of being actualized by its parent.

The past of OS is represented by a particular solid thread connecting all our past **actualized** states as it meanders through the matrix of all past possibilities. Our perceived past or history can also be described as a specific sequenced subset (previously actualized states) of the all-past-states database. Likewise, our probable future is represented by a dashed thread snaking through a vastly smaller database of future possible states, picking out only the states with the highest likelihood of actualization, as it moves sequentially from generation to generation. This future thread represents the most probable reality or probable future surface of OS (see Topics 1 and 8 in Chapter 79 of this book).

We have described and generated a set of state vectors representing everything significant that can happen (including everything significant that might have happened and everything significant that might happen yet) through **M** generations beyond the original common ancestor OS. All this is calculated between each DELTA-t. Because this concept is complex, let us summarize quickly before continuing.

Previously, we developed probable reality surfaces for **M** sequential simulated increments of DELTA-t. These probable reality surfaces only described **everything most likely to happen**. Now, we have broadened that concept by describing **everything significant that could happen**. The process starts with OS at a particular DELTA-t and projects (simulates) **M** generations of **possible** future significant states of OS. This is accomplished between successive **actual DELTA-t** increments (real-time is standing still in OS). It is accomplished by incrementing the $NPMR_N$ delta-t loop, which, at each iteration **m**, generates **all** the significant possibilities (the OS_i) for **each** of the OS alternative states previously defined. This delta-t process continues through **M** iterations, progressing and expanding to project or simulate **everything significant that could possibly happen** during the next **M** iterations of DELTA-t.

During the next **actual DELTA-t** (real-time in OS moves forward one increment), one and only one of the OS_i states, is actualized to become our present moment. The actualization of only one state leaves a large set of unactualized past possibilities. Every state not connected upstream to our newly actualized present state becomes an unactualized past possibility. In other words, only those relatively few states that are descended from the just actualized (our new present) state now make up our future possibilities. TBC always maintains a calculation space of **M** generations beyond the present. (We will generalize the concept of **M** later but it will serve us well in the meantime to think of it as a fixed integer.) As this process continues, redundant states among the unactualized past possibilities are collapsed. Entire branches of this family tree may cease to expand for lack of further significant unique possibilities.

The initial massive calculation (running the delta-t loop to generate every significant possible future state through **M** generations) must be done only once (say during the first increment – the beginning of time for OS). Other than a few relatively minor adjustments that may need to be applied to the previously generated states (allowing for unforeseen changes in initial conditions and imperfections in TBC's evaluative and predictive software), all that remains is the creation of the newest generation of children states.

Every actualized reality system state vector, flourishing and evolving within its own dimension, represents a **dynamic open** (entities and objects can come and go) reality with an active copy of you and everyone else (including all the objects and energy) it inherited from its parent (along with all the pending potential choices, interactions, and conditions).

It may be helpful here to point out that dimension is to TBC as a line on a sheet of paper is to us. Or better yet, as the text-line on a computer screen is to us – simply press the enter key to get a new one. Those analogies are not perfect. Perhaps a better one would be that dimension is to TBC as a saved file in our computer is to us. You get the idea. TBC spins off a new computationally alive dimension within Our System's multidimensional reality for every uniquely significant reality system state vector it generates.

In a bigger picture, each dimensioned local reality (the various PMR_k for example) describes a diverging, branching, set of uniquely dimensioned potential worlds. Think of saved files that may contain sub-files – folders within folders – with each folder or sub-file containing an executing piece of the overall simulation. Each reality exists within a unique dimension, folder, or memory space within TBC. All these realities existing within their various dimensions or dynamic folders are computationally alive (can be modified) subsets of a larger simulation, expanding into their potential futures by the beat of their **own** time artificially constructed or simulated by successive increments of DELTA-t_k. OS is one of those local realities – the one that we sentient beings in OS have collectively chosen to actualize. I will discuss this subject again from a slightly different viewpoint in Topic 3, Chapter 83 of this book where we will again contemplate an Even Bigger Computer (EBC) and multiple PMR_k.

We have thus constructed a process to support everything significant that can happen, in fact, does happen – at least in TBC calculation space. Nevertheless, everything that can happen is not actualized. If, in the rare instance (allowing for imperfection in TBC's software) where the state that is actualized is not (to a significant degree) one of the previously generated OS_i, then it is simply added to the set of OS_i.

Although TBC's software can be exceptionally clever and efficient, it does not have to be perfect. Perfect processes, like infinite processes, are unnecessary to the development of this model. We are talking about real processes here, processes that are imperfect and finite. There is a finite number of beings, objects, and energy states among objects, and each of these has a finite number of choices and ways to change. All the significant permutations and combinations of all the possibilities through all **M** generations is probably an extremely large number (especially from our PMR perspective), but it is finite and consumes only a tiny fraction of the capacity of an apparently infinite (but actually finite) AUM.

81

■ ■ ■

A Model of Reality and Time
Viable Alternate Realities
and Histories

■ ■ ■

The previously described process of progressing OS through **M** generations beyond the present (in order to represent everything significant
that can happen) began to include PMR with the initiation of the Big
Digital Bang, or the Big Set of Initial Conditions, or the Big Start of the
Simulation. Our scientists use the word "Big" because we are such a tiny
part of the entire universe. However in a bigger picture, our physical universe is neither big nor exclusive. In fact, it is every bit as small and insignificant from the Big Picture view as it is grandly immense from our little picture (PMR) view. Nevertheless, "Big Bang" is a catchy and likeable term
even if it is not that big and even if the bang is only a digital one in calculation-space within the space-time simulation section of TBC. OS has incremented its way forward in time within TBC ever since the time loop driving the Big Bang digital-mind simulation started incrementing our PMR
forward by applying the space-time rule-set to the given initial conditions.

When the state of PMR (within the simulation) was mostly rapidly
changing matter and energy, alternative space-time realities were generated. Some were produced by sampling over the randomness (choices)
within material and energetic processes. As always, the Fundamental
Process was in the driver's seat; all possible states were generated but only
those universes supporting PMRs that were uniquely useful, productive,
or significant were propagated forward. One of them became our universe while others evolved, in their own dimensions (sub-sets of TBC), to
represent some of the other PMRs mentioned earlier (see Chapter 76 in
this book).

Nothing could be more elegant or efficient than to initiate the virtual material part of AUM's continuing experiment in consciousness, than beginning with a huge virtual space-time energy (remember $E = mc^2$ or equivalently, $m = E/c^2$) at a point (or within a small volume) and... Bang! Or rather, Big Bang! Wow, look at that stuff go (evolve) – one DELTA-t at a time – all according to the space-time rule-set. Our scientists have some simulations like that, but I am sure they are not as good as TBC's. The previous sentence should not be construed to denigrate our brilliant, bushy-headed scientists, but is simply attributable to the fact that TBC has a much better computational capacity and throughput as well as the **complete** space-time rule-set.

Recalling that simulation time can run much slower or faster than real-time, and that any result or intermediary state may be modified, saved, or deleted, gives us an appreciation of the control AUM has over the eventual development of our local PMR reality and the evolution of the space-time rule-set used to define allowed interactions (within this book, see Chapter 77 and Topic 2 in Chapter 79). AUM may have initiated, run, and rerun this simulation process many times before all the initial conditions, parameters, rule-sets, and relationships had evolved to the point that the results met all the specifications of a top-notch learning laboratory capable of effectively supporting the consciousness cycle. This is how AUM and TBC generate the PMR virtual reality set that we call home – likewise, for all the space-time reality sets that other sentient beings call home.

When a subset of sentient beings within $NPMR_N$ was able to use (inhabit) the evolving virtual physical reality we call PMR profitably, their free will choices driving learning, growth, physical evolution, and spiritual evolution became interactive with the virtual objects and energy of PMR as well as with each other. These interactions generated the OS probable realities and alternate reality state vectors that were the predecessors of our present OS. Some of the sentient beings in $NPMR_N$ that existed before our PMR's digital Big Bang were a part of OS by virtue of their involvement in planning and executing OS's origins – the seemingly mystical beginnings of our universe – yet mystical only from a PMR perspective.

A still living and still potent history of OS (actualized and unactualized) continues to exist in TBC where any saved state is able to calculate or simulate more children if something unique and significant is added or changed. This easily animated living history is available to us from $NPMR_N$. For example, any state vector that lies on our history thread, regardless of where it is located on the thread, is available not only for inspection but also for a what-if tour of the probabilities. The accumulated database of everything that could have possibly happened since the PMR virtual reality was created provides a huge resource for data mining.

There are many unactualized states that can be strung together into continuous threads that represent significant choices that could have been made differently; such traces represent an unactualized history that is different from ours. Any number of unactualized history threads may have shared a common state with our actualized OS history at some particular DELTA-t, but subsequently have gone their separate ways. These unactualized and usually less probable history threads represent a particular sequence of events that did not happen. Nevertheless, the study of such what-if world-lines or sequential traces through the unactualized state vector database can reveal the long-term impact of decisions, choices, and intents – an excellent educational and evaluative tool.

Recall that each state vector has an associated expectation value that properly fixes its likelihood of happening given the initial conditions of the parent. One can sequence or thread his or her way through unactualized consecutive reality states (past or probable future) that are OS derivatives in a huge multitude of ways that are intentionally controllable and definable only by a steady, low-noise, focused aware consciousness.

These sequences of related states are (from our view) called parallel, divergent, or alternate realities. These alternate realities are computationally viable in TBC and are populated by mathematical representations (statistical simulations) of each sentient player as well as mathematical representations of the objects and energy. It is relatively easy for a novice explorer of the RWW network to get lost or confused in these parallel worlds. The explorer comes back with information that seemed to be correct and clear at the time, however, later analysis often indicates that paranormally derived information does not make sense, has little PMR value, or is shown to be false or inconsistent by subsequent events or information within PMR. It is easy to end up wandering unknowingly in the unactualized data matrix of TBC because it (the interactions and the experience) appears as real as any experience can be. The experience of an unactualized computational reality is like seeing a movie where you are simultaneously an actor (exercising free will to drive the story line) as well as a member of the audience. An unintentional experience of an unactualized computational reality is indistinguishable from the experience you are having now – except that it disappears when you shift your focus back to PMR (as PMR tends to disappear when your awareness becomes preoccupied with a mantra, a problem, entertainment, or befuddled by drugs).

The set of state vectors comprising the history thread of OS, the OS present, and the set of OS probable futures is no more or less real or dynamic than the set of state vectors that comprises the unactualized past. All are of equal reality and validity. All share the same properties and are similar data types within TBC. The only difference is that some were actualized by our choices while others were not. The only difference is us – what our individual quality expressed as a free will intent actually did. The set of actualized OS state vectors represents the collective quality of our consciousness within OS, expressed as a unique path through the possibilities.

Previously, I have referred to choices and changes in objects and energy as drivers of actualization. The possibilities that may be actualized are derived from the possible free will choices of sentient beings, the constraints of the space-time rule-set, as well as from the randomness associated with objects and energy (material movement, changes, interactions). Physical randomness and the constraints of the space-time rule-set drive the actualization of natural events such as earthquakes or thunderstorms. Physical changes may be expressed as the choices that inanimate objects and energy make.

In that sense, objects and energy make choices as they evolve toward minimum energy states along the path of least resistance. Objects and energy must make their choices in strict obedience to the space-time rule-set – the consistency of space-time physics is important to the usefulness of PMR. These choices are not the moral choices of a free will and do not reflect the quality of consciousness of the chooser. To avoid confusion, let's call the results of physical or natural randomness (such as radioactive decay) rule-driven possibilities, as opposed to the choice-driven or free-will-driven possibilities that are sometimes made by conscious, aware, sentient entities.

Sentient beings may contain small random components within their intent, and larger random components within their perception, interpretation, and reaction to their internal and external environments. Random components play a part in what sentient beings experience, and how they react to, and interact with, that experience. Consider the random components influencing experience as spice, not the main ingredients. Randomness occasionally dominates the flavor of existence for some entities some of the time (usually short-term), however, the quality of your choices are a much bigger driver of eventual long-term outcomes than randomness.

On the upside, randomness helps to deliver an interesting and ever-changing array of experiential opportunities to which we interactively apply our free will intent at whatever level of quality we have thus far generated.

It is these experiences, these opportunities with their sometimes-random components that enable us to evolve our consciousness. When free will meets an opportunity to interact with our internal or external environments, consciousness has the opportunity to grow, stagnate, or degenerate. Do you see why a virtual physical PMR-type learning lab is so useful? Straightforward interactive opportunities with clear results and feedback abound everywhere we focus our attention in PMR. PMR is constrained to be a highly interactive experience at the most basic level of relationship. Interacting with the other PMR players (an interaction of free wills) is the main thing we do here in PMR: This interaction is the basis of our opportunity to evolve the quality of our consciousness.

The set of possibilities that may be actualized as a result of many individual actions taken by a group such as a club or organization (or perhaps a government) is defined by a superposition of all the free will choices of all the sentient beings in OS that have some influence or impact on the actions of that organization. Thus, we also describe group choices as choice-driven possibilities. Recall that a motivation and intent that has large random components represent a less developed consciousness with large entropy. This is true of group consciousness as well as individual consciousness.

Possible events, with any value of probability, may be primarily rule driven, primarily choice driven, or a combination of both. Thus, for example, some relatively clear (large signal-to-noise ratio) future events (see Chapters 75 and 79 in this book), such as potential coming earth changes, gain their strong signals or large stable and growing probability (expectation values) from an interactive mix of rule-driven and choice-driven probability calculations.

From Chapter 79 in this book, recall that the probabilities of actualization, calculated for each successive value of **m**, define the probable reality surfaces. Also recall that peaks or lumps on the probable reality surface represent the expectation value of a particular event. Do you see how choice-driven and rule-driven actualization creates uncertainty in the expectation values of probable future events? Do you also see that a dynamically evolving free will that is bound to directly express the true quality of its consciousness, along with ubiquitous random components that are typically small and short-term, conspire to maintain a rich mixture of uncertain yet personally focused opportunities within PMR?

Also recall that it is uncertainty, according to the psi uncertainty principle, that allows good planning and manipulations from $NPMR_N$ to influence PMR happenings without being obvious – without stepping outside the space-time rule-set and PMR causality. Consider the natural uncertainty and small random components modeled within the digital consciousness system as the lubricant that helps make the system work.

▶ Let us take a break for a minute in the form of a short aside to gather up some ideas and concepts that fell on the floor while we were explaining the mechanics of reality system state vector generation, propagation, and uncertainty.

It is worth reiterating at this point that there are similar consciousness and physical-matter reality systems working through similar evolutionary processes in other space-time PMRs within $NPMR_N$. These realities can be very different from our PMR and exist independently of OS. TBC can be thought of as representing a dedicated server (or calculation space allocated within a larger AUM mainframe process) for the PMR_k within $NPMR_N$. All PMR_k are part of (accessible through) the same NPMR communications net. All are available on the RWW for inspection and participation.

Another point worth mentioning is that the clever evaluative software that defines "significant" by filtering out non-productive, insignificant, redundant, meaningless, or useless reality states does not need to be any more mystical than our artificial intelligence or expert systems software. Like its PMR cousins, it utilizes evaluative criteria, pattern recognition, history files, statistics, and predictive algorithms. Its accuracy is dependent upon an extremely detailed and unusually complete knowledge base.

I have evoked (assumed) nothing fundamentally new or spooky here – only bigger, faster, better. We humans have a tendency to see ourselves with inflated significance. There are perhaps many fewer significant states than you might assume at first glance, making the calculation problem somewhat less monstrous in volume.

Please make note of the fact that thus far, I have not evoked, nor will I evoke, either perfect or infinite processes. OS contains finite numbers of beings and objects, with finite **significant** choices and finite **significant** random events that impact, influence, or interact within a finite universe of objects and energy capable of moving or changing in a finite number of **significant** ways. Thus, though it may seem overwhelmingly large, there is a **finite** number of unactualized past and future OS reality-system state vectors. Recall from an earlier discussion that non-interactive subsets of our larger physical and nonphysical realities can be handled separately. We only have to include that part of $NPMR_N$ and our physical universe that is interactive with our OS's super set of everything that can happen in order to fill out our OS data matrix (see Topic 8 in Chapter 79 of this book.)

Some may think that the data storage requirements and processing requirements are too large to be workable. If you believe that, it is because your vision of what is possible is stuck in the PMR of the twentieth century. To an apparently infinite, highly evolved, aware digital consciousness that employs processing speeds and storage capacity trillions upon trillions upon trillions of orders of magnitude greater than what you could ever imagine in your wildest theoretical dreams, the computational burden is almost trivial. The computational requirements are significant, but they are also finite and well within the capacity of TBC.

On the other hand, you might believe that the mechanics of generating, storing, and progressing OS state vectors as described above is a mystical or supernatural process. It might seem that way to some, but remember what we learned in Chapters 18 and 20 of Book 1 – that what appears to be mystical is relative to your experience, and dependent on your perspective. From the perspective of *My Big TOE,* it appears to be a straightforward bookkeeping and computer science application focused on a relatively small (about the size of a Petri dish) and relatively simple (like raising guppies in an aquarium) problem. ◀

82

■■■

A Model of Reality and Time
Multiple Copies of You –
Creating Your Reality

■■■

Previously, I defined everything significant that could have happened or that might yet happen. Let us now focus on a piece of our history thread that contains you, as a sentient being, in it. From any particular state that contains your existence, a multitude of dependent child states is generated, which in turn have children that have children through **M** generations. The phrase "everything that can happen does happen" implies that each of the possible representations of you – each virtual you – contained within each **possible unactualized** OS state vector represents the array of possible choices you **could** make at any particular DELTA-t.

You are the entity that made the free will choices you made. There are representations (models) of you that populate all the possibilities that you could have made but didn't within the unactualized past. There are also models of you populating the unactualized possible future states. You, the evolving sentient unit of individuated consciousness, exist only in the present as you make free will choices that actualize one possibility over another and thereby modify the quality of your consciousness. The models of you we are referring to are simply an enumeration of all the possible choices you could have made, or could yet make, with an assessment of the probability that you would actualize each choice. Thus, the model of you is simply a probabilistic estimation of the significant choices you are likely to make under each of the unique circumstances defined within each state vector. This model of you is updated and improved with each set of choices that your conscious free will intent actualizes. Your model represents your quality – it is the result of you – the result of every thought, feeling, desire, intent, choice, and action you have ever made.

Although this simulated or virtual you makes all the same choices you are likely to make in a given situation, it is incapable of learning and changing its representation of you. It is not making free will choices. The replicated you that exists within every unactualized (past and future) state is merely a statistical representation of all the choices you could possibly make and the likelihood (probability) that you would make each one. Thus, TBC populates **all** alternate states with a probabilistic model of you that knows (from your quality and from your history) the likelihood of your making each possible choice (given your past choices and the current array of options). The virtual you is a reasonably accurate simulation or projection of the actual you. (Keep in mind that the actual you we are talking about is you as an individuated unit of consciousness, not you as a PMR body). Think of it as a duplicate of you, a representation of you, **without** the ability to choose by exercising the quality of a freely willed consciousness.

Each alternate or replicated you that inhabits every saved state vector (including the ones you have previously actualized) is actually a representation (predictive model) of your primary energy in $NPMR_N$. Each virtual you represents the likelihood of you making the given significant choices according to the statistics and probability functions that currently model the quality of your consciousness. Though this model of you is an accurate high-fidelity model, it is only a model. It does not represent a sentient consciousness with free will. It is not expressing your intent and is not directly helping or hurting your effort to increase the quality of your consciousness.

The what-if scenarios (divergent reality branches) that you can access in the unactualized past data matrix, for example, are populated exclusively by these modeled individuals who will interact with you (the new inputs you create) according to their own modeled behavior. The adventures or experiences you may have in these or other parallel realities are as real as what you are experiencing now. They are run in what appears to you to be your real-time, are typically not saved, but can be regenerated anytime. Think of it as querying a multidimensional database with your intent where the result of the query is a movie that you both act in and watch.

It is possible (though unlikely) that your consciousness may be integrated with or inhabit more than one free will motivated choice-making entity at the same time. Whatever arrangement best optimizes your growth potential can generally be implemented; nevertheless this arrangement is not often put to use. Because of the strong coupling between lessons learned from previous physical experiences and probable success with the present physical experience, it is usually a better strategy to plan and execute these learning experiences serially instead of in parallel. Parallel space-time (physical) projections may be in the same PMR at the same time or be scattered among any of the PMR_k. The more developed the consciousness, the more easily it can handle complex arrangements.

Your totality is the sum of all of your projected consciousness fragments that have ever exercised free will. Your primary consciousness energy, sometimes called the oversoul (where the results of your efforts are accumulated) remains always conscious in $NPMR_N$. Think of the oversoul as your main personal data folder. In general, here is how it works: Individual subsets of your sentient (free will possessing) being or consciousness-energy are projected into various frames of reference (dimensions or realities) where their interaction with the challenges and opportunities that they find there produces a change in the quality of your overall consciousness.

The fragment of you (who you think you are) in this PMR may be no more real, special, significant, or successful than other free will expressing fragments of you that have been, are, or will be spread around in various realities (past, present, or future).

Your present (in this reality) personal identity or awareness (the fragment that is currently the brilliant, good-looking, and lovable being that you know you are deep down inside) may be the grandest of the bunch – or it could be the worst, or somewhere in between. Your personal identity is simply the fragment of consciousness (entity) that has made all the choices you have made in conjunction with all the people and objects with whom you have interacted. That statement is also true for all the other past, present, and future free willed projections or fragments of you. Though your oversoul is immortal (disregarding a few rare exceptions), your **present** personal identity will only persist as long as your oversoul finds it profitable to maintain it. The Fundamental Process urges the oversoul to maintain and progress only what is profitable.

Some people refer to their past fragments projected into the PMR reference frame as past lives. That is a good descriptive term, but carries a PMR-only connotation that is too small to fit the bigger picture of all the potential sentient fragments (physical within any of the PMR_k and nonphysical within any of the $NPMR_n$) that your oversoul might contain. On the other hand, many sentient entities prefer to specialize in one, or sometimes a few, similar local realities in order to optimize their learning efficiency by narrowing their focus. Thus, an individual might specialize in learning within the context of OS, for much the same reason that a physician might specialize in dermatology. Most entities initially specialize in a single reality system, eventually broadening their experience-base as they develop the quality of their consciousness.

It is possible, though somewhat unlikely, that you have more than one fragment (sentient being belonging to your oversoul) existing in the same PMR at the same time, but you do not need to worry about competition among the various fragments for energy. Such energy is abundant in a digital consciousness reality – from our view it appears to be infinite.

Think about it: The total energy consumed by hundreds of thousands of players in a large war game is nothing more than organization, rules of interaction, goals, memory, and digital processing power. Our concept of energy as fuel (a limited animator of action) is a reflection of our specific space-time rule-set and does not apply to the larger reality. In virtual digital realities, virtual energy fuels the virtual action. There is unlimited energy awaiting your command in NPMR.

Primal Energy in NPMR is like air on earth. The supply is finite, to be sure, but it's there for all to breathe as needed – there is no competition, every land based critter on earth consumes as much as he or she wants or can use. Digital consciousness energy is simply intent manipulating bits – it does not depend on a limited **outside** source that imposes practical usage or conservation limits on us. In a practical sense, your access and use of nonphysical energy is limited only by your personal quality.

Your ability to apply energy to PMR from NPMR (induce paranormal events) by focusing your intent may be limited by your beliefs, ignorance, access privileges, and the psi uncertainty principle, but it is never limited by an unavailability of basic energy. Light workers (a term among New Agers for those who have learned to manipulate consciousness energy between $NPMR_N$ and PMR) never run out of light – an endless supply is always available. If you know how, you can affect a virtual reality from the outside by applying virtual energy (modifying expectation values by intent), but only in accordance with the rules (such as the psi uncertainty principle) of the next larger (parent) reality. To affect a virtual reality from the inside, you must work within the defining rule-set (like putting gas in your tank).

Each fragment projected (in serial or parallel existences) by your oversoul is given as much energy and support as it can profitably use to maximize its learning opportunities. Situations, conditions, and environments are often chosen or planned to provide the kind and type of opportunities that are needed most or that will most effectively improve the whole.

Do not misunderstand me and immediately jump to the conclusion that you are the CEO and chairman of the board of your oversoul. Think of yourself more as a member of a team. You may be on a specific mission to accomplish a specific task that has to do with your personal growth and perhaps the growth of others, or your mission may be non-specific, but more importantly you are adding to the growth of the whole of you while you are focused in this particular PMR experience.

The proper (most productive) perspective is that you are here to develop the quality of consciousness of the fragment you presently represent, not as an agent of your oversoul. The best thing you can do for the whole is to improve yourself by effectively utilizing the opportunities you have. It is that simple. An oversoul that successfully lowers its entropy contributes that benefit to the entire consciousness system.

Thinking that, in your case, a mistake **must** have been made and that your opportunities are more difficult, less fruitful, or contain less potential than they should, or that others have not done their part, or that you are not getting enough help, is a serious mistake. If you think that this might be the case, you don't understand how things work. Trust me; you have everything you need to succeed and the free will to make success happen. You are not in a disadvantaged situation that has been imposed upon you from the outside. If you believe that you are in a disadvantaged situation, you are indeed in a disadvantaged situation, but one created by your beliefs, fears and ego – a situation that only you can change. The success that I am referring to is the success of improving the quality of your consciousness, not making lots of money in the stock market.

You have access to all the quality and understanding that belongs to the whole. If other fragments (in serial or parallel existences) of your fundamental energy make great leaps forward or backward, the quality of your personal fragment of awareness is immediately affected. Great leaps in either direction do occasionally happen, but do not worry too much about the downside. Typically, a fragment of your fundamental energy changes the quality of its consciousness by only very small steps and not by great leaps – that fact is a double-edged sword.

For the most part, personal growth, the evolution of your consciousness, is a slow and steady process that cannot be easily hurried or advanced by taking shortcuts. The various individual fragments (in serial or parallel existences) of your oversoul are all independent initiators of free will choices.

The individual specific experiences of the various fragments of awareness are kept more or less separate (filed in their own folder within the oversoul folder so to speak) and do not, for the most part, become shared experience. However, the spiritual growth resulting from those experiences is immediately available and shared. You can access (and sometimes interact with) the experiences of the various individual fragments of your oversoul after becoming sentient and in control of your energy within $NPMR_N$. This remains true even if the fragments are not concurrent because all reality states are saved in TBC and remain potentially viable.

Keep in mind that your personal awareness fragment represents only a tiny portion of the free will choices and changes in material and energy that define a present PMR OS state at the end of some DELTA-t increment. Your local reality is a shared reality that represents a vector sum of all the free will choices of all the interacting players. How the virtual ball bounces and the virtual cookie crumbles is the result of a shared effort. All the sentient entities existing upon the earth (Team Earth) are in a collective dance of creative interaction. We get to live with what we create from the opportunities we have. Each individual is an important part of, and affects, the whole. There is no way to cheat others without cheating yourself. Team Earth succeeds or fails together.

Recall that probable futures were generated based on the calculated probable choices of the virtual you (a simulated you). Free will choices were subsequently made which caused one of those probable futures to be actualized. Keep in mind that the things that did not happen (were not actualized) became part of the set of state vectors that defines the un-actualized past possibilities. The data within TBC defines what could have been and what is, as well as the part you played in generating that difference.

The choices you made, as well as the possible choices you chose not to make, remain part of the living viable record within the memory (mind) of TBC. All information remains accessible indefinitely on the net within the thought-space or mind-space of $NPMR_N$ where TBC resides. The same is true for all fragments of you. All the information is there to be utilized to maximize the learning and growth of the quality of your consciousness. (Remember that most embarrassing moment of yours – the low point in your quality of being – it's all there, every last minute detail, everything you said, did, and thought! There are no secrets.) This school has one bodacious library! You simply need to know how to apply (yourself) for a card.

The concept of everything that can happen does happen represents an impressive array of information. The tricky part, common to accessing any huge information source, is to find the particular data you want. Knowing what is available and how it is categorized is crucial to developing the skills to access selected data efficiently. Over the centuries, many people have figured out how to tap into a portion of the available database. In general, the smaller your reality, and the less you understand the Big Picture, the easier it is to misinterpret what you find.

Let us peek into the OS file folder and see what data are in there. Thus far, we have discussed the computationally viable states of OS that lie on our collectively actualized history thread as well as the unactualized past states that represent all the possible significant choices you and everybody else could have made (but didn't), and the unactualized future states that represent that set of collective choices that may yet be made. All this information is available for you to experience or to study. The data are not volatile, and the database remains always available. Everyone in OS, physical or nonphysical, participates in generating the OS database.

Others can access your data (every intention, feeling, or thought you have ever had), but there are rules, and only those who have learned how can do so with full awareness. It is not likely that this information will be used for anything other than to help you grow up. However, if you break a rule or get pulled in before a judge, every detail will be scanned to determine the truth of the matter. It is impossible to hide or bluff. Everything is there. Everything! Yes, even that!

The unactualized past is not only recorded, but is alive and kicking in the form of an interactive digital simulation where every player is modeled by a set of expectation values relevant to the possible significant choices. The unactualized past, as well as the states on our history thread, continue indefinitely within the memory of TBC and maintain their interactive potential as unique simulations that are executed according to TBC's statistical models.

The unactualized future states work much the same way except that some of the time (for some individuals under given circumstances), access to certain files is denied. Access is typically denied when the information sought would decrease the probability that an individual will improve the quality of his or her consciousness. Access limitations to the probable-future information are applied for the same reason that we protect small children from hazards (such as loaded guns, dangerous animals, or toxic household chemicals) they are ill equipped to handle.

Many, if not most, people do access some of the data that exist in these probable realities. We call the results of these vaguely intentional queries our intuition. Some people are better than others at accessing the data that are available to their intuition. A developed intuition is a developed awareness of how to purposely access specific data residing in the probable reality and actualized historic databases. Some get good enough, by

combining a natural temperament or attitude with practice and technique, to amaze their friends and sometimes even eke out a living.

Typically the lack of Big Picture understanding, as well as fear, beliefs,
ego, and natural access limitations, leads to random and systematic errors.
These errors and limitations erode confidence and credibility and prevent
individuals from making a big splash among their skeptical reality-mates,
though some of their little individual splashes can occasionally be impressive.

You can project your sentience into any of these states and simply look
around. Or you can follow a sequence of connected states that produces
an effect that is like following a movie frame by frame. Run the frames by
fast enough and the action seems realistic – an experience – exactly like
being there. Because all possible choices are available, you can theoretically
make the story line run any way you want to, but it is much more useful
to follow variations of your personal choices and let all the other virtual
players make only their most probable moves given the new conditions.

By carefully using your focused intent, you can design and create
almost any query-filter you can imagine, apply it to this database, and
experience the results. The experienced end product is exactly like being
there. Imagine a 3D holographic movie where you control the conditions
or filter-sets that define the possibilities and probabilities of the action.
Lack of specificity adds random or nominal components to the filter definitions intended. If you do not know enough to define and control the
process precisely, the results appear to be jumbled and have a mind of
their own. In any case, your experience of this data (filtered by your clear
or jumbled intent) appears as if you are viewing, and participating in, a
reality populated by fully sentient beings with free will. That it is a simulation – a calculation space populated only by probability models of actual
beings with free will – would be transparent and thus unnoticeable.

Simulated or emulated digital consciousness does not seem artificial to
an individuated digital consciousness with free will. The only difference
between you and the emulation of you is that you have the free will necessary to modify the quality of your consciousness. The expectation value
driven choices of each modeled player are generally an accurate and realistic representation of that player's quality and intent.

Because of the way errors propagate through a system of unactualized
states, the further out (in time) that you go beyond a particular event, the
less accurate or meaningful are the results based upon that event. Most
errors occur for the following two reasons: Individuals, because of their
free will, sometime perform unexpectedly; and, sometimes there are two

or more extremely significant (potentially divergent) choices that have nearly equal probabilities of being actualized.

Now that we better understand the attributes of probable reality state vectors and their interactive player probability models, I can be more precise about the process of generating probable realities. Because of natural errors, the value of **M** can only get so large before it no longer makes sense to propagate further generations.

M does not have to remain forever fixed – it is simply easier to initially explain and understand the process that way. Where the probability distributions of discrete events become low and flat (fuzzy results), no further generations (no more child states) are calculated. However, where the distributions remain clear and crisp (high, distinct, peaked, and with individually resolved events), generation after generation is computed until clarity and resolution is lost. As a result, there is a statistical reasonability criterion that determines the number of generations that are propagated forward for any given set of circumstances with **M** being a nominal value.

> ▶ This short aside will help relax the intensity of the thought (or is that fog?) around your brain: With a little imagination, we could search out a sequence of contiguous states (thread) in the database of OS's unactualized state vectors that might be especially interesting. One interesting pair of threads to compare might be those leading to the Allies winning and also losing WW II – with an alternate (but unactualized) reality movie created by quickly flipping through the state frames of the most likely consequences of different choices. Our history has recorded the terrible price that was paid; now we could also ascertain the value of the benefits gained or lost as a function of time.
>
> Somewhere there is a virtual reality where Saddam Hussein and his army were pursued and destroyed completely during Desert Storm (the US vs. Iraq, Gulf War of 1991). Do you wonder what difference, if any, that might have made? What if John Kennedy, Martin Luther King, Mahatma Gandhi and Jesus had not been killed – where would that leave us now? What if the quarterback hadn't fumbled and you hadn't forgotten your ex-wife's birthday five years in a row? What difference does any particular event make in the long run? I am sure that you can think up many interesting questions relevant to your life.
>
> Your sense of the possibilities is your only limit as to what might be discovered by following a virtual thread along in OS's non-actualized database (past or future) within TBC. You could go there yourself and learn how various historical or personal what-if scenarios would have (most likely) turned out – a great source for writing fiction. In fact, much of our imaginative creativity and inspiration has its source in an intuitive

connection to the OS database within TBC. More importantly, it provides a great resource for understanding the impact and importance of your personal choices. As such, it provides an excellent tool for helping you design future experience packets, as well as learn from, and maximize the growth potential of your experience.

A word of caution to those who know how to at least partially open the door of their mind: Do not spend too much energy or time in nonphysical realities or you may neglect or waste the precious time you have in the one reality that is most important to your personal growth – physical reality. Much is perhaps interesting, but only a small subset is pertinent to your personal spiritual growth and therefore important. Do not invest too much time in what is not important. ◀

Back to work. A projection of a fragment of your sentient consciousness into a space-time reality could get spiritually dead-ended in a situation where you **believe** that you do not have significant choices left – you feel that you cannot produce uniquely useful or significant choices during future increments of DELTA-t. Under those circumstances, your inability to find a way out of the perceived dead end will typically cause you to find a way (perhaps a fatal disease or accident) to check out of, or shed, that particular reality. Though most humans do not check out because of being dead-ended, a significant minority do. There are usually many ways out of an imagined dead end, many good choices (other than termination of the experience packet) are typically available, but they remain invisible to a self-referential consciousness limited by belief, fear, and ego.

It should now be clear that because of the choices you make and the objects and beings you interact with, you have considerable influence over the reality in which you develop and evolve your being or consciousness. Interacting significantly with virtual players (mathematical models of yourself and other related beings, objects, and energy) within parallel realities is roughly analogous to playing against the computer in an educational computer game or performing parametric analysis within a simulation. These tools are there for your use, but do not become enamored of them, get unproductively lost in them, or become overly dependent upon them.

Abuse is not widespread because for the most part if you are wise enough to use the available tool sets coherently and powerfully, you are wise enough to use them wisely. The abuse that does occur is usually at low power or of a singular (as opposed to widespread) nature. The entire structure of $NPMR_N$, including the space-time part, is designed to provide you with the experience and opportunity you need to optimize your learning. The experiences, situations, and opportunities that confront you in PMR are for the most part the ones your evolving individuated consciousness needs in order to learn what is critical to its continued growth.

The path that you are on is your path of maximum opportunity as designed by you with the help, advice, and cooperation of others. You are involved with your lessons, opportunities, and choices as they define your spiritual evolution and growth within the reality that you and your co-conspirators (others who are significant to your existence or being) are producing.

What can you or others do that is significant enough to effect a change in the quality of your consciousness? Significant, within this context, means potentially capable of path-altering and life-changing effects. If the overall results of your **potentially** significant choice don't actually make a significant difference to you or anyone else (who might be affected by your choice), then that choice contains little power to effect a change in the quality of your consciousness.

When some guy, for example, makes an important choice, perhaps to marry and have children with that lovely woman he has been dating for the past five years, popping the question produces choices for her, her family, her roommate, her landlord, his mother, his other girlfriends, and so on. All affected are in an interactive dance that alters everyone's personal reality as each make their choices. If the same man makes a choice to scratch his head with his right hand instead of his left hand, as is his habit, nothing significant is either affected or effected – no opportunities for growth are likely produced – it is an insignificant act that does not have to be tracked.

▶ Keep in mind that significance in the Big Picture is about changing the quality (entropy) of your consciousness while significance in the little picture is most often related to satisfying the needs, wants, desires, and expectations of your ego. Trivial choices with a low probability of Big Picture significance can be safely ignored. Calculations of the probability of significance do not have to be perfect – if they are consistent and reasonable, a working virtual reality can be functionally defined. Given the interactive complexity and purpose of PMR, consistency and the continual availability of a rich array of significant choices are much more important than accuracy in every minute detail. By ignoring choices that are expected to be trivial, the requirement for computational resources to compute the likely consequences of virtual reality interactions is greatly reduced. Likewise, the computations of consequences do not have to be perfect. Absolute precision of each event and its interactions and eventual outcome (as required by determinism) is incompatible with a finite, statistics based digital virtual reality.

Extreme precision in defining and progressing each event is not important to the purpose of PMR. As long as the psi uncertainty principle constrains the mechanics of reality to present the appearance of an interactive objective causal reality to the individuated units of consciousness evolving within PMR, the functionality, value, and purpose of PMR is adequately maintained. Physicists working from a deterministic model of reality have been tripping over this conceptual error ever since Isaac Newton uncovered a few significant pieces of PMR's space-time rule-set. Their insistence that all phenomena be forced into a deterministic causal model has blinded them to the true nature of the reality they are trying to understand.

Even statistics based quantum mechanics is coerced to exist within a deterministic straight jacket; the statistical description at the root is seen as only a nebulous intermediate step before the final states collapse to some measurable physical reality. The joke is: The statistical description represents the actual reality while the resultant final physical state represents only a virtual shadow of the more fundamental statistical reality. Getting it wrong is normal enough, but believing it to be the exact opposite of how it actually is adds a touch of humor to the sanctimonious recitation of scientific dogma by the high priests of science. Belief traps and narrow paradigms don't make you stupid; they simply limit your capacity and ability to understand. ◀

The "everything significant that can happen does happen" process would seem to be an excellent experimental design for a spiritual $NPMR_N$. It involves everyone in OS. PMR entities are engaged in a physical experience that constitutes a self-designed consciousness quality increasing, entropy reducing, training class called life that takes place within a learning lab called PMR. In the class of life, within the earth frame of reference, what you do (the choices you make) and the shared reality you live in accurately and inexorably reflects the quality of your being. Additionally, it offers you the specific opportunities (potential choices) you need to enhance the growth of your individuated consciousness.

The meaning, purpose, and direction of all individual and collective consciousness (life) within $NPMR_N$ (of which our PMR is a subset) is focused and animated by the evolutionary requirements of the consciousness cycle and motivated by the purpose of the larger consciousness ecosystem. The fundamental driving imperative is: Evolve! Grow! Become more! Seek out profitability! Reduce the entropy of your consciousness, become animated and motivated by love. To accomplish this evolution revolution, at every level of existence, the Fundamental Process promotes profitability while eliminating failure. If a being is highly sentient, successful evolution requires that this being use its energy to decrease the entropy of its individuated consciousness. By doing so, it decreases the entropy of the entire AUM consciousness system as it decreases the entropy of its own oversoul.

Do you now have some faint notion of the immensity of this reality, and how minute, by comparison, OS and our history are? Given that OS and its history represent a miniscule subset of the possible content that is interacting or stored within the Big Fractal Picture, how much more minute, by comparison, is our physical universe, our solar system, earth, humanity, and you? Small, yes! But, we are still significant. Very significant indeed! Notice how small a virus or a bacterium seems to us and how viruses and bacteria are not only significant, but also vital to our physical existence. Imagine how small a hydrogen atom must seem to a virus – are hydrogen atoms significant and critical to our continued existence? Imagine how small a neutrino must seem to an atom – are neutrinos significant? Yes! All are very significant! You have now seen how the Big Picture of your immediate (local) reality works – this is not necessarily the mechanics of the biggest picture, but it is extremely big relative to our typical PMR view all the same. Is your view of our immediate larger reality like the atom's view of a human? Maybe.

If your sense of personal importance or significance is shattered by the sheer size of the bigger picture, you should concentrate on playing your part (your important part) to the best of your ability. You should focus on maximizing every opportunity to grow the quality of your being – the success of which is reflected by the choices you make. Be your best and don't worry too much about the small stuff, or the big stuff. Just be. That's what the neutrinos, atoms and viruses do – and hey, who ever heard of an unhappy neutrino, a neurotic atom, or a depressed virus?

83

■ ■ ■

Ramifications
Changing the Future,
Changing the Past, and
the Quantization of Time

■ ■ ■

Let us pull together and summarize what we have learned thus far. This short condensation and solidification of the concepts presented in the previous several chapters will get us ready to explore the ramifications of existing and experiencing within a virtual digital reality. This is where the fun starts, but first we need to make sure that we have a firm grip on the basics.

1. Future Probabilities are Non-Binding – Things can Change

We can say that the past, present, and future all occur and exist within TBC simultaneously – if by future we mean the probable un-actualized (potential) realities. The future is not a done deal; it exists only as a complete set of probable futures with varying probability densities (peaks on the surface) forming and changing around possible future events. All these states exist within TBC at the same time. All are fertile and capable of producing new child states given a change in initial conditions. All can be visited and experienced by an appropriately aware consciousness. Consider the many software applications, simulations, and folders sitting in your computer. They are all extant at the same time, all are ready to do whatever they do, some are dynamically executing (running, changing, operating, incrementing local time) while others are idle awaiting input or simply storing results.

Some probable reality surface peaks (within a dynamically executing reality) may be narrow, indicating a more precise probable time of potential actualization, while others are broad, indicating larger uncertainty in the time of potential actualization. These probability densities (peaks) may vary like the topology of any other surface function, in amplitude, width, shape, and area under the curve. As DELTA-t increments and our PMR time moves on, choices (random and sentient) are made and one of the potential future states is actualized into the present. This creates a new set of probabilities and expectations about what might happen next. The entire set of future possibilities is recalculated after each DELTA-t.

Because our future is both dynamic (updated every DELTA-t) and statistical, it exists only as a set of uncertain probable realities; consequently, it has a fundamentally different nature than our past (which is the set of stored reality system state vectors that was actualized by the beings, objects, and energy of OS). Although many future events are accurately predictable (which is the basis for precognition) because of high expectation values, they have not yet been actualized or chosen into the OS present – and things can and do change. This is why Eastern traditions talk only about past lives and not future lives and why karma is attached only to past events and situations, not future ones. We have not yet actualized our future – it could turn out to be any one of a large set of possibilities, some being more likely than others. We turn the probable future into our present by making our collective choices and thus removing all uncertainty.

The probable future is ours to peruse within limits set for our own good. An individual, a group, or an event can have its future probable reality surface calculated – it simply depends on how you want to filter or query the probable future database and whether the circumstances support your being granted access to the data. Access must be both earned by you and granted by the system administrator.

Having someone else retrieve the data for you will not get you around the access problem. Access is granted by the system administrator based on the probable use of the data, not on who retrieves it. There is no way to cheat, trick, or force the system into providing information that will not produce a positive spiritual benefit for all who might be affected by it.

One can modify the probable future surface with focused intent, thereby raising or lowering the expectation value of a future possibility. With a properly focused mind, new peaks can sometimes appear and others of long standing may slowly disappear from the probable reality surface representing a given entity. This is generally how the mind utilizes nonphysical mental energy to manipulate future events within PMR without violating the psi uncertainty principle. The psi uncertainty principle tells us that fuzzy short peaks are easier to modify than tall sharp ones and that events become increasingly more and more difficult to manipulate the closer they get to the present.

▶ I said "slowly disappear" in the previous paragraph out of practical considerations not because slowness is theoretically necessary. The organizational energy that created that peak in the first place must be dissipated in order to remove or reduce that event's expectation value. If an event's expectation (future probability of being actualized) still has energy being poured into it, that energy must be dissipated as well if its associated peak is to be eliminated. Your mind, depending on your quality, interest, and instantaneous access privileges, can only apply so much power (do so much work per unit of time) to reorganize the digital energy that represents the expectation you are trying to change. Your ability to manipulate bits is limited and must compete with the ability of others to intentionally or unintentionally manipulate those same bits.

In Chapter 47, Book 2, we used the term "force of being" and said:

*Constraints come in many forms. For example, if someone who has the ability to manipulate nonphysical energy is asked to remove (dissipate or dematerialize) a tumor (noticeable lump) from someone's PMR body, the energy required is dependent upon, among other things, the degree to which this tumor is connected to PMR reality – its force of being in PMR. Quantitatively [force of being is associated with] the expectation value of the future event. If the body's owner and a few others are mildly distressed about the **possibility** of a malignant tumor, the removal energy may be relatively low (easy to accomplish). In this situation, much uncertainty exists within PMR – but not necessarily for the one viewing and manipulating the body's nonphysical energy from within NPMR – to that person, both the physical and nonphysical properties of the lump may be clear.*

If, on the other hand, four doctors and half the residents at the local hospital have looked at the CAT scan or felt the lump and are relatively certain the tumor is malignant, the removal energy is somewhat higher. When all of the above get the biopsy report confirming a fast growing, incurable, always-deadly malignancy, the required energy increases. The more firmly the malignant tumor's existence and likely outcome (degree of causal certainty) is held and shared in the minds and expectations of credible sentient beings, the taller and denser its probability function becomes. The uncertainty of the outcome dissipates and the probable event of dying in PMR from this cancer becomes much more difficult to change by manipulating energy in $NPMR_N$. Fortunately, the intent and attitude (mental focus) of the body's owner has the greatest potential impact. Unfortunately, this attitude is often driven by its fear and the opinions and fears of others.

Now, by definition, it takes a miracle, where before no miracle was required. The confident knowledge of the doctors, which is based on test results and historical precedent (mortality statistics), actually affects (decreases) the probability of actualization of other alternative possibilities such as the cancer spontaneously going into remission, or the tumor turning out to be benign. As in quantum mechanics, performing the measurement forces the result to pick a state compatible with PMR causality (compatible with the PMR space-time rule-set). Typically, the most likely state at the time of the measurement is picked unless there are several states of equal probability, then one outcome is picked randomly from among the set of outcomes (future states) that are all most likely. The individual with the tumor, along with his or her friends and loved ones, can inadvertently help drive the final outcome to an unhappy ending by causing the probability of a fatal outcome to grow, and the probability of a non-fatal outcome to shrink. Beware: you can be easily drawn into a fatal dance of expectation with those connected to you, or to your condition. The best time for intervention, whether from PMR or NPMR, is long before "fatal" becomes a near inevitability.

Now you can understand what was being said from a much broader perspective. The terms "density" and "force of being" refer to the cumulative amount of digital energy of organization invested in a given expectation or peak on the future probable reality surface. The more digital energy that is collectively and individually invested in a given outcome, the more resistant its expectation value (likelihood of future actualization) is to change.

Digital energy can accumulate just as physical energy can (charge in a battery or water being pumped into a tall tank). Just because you know how to reorganize bits and manipulate digital energy within the larger consciousness ecosystem, doesn't mean you can do whatever you want. Even if you obtain the necessary access and permission (which are not always easy to get and are granted on a case-by-case basis), you still have a limited ability to change events that exhibit a strong force of being. Yours is not the only intent and organizing force active within OS. There are limits to what you can do in NPMR, just like there are limits to what you can do in PMR – and for much the same reasons.

All realities, all virtual existences, operate according to their own causality (rule-sets) and you must live and operate interactively within those rules. Forget about your ego-dream of becoming superman or a miracle worker – those represent an exceptionally poor choice of goals. Existence is not about having it your way or impressing your friends. Nor is it about helping your fellow humans to be healthier, happier, and more comfortable. Given the right intent, those may be very worthwhile activities that help you focus and improve your learning within PMR, however, your existence is about **you** individually increasing the quality of **your** consciousness and, where possible, facilitating others to do the same. Spiritual growth is a personal accomplishment that must be achieved individually by each unit of individuated consciousness whether that unit is you, your puppy dog, or the Big Cheese. Helping others is necessarily on the path to that accomplishment but it is a byproduct, not the final goal or endpoint. ◀

State vectors are not closed systems; outside energy, beings and objects can interact at any time. TBC simply keeps up with all the changes and their effects. Our reality is dynamic, interactive and uncertain because that is the nature of individuated consciousness.

There is no conflict between free will, and predestination or precognition. Visions of the future are simply visions of probable realities. The past (all stored reality system state vectors, whether we have actualized them or not) is alive, vibrant, available, open to us (for inspection or what-if analysis) through $NPMR_N$, and fully capable of calculating new realities based upon new initial conditions. Future probable reality surfaces are generated (for each state of OS, each DELTA-t) by looking at only the most probable states that exist downstream among our unactualized future possibilities.

State vectors can be perceived either singly or strung sequentially together as frames of a movie. You can experience probable future states (as a nonphysical interacting observer) by focusing your intent while sentient in $NPMR_N$ (like experiencing a movie from a god's eye view). The probability, likelihood, or expectation value of each and every potential future state is known, but it may or may not be available to you (along with other details of the future).

You may or may not be blocked from knowing the probable future by information filters that reflect a concern with the impact that this knowledge might have on the evolving quality of your consciousness – just as an adult would prevent children from playing with fire, or accessing inappropriate material on TV, or on the World Wide Web (WWW). It is **your** quality, **your** degree of consciousness development (lack of entropy), which determines what you can and cannot clearly experience – regardless of who is receiving the data. If future probabilities are shut off from your knowing, the problem may be one of simple ignorance – you simply may not know how to access the information. Or perhaps a limit has been placed on your access to knowledge by wiser entities who are aware of existing intrinsic personal limitations, and who are looking out for your best interests. For example, you may cause more harm than good to the process of developing the quality of your consciousness if the motivation for gaining the information is in the service of your ego, for personal material gain, or likely to feed or encourage other high entropy intents.

2. Changing the Past

Our past (the history of OS) can be thought of as a series, or thread of actualized states. Because TBC stores **all** results at the end of each

DELTA-t interval, our past (as well as other non-actualized past possibilities) continues to exist and is accessible through $NPMR_N$. You may decide to modify the actualized past by projecting a fragment of your consciousness into a reality described by a time now past but still defined in TBC with its associated DELTA-t. (You may make the trip by decrementing DELTA-t along our history thread to that past point in time or simply specify the time, place, and reality frame). At this point, you may use your focused free will intent to follow the thread of a specifically different set of choices and a specifically different set of conditions than those that were actualized. The resultant possible but unactualized past that you create is experienced as an entirely unique and realistic set of sequential happenings with which you may actively interact. However, because you can process only so much data per unit time, you will be aware only of those interactions that you specify as being of interest to you.

Relative to the participating observer, it **appears** as if the initial conditions of the beginning state vector have been changed to reflect the observer's new choices and conditions, and that all the various players (represented by predictive mathematical models) interact with those changed initial conditions by modifying their choices, intents, and actions. The observer sees a new reality, based on the new initial conditions that he or she brought to the starting point, branch from the original. The process repeats and continues as long as the observer maintains his focus and is interested in seeing what happens next. Interacting or experiencing within this new series of events appears to be identical to interacting or experiencing within the present OS reality.

That the other players are mathematical models that do not exercise free will is entirely transparent and undetectable because the database contains **all** significant possibilities ordered by their expectation values. It simply appears to the participating observer driving this re-write of history that a new reality has branched from the historical beginning point (state vector) where he first changed the initial conditions based on a modified free will input to that system. Another way of looking at this is that the observer is implementing a specific series of dynamic queries of the TBC database by his intent.

The old thread (our history) is not changed or affected in any way. It continues to increment (move forward in time), unperturbed by the apparent new branch diverging from that old (past) DELTA-t. From the perspective of TBC, no new reality is created – previously stored data from the everything-that-could-happen-does-happen database is merely being accessed and ordered differently. As you modify the input data (choices) within any given state vector new probability configurations and expectation values (for yourself or for others) are automatically computed. No new files or reality state vectors are created or stored; you are simply exploring the possibilities within a virtual simulation. This virtual simulation allows one to explore probable event-chains or causal threads within a higher order virtual simulation. Having a simple virtual simulator within a more complex larger virtual simulation might seem a tad confusing, but that is simply the nature of digital reality. Employing simulations within simulations within simulations is similar to utilizing nested subroutines or sub-simulations – a common programming practice within PMR and NPMR.

The OS database of unactualized possibilities within TBC may be viewed, queried, explored, experienced, or lived in unique new ways based upon the intent of the participating observer. What you do **outside** your reality (such as exploring and interacting within the OS database of unactualized possibilities) **may** create the experience of new realities, but it imposes no new data on TBC, nor does it affect your home reality (OS) except in as much as it changes the quality of your consciousness (providing the experience precipitated learning and growth).

Only what you do with your free will choices **inside** your larger reality from either PMR or NPMR can change that reality directly (for the one who is making the changes as well as for others). Individual interaction within any past or probable future set of states within TBC's calculation-space takes place outside your operational reality and has no effect on any sentient being (consciousness with free will) other than you. One might say that the participating observer executing the what-if analysis is working only with a volatile scratch copy of the original state file so that the changes he makes provide the full experience of an operational reality but leave no permanent record. From another equivalent viewpoint, it is clear that running a dynamic query filter does not alter the original data set. Though your what-if analysis leaves no permanent trace, a generated experience (unique tour through the data) can be reconstituted from the original data anytime.

Our PMR past (as well as unactualized past possibilities) is always alive and well in TBC. The **experiencing** of new pasts, represented by either the reality branching or database querying viewpoints previously described, are always available and accessible via the RWW. Whether you see this process as a database query or the branching of a new reality is relative to your perspective. Both views are identical – it is only a matter of your frame of reference. Thus in summary, you may create and experience a new reality that branches from some past or probable future state vector by introducing new initial conditions into the beginning state, or equivalently, you may trace a unique path through TBC's database of unactualized past or future possibilities.

Your present fragment in this present reality (what and where you think you are now) is not necessarily the main one. Mostly there is no main one – all your various projections into various present states within various reality systems containing beings with free will are merely different. All these divergent realities (sometimes called parallel realities because they are initially similar to each other or may share a common state) exist simultaneously in $NPMR_N$ as books all exist simultaneously in the library or files and folders all exist simultaneously in the computer. Imagine every OS state vector since the beginning of space-time time, as well as all those projected $\mathbf{M} \cdot (\text{DELTA-t})$ into the future, existing within a huge on-line database in TBC. The past, present and probable future of OS all exist simultaneously within one database – a database that is accessible by the queries you design and execute with your intent.

The present is unique because it contains the actual ongoing virtual reality game or happening in which the players exercise free will. It constitutes the learning lab and generates the direct experience that provides custom-designed original opportunity to learn and to grow the quality of your consciousness. Everything else is the result of calculating and storing the possibilities – data manipulation built upon the analysis of previous free will choices. Though these manipulations produce some fine experiential educational tools, life's game of consciousness evolution is played, for the most part, in the present moment.

New potential realities are constantly being generated from present choices. Likewise, modified histories are available for you to experience as your intent follows any number of possible threads winding their way through TBC's state vector database. As free will choices are made, particular possibilities are actualized into the present. The present becomes the past when the next DELTA-t time increment is initiated. This actualized history is generated by the beings, objects, and energy that interactively coexist in a particular present reality state at a given time increment DELTA-t. Every actualized and unactualized reality state vector, representing all variables, possibilities, and probabilities for each DELTA-t, is saved in the state vector database, so that **all** possible realities, beginning with the first DELTA-t, simultaneously coexist within TBC. These saved state vectors are open to the possibility of a change of variables or initial conditions so that additional learning can be squeezed from a what-if analysis that illuminates the long and short-term impacts of a given choice or series of choices.

If AUM thought that a particular state vector was particularly interesting or held particular promise, it could use that as a pattern to generate another (or many) evolutionary experimental branch populated with entities with free wills. Because we and all else (including AUM) are digital, it could simply start with a copy of the original (such as OS) and then make modifications as required before setting it off to evolve on its own. Thus, there is another level of parallel realities that evolve on their own, are populated by entities with free will, and are not merely uniquely filtered output reports of existing data. Some of the various independent PMRs share common states on their history threads (one is a branch of the other).

You (your present conscious awareness) can visit any state of any reality system anytime as an observer by focusing your intent while sentient in $NPMR_N$. If you significantly interact with any of these states while in $NPMR_N$ or become experientially aware in them and have a significant potential to direct the action with your free will intent, you will have an opportunity to learn and grow from the experience.

The generation of a unique reality branch does not modify or change other previously existing reality states or sequence of states. What was, and what is, continues undisturbed on its merry way, unaffected by either the creation of what-if branches within the calculation-space of TBC, or the creation of new niches within the greater consciousness ecosystem fractal.

A common sci-fi plot element requires changes in the present and future to follow instantaneously from inadvertent or purposeful changes to the past (usually made by a nefarious or bungling time traveler). The idea that any change affects everything downstream is based on the erroneous assumption of a continuous reality instead of a digital one and a fundamental misunderstanding of the functional relationships that interconnect the past, present, and future. Fiddling with OS's unactualized or actualized past has no effect on the existing OS data within TBC and no effect on the present choices being made. The one extremely important effect it does have is to provide learning opportunities to individuated units of aware consciousness that result in a decrease of the units' entropy. Simply running a query that results in a custom build-as-you-go interactive holographic movie output report does not change the data in the database – but it may change you. And you may change future intents and choices that will define future states.

3. The Significance of Quantized Time and the Concept of an Even Bigger Computer (EBC)

Let us re-examine quantized time and TBC from the bigger picture perspective of multiple PMRs each within their own OS and progressing along by incrementing their own time increments. I will let the subscript k differentiate and enumerate these various other OS-PMR-DELTA-t type systems. Time will accumulate in these systems by adding up all the DELTA-t time quanta that have been sequentially incremented within that system.

Using this notation, our OS is only one of multiple OS_k simulations in TBC, each having its own PMR_k. The total elapsed time from the beginning of time (the first DELTA-t_k in each PMR_k) is T_k. Here, $T_k = K_k$ (DELTA-t_k) where K_k is an integer, specific to each PMR_k. DELTA-t_k is defined as the time quantum of OS_k, and is the smallest discrete time increment within PMR_k. Time in any PMR_k seems continuous because DELTA-t_k is so small. Each virtual physical reality represented by a PMR_k runs on its own local time defined by incrementing its own DELTA-t_k. As far as I can tell, all the DELTA-t_k are about the same size. However, different PMR_k may have different values of K_k, and therefore different values of T_k. Among the PMR_k within their respective OS_k, there are often shared beginnings with major decision points creating branches where OS_k and OS_{k+1} go off their own way – an attribute that only a digital reality can easily manage.

It is quantized time and the storage medium in TBC that allows history to remain alive and vital as a seed for additional virtual realities and as an educational tool set.

It is quantized time and digital processing that allows independent branching of realities so that a change to a past state does not cause subsequent, automatic, and instantaneous change to occur in the future belonging to that past.

It is quantized time that allows discrete and independent reality system state vectors, each with a probable future.

Quantized time allows the probable future to be progressed and tailored efficiently to each discrete state vector.

Quantized time gives TBC control and flexibility because it can stop incrementing (pause) T_k after any particular DELTA-t_k increment for as long as it likes (time literally stands still) and then start it back up again without noticeable effects. AUM-EBC-TBC, due to the characteristics and properties of digital simulation, can at any time take (copy) any reality state vector, modify its initial conditions and let it be the seed for a new virtual reality experience simulator learning lab. That is analogous to taking the best or most interesting bacteria found growing in all the Petri dishes and using them as the basis for a new set of experiments.

Quantization and the digital nature of TBC also allow TBC (to the extent that TBC is fast enough) to speed up or slow down any specific reality PMR_k by incrementing the $DELTA-t_k$ as fast as it wants (not to exceed as fast as it can) relative to $NPMR_N$ real-time by adjusting the size of ($DELTA-t_k$) or modifying the time base defining the fundamental **simulation** time quanta $delta-t_k$.

Because of the properties of digital simulation, AUM and TBC could rewind our reality movie or do any number of tricks (similar to the magical things we do by manipulating digital audio and video in our computers). Fortunately for us, tricks for the sake of tricks eventually become boring and disturb serious science projects. This type of meddling in ongoing consciousness systems is not often done. AUM is not merely a big kid playing computer games. Sure, AUM might create a reality-dimension or two for amusement or gaming, but it has nothing to do with our local reality (OS) – unless of course, we **are** the fun one. I am having fun, aren't you?

Because of quantized or incremented time, TBC can run our play, our reality movie, in fast forward or reverse as well as in slow motion or pause it (relative to time in $NPMR_N$). Our sense of time, our PMR time, is thus an **artificial construct** relative to time as it exists in $NPMR_N$. Likewise, $NPMR_N$ time is an artificial construct relative to time as it exists in NPMR. AUM is the dude with the smallest time quantum – that fact puts it in charge – as the originator of the fundamental beat that everyone else must dance to. Think of PMR as a subroutine in a larger simulation that is incremented by DELTA-t because the objects, energy, and beings it models only change infinitesimally relative to one DELTA-t.

Quantized time allows AUM, or whatever part of AUM is designing the software for EBC, to control the experiments or the consciousness cycle for optimal results. The Space-Time N-Division Management Team can collect and further manipulate the data it generates. It can easily terminate unproductive scenarios and rerun particularly interesting ones with or without new input data or new random number seeds. Because consciousness and thus reality is digital, N-Division's Management Team can easily update hardware or software, and modify experimental design on the fly without disturbing the experiments. AUM's experiments are flexible and optimized to capture all possible outcomes – they create their own complete set of result statistics about evolving consciousness. All such digital machinations are completely transparent to the players – they, for the most part, remain completely clueless within a reality that, by design, always appears physical and continuous.

Quantized time and digital processing allows for the simultaneous discrete existence of a variable number of multiple projections of you. It allows TBC and other sentient beings that are aware in $NPMR_N$ to run what-if analysis by providing a dynamic representation (mathematical model) of all the players and all the possible significant conditions.

The time in $NPMR_N$ appears to be fundamental and continuous to its residents. Nevertheless, it is merely a higher-level loop in the simulated time construct that is defined or manufactured within EBC and TBC. Although we know that the passage of time in $NPMR_N$ is created by a series of even smaller quantized time increments, you would not be able to experientially notice or measure that small a time increment from within $NPMR_N$ because of your local perspective in that reality. Similarly, you cannot be aware of $NPMR_N$ from the limited local perspective of PMR. To become aware within the reality that is **parenting** $NPMR_N$, you must gain the perspective of the next higher level in the reality onion. You have the proper vantage point to analyze the dynamics, content and structure of a reality only from the perspective of the next level of generality.

From an $NPMR_N$ perspective, NPMR and beyond represents a reality that appears mystical and lies beyond the logical causality of the residents of $NPMR_N$. A detailed exploration of the mechanics of NPMR would require a sentient perspective from the next higher level of organization **beyond** the various $NPMR_n$. All the various $NPMR_n$ seem to run on independent, though similarly derived, time bases.

From my experience, I can clearly **infer** a larger system (run within the EBC – Even Bigger Computer) representing another higher level of organization within AUM wherein TBC is either emulated within an outer loop or simply a subset of the EBC. This larger system is dynamically driven by a time quantum that is much smaller than the one that animates OS.

I have not yet **directly** experienced such a reality as a well-defined understandable place with well-defined rules. I do not yet know how to interpret those experience data within my limited NPMR and $NPMR_n$ level of awareness. When dealing with phenomena that relate to reality systems beyond $NPMR_n$, I am like a PMR scientist working with atoms and electrons that he or she can infer, but cannot directly experience. Such a scientist can experience only the effects of atoms and electrons, but not the atoms and electrons themselves.

That gives me something to work on in my spare time. I have been exploring NPMR, $NPMR_N$, OS, and PMR for only thirty years – there is much that I have not yet seen and experienced. Perhaps the $NPMR_n$ represent the outermost layer of our **practical** (operational) reality onion, leaving NPMR as a simple $NPMR_n$ container or media – and only AUM beyond that. One must eventually run into the outer edges of the greater consciousness ecosystem, the boundary of the AUM-reality itself.

That AUM is perhaps a single cell populating the lower intestine of an AUMosaurus is probably far beyond the reach of AUM's awareness. You and I should not feel too bad if we do not get that far.

84

■ ■ ■

Ramifications
Communications, Time Travel,
and Teleportation

■ ■ ■

Communications

Neither apparent PMR distance nor spatial separation affects signal quality on the RWW net within $NPMR_N$. For example, an earthbound individual would experience no noticeable time lag between questions and answers if he were telepathically carrying on a conversation with someone located near Alpha Centauri (a star 4.4 light-years away from the earth). Transmission time and variations in signal-to-noise ratio are likewise not noticeably affected by a perceived separation between the sender and the receiver even if each is in different PMRs or in different $NPMR_n$ – much as a web page (as seen by my browser) hosted on a server in Australia is as clearly and quickly represented on my monitor as a web page hosted on a server in the next room. It would appear to take no longer to access one grouping of data (dimension) within TBC as any other – distance is only virtual and has no meaning outside of the virtual PMR_k.

Time Travel and Teleportation

Energy (the ability to rearrange bits and affect organization, and thus content, within a system) from $NPMR_N$ can be directed to any PMR_k at any point on its thread of actualized history, including our PMR as a particular instance of PMR_k. Because energy can propagate between any PMR_k and any $NPMR_n$, (as well as between everything else that is on the NPMR RWW net), there seems to be no a priori reason why **what we perceive and experience** as physical matter could not be transported either directly, or by using NPMR$_N$ as an intermediate hub or router, between realities within NPMR$_N$. The same goes for conscious embodied beings.

In fact, because physical reality is only a virtual reality (a rule-set that defines the experience of individuated consciousness within the space-time subset of AUM), jumping or teleporting from one experience-set to another one in a different time and place would not seem impossible. Dimension hopping between realities is not so difficult. Because all is consciousness, it would seem that teleporting through what appears to you as time or space would merely be a matter of getting around within TBC.

How could an AI Guy in a WWII war game simulation, get into the WWI or Viet Nam war game simulation running in the same computer? He would need to know, understand, and be able to operate within the larger computer reality. He would need to be able to copy his algorithms, memory and data (himself) into the other simulation and insert himself properly into the appropriate time loop drivers. He would need to understand the content and mechanics of the new simulation enough to modify the code of the new simulation host to integrate himself (share data) into that simulation in a useful way.

Do you see the levels above his local reality in which he would have to operate in order to teleport into a totally different simulation? But what if he wanted only to teleport around within his own simulation – perhaps be two places at the same time? That's much easier but he still needs to be an exceptionally aware and savvy AI Guy in order to modify the data that drives his simulation and to avoid violating the psi uncertainty principle. The same applies to us.

Teleporting a virtual body, which **appears** to be physical, between different local realities is analogous to switching between two virtual reality games – say going from Jungle Safari to Alpine Adventure – entirely possible, but you need to be good at parallel processing and interacting with (programming) the computer. Teleportation is only a matter of switching your consciousness to a different energy packet exchange interface (see Chapter 66, Book 2) after establishing the appropriate data links so that all interactions with other players are properly taken into account. That may seem challenging but is not as difficult as it sounds. The user interface between an individuated consciousness aware in NPMR$_N$ and TBC simplifies most of the required actions.

Fortunately, there is little advantage, need, or incentive for you to take your PMR body along as you explore the best sites on the RWW. Keep in mind that what is physical is relative to the observer. We appear to be nonphysical to other PMRs as they appear to be nonphysical to us. Thus, the idea of dragging a **physical** body to other reality frames makes no sense. There is no physical body, except in your mind. Teleporting a **physical** body has meaning only within your own local reality where you must figure out a way around the psi uncertainty principle. It is much easier and more practical to manifest an **additional** body appropriate to the place visited. Possessing multiple bodies in multiple realities is not a problem.

Why bother teleporting? There is little advantage and no point to dragging a unique body around with you wherever you go. Being either apparently physical or nonphysical in several reality dimensions at one time is not difficult if you can parallel process sufficiently well. Having one unique identical body at two places at the same time within the same reality stresses the psi uncertainty requirements more than having two physically different bodies at two locations at the same time. Having one physical body and one or more nonphysical bodies in the same physical reality at the same time does not stress the psi uncertainty principle.

Actually, I cannot imagine a situation where anyone's spiritual growth would be enhanced by teleporting a physical body around within a single physical reality. It is simply not important, and not worth much effort. On the other hand, though it is entirely irrelevant in the Big Picture, it certainly would be lots of fun – imagine all the great gags you could pull. Also, it would be convenient (like being rich) as well as save transportation costs and travel time. If great gags and convenience are your goals, get working on it. Beam me up, Scottie.

Teleporting your awareness (nonphysical body) is less complicated because the objective physical experience connection to the space-time rule-set (your physical body) is left behind undisturbed. Traveling with only the consciousness is simpler than manifesting an additional body because local psi uncertainty principles are more easily satisfied. The mind is free to go, be aware, and experience anything anywhere on the Big Picture reality net (with a few access restrictions). Many PMR residents do not comprehend this; consequently, they never go anywhere – they simply stay at home and baby-sit their bodies. They have no idea there is a wholly new and magnificent reality out there to explore – or how important it is to the success of their mission – or for that matter, that there even is a purpose or mission to their life. They have their you-know-what stuck in a belief trap. Contrary to popular opinion, ignorance is not bliss, it is just ignorance.

Unique addresses exist for any 3D space-time point or being in any PMR (the same goes for points and beings outside space-time). Conscious beings, embodied or not, can use their minds as a vehicle to directly, or by using $NPMR_N$ as an intermediate hub or router, transport their awareness between any two points within our PMR (our universe) or within the larger reality. Think of it as hopping between different clumps of memory cells or locations in TBC. Using local coordinates that uniquely define a particular state vector, your awareness can end up anywhere you want it to in the reality of your choice.

Thus for example, you can make a custom designed worm hole from a point in your present space-time-reality to a point on our shared history thread, or to any other space-time-reality point in any other reality (our PMR or not, our history thread or not). This process is successfully and commonly used by beings (embodied in PMR or not) that are sentient in $NPMR_N$. Navigation of your awareness is consciously directed by focused intent over a low noise background and is limited by the extent of your knowledge and your experience. As long as there is a unique destination URL or address (and you know what it is – or know someone who knows), you can navigate to any dimension, reality, place (local coordinates), person (unique consciousness ID), or thing.

The address reads as any other, from the more general to the more specific. Say you want to check out the healing process of a stomach ulcer that is bothering a good friend of yours who lives in some other PMR, the address might be as simple as: reality, individual, stomach, ulcerated cells – as long as the individual was unique within that reality. You mail your energy or consciousness by intent (see Topics 5 through 9 in Chapter 76 of this book). Essentially this is how remote viewers (RV) and out-of-body experiencers (OOBE), as well as other mind travelers, get to and from wherever they go, whether they express it like this or not.

Clear and complete intent is critical. Knowing you need a city and street but being unaware that a state and house number are also required to find a particular resident creates a large **set** of possible solutions because some cities in different states may have the same name and many cities have the same street names. If intent is not clear and informed about how to specify the uniqueness of an address (at all levels of the address), those parts of the address that are not sufficiently specific are often filled in with a random selection from, or a smear across, (depending on the circumstances) the possible solution set. Unfortunately, the explorer often never knows that randomness or inaccuracy has crept into his vision. If the intent is foggy or imprecise, the results might be foggy (smeared across the possibilities) or they might be clear but inaccurate (random selection).

Given the numbers of multiple realities and given that similar reality branches could be almost identical in certain areas (unrelated to the uniqueness that created the branch in the first place), it is easy to understand that getting the right address (knowing what you are doing) becomes more and more difficult the farther you stray from home (going beyond your comfortable understanding). This is, and has always been, a common problem of all explorers – from Christopher Columbus to Lewis and Clark to Captain Kirk. The solution? Carefully and continually turn very small parts of what is unknown into what is known. It is a slow, time consuming process, and in NPMR, must be accomplished by each explorer by and for himself.

No one can give you a detailed map of the territory, but they can give you specific addresses. Successful travelers (those who have controlled out-of-body-experiences or who successfully practice remote-viewing) understand the importance of a clear address, the need to suspend the noisy ego, and the necessity for many hours of developmental experience to guide the way.

Once you have projected your awareness from NPMR to any reality-system state vector (past or probable-future, actualized or non-actualized, OS or otherwise), you can easily specify your intent and let DELTA-t run forward or backward from that point of entry to make a realistic movie from the sequential frames. However, you must be careful at branch points to not get sidetracked into a reality other than the one intended. As you might imagine, it is easy to get lost and turned around – your intent must be unique (which is sometimes difficult because often you are not aware of all the variables), steady, clear, and complete. Anything else may well invoke that old computer-programming truism: Garbage in, garbage out. Awareness of the potential difficulties and plenty of carefully evaluated experience are the only remedies.

Flying Faster than a Speeding Photon, and the Art of Generating Multiple Bodies That Simultaneously Belong to Your Present Personal Awareness.

Warp speed is perhaps not important if you can simply materialize yourself and your spaceship from any space-time-reality to any other space-time-reality. It would be like sending a matter-gram of yourself over the reality net by plugging your conscious awareness into some other reality's experience simulator. If you choose the same PMR reality, you will merely teleport around within the same universe – again, uncertainty (as required by the psi-uncertainty principle) must obscure problematical rule-set violations at both ends and within all PMRs. When you realize your body and your spaceship are only rule-set based experiential delusions of mass by a hallucinating nonphysical consciousness, the concept of warp speed, the necessity for spaceships, and teleporting take on new meaning.

Splitting, duplicating, or fragmenting is a natural thing for your consciousness to accomplish. There is no rule or law that demands only one **apparently physical** body at one time at one location in one space-time-reality. As long as the rules of interaction are obeyed and the psi uncertainty principle (described in Chapters 47 and 48 of Book 2) is met, AUM and the Big Cheese are happy campers.

The idea that you must take the body-energy with you by dematerializing or disengaging from your shared experience here (from the viewpoint of others here) and then re-materializing or engaging a shared physical experience somewhere else is based on a misunderstanding of the nature of reality. For some reason, it seems intuitive to citizens of PMR that a body must first disappear (dematerialize) from where it is before it can reappear (materialize) somewhere else. Because your body is only a virtual one to begin with, all this materializing and dematerializing is silly. It is much simpler and more reasonable to create an appropriate form (body) in whatever reality you happen to be in. Fortunately, it does not take much energy to materialize an additional virtual physical body in most virtual simulations within $NPMR_N$. Actually, going bodiless is difficult to do. It would appear that some sort of body that defines the boundary of your individuation is automatically attached to your being whether you intend it to be there or not.

In the non-space-time portions of $NPMR_N$ (outside those reality dimensions that operate under a space-time rule-set), everything travels faster than warp speed; accordingly, travel time always **appears** to be instantaneous and never becomes an issue.

There are rules that dictate how, and to what extent, you can interact in other reality systems – one must observe all local laws including local rule-sets and psi uncertainty.

Dematerializing, materializing, and going faster than light-speed are merely local PMR issues that make little sense in the Big Picture and are not important to your personal growth.

Leaving the PMR Body Behind

We may, in the end, be like old-dog executives who get their secretaries to make a hard copy of their e-mail before they will read it. In the bigger picture, we resemble the people who do not like teleconferencing because they can't shake hands and do not like telecommuting because they cannot continually watch or monitor their people at work. It may be that we are so used to hauling our bodies around with us that we simply cannot imagine going anywhere without them. Unfortunately, what we cannot imagine becomes impossible to accomplish.

Perhaps in the future, dragging one's body about won't seem necessary or desirable – especially if everything but the virtual body you experience within PMR can easily make the trip. Travel within NPMR is governed entirely by intent. Nonphysical travel within PMR (with all the attendant physical sensations of being there included) has the potential to become more and more like using the telephone, telecommuting, teleconferencing, and communicating via e-mail and the WWW – all concepts of disembodied communications that we are slowly getting used to, and finding extraordinarily efficient.

85

■ ■ ■

Ramifications
The Fractal-Like Patterns of Reality
The Seed of the Universe
in the Eye of the Gnat

■ ■ ■

Interestingly, the basic mechanism that I have described as the generator of Big Picture reality seems to repeat itself at all levels throughout
the entirety of our reality. Everywhere we look we see the same simple
pattern repeated. From multiple branched diverse actualized and unactualized history traces, to our PMR universe, to the earth's multitude of
diverse species. From our souls, to our cells, to our technology, cities,
and businesses, we see the same Fundamental Process. The Fundamental
Process of evolution is applied repeatedly in the same way that a fractal
generates a higher order representation, a picture or design, by repeti-
tiously applying a pattern at differing scales through a simple relational
mechanism applied recursively. The basic pattern in reality dynamics is
not geometric, but rather one of process, of becoming, of actualization
and evolvement.

The Fundamental Process is as follows: It starts from any point (any
level) of existence or being, spreads out its potentiality into (explores) all
the available possibilities open to its existence, eventually populating only
the states that can maintain significant profitability while letting the others go. The successful states are progressed to their logical conclusions by
iterating the Fundamental Process. Additionally, new or intermediate
states can generate new states. States that no longer hold potential for
profitable growth either disappear or are recombined with others with
which they are redundant. Potential is maintained, in the event new initial or environmental conditions appear.

We saw in the two previous books how this simple process applied to the potential of consciousness generates synergy that develops into low entropy high quality consciousness, individuated consciousness, NPMR, time, space-time, OS, PMR, and the entire Big Picture of our reality. In this section we found that the same process enables state vectors, probable future reality surfaces, and everything significant that can happen to happen. Additionally, we have seen how this same process generated our universe, among others, through the Big Digital Bang. Consequently, it should not be surprising that every level of existence, including physical matter and our carbon-based biology, follows a similar evolutionary prescription.

Consciousness, the space-time rule-set, and the Fundamental Process together set the boundaries and define the content and dynamics of our local reality. The natural uncertainty expressed by quantum mechanics, psi uncertainty, interactive personal relationship dynamics, as well as the randomness found in nature serve to further stir the pot of possible choices and possible outcomes.

The Fundamental Process of evolution constitutes a reality-generating pattern of applied process. You can notice it everywhere you look. It should be obvious that Darwinian Mechanics followed this same process to successfully populate the earth with diverse life-forms – each with uniquely specialized features (such as the human's brain, the frog's tongue, and the multi-faceted eyes of flying insects).

In a more mundane sense, this is also how we humans pursue our daily lives – projecting and calculating potential value into all our known options, choosing the best states or outcomes, and then progressing those forward by our actions and choices. Technologies, businesses, cities, political or financial empires, complex computer software, capitalist markets, corruption, and crab grass, as well as consciousness and sentient beings, all grow, expand, and evolve into the available possibilities by an identical process.

The cells (biology), molecules (chemistry) and atoms (physics) in our bodies (or anywhere else), execute their own version of that same basic process within their given potential or possible states. From the individual particles of high-energy physics, to all PMR and $NPMR_N$ within TBC, we see this simple pattern repeated. Could the Fundamental Process be the mother of all fractals? It is an interesting thought. If true, perhaps we can find the father of all fractals as well.

To be sure, the Fundamental Process does not generate a **geometric fractal**; instead, it produces a **process fractal**. You may need to generalize your concept of fractals, but the similarity to fractal dynamics and structure is obvious. When we look at our reality, we see the results of the evolutionary process repeated at various scales and levels generating intricate convoluted patterns. We see the digital (virtual) energy of synergistic organization creating a complex ecosystem that employs the Fundamental Process to recursively iterate layer after layer of interactive process. Each layer becomes the foundation for the next.

Together, consciousness and the Fundamental Process evolve an ever growing, monstrously complex reality system – a system where every part or entity at every level of existence explores its full potential while populating only those states determined **by itself** (often using criteria germane to the next higher level) to be significant and useful **to itself** (often for a higher-level purpose or activity). Solar systems, galaxies, human bodies, insects, and consciousness all evolve through the same pattern (as does everything, including AUM). Why is that simple consistency not surprising?

Beauty and power expressed within elegant simplicity! Does that not appear to you to be a common blueprint or theme from Mother Nature? Absolutely, Occam's razor is based on the truth of it, and most mathematics and natural laws clearly exhibit those characteristics. Mother Nature analytically decomposed exhibits an applied fractal process at the root. Elegant simplicity driving powerful results is the obvious consequence of good design and good programming within TBC.

Evolving a powerful, elegant, and simple fractal-like process to be the engine at the core, the motivating principle of reality, is the natural result of the Fundamental Process iterating toward optimal system solutions. It is the nature of all successful, large, and complex self-modifying systems to be constrained at the top level by elegant simplicity, otherwise chaos would destroy their viability. AUM and Mother Nature cannot help but reflect the fractal property of awesome complexity generated by the recursive application of elegant simplicity because that is how they themselves are constructed – they work as they are. Elegant simplicity is their secret, the key to evolving stable and productive complexity. For example, large social systems often fail (exhibit high entropy) because they lose the values (rule-set or constraints) that must provide a viable foundation for stable and productive evolution.

Toward the end of Chapter 54, Book 2 we discussed the concept of consciousness as a fractal system. There we saw that the **content** (as opposed to structure) of reality was derived from consciousness within consciousness within consciousness – computers within computers. From AUM to EBC to TBC to personal individuated consciousness in NPMR and PMR to human brains, to silicon-based computers designing better silicon-based computers, all reality appears to be populated and regulated within the form and structure of a giant fractal consciousness ecosystem. Each level of existence is derived from the consciousness above it by repeating the same basic attributes of consciousness on different scales and with differing form-functions. All entities are chips from the old AUM block at various levels and forms of being and awareness.

Digital consciousness, with its multitude of recursive interdependent expressions, is the engine at the core of reality – a fact that leads us to characterize consciousness as the father of all fractals.

The Fundamental Process of evolution permeates all reality as process within process within process, while consciousness provides the substance or self-organizing media upon which evolution's process operates. Gazillions of reality cells provide a conceptual digital medium that can organize and reconfigure itself to lower its entropy, or equivalently, increase its useful energy. The Fundamental Process converts the potential for self-organization (potential digital energy) into aware brilliant love-consciousness. Our two basic assumptions given in Chapter 24, Book 1 – consciousness and evolution – can now be seen as the fundamental **substance** and **dynamics** of the larger reality – Father and Mother of All That Is.

Consciousness brings content, substance, value, potential energy, time, organization, and entropy while the Fundamental Process brings dynamics, structure, motion, change, profitability, and a process to convert potential mind into active mind by lowering entropy. Because that is all it takes to form reality, reality is nothing other than that. In all of its gazillions of forms and functions, from the most intricate superfine detail to the most global of overarching concepts, variations of these two fundamentals recursively repeated and applied at every possible scale, one level building upon the other, have produced All That Is. As in all fractals, a repetition of basic patterns and simple rules for change, applied recursively, yield a large, detailed, and complex result.

No doubt you have seen pictures of fractal images. Consider that the Big Picture is a Consciousness-Evolution fractal image. Think of our larger reality as an evolving mind-fractal. Do you see why **3D geometric** fractal images closely describe the natural objects within our **3D geometric** space-time reality? They are of a similar fractal nature and similar fractal type. Would not a fractal best and most accurately describe another fractal? It makes sense that a fractal reality constrained to **geometric** space-time (like PMR) could be described by geometric fractals.

Reality has a fractal nature because that is how it is built (evolved). Reality is a fractal, the result of a fractal process applied to the self organizing capacity of consciousness. AUM's conscious awareness, TBC, the space-time rule-set, and our beloved local PMR are generated by a recursive application of the Consciousness-Evolution fractal process. AUM, consciousness, and all reality are the result of the Consciousness-Evolution fractal propagating its way through the possibilities.

You not only live in a fractal reality and are a piece of a large digital fractal system, but you are a fractal component! Both the structural and the dynamic aspects of your individuated consciousness are part of a larger interactive pattern that we have called the greater consciousness ecosystem. This ecosystem (interactive interdependent system) is a complex consciousness-evolution fractal that is continually energized by applying the consciousness cycle to its self-evolving, self-organizing components. The virtual reality we experience as physical reality is simply a piece of that same fractal pattern. Carbon based biology; social systems, business, and political organizations; technology; and the non-sentient physical matter of PMR (trees, mountains, lakes, clouds) all express geometry-limited fractal characteristics in both form and content. We are all individuals, and are likewise made up of individual parts, but we, as well as our parts, are of the One Pattern – One Evolving Consciousness – all part and parcel of the Big-Fractal-Picture of Reality.

Once you get the idea of how one might generalize the fractal concept to include process and organization as well as geometry, it becomes clear that a reality based on the Fundamental Process of evolution working upon digital consciousness must necessarily result in a consciousness-evolution fractal, where consciousness is the reality medium (the malleable substance to which the process is applied) and evolution is the process. PMR, being a component of such a reality, should have the word fractal written all over it – and it does. Governments, societies, cultures, businesses, economies, technology, people, critters, plants, mountains, forests, and rocks (as well as the ecologies each generates to sustain itself) are all created and driven to their present state through a fractal process of simple rules applied to a self-modifying complex system.

An evolving complex interactive system of consciousness is necessarily implemented as a process-fractal because consciousness can only interact with itself. Consciousness acting upon consciousness – pulling itself up by its own bootstraps through the replication of a simple process applied recursively at all available levels and scales where each new layer is built upon the previous one.

For every unique reality dimension, the profitability equation that drives its evolutionary process must reflect the rule-sets that constrain the possibilities within that reality dimension. Thus within OS, profitability requirements push consciousness toward lower entropy states while pushing PMR virtual matter toward lower energy states.

The Fundamental Process is the fundamental process; its profitability equation defines success; the environment applies the form factor; and space-time ensures consistency through the constraints defined by its rule-set. Realizing why geometric fractals accurately describe our 3D environment is just the beginning of understanding the fractal properties of our reality. Understanding the close connection between process fractals and ecosystems will eventually allow us to optimize our social, economic, cultural, organizational, and technological creations and institutions.

One needs to expand the limited concepts of ecology and ecosystem dynamics to include all systems of complex interdependent human activity (governments, societies, cultures, businesses, economies, technologies, and others). With a better understanding of process fractal dynamics we should be able to squeeze much entropy from the products of human organization. A better understanding of fractal ecosystems and the fractal nature of our reality will help create and define order within chaos.

To see the picture that I am painting one must expand their concepts of energy and evolution and realize that digital energy (the energy of organization created through entropy reduction) is not resource limited or limited to mass and fields, and that evolution is a fundamental process that applies to all self-modifying complex systems.

Once the Fundamental Process is better understood, it will become clear that we are surrounded by many critically important evolving systems that all follow the same simple process. Similarly, once the fractal nature of evolving systems is better understood, chaos will melt away into manageable and profitable processes. Furthermore, once the ecological model of evolving systems is better understood, human activities can become more efficient, cooperative, and productive on an ever greater scale.

To summarize, evolution defines the fundamental process, process fractals define the primary construction mechanism, and ecology defines the interactive structure of profitability. The techies in the audience should be reaching for their pencil sharpeners. Before them lies the opportunity of discovering the science of applied ecological fractals – a Big Picture approach to complex interactive systems analysis – which is timely and important because in the twenty-first century we all live, work, and play in a complex interactive world whose supporting systems are becoming more complex and interactive every year. What now appears random and chaotic from our little picture view is actually buttoned down and well behaved from the view of Big Picture fractal science.

Let's toss a TOE bone to the techies: Go boot up your computers and be the first on your block to invent a process fractal as well as a new academic discipline and an important new applied science, and you will be granted the rights to universal fame and three additional gold stars.

Rudimentary process fractals are already being applied by scientists investigating social and cultural dynamics in an effort to understand patterns of crime, ethnic segregation, and why the Anasazi abandoned northern Arizona 700 years ago. Although these elementary process fractals are constrained by exceedingly limited rule-sets, the artificial societies they model represent extremely complex results. (The simpler and shallower the rule-set the more likely it is that the process fractal will quickly converge to a steady state condition.) Using a process fractal to model a digital system dedicated to reducing its overall system entropy through evolution (self-improvement) will provide a rough simulation of our larger reality wherein the consciousness cycle will evolve as a winning strategy. Then, if the constraints of space-time are imposed to improve the efficiency of the consciousness cycle, we will find ourselves to be the result.

These are incredibly powerful and cool concepts. If you are unfamiliar with fractals, go look them up and find out what they are. After doing so, the Big Picture of our larger reality will suddenly make more intuitive sense. From the eye of a gnat, to a supernova, to the mechanics of NPMR, to AUM itself – all are expressions of a relatively simple consciousness-evolution interaction that repeats, copies, and clones itself at many levels and scales into a single magnificent fractal mental-image of digital existence. We are thus a simple but convoluted and recursive repetition of All That Is within All That Is. The greater consciousness ecosystem is a naturally occurring fractal system where each separately dimensioned habitat (virtual reality or dimension) reflects the fundamental pattern of evolving consciousness and is interconnected to, and dependent upon, the whole. Now, that's a Big Picture, but one that is within the grasp of our understanding.

Had I told you on page one that you are an individuated portion of a larger fractal pattern that constitutes All That Is within a digital virtual reality based upon evolving consciousness, you would have rolled your eyes and put the book back on the shelf. Hopefully, after you have finished rolling your eyes, you will at least allow this elegant concept to reside somewhere in a remote corner of your skeptical but open mind as you search for the personal experience-data that can confirm or deny its value and applicability.

You are a sentient entity designed primarily to evolve the quality of your being – the quality of your consciousness – because that is what one does in a Consciousness-Evolution fractal. That is the pattern of which you are a part. What else could you do as an individuated consciousness-evolution pattern subset existing within a consciousness-evolution fractal? Trying to do anything else, trying to break out of the pattern is futile – you are the pattern, the pattern is you – it is how you are defined. If you are unaware of this fact, you may be missing the point of your existence. You may be spending your time and energy in ways that are not important to your evolution, to your larger purpose within the larger reality. You are what you are – there is no point trying, or wishing, to be something else – you might as well learn how to play the game that you are in. There is no other game. If you get good at it, it is more enjoyable and more fun. Hanging out, clueless, in the middle of the playing field while others are fully engaged, having a blast, and making progress is a sad waste of your opportunity and of your potential.

The larger reality is not primarily a place defined (and bounded) by n-dimensional geometry where we live and objects exist, but rather a **process** for **becoming**, a process containing progressive (evolving) states of being – minimum energy for artillery rounds and elementary particles; minimum entropy for bright sentient entities such as you. The dynamics of the larger reality are driven by a process designed to facilitate the evolution of consciousness where individuals and the larger consciousness system mutually seek and find new end-states of higher profitability. The larger reality provides the opportunity for personal growth: the possibility of being all that you can be and more than you ever thought possible.

The N-Division piece of the larger consciousness ecosystem, reflects the fundamental nature of $NPMR_N$, requires local experiential subsystems (such as our PMR) of mass, space, time, and limited consciousness to serve as tools we can use to provide the opportunity for us to evolve through exercising free will choices. Why us? Why are we like that? Because consciousness evolving through the application of intent to free will choices reflects the repetitive pattern of our reality fractal. That is how consciousness and the Fundamental Process interact. We are provided the opportunity to apply the pattern of evolution to ourselves as an integral part of a larger process.

As with any seed, hologram, or fractal, the design of the whole is captured, expressed, and implemented within each part. The tiniest part (a neutrino, electron, or perhaps the intricate eye of the gnat) expresses the same Fundamental Process that represents and describes the universe, and beyond. This fundamental process of systems evolution is a process, within a process, within a process ... each tiny part containing the blueprint that drives, as well as explains, the whole.

You, both the physical and nonphysical you, are one small individual part that contains the essence, the pattern of the whole – an evolving individually-conscious piece of the larger evolving consciousness. You repetitively apply the Fundamental Process of evolution to yourself as you interact within your internal and external environments. You do the process, the process does you. You are a creator that exhibits, uses, and manipulates the evolution of consciousness and you are the result of that same process at both the nonphysical and physical levels. The Fundamental Process provides the dynamic principle that builds the structure, while consciousness provides the medium – the content, the fundamental energy, and organizational potential – that the process works upon. At the core is elegant simplicity, the hallmark of Big Truth.

No doubt there is much left to do and much left to understand. Nevertheless, is it not remarkable that one simple process and one simple energy form are all that is required to produce All That Is? Is it not remarkable that *My Big TOE* has been entirely and logically derived from only two simple assumptions – evolution and the potential energy of consciousness? Is it not remarkable that from these two assumptions a theory exhibiting elegant simplicity in form, structure, and application has seamlessly combined physics, metaphysics, ontology, epistemology, and cosmology to answer many questions of science, philosophy, and theology (both ancient and modern) that have gone begging for answers for years? Is it not intellectually and emotionally satisfying that the full and complete answers to these historically unanswerable questions turn out to be relatively simple, straightforward, and logically concise once old paradigms are generalized to provide a more accurate perspective of the nature of our reality?

Do you see how the application of this Big TOE makes the fundamentals of life, science, philosophy, and metaphysics easier to understand and more obvious to apply and use, while driving nothing to greater complication or obscuration? Big Truth simplifies issues and understanding – it never complicates. On the other hand, beliefs (as do lies) require an ever greater complexity to support them as they spread and grow. None of what we scientists have worked so hard and long to understand has to be discarded – only relegated to a subset of a larger, more comprehensive understanding.

You, I, and the dog next door are of one consciousness, one source, one connected evolutionary pattern. Nevertheless, every level of existence has its own mission and purpose. We sentient beings in OS have our mission, our reason and purpose for being. That mission requires that we have free will and the ability to interact freely with other units of individuated consciousness within the constraints of our experience and the quality of our unique personal consciousness. This arrangement provides us with the maximum opportunity to evolve our consciousness, our being, and our quality according to the pattern of our larger reality.

> ▶ "Hey Jake! Put that stupid toe-book down and bring in the cooler while I turn on the tube! Wrestling's coming on! Can you believe it? The 'Mad Menace' is going up against 'Killer Mc Bee' in a grudge match – that ought to be really good!
>
> If 'The Bee' gets his stinger out, man, there will be blood everywhere. That would be so cool! The last time he got banned for three months – remember? And Jake, listen to this – hey, pop me a cool one, pal – after the main event they are going to let the women wrestle. I don't know their names but I got a good look at 'em on the pre-match interviews and one of them has really big…" ◀

Improving the quality of consciousness, advancing the quality and depth of awareness, understanding your nature and purpose, maturation of soul, manifesting universal unconditional love, letting go of fear, and eliminating ego, desire, wants, needs, or preconceived notions – these are the attributes and the results of a successfully evolving consciousness. What do the facts of your life, the facts of your existence, and your results say about the quality of your consciousness, the effectiveness of your process, or the size of your picture?

Progress is measured only by results – clear obvious results. There are no quality points given for good effort or nice try or for believing or not believing anything. You are an in-charge (of your opportunities) responsible adult individuated consciousness. You cannot bribe, cheat, fake out, or hustle the system. There is no spiritual welfare system. There are no shortcuts. You have to do it 100% by yourself and your friends, connections, and cash cannot help you. Excuses earn zero credit irrespective of how well justified they are. Only results earn credit.

On the other hand, there is no practical time limit – you can take as long as you want or need to. There is generally no such thing as absolute failure – failure is relative and simply equates to a lack of progress. There are no exits, no escapes, no way out or around, except through personal growth.

If you think that you may have gained some new knowledge or insight in the process of comparing Big TOEs with me, I must remind you that with knowledge comes responsibility.

I thought you would want to know.

86

■ ■ ■

Section 5 Postlude
Hail! Hearty Readers, Thou Hast
Completed Nonphysical Mechanics 101

■ ■ ■

If you have made it this far without skipping over too much of the more difficult material and have let the concepts presented in Section 2, 3, 4 and 5 intermingle and bounce around inside your skeptical but open mind, you have earned my respect and appreciation regardless of what you think about what you have read.

Now that you understand the Big Picture evolutionary process of improving the quality of your consciousness by making choices based on right motivation, simple things such as the point of your existence and the meaning of your life should be obvious. You should, by this point in our journey, have a good idea of where you came from, where you are, where you need to go, and how to get started on the journey to get there. Hopefully, you will remember not to confuse reality with the model of reality.

No doubt, a few readers are struggling to justify their old and comfortable paradigms. For the most part, their justifications, though familiar and traditional, are neither logical nor objective, but appear, from their view, to be both. Wrestling old paradigms and beliefs to the ground is exceptionally difficult to do. To be a lonely warrior in a strange land requires more courage and personal strength than is commonly available. By having almost completed this trilogy, you have proved yourself to possess uncommon fortitude, strength, and courage as well as plenty of extra time.

For some, stepping off the well-beaten path brings up deep seated anxieties – the unknown, the non-conforming, and the unaccepted strike fear deep into the hearts of those in need of reassurance. So it is with all social (herd) animals – and so it shall always be – it is their fundamental nature to seek safety in numbers and in conformity to norms. Beware: In a fearful stampede to save themselves from the terrifying menace of an original idea, the herd can become a mindless destroyer of the light. This is simply how it is: **No fault or error is implied**.

For bravery, courage, and outstanding stamina, I do hereby grant you permission to paste four additional shiny gold stars in your book. Congratulations! Thy treasure of unearthly gold doth greatly grow.

Unfortunately, many readers find at this point that they need to read Sections 2, 3, 4, and 5 again to pull it all together in their heads. At first reading, the introduction of so many new and unusual concepts keeps your head spinning and often causes philosophical discombobulation, existential confusion, metaphysical vertigo, as well as belief-trap withdrawals. These in turn inhibit Big Picture conceptual coalescence. It is a sad fact of life that much of what you read in earlier sections will make more sense now that you have completed Section 5.

Isn't that how it always is? You are properly prepared to take a difficult course – and thoroughly learn the material – only after you have finished struggling through it for the first time. Unfortunately, we almost never retake a course because we are in too big a hurry to begin struggling through the next one. I am reminded of the office maxim: There is never enough time to do it right, but always enough time to do it again. That rings especially true when your view of time spans multiple lifetimes.

The next and final Section (Section 6) is a short wrap-up that will help you put what you have been exposed to during the previous four sections into a more personal and balanced perspective. You are now on the downhill slope of the Big TOE thrill-a-minute ride into your consciousness – only a little more effort is required and you will be entitled to brag to your friends that you were able to get through the entire thing – from shaft to knuckle to nail. Though they will, no doubt, be envious of your new twinkling stars, you should begin thinking about what you are going to tell them when they ask you what these books are all about, what you learned, and whether or not contemplating your Big TOE is as fun, interesting, and useful as it sounds.

▶ At the very least, if you can think of nothing else to say, you can change the subject by firmly asserting the undisputable fact: that during meditation, your big toe is always easier to see than your navel. With sustained concentration being an immensely difficult thing for most people, your navel challenged friends will immediately see the advantage and think you clever to have penetrated the infamous belly-button barrier and obviated the nuisance of propping up mirrors all in one ingenious stroke. Ahhhh ha! So that's why they pull their feet up on their knees like that! Now I get it! It's the big toe, stupid!

Shhhhh, whisper! This is a great secret of the Mystic Inner Circle, don't blab it all around. Nobody is supposed to know why we sit like that and wear open toed sandals year-round. Think of it my friend, through your clever choice of offbeat dubious reading material, you have discovered the ancient mystical key to the door of enlightenment – your friends will no doubt be mightily impressed. One more thing, please, do not tell anybody that I told you this. If you must tell, use someone else's name – I knew I could count on you – thanks. If it is found out that I have let this secret slip out, ancient dis-embodied Tibetan warriors will come to the foot of my bed at night and tickle my feet. Oh jeez, a fate worse than life! ◀

We will regroup at the beginning of Section 6 in the morning where you will receive your final set of briefings before going off on your own into whatever reality you have thus far created for yourself. It is time to think about what comes next – and what, if anything, reality has to do with you. The most significant questions for you to ponder are: Just how real are you really? How aware are you of awareness and how large a larger reality can you stand to understand?

In Section 6, I provide the perspective with which you can sum up everything you have been exposed to on this trip through Big-TOE-land. Perhaps you will begin to formulate a strategy for where you want to go next – and how you might get there. After Section 6, the ball is in your court – and you will need to figure out what the game is, and how to play to win. This is the last inning pal, get some sleep and be ready to go at sun-up – this time we will not wait for stragglers.

Section 6

■ ■ ■

The End is Always
the Beginning:
Today is the First Day
of the Rest of Your Existence

■ ■ ■

87

■ ■ ■

Introduction to Section 6

■ ■ ■

You can secure a major advance in Big Picture understanding by merely appreciating your ignorance within a larger point of view, even if your experience allows no perceptible conversion of that ignorance into knowledge. Learning begins when you are capable of appreciating your ignorance enough to ask a question and care about the quality of the answer. When you know or believe you know the right answer, learning is not possible.

We are not children. Simply asking the question is not enough. There is no Mom, or Dad, or organization capable of supplying us with an easy answer that is also a quality answer. A quality answer is one that is part of your personal solution without also being part of your personal problem.

One of the most obvious Big-questions left hanging in thought-space after reading *My Big TOE* is: What now? I think that is a good question – it may be less obvious, but I have that same question. What, if anything, are you going to do with the ideas and concepts you have encountered in *My Big TOE*? That is my question. Perhaps you will be willing to share your answer with me and others at: **http://www.my-big-toe.com**.

Unfortunately, if you are to reach significant conclusions about the content of *My Big TOE* you will need to either build some bridges or jump to them across a chasm of ignorance and ingrained belief. How many bridges and how much distance they must span is dependent on the breadth, depth, and quality of your experience data and the size and shape of your personal chasm.

Building logically and scientifically sound bridges from carefully evaluated experience is always a more difficult methodology for reaching conclusions than jumping to them, but if you want to progress beyond chasing your tail it is absolutely necessary. We think puppies and kittens are funny when they are in full and ardent pursuit of their tail for a few seconds at a time. Imagine how funny a human must look dedicating its entire life to a self-referential endless loop of non-productive tail chasing. Actually it is more sad than funny, but these are the typical views when looking from the perspective of the Big Picture.

To help you decide what to do about the ideas and concepts you have encountered in *My Big TOE*, I have compiled six suggestions for your considered action in reverse order of difficulty. 1) Do not throw good after bad – drop these ideas out of your mind like hot (or rotten) potatoes – use the books to warm yourself this winter, or give them to some mushy-brained acquaintance you wish to irritate. 2) After some initial excitement, forget about the Big Picture – it is too big to do anything about anyway. The ideas presented, though intellectually interesting, will naturally drift away as you pour yourself into the next activity, and the next one after that. 3) Because you are reading it only for the t-shirt, go to the **http://www.my-big-toe.com** web site and see if you can procure an official Big TOE t-shirt. After all, you did read the entire book and have the empty aspirin bottles and 17 shiny gold stars to prove it. For that alone, this renegade wordsmith owes you – big time! 4) Send a fiery letter full of indignation and well-placed exclamation marks to the publisher requesting them to cease and desist from further polluting the accumulated knowledge base with mindless drivel. Insist your letter be forwarded to the author at his home in Antarctica. 5) In a bubble bursting with fresh enthusiasm, decide to get someone (book, class, seminar, guru) to help you develop the results of your personal Big Picture understanding before sooner or later returning to 2) above. 6) Do not throw good after bad – commit serious time and energy to a lifelong skeptical but open-minded personal pursuit of Big Truth – and go energetically wherever that path might lead. Become a pudding-head on the path to true love.

Any of the above six choices (and countless others) are acceptable: you must move in a direction that **appears** to be forward from wherever you are and from whatever perspective you have accumulated thus far. You make choices, and the choices you make, make you. Free will is the driver. Choices and consequences, reaping what you sow, getting what you deserve, and deserving what you get – that is how the game is played – lose or win. All individuals are equally important and valuable, and all choices are valid steps along your path of being, whether you know what you are doing or not.

My advice to you and hopes for you have nothing to do with which of the six choices you settle on – whichever choice represents who you are today **can eventually** take you where you need to go tomorrow. My advice and hope is that you will choose only after careful consideration and that you will constantly reevaluate whatever choice you make as the evidence of your lifetime dribbles in.

88

■ ■ ■

I Can't Believe I Read the Whole Thing
Uncle Tom, Pass the Antacid Please

■ ■ ■

My Big TOE was written to allow me to share the tentative results of my explorations with you. It is in the style of a personal conversation between you and your eccentric Dutch uncle – strange old Uncle Tom. The informal tone and use of humor reflects how I normally relate, one-on-one, with good friends who want to know, and whose eyes have not yet begun to glaze over. This approach is calculated to maintain interest and to minimize the wow and gee whiz effects.

I could have made a more serious, formal, erudite presentation that would have been much more intellectually impressive – like the typical book authored by some hot-shot scientist from the prestigious University of Tough Love with an impressive string of cryptic letters after his name. Something like this: Dr. Uncle Tom, B.S., I.M.S., Ur.D. where BS is self-explanatory and S and D imply smart and dumb respectively – at least that's the message intentionally left lurking between the lines. Trying to be impressive or worse yet, seeing yourself as actually being impressive is, in the Big Picture, equivalent to having a lobotomy.

Contrary to popular belief, you do not have to be a doctor, lawyer, executive, high ranking military officer, government manager, politician, or university professor to get a lobotomy – but if you are one of these supremely impressive individuals, you get to go immediately to the front of the line, no questions asked. Don't giggle, that is no small advantage – there is always a very long line at Lobotomy Central.

Appearing impressive is something you should always carefully **avoid** – even if you must act a little silly sometimes. Silly mixed in equal measure with down-home and funky is always a sure winner in the "oh, it's only you" sweepstakes. Avoiding actually being impressive is exceedingly simple for most of us while avoiding the appearance of being impressive requires a more concentrated effort on our part.

It is important that you do not take the attitude that serious progress along the Path of Knowledge is only available to the special few. That it is beyond your practical reach – only for robed gurus who dedicate their life to the pursuit of spiritual perfection or Ph.D. physicists who are so far out in the esoteric ether that even they have no idea what they are talking about. Nothing could be further from the truth. You can succeed superbly without changing your outward life-style very much. You should know that I am a regular down-to-earth guy with a job, a wife, young kids, multiple dogs, birds, marsupials and snakes; one cat, two old cars, a whopping mortgage, and last but not least, a very strange sense of humor.

Personal spiritual progress does not require disengagement; it does not require a high and lofty demeanor, or a serious superior-to-thou attitude. Clearly, those attributes indicate a **lack** of spiritual progress, a **deficiency** of consciousness quality. Unfortunately, people have a natural tendency to place those they think know more than they do on a pedestal – and denigrate their own worth or progress by comparison. This is a belief trap that makes progress much less likely. You will evolve more quickly by tossing out the Great Guru and seriously working on your own, than by hanging out with the Great Guru in lieu of doing serious work on your own. Quality of consciousness is, for the most part, not gained through association or by osmosis.

Placing others on pedestals is nonproductive and makes **you** feel less competent by comparison. Do not fall into this energy and incentive destroying belief trap.

Natural and simple processes usually work the best. It is my hope that being "one of us" in language and style will reach more folks than informal and corny will lose. I am betting on the likelihood that many people will listen more openly and less defensively to their Dutch uncle. More importantly, it is clear that the minimum number of readers will get trapped by the wow effect if the delivery is straightforward, humorous, and informal. That gives the uncle-dude a more effective place from which to communicate and share whatever I think might be of value to you.

Those turned-off or lost for want of a serious, intellectual, top expert demeanor are probably too far gone into the depths of ego and self-referential belief traps to get much from *My Big TOE* anyway. It all works out. What you end up with is: Funky Physics for the Rest of Us – a personal tour through reality with your Dutch uncle. I hope the style, tone, and humor was as good for you as it was for me. Self-important professional stuffed shirts and pompous professor types will need to descend to a more common level to ferret out the few Big TOE golden nuggets that happen to be color coordinated with their egos and career goals. If that is too painful a condescension, they can always watch TV instead.

Looking and acting normal and being normal can be two separate things. The point is: You can seriously and successfully pursue the Path of Knowledge, the path of consciousness quality evolution, of spiritual development and continue to carry on with your normal life. Sure, there will be lots of changes, but none that require you to step dramatically out of your lifestyle as an **initial** investment in future growth. The quality and evolution of your consciousness is not dependent on the **form** of your physical existence. However, the **quality** of your physical existence depends absolutely on the quality of your consciousness.

Because most, if not all, of the people we know who are successfully pursuing a spiritual path or path of personal growth are not normally immersed in our Western techno-culture, we **believe** incorrectly that family, jobs, and mortgages are incompatible with a serious bid for enlightenment (understanding and living the Big Picture at a deep level). Likewise, because we have a small, half-formed picture of what consciousness quality or spiritual progress means, we **believe** that dropping out of the mainstream (becoming monks, priests, nuns, living in solitary places, joining spiritually focused organizations, performing rituals, wearing funny clothes, or severing material connections) must be required: They are not.

The **form** of your material involvement with the local reality is not important. How you interact with that environment, how you relate with others, what you make of the available choices and opportunities – that is important – and that can be done anywhere under a wide range of circumstances that includes Western culture and lifestyles. Do not let these two belief traps (the wow-effect and the drop-out blues) discourage you from setting out on a quest to evolve the quality of your consciousness.

When you hear others use these erroneous beliefs as excuses for why they cannot successfully pursue a more spiritual path, first shout: Copout!, and then, Bullpucky! Next, apologize for your rude uncontrolled outburst. Finally, explain with great empathy and compassion why nothing is required to change but them, and that creating the right environment **first** typically retards progress by inadvertently putting the cart in front of the horse, causing energy to be focused on issues of minor importance. Successful learning does not flow from the right environment – the right environment flows from successful learning. The environment that is most important with regards to your growth is the one in your head. The appropriate physical environment will form on its own.

The West, unlike the East, has not evolved its own distinct cultural processes to support the evolution of individual consciousness. When and if it does, dropping out of the mainstream will not be a requirement. In the West, neither our religious nor our secular processes are designed to help individuals outgrow the belief traps that dramatically limit their potential growth. To the contrary, our Western institutions foster, support, and demand, an array of limiting beliefs with great gusto. The same is true in Eastern cultures.

Westerners are especially proud of their logical and scientific ways. That Western attitudes and processes are often belief based and not as logical and scientific as they appear to us through our tinted cultural glasses, is obvious to many who live outside Western culture.

Nevertheless, the Path of Knowledge, the warrior's path, the relentless pursuit of truth, the application of logic and the intellect in the service of consciousness evolution are right down our Western alley. Progress on the Path of Knowledge depends upon rational process. Rational process is where the Western worldview supposedly shines – the one thing Western culture prides itself on more than any other. We of the West thrive on rational process. In fact, we are so obsessively committed to rational process that everything we do is construed to be the result of a rational process whether it is or not. We simply define it to be that way! If we do it, we can somehow justify it as a rational and logical process; at least the men can, the ladies often know better.

Ahhh...so close, yet so far away! A decided spiritual advantage lies unseen and thus unused. The Path of Knowledge is not well worn by Western feet. We of Western attitude live with and enjoy the material advantages, regret or do not see the spiritual disadvantages, and for the most part entirely miss the spiritual advantages. Every culture has its advantages and disadvantages. Because you must live with **all** the disadvantages of your culture, it is a shame to allow self-imposed belief-blindness to deny you access to some of your culture's more significant advantages.

Hopefully, *My Big TOE* has occasionally prodded you out of your habitual no-brainer comfort zone, or at least challenged your mind with a few new concepts. Perhaps you have taught yourself a thing or two in the process of struggling with the unusual ideas found here. That you end up agreeing, disagreeing, or better yet undecided, is not important. That there was some mental motion, some stirring about within an open mental space – **that is** important – and if the motion, the seeking, continues for a long time, that is even more important.

Whether or not **your** Big TOE is correct or flawed, diminutive or great, is also not important as long as it is moving you in the right general direction. Because an individual's Big TOE must be perpetually in a state of growth, improvement, and evolution, the motion part is easy. Maintaining your heading in the correct general direction is also easy for committed, honest pudding tasters. If you start growing **a** big TOE and continually progress it in the right general direction, you will eventually end up with **the** Big TOE; that is why it is more important to get serious and get going than it is to make sure you are doing it absolutely right. From the perspective of PMR, the process of reducing entropy within your consciousness has many valid approaches. It is a naturally convergent process that will take you from wherever you are to where you need to be. Regardless of what direction you initially take, if you are serious and persistent, you will eventually end up with the same result. Some paths are simply more or less efficient depending on the individual.

The important thing is to start. Your understanding does not need to be perfect as long as it is constantly and consistently converging on rightness. That your Big TOE enables **you** to focus **your** energy on the task of growing toward a higher quality of consciousness is the critical ingredient of eventual success. Your constantly updated Big TOE should be the central part of your continual process of converging on perfection, correctness, enlightenment, and love.

If reading *My Big TOE* has caused you to grapple with your beliefs in the context of a larger reality and induced you to come to **any** helpful tentative conclusions about the quality and purpose of your life, it has been a great success from my view. However, if you agree completely with everything in *My Big TOE* but do not gain momentum or understanding that helps you to improve the quality of your consciousness, then this trilogy has helped you focus your experience, or simply entertained you; in either case there is little significance. If that is your case, I can only hope that I have left a spark of interest or understanding that will produce flames later on. For those who agree and disagree with *My Big TOE*, if no spark of greater knowledge or no flame for seeking truth is carried away, I have wasted your time. The information in *My Big TOE* is meant to be news that you can use, not chatter splatter for New Age groupies or PMR apologists.

I have had loads of fun hanging out with you for the last several weeks or months, or however long it has taken you to get this far. We have covered a lot of new conceptual ground together and have had some laughs in the process. If you have read the entire trilogy and are still hanging with me in this quest to see the Big Picture and develop a comprehensive model of reality – and to find your place in it – I expect that you have what it takes to go the distance. You are among the few to whom I hope to make a significant difference.

You may find this difficult to believe, but some who started on this journey with us, including a few who approached *My Big TOE* with serious open-minded intent, became and remain frustrated to the max. Many of these didn't make it past Section 2 or Section 3 because of the high levels of frustration – a natural filter that you passed through with flying colors.

Why all this frustration and angst? Because I emphasize repetitively that it is up to you to go out and get evidence for yourself and come to your own conclusions. Add that to the fact that I constantly reminded you that your growth is critical to the quality of life you experience and to your evolutionary progress. Furthermore, I am forever pointing out the purpose of your existence and emphasizing the importance of your success (in the little picture and the Big Picture) in fulfilling that purpose and that you alone must assume full responsibility for your success or failure.

Additionally, I make it clear there is precious little help within PMR to aid your success, and that a long-term focused effort is required to pull yourself up by your bootstraps one tiny increment at a time. On top of all this tough love I am continually insinuating that you are probably not as spiritually evolved as you think or hope that you are; that you are more than likely driven by ego, wants, desires, and fear; and that your existence is probably not as close to the center of the reality-universe as you may have imagined. Wow! What a trip ... and that's the good news!

Frustration accrues because although I provide some technique and direction on how to get started, there are no guarantees of immediate success and few hints about what to do after you do get started. I am not holding out on you, or being vague to get you to read my next book, or avoiding the issue because I don't know the answer – it is simply that growing-up is something that no one else can tell you how to do.

Your mother and your boss can tell you how to **act**, but only you can decide how to **be**. Unfortunately, acting properly (exhibiting proper behavior and intellectual understanding) without proper being, though civilizing, is of little value in raising the quality of your consciousness. Getting into a learning process that is significantly ahead or behind your grade level is more counterproductive than helpful: Boredom and frustration are terrific inhibitors of actualizing potential. Mass marketed materials are not the right media for effectively guiding an individual's personal growth.

I could, if I had little understanding and even fewer scruples, try to convince you that I possess the knowledge of a sure path to greater spiritual quality and then sell it to you on the side in the form of books, tapes, lectures, seminars, and training courses, but that would be more marketing bullpucky than truth. There is no shortcut that significantly minimizes the work you must do on your own. There are many equally valid paths. You should choose one that fits your demeanor, style, and situation. None of them can give you anything but an **opportunity** for you to do the work required to improve yourself. Promises of anything more are empty and generally made by those more interested in the quality of your bank account than in the quality of your consciousness.

My Big TOE is not a how-to book, nor is it an advanced tome on better living through physics and meditation – it is simply a Reality 101 survey course – just enough information to get you started on your own adventure of discovery. How-to instructions, to be effective, must be individually focused and personally delivered: A one-size-fits-all book is not a good medium for teaching individuals how to experience the larger reality. A book may teach you useful meditation techniques (see Chapter 23, Book 1), but it intrinsically has a difficult time helping you interpret and understand (guide) your personal meditation experiences. How useful is the former without the latter? For the large majority, not much. Typically, such a book will serve only to shift the point of terminal frustration from a pre-meditation, "I do not know how to get started," to a post-meditation, "My experiences are indefinable, uncontrollable, and without specific meaning."

The first type of frustration: Enthusiasm constrained by ignorance, has a long half-life, and may eventually lead to real progress if the enthusiasm can overcome the constraints of the ignorance before it decays. The second: I've done that and it doesn't appear to lead anywhere, has a much shorter half-life, rarely leads to real progress, and makes it less likely that the individual will ever make real progress. The second type of frustration is not always, or necessarily, the result of an unguided effort. Certain bright, dogged, robust individuals, who are ready to learn, dive right in and, with little to no guidance, become great swimmers in the crystal clear waters of the larger reality. Unfortunately, these individuals are rare; the rest essentially do a belly-flop, spend their life doggie paddling around in circles, or get out of the water permanently.

An immediate concern of mine is that this trilogy, because it is targeted at a broad base of readers, might do more harm than good by encouraging an epidemic of the belly-flop blues. That is a downside risk I accept in order to offer the upside potential of stimulating significant new growth.

The perseverance, effort, and lack of ego required for dramatic success act as a natural filter to cull out those who are not ready to access the greater power and responsibility of an aware consciousness. Balance and stability are inherent to the process.

The rare few who are powerfully driven, sans ego, to find answers will find a way – they always do – they need only to see the possibilities; it is never easy, but for them, easy is not required.

In terms of learning about being, we in the West have a learning disability built into our cultural values. Our model of education employs an intellectual process designed for stuffing facts into the student's head. Learning to be is necessarily an experiential process, not an intellectual process. Much of our frustration derives from wanting to achieve the latter (learning to be, growing-up, evolving our consciousness) by employing the former (intellectual process). We push and push on that intellectual string until we give up in exasperation and conclude that the larger reality is either inaccessible to us or simply does not exist.

Babies do not learn to walk or speak by thinking about it, being told about it, logically analyzing it, or having it explained to them – it takes only a **dim glimmer** of the possibilities and great courage and determination. *My Big TOE* is trying as hard as it can to supply the dim glimmer; you must supply the courage and determination.

Feeling frustrated because the solution is not one you can get an intellectual grip on or intellectually master is futile unless the frustration drives you to continue trying until you become a consciousness toddler. A valiant and ardent search for the right intellectual understanding, the right intellectual tools, process, procedures, directions, recipe, or prescription is a search in vain. The key you are looking for does not exist in intellectual or physical form. There is no recipe, syllabus, outline, book, or set of instructions that holds the answer for you because this is not an intellectual exercise. This fact is extremely difficult for a product of modern Western education to understand.

Finally, if you are discomforted or irritated by hearing me continually infer that you probably have a long way to go and a lot of work to do, consider the use of the carrot and the stick. Providing the dim glimmer is a carrot and stick process; nothing else works. Carrots without sticks are less irritating than sticks without carrots but neither approach will produce any smart mules. To encourage your personal journey of discovery, I have scattered a few enticing carrots here and there throughout the previous five sections. It is also possible that you received a whack or two with a humor-stick that was hand crafted from your ego. It is my intention that the application of sticks should be more of a gentle poke or helpful nudge than a whack. However, I am fully aware that one person's nudge is often another person's whack. A sudden or forceful growing up typically appears to be harsh.

If *My Big TOE* has left you feeling frustrated or disconcerted, it may be because the truth often does not fit comfortably into our established routines and belief systems, and that we often do not want to hear what our intuition is trying to tell us. Not knowing for sure whether Uncle Tom is absolutely correct, nuts, or hopelessly lost, is irrelevant to your personal growth. Not starting on a journey of discovery to find answers because you lack answers is a self-defeating deception of the ego aimed at denying fear and reducing anxiety. Embrace uncertainty: You cannot know where your path will take you. Plan only life's journey, not life's destination. Accept that you are always more ignorant of your ignorance than you think. Do not let the heights of your uncertainty, the depths of your ignorance, or the importance of your mission intimidate you.

I have discovered that most readers like Section 1 the best. Most like my descriptions of NPMR, comments about the Big Cheese, and at least a few of the jokes. Many also enjoy Uncle Tom poking fun at self-important stuffed shirts and iconoclastically tweaking the nose of brain-dead dogmatists. In general, people like to hear about the critters I've met, non-physical sex, gender identification in NPMR, and stories of the epic adventures and conflicts that are taking place within the larger reality. Some are fascinated by how things work in NPMR and the interaction between PMR and NPMR. These stories, anecdotes, and descriptions are interesting to most everyone because they are not challenging (no conceptual heavy lifting is required) and because they have great entertainment value.

One attribute that makes entertainment so popular is that it occurs at a distance – the one being entertained does not have to participate or get personally involved. Being entertained is a pleasant, safe, secure, comfortable intellect-at-a-distance activity that requires little effort, carries no responsibility, and therefore generates no fear, guilt, potential for failure, or personal growth. By vicariously reading about (or watching) others doing amazing or cool things, we are able to share and confirm the hero's reality in terms of our own reality without the effort or risk. Our own possibilities are thus **theoretically** expanded. That's not all bad, but the limitations are obvious.

I could spin stories of adventures, battles, and intrigues in NPMR that would keep you interested and spellbound until you decided that I must be nuts. This is what most people want me to talk about most of the time – all useless, except for its entertainment value. Entertainment value is highly coveted in the marketplace, but it is better at developing cash-flow than it is at developing consciousness.

▶ News or 'truth' as entertainment – that is the tabloid approach – and the rest of our mainstream news producers are in that same boat or headed there in a big hurry – they are merely less obvious. We the people are driving the content of all mass marketed information to that same sad state of affairs. The dynamics are simple: 1) Information distribution is a business and must sell to remain competitive and profitable; 2) People prefer to be entertained; and, 3) The truth is not in big demand.

Likewise, people often prefer a tabloid approach to reality as well. (See the aside within the aside near the middle of Chapter 24, Book 1.) Unfortunately, little to no improvement in the quality of your consciousness can be achieved by listening to other people's experiences because making meaningful improvements (consciousness evolution) is **not** an intellectually-driven process. Quite the contrary: Personal growth is an intent, free will, choice making, choice executing, consequence assessing, pudding tasting, experience-driven process.

Gather 'round folks, I am about to blab a rare heart-felt confession. I am not interested in entertaining you with my experiences in NPMR because there is little potential value in it. Actually, it is worse than that; the downside is significantly greater than the upside. Interesting and exciting tales from the Big Picture serve mainly to distract your focus from what you ought to be paying attention to. Spiritual growth through entertainment sounds as dumb as it is. Improvement in your consciousness quality must flow from your own direct experience. Intellectual analysis cannot generate new experience. Enough said. ◀

The experience of others can provide only a tentative bigger view of the possibilities – it may provide some direction, but it cannot move you forward toward greater personal quality by even a single small step. You must learn to recognize what is mainly entertainment and not confuse it with those experiences that could directly lead toward personal growth. It is not always as easy as it initially appears to make that discrimination, but until you can tell the difference and act accordingly, you will remain in square one.

Entertainment is undeniably what the marketplace wants, but that is not my interest, function, or intent in creating this trilogy. You already have a huge array of choices if entertainment and quick spiritual fixes are what you are pursuing. Here, you are expected to do your own serious iterative thinking and experiencing in order to come to your own tentative conclusions.

You now know some of the considerations and dynamics that shaped the writing of *My Big TOE*. I made you wait until almost the end so you wouldn't be tempted to second-guess my presentation and delivery. The inside scoop divulged in this chapter is reserved for my favorite readers – those who were tough enough, interested enough, and determined enough to make it (almost) all the way to the end.

▶ Now you know: *The Untold Secrets Behind My Big TOE.* (How's that for an oxymoronic tabloid headline?) Or, how about this: *Uncle Tom – Exposed!* People would love it, and sadly, find it more believable if Uncle Tom turned out to be an ancient Egyptian wizard accidentally raised from the dead by leaky radioactive waste. What is the real story behind *My Big TOE?* When a senior editor at a small publishing house read an early version of the manuscript, he called me to ask, "Where did you get all this stuff?" Would you believe I teleported here from a spaceship that remains hidden on the dark side of the moon just to implant these words of wisdom into the collective consciousness of the pathetically inferior but nonetheless deserving earthlings, so that eventually they might become worthy of contact by my people?

OK, OK, this is the real, true, truth: My name really is Tom, and I am (or at least used to be) The Big Cheese's favorite Uncle. The unvarnished truth is that The Big Cheese sent me here because I beat him three times in a row in the heavyweight lightning bolt hurl. That is a fact – three times! Nobody had ever done that before. I know in my heart of hearts that he sent me to this clueless humanoid-hell-hole to get even, but what he said was…let me give you the exact quote… "Uncle Tom, see if you can help those bone-headed earthlings get their shit together." That is exactly what he said, word for word, honest! Jeez, I worked on designs for outhouses, flush toilets, and sewage treatment plants for almost 2,700 years before I figured out that I was supposed to write these damn books!

This is tons of fun, and I am tempted to go on and on, but let's get to the point. Many people would find *My Big TOE* more credible, much easier to take seriously, and less threatening if its origins were fantastical (as above), accidental (the result of being bonked on the head by a meteorite), or the words of a mysterious and mystical individual – perhaps a monk that has been in a cave for thirty years or a super-duper guru from a foreign culture. Why? Those sources are easy to keep at an ego-safe distance because they have nothing to do with us personally; they imply no responsibility on our part. ◀

▶▶ "If this so-called enlightened man is just like us, then he couldn't know that much more than we do. Right? If he is, and he does, we should be able to do the same thing. Right? Well, then, why haven't we?

"The unmistakable conclusion is that either he is non-credible, or we are failures, or at least have come up very short in actualizing our potential. Intellect, please tell me which possibility is the one I should believe.

"My ego and intellect agree that I have done all I can reasonably do. Whew, I feel better already. I knew that he was too ordinary and just like me to know that much more than I do.

"There for a minute, I felt small, fragmented, and insecure; like I needed to refocus my life completely, but now I am whole and in charge again.

"Too bad Mr. Know-it-all didn't pan out; I wish that he possessed the truth and insight that I have been seeking these many years. He is, it turns out, at worst a hustler, or at best simply confused – what an ego in either case! The world is full of people like him who think they know **the** answer. I have listened to them all, but none of them have ever **done** anything for me. I remain the same. Perhaps a few of the more popular ones have part of the answer, but none seems to have the entire answer. I am beginning to think that knowing the entire answer is impossible, that life is meant to remain mysterious, that we can never know the larger truth. If nobody really knows, or can know, the answers to life's hardest questions, then I am doing as well as can be expected and better than most.

"Surely, we all agree, one must be exceedingly careful about who and what to believe.

"Nevertheless, I must continue my search for The One who knows the truth, The One that will share the truth with me so that I too can become enlightened. Perhaps the next Mr. Know-it-all (they seem to come and go all the time) will be the real one – the one who will set my spirit free!

"I have tried it all. My hopes soar with every new process or teacher. Nonetheless, my spiritual quest seems to wander aimlessly with little real progress. All the enlightened gurus are totally inaccessible. I can't quit my job and family to hang out with some guru – that would not be responsible or right. Maybe there is no way to get to there from here?

"I wonder if I am doing something wrong. If I could only meet an enlightened person who was accessible, who could relate to me, who understood my needs and commitments – someone more like me." ◀◀

▶ Is that guy lost or what? His almost total irrationality is derived from belief trap piled upon belief trap until only the twisted framework of a bizarre self-apparent logic remains. I counted at least twenty-eight unique belief traps piled one upon the other – how many did you see? Unfortunately, not much forward progress is likely for this particular tail-chasing seeker of spiritual truth until he learns to break out of the endless loop of Catch-22s he has created for himself.

It is a simple fact that everyone has the **potential** to learn what I have learned, experience what I have experienced, and do what I can do. It may take significant effort and dedication, but it doesn't have to disrupt your life very much. This fact is deeply discomforting to the ego. Rather than jump for joy at the possibilities, a much more common response is to shrink from the personal implications and to deny personal responsibility. One always finds it easier to generate excuses than to generate results.

If my knowledge were the result of being hit by tiny meteorite fragments (dust), contact with an alien, falling off a ladder, or channeling the material directly from god himself, you would have a ready-made excuse. The material is clearly beyond your practical knowing. You can do no more than choose whether to **believe** it or not. There can be no blame, no self-expectation, and no responsibility or guilt; your interaction with the material is easily constrained to an intellectual exercise that is kept at a safe distance from personal involvement. Now **that** is entertainment. Ahhh, this feels much more comfortable, and thus more credible. Our ego's job is to make us feel comfortable by redefining and interpreting our perceptions and conclusions to calm our fears and suit our needs. The typical ego can be convincing, cleverly subtle, and is extremely good at its job.

On the other hand, if you do have the energy and drive to find out for yourself, getting the initial kick-off briefing from your Dutch uncle who has been there and done that should greatly reduce the threshold to getting started. The technique is found in Chapter 23, Book 1 – the rest is found in your intent. ◀

Although reading this trilogy demonstrates great effort on your part, digesting it is another matter. Digestion is a much slower, more complex, and more significant process – it determines what you will absorb as opposed to what you will excrete. Being somewhat of an expert in the field of excretion processes, let me offer this advice: While *My Big TOE* is **slowly** digesting, you should take the opportunity to contemplate **your** Big TOE (you have one whether you know it or not); that will make the absorption process as efficient and effective as possible.

The antacid is on the top shelf right next to the extra strength aspirin.

89

■ ■ ■

You Must Climb the Mountain
to Get a Good View

■ ■ ■

There is no such thing as heresy if there is no such thing as belief. Likewise, you cannot be sucked into someone else's belief trap if you do not need a soothing belief. Replacing one belief system with another is of relatively little value. Perhaps there is some value if the new belief system is less limiting than the old one, but the **apparent** lack of limitations won't do you any good if you do not stretch out and take advantage of the new-found freedom. A man who will not leave his room because he does not know how, or is afraid to open the door, is trapped just the same whether or not the door is locked.

Now that you have seen the details of the overall model of reality contained within *My Big TOE*, it may be a good idea to re-read the foreword to this trilogy to regain the proper perspective for absorbing and interpreting what you have read. Additionally, look at the last several pages of Chapter 45, Book 2 (begin just after the free will aside) and at the discussion of fractal existence in Chapter 85 of this book. Reading these top-level high altitude summaries will prime you to get the most from the upcoming discussion that integrates *My Big TOE* into traditional Western science and philosophy.

There is a good chance that you will find something there that will help put the Big Picture into a more usable, or at least more comfortable, perspective. Such a review may dramatically facilitate the process of coming to your own conclusions – which is, by the way, your next challenge.

It is always interesting and sometimes informative to take another look from the mountaintop after extensive exploration of the valley. While you are there, you can assess whether or not I made good on the up-front claims that were used to encourage you to buy and read this book when you could have been more comfortable and better entertained watching TV.

If you want to communicate with me directly, go to the My-Big-TOE web page at **http://www.My-Big-TOE.com**. This web address is **not** case sensitive, but **the hyphens are necessary**. You can send email to me and the publisher from this web site, as well as acquire all Big TOE books and other paraphernalia. Here, you can keep up with the latest in Big TOE info, happenings, and discussion groups. By all means, share your complaints, praises, and comments with me by way of the web. It could be fun and educational, help you integrate what you have read, and enable you to be more connected and engaged with others. Best of all, you will find a public place to express your politely worded and erudite opinions, experiences, and learned thoughts – however pro or con they may be. A constructive sharing of ideas, experiences and feelings may go a long way toward helping you develop and grow your own Big TOE. Your experience and insight may present a valuable learning opportunity for me and for others.

You may or may not raise your consciousness and lower your entropy while dancing the Big TOE boogie with kindred spirits at the My-Big-TOE web page, but it might make you feel better to share your feelings and knowledge and vent your angst. You may even bump into an unforeseen opportunity to learn something of value.

My Big TOE is a model that simultaneously makes sense of both your subjective and objective experience as well as defines the ultimate significance of that experience. If you do not bring your experiential data to the table (before, during or after reading this trilogy), you will not get as much as you could from Sections 2 through 5. As long as *My Big TOE* remains only a pile of words that you can either choose to believe or choose not believe, it cannot effectively serve as a catalyst to your personal growth.

This Big TOE trilogy represents my best effort to facilitate the process of you growing your own big-as-possible TOE based on your personal experience by gently tugging at the constraints that may be limiting your vision. My intent is to set your mind free to find truth, not to pile on another layer of belief on top of what you already have, or replace one of your current beliefs with a new one. Freedom – spiritual, emotional, and intellectual freedom – provides the necessary environment for learning. Open-minded skepticism is the primary tool you will need to maintain a free mind capable of significant evolutionary progress.

Don't forget that other models can be different from *My Big TOE* and still be correct within the context of their specific conceptualization. My view is not the only useful or correct view – it is not an exposition of the one truth. If you are looking for the one correct expression of truth, you do not understand the cultural characteristics of expression, or the nature of truth.

A major purpose of this trilogy is to serve as a Big TOE construction kit – a way of looking at reality that helps you make sense of your experience within a larger systems view. Its value lies in its ability to describe all the data from all sources within all cultures in terms of a Western scientific-philosophical context. *My Big TOE* is designed to be most easily comprehended by the Western mind-set; other mind-sets may better understand and get more from other models.

With the *My Big TOE* trilogy, I have made an effort to develop a perspective that loosens up self-imposed constraints while providing a rational structure that will serve as a tower or mountain from which you can leap – soar off on your own – like a Jonathan Livingston Reader.

It is intrinsically difficult to leap and soar from the bottom of the PMR pit where most of us live; a little altitude can be invaluable. With no altitude, we need a long running start, which is much more difficult to achieve and often leaves us flapping furiously yet not getting off the ground.

Grounded! Caught in a belief trap, flapping about pathetically, unable to take flight into the reality that lies within. Can you relate to this pathetic bird-rat-being-pit-dweller-person?

You must climb the mountain to get a good view. You must look for the conceptual high ground from which to launch your vision of a personal Big TOE. At least, you must be able to differentiate up from down; if you can do that, and persevere in your effort, the rest will follow.

One note of caution. There are those who would entice you to get into their psychotropic drug cannon. They will offer to launch you into inner space like Bozo, the Human Cannonball Clown at the circus. **Do not take the trip**! If you do, you will no doubt end up a clown like Bozo except there will be no net and no applause. Consciousness altering drugs do **not** constitute a viable path to an increased quality of consciousness and are a thousand times more likely to be a serious problem than a serious solution.

If you cannot fly on your own, being shot out of a cannon to imitate flight constitutes a short irrelevant trip to nowhere ... and then ... SPLAT!

Being shot up into inner-space like a Fourth of July bottle rocket cannot teach you anything at all about flying, navigating, interpreting, or effectively utilizing thought-space; the view is short-lived and the trip can be deadly to your spiritual progress. In fact, using a chemical rocket to provide access to inner space almost always makes learning to grow and soar on your own more difficult.

It is said that misery loves company. Do not join the losers who are looking for a shortcut or a good time. Traveling down the path of chemically induced altered states of consciousness eventually and surely will take you to the opposite of both.

Because we live in an environment that is constantly trying to sell, manipulate and control, we are conditioned to defend our ego's constructs and to argue with and resist what others are constantly throwing at us. Eventually, by force of habit, this conditioned response hears me say – "I know more about these things than you ever will, simply believe what I tell you until you can figure it out for yourself" – despite what I have said to the contrary. Do you ever feel as though you are living in a grade-B reality where the zombies have taken over the world?

I have some good news and some bad news. The good news is that absolutely nobody but you can block your path toward quality consciousness. The bad news is that absolutely nobody but you can unblock your path toward quality consciousness. Actually, the bad news is not that bad. It simply means there are some things that money, status, good looks, sex appeal, and political power cannot buy – some things you either have to do for yourself or do without. Opportunity is everywhere; it is only gumption and vision that are in short supply.

90

■ ■ ■

The Significance of Insignificance

■ ■ ■

If improving the quality of your consciousness seems to be the point of existence, does that imply that worldly (PMR) pursuits, such as studying mathematics, physics, philosophy, basket weaving, digging ditches, or becoming an artist, are a waste of time? Absolutely not! You should do something besides contemplate your Big TOE. Everyone should have something externally productive to do with his or her time – something that is at least of some value to others. It is not what you do that is important – nor is it how valuable it is to others that is important: It is what you learn in the process of doing it that is important. For this reason, almost any job or activity will suffice because it is your interaction with yourself and others that creates many of the best learning opportunities.

If, after reading *My Big TOE*, everything in PMR seems less interesting and less significant, you have missed an important point. It is the experience we have in PMR that gives us the opportunity to learn by making choices that are precipitated by interaction with others. All sorts of activity in PMR – butcher, baker, candlestick maker, doctor, lawyer, Indian chief, housewife, teacher, physicist, bum, executive, or guru – all provide the opportunity you need to learn and grow the quality of your consciousness.

It is not that any profession or life's work has become less important because you now have a bigger picture. With a bigger picture there are other things that are also important that did not exist in the previous little picture. Physical possessions, family, friends, or your career, for example, are not any more or less a part of your life than they were before you grasped the Big Picture; it is you who may change the way you value them as your ego, knowledge, wisdom, or perspective changes. From a new

774 | *My Big TOE* | The Significance of Insignificance

larger perspective, as the ego value of what you do subsides, the real value (interactions with yourself, your tasks, and with other people) grows. Your career, or whatever you do, may seem less important in the Big Picture, but the potential value its interactions have for you should seem more important from your larger view.

What you are interested in is a function of what turns you on, and where you invest your energy. Again, a bigger view adds additional interesting things into the mix of possible things in which you can invest your energy. As you evolve your being (grow the quality and lower the entropy of your consciousness), the focus of your energy investment constantly shifts because a broadened reality includes new awareness that changes the potential investment mix.

Simply look at your life – say from age five to your current age – to see the truth of that statement. As you have grown up, the things that are important to you constantly change as your awareness broadens. The process of changing your investment focus continues as long as you continue to grow. If you are able, interested, and willing, the rate of growth accelerates after thirty-five – the growth rate curve can continue to be exponential for some time, and does not peter-out and go asymptotic (to the time axis) as most people think (imagine a plot of growth rate versus time). Your growth rate becomes asymptotic (slows to a crawl) only when you believe you know almost everything important.

That sense of knowing it all, (at least everything significant) usually blossoms at two, then again during the teenage years, and then gels, or becomes permanent, for most people around the age of forty. At two it is called "terrible," during the teen years it is called "self-absorbed" and "wild"; at forty it is called "mature" (obviously it is the older group that makes up the names). Knowing almost everything important is seen as the end point of growing up. This is like a sheltered (doesn't get out much) four-ounce juice glass half full of prune juice thinking it contains all the fluid on the planet. It knows nothing of the world's great lakes, oceans, rivers and seas, and being secure in its ignorance, feels self-important and complete in its mastery of fluids. Oh my! Poor little delusional juice glass!

If *My Big TOE* leaves you feeling deflated, it is because you are seeing the juice glass as being half empty instead of half full. Where you are now, what you are presently doing is almost certainly a great place from which to learn and grow. You do not need to make changes to your environment; you need only to make changes to yourself.

If you continue to feel deflated because you are not making sufficient progress in changing yourself, because you do not know how to start,

| http://www.My-Big-TOE.com |

cheer up – it's easier than you think. Not starting because you do not know how is like a one-year-old baby believing it will never walk or talk because it does not know how. Bullpucky! That is a copout attitude laying down a defeatist rap. Try this rap instead:

Just go do it!

Do it now!

That is all there is to it.

Wow!

Just go do it

You don't need to know how!

If you can sing that to a funky beat while you play the imaginary bongos on a tabletop you are on your way! Go! Go! Amigo!

What? You do not want to look nuts in front of your friends? I understand, you are new at this sort of thing – tell your friends that this mind-rap is actually a secret Tibetan chant recently translated from a primitive Sanskrit text found etched at the bottom of an ancient "Monks Do It Better" bumper sticker. Also remember, actual sanity flows from freedom while only pretend sanity flows from conformity.

Additionally, you might want to work on the following: Begin to understand your beliefs and their limitations. Inspect your ego regularly to see if it is growing or shrinking and expose some of its more blatant fantasies to yourself, then to your loved ones, and finally to your friends. Dedicate some of your energy every day to finding and verifying the truth. Become aware of your motivations and intents. Turn off the TV and get acquainted with your mind. Learn to meditate (see Chapter 23, Book 1). Be kind and loving in all your interactions with others. Stop thinking about yourself and your wants, needs, and desires. Figure out what your fears are and outgrow them. And above all, continue to objectively taste that pudding to see how you are doing. Only real measurable, bona fide objective results are acceptable. If you do not get results that you, as well as others, can easily see after a serious six-month effort, do something else. Be patient, real progress takes serious dedication over a long time. Have fun always.

Just go do it. You don't need to know how!

91

■ ■ ■

Traveling in Good Company

■ ■ ■

Since Big Truth is the same for everyone for all time, I figured that there must be lots of other similar material published somewhere. After I completed Chapter 90 of Book 3, I decided it was time to do a literature search and find out what science and philosophy had to say about Big Truth. I know, I was supposed to do that first – however, *My Big TOE* is not meant to be a scholarly work that merely adds an original twist to the work of others. This Big TOE is between you and me, dear reader: Its value to you defines its significance to me and what anybody else thinks about it is not relevant. My goal is to point out and explain the logical pattern that I have noticed in my own experience. Since I am not trying to convince you of its correctness, I don't need much help. After all, it is the results of my experience that I am writing about and it is the results of your experience that judges the value of my effort. If you and I are not the world's top experts on the meaning and significance of our personal experience, who is?

What I found out in my search – restricted to the processes of science, mathematics and logic – was that the ideas presented in *My Big TOE* are ideas that many individuals, judged by history to be great thinkers, have touched on. Even if many of these stellar individuals did not have the entire picture, they often saw a significant part of it.

I am not, I must confess, a particularly widely read individual. The upside is that what I have written in *My Big TOE* is uniquely mine and has not been influenced very much by others. This is no intellectually slick term paper laced with expert references and quotes to prove the point; there is no point to prove. Fortunately, the truth is the same for everybody, and all seekers who find any particular Big Truth will more or less reach the same conclusions. I expect that there are hundreds of books that share many of the concepts laid out in *My Big TOE*. No one has the market cornered on truth. Wisdom (both pseudo and real) has found creative outlets throughout the ages and will continue to do so.

On the other hand, the downside is that I cannot very easily come up with intellectually satisfying, cool-sounding quotes from big intellectual guns that if sprinkled throughout my text would lend credibility to my discourse (particularly at chapter headings, as is the current fashion). I have often thought that when the quotes are actually relevant, this can be a clever and effective literary device.

Not being that clever, I am forced to find other ways to invoke the knowledge of others and thus give the reader the illusion that there is indeed safety in numbers – that thinking big ideas that lie off the beaten path is not such a dangerous and crazy thing to do.

The only real danger is that once you find new knowledge, you automatically gain the responsibility for that knowledge. With both new knowledge and new responsibility, your growth rate will begin to accelerate dramatically. Before long you are a changed person! This, without a doubt, is the main danger associated with traveling the Path of Knowledge: that you might grow up as the quality of your consciousness improves. That may not sound too bad, but if those around you do not share your new broadened perspective, traveling through inner-space can be a lonely trip. Of course, you and Peter Pan can avoid that result if you want to.

The scientific literature is full of material that supports many of the concepts in *My Big TOE*. From physicists of the caliber of Albert Einstein and David Bohm we find several quotes that clearly show their sense of the larger reality.

"I wished to show that space-time is not necessarily something to which one can ascribe a separate existence, independently of the actual objects of physical reality. Physical objects are not in space, but these objects are spatially extended. In this way the concept of 'empty space' loses its meaning."
— ALBERT EINSTEIN JUNE 9TH, 1952 NOTE
TO THE FIFTEENTH EDITION OF *RELATIVITY*

Here Dr. Einstein is trying to explain that space-time is not something that physical objects exist within. He thought that space and physical objects were of the same substance – all were part of a single unified field. What appear to be solid objects are merely regions of higher field density than what appears to be empty space. What you perceive as reality is merely your experience of the various interactions within the unified field. The bottom line, according to Einstein, is that there is no such thing as physical space. Physical space (what you think you live in – what you **believe** constitutes your reality) is merely an illusion. Without physical space, there can be no physical reality. Einstein's deepest and most intuitive understanding of reality was expressed by his effort to develop a **nonphysical** TOE that he called "Unified Field Theory."

> *"If we think of the field as being removed, there is no 'space' which remains, since space does not have an independent existence."*
> —ALBERT EINSTEIN, GENERALIZATION OF GRAVITATION THEORY

> *"Reality is merely an illusion, albeit a very persistent one."*
> — ALBERT EINSTEIN

The following quote, attributed to "one of Einstein's letters" by Rudolf v. B. Rucker in his book *Geometry, Relativity And The Fourth Dimension* (p. 118) and also found in *Quantum Reality, Beyond the New Physics*, p. 250, captures Einstein's larger sense of reality and purpose.

> *"A human being is part of the whole, called by us 'Universe,'*
> *a part limited in time and space. He experiences himself, his thought*
> *and feeling, as something separated from the rest – a kind of optical*
> *delusion of his consciousness. This delusion is a kind of prison for us,*
> *restricting us to our personal desires and to affection for a few persons*
> *nearest to us. Our task must be to free ourselves from this prison by*
> *widening our circle of compassion to embrace all living creatures and*
> *the whole nature in this beauty. Nobody is able to achieve this*
> *completely, but the striving for such achievement is in itself a part*
> *of the liberation and a foundation of inner security."*
> — ALBERT EINSTEIN

It is clear from Einstein's many writings that space-time is not the place where we live, but rather a field of which we are a part. Mass (including our bodies) is merely a higher density portion of that pervasive field – matter bumps on the lumps in the space-time consciousness sheet. That this fundamental field was nonphysical and associated with consciousness was not as clear to Einstein as it was to his friend, associate, and fellow physicist, David Bohm.

> *"To meet the challenge before us our notions of cosmology and of*
> *the general nature of reality must have room in them to permit*
> *a consistent account of consciousness. Vice versa, our notions of*
> *consciousness must have room in them to understand what it means*
> *for its content to be 'reality as a whole.' The two sets of notions*
> *together should then be such as to allow for an understanding*
> *as to how consciousness and reality are related."*
> — DAVID BOHM FROM THE INTRODUCTION TO
> *WHOLENESS AND THE IMPLICATE ORDER*

From Max Jammer's book *The Concepts of Space* with an enthusiastic foreward and endorsement by Albert Einstein, we have further clarification (p. 171):

> *"Hence it is clear that the space of physics is not, in the last analysis,*
> *anything given in nature or independent of human thought.*
> *It is a function of our conceptual scheme [mind]. Space as conceived*
> *by Newton proved to be an illusion, although for practical purposes*
> *a very fruitful illusion – indeed so fruitful that the concepts of*
> *absolute space and absolute time will forever remain in the*
> *background of our daily experience."*

From the same source (p. 175) we find the great mathematician and physicist Karl Friedrich Gauss *"considered the three dimensionality of space not to be an inherent quality of space, but as a specific peculiarity of the human soul [consciousness]."* Also we find out from Dr. Jammer that time is the fundamental quantity and that space is a derivative of time (p. 169). Dr. Jammer says, *"This fact is of extraordinary significance because it proves that space measurements are reducible to time measurements. Time is therefore logically prior to space."*

I am sure a high-frequency state-changing spaceless nonphysical consciousness like AUO was very happy to get that news. If you are trying to remember where we first derived that conclusion, look in Chapter 31 of Book 1.

Eugene P. Wigner, a Nobel Prize winner and one of the leading physicists of the twentieth century, wrote a paper entitled: *The Place of Consciousness in Modern Physics* wherein he discuses the future of quantum physics. Dr. Wigner said: *"It will remain remarkable, in what ever way our future concepts may develop, that the very study of the external world led to the* **scientific** *conclusion that the content of the consciousness is the ultimate universal reality."*

Unfortunately, the non-objective (from the view of PMR) process that is required to access the science of consciousness ("the ultimate universal reality") was not fully appreciated. It may seem obvious to you that consciousness belongs to the individual and therefore is personal and cannot be studied in the same manner as a moon-rock, but at the time, and even today, this is not clear to everyone. However, the objective scientific importance of subjective experience was not overlooked entirely.

Willis W. Harmon (Ph.D.), noted futurist, forward thinker, Director of the US Educational Policy Research Center at Stanford University, in a 1969 paper entitled *The New Copernican Revolution* discusses the coming science of subjective experience.

> *"The science of man's subjective experience is in its infancy.*
> *Even so, some of its foreshadowings are evident. With the classification*
> *of these questions into the realm of empirical inquiry, we can*
> *anticipate an acceleration of research in this area.*
> *"Young and incomplete as the science of subjective experience is, it*
> *nevertheless already contains what may very well be extremely*
> *significant precursors of tomorrow's image of man's potentialities."*

At this point Dr. Harmon provides a discussion of what he terms "...an impressive amount of substantiating evidence." Next he describes the ongoing research that supports his projection. He then goes on to say:

> *"Assuming that the evidence substantiating these propositions continues to*
> *mount, they have the most profound implications for the future. For they*
> *say most powerfully that we have undersold man, underestimated his pos-*
> *sibilities, and misunderstood what is needed for what Boulding terms 'the*
> *great transition.' They imply that the most profound revolution of the edu-*
> *cational system would not be the cybernation of knowledge transmission,*
> *but the infusion of an exalted image of what man can be and the cultiva-*
> *tion of an enhanced self-image in each individual child.*
> *"It is perhaps not too early to predict some of the characteristics of the new*
> *science. Preliminary indications suggest at least the following:*
> *"Although we have been speaking of it as a science of subjective*
> *experience, one of its dominant characteristics will be a realizing of the*
> *subjective-objective dichotomy. The range between perceptions shared by all*
> *or practically all, and those which are unique to one individual, will be*
> *assumed to be much more of a continuum than a sharp division between*
> *'the world out there' and what goes on 'in my head.'*

"Related to this will be the incorporation, in some form, of the age-old
yet radical doctrine that we perceive the world and ourselves in it
as we have been culturally 'hypnotized' to perceive it. The typical
commonsense-scientific view of reality will be considered to be a valid
but partial view – a particular metaphor, so to speak. Others, such as
certain religious or metaphysical views, will be considered ... equally
valid but more appropriate for certain areas of human experience.
"The new science will incorporate some ways of referring to the
subjective experiencing of a unity in all things (the 'more' of William
James, the 'All' of Bugental, the 'Divine Ground' of Aldous Huxley's
The Perennial Philosophy).
"It will include some sort of mapping or ordering of states of
consciousness transcending the usual conscious awareness (Bucke's
'Cosmic Consciousness,' the 'enlightenment' of Zen, and similar concepts).
"It will take account of the subjective experiencing of a 'higher self'
and will view favorably the development of a self-image congruent with
this experience (Bugental's 'I-process,' Emerson's 'Over-soul,' Assagioli's
'True Self,' Brunton's 'Over-self,' the Atman of Vendanta, and so on).
"It will allow for a much more unified view of human experiences
now categorized under such diverse headings as creativity, hypnosis,
mystical experience, psychedelic drugs, extra-sensory perception,
psychokinesis, and related phenomena.
"It will include a much more unified view of the processes of
personal change and emergence which take place within the
contexts of psychotherapy, education (in the sense of 'know thyself'),
and religion (as spiritual growth)."

Dr. Harmon's vision and understanding of the fundamental scientific and individual importance of individuated (subjective) consciousness was right on the money but considerably more ahead of his time than he suspected. Never underestimate the power of the visionless center of objective science to maintain the conceptual status quo. Why did the vast majority of Western thinkers believe the earth to be flat when it had been demonstrated and logically confirmed to be spherical by several well-known and well-respected scientist-philosophers hundreds of years earlier? The political correctness of scientific **belief** weighs heavily at the center. Sometimes brilliance at the conceptual edge must be patient before the less flexible conceptual center is capable of seeing the light.

Let's return to the discipline of physics to see the next piece of the reality puzzle fall into place. Albert Einstein (Unified Field Theory) asserted a nonphysical field as the basis for both matter specifically and reality in general, thereby moving science closer to the truth, but he did not appreciate the discrete digital nature of both space and time or the role of consciousness (instead of space-time) as the primary energy field. Einstein's student and colleague, the great quantum physicist David Bohm (along with a few of the best quantum mechanics theorists such as Niels Bohr, Werner Heisenberg, and Eugene Wigner) made the consciousness connection but missed the digital connection and the Big Picture.

> *"One has to find a possibility to avoid the continuum (together with space and time) altogether. But I have not the slightest idea what kind of elementary concepts could be used in such a theory."*
> — Letter from Albert Einstein
> to David Bohm October 28, 1954

Physicist Dr. Edward Fredkin of MIT, Boston University, and Carnegie Mellon finally made the digital connection. In 1992 Dr. Fredkin published two papers: *A New Cosmogony* and *Finite Nature*. In these formal scientific papers, presented within traditional scientific forums, Dr. Fredkin developed rationale supporting both quantized space and quantized time, along with a description of reality as fundamentally digital. The science of information theory and mathematics led him to postulate an "Ultimate Computer" as the basis of a digital reality that computes our physical existence.

> *"If space and time and matter and energy are all a consequence of the informational processes running on the Ultimate Computer then everything in our universe is represented by that informational process. The place where the computer is, the engine that runs that process, we choose to call 'Other'.*
> *"Where did Other come from? This question is actually quite easy to fence with. The nature of systems of laws that can support computation is very much broader than the nature of systems that are limited to the physics of our universe. In other words, many of the properties of our world that are necessary for our world to take the form it has are not necessary for other kinds of worlds that can support universal computation. Universal computation, the kind that can simulate other general-purpose computers, is even a property of all ordinary commercial computers.*
> *"There is no need for a space with three dimensions; computation can do just fine in spaces of any number of dimensions! The space does not have to be locally connected like our world is. Computation does not require*

conservation laws or symmetries. A world that supports computation does not have to have time as we know it, there is no need for beginnings and endings. Computation is compatible within worlds where something can come from nothing, where resources are finite, infinite or variable."

Dr. Fredkin goes on to prove that "other" must be other than physical, that is, nonphysical from the PMR point of view. It would appear that Dr. Fredkin is talking about TBC and the space-time rule-set; yet Dr. Fredkin's work is wholly rooted in modern math, information theory, and contemporary physics.

Dr. Fredkin continues to analyze ("intelligent speculation" as he puts it) the beginnings of our universe.

"If we assume that the Ultimate Computer was purposefully constructed in 'Other,' we can immediately answer the puzzle of the origin of the Universe. It's simply a matter of the following process taking place in 'Other.' The initial conditions are set into the engine and the engine is set into motion; it starts to compute. Those two steps are outside the domain of physics."

That passage from *A New Cosmogony* should remind you of our description of The Big Digital Bang – The Big Simulation – discussed in Chapter 63 of Book 2, and in Chapters 73 and 81 of Book 3. Dr. Fredkin continues to explore the nature of "other":

"As to the question 'Why didn't the thing in Other just do it in its head?' The answer is quite straightforward: Doing it on a computer is exactly the same thing as doing it in one's head [with consciousness]. Both are examples of using an informational process to get to the answer.
We are not referring to the thing in Other finding an analytical solution in its head (the speedup theorem forbids such solutions) but rather to it imagining each step of some cellular automaton in its head. Strangely enough, that's exactly the same as doing it on a computer."

Is it not obvious that digital consciousness simulating OS nicely fits the form, function and nature of Fredkin's "other"? What Dr. Fredkin did not understand was that consciousness **is** the computer. I find it interesting that information theory, physics, and mathematics, once again (this time from an entirely different starting position) point to consciousness as the ultimate reality, the fundamental source of All That Is.

Dr. Fredkin seems to have started something with his concept of a calculating (simulating) digital reality based in a nonphysical computer. Today there are a growing number of physicists, computer and information scientists, and mathematicians scattered around the world that pursue research in what has come to be called digital physics. Digital physics, like relativity and quantum mechanics, is serious hard-core science that is, for the majority of people, still way out on the edge.

▶ Let's pause a moment to collect together what some of the most respected scientists of our time have thus far jointly discovered. 1) What we perceive as our local physical reality is actually a nonphysical virtual reality, a subset of a larger more fundamental reality. 2) The doorway to experiencing this larger reality is through the individual's subjective mind. 3) The larger reality is based upon consciousness, which is the basic substance, energy, or media of existence. 4) The larger reality is discrete and must be the result of digital computation: The consciousness of (3) above is a digital consciousness and that the virtual reality of (1) above is a digital simulation.

Ladies and gentlemen, are you astounded to find out that *My Big TOE* has explained, derived, and pulled together, within a coherent theoretical framework, some of the best theoretical efforts of contemporary modern science?

Jeez, I don't know about you, but it astounded the hell out of me!

Notice how the seemingly wild-eyed conjecture of one old Dutch Uncle-dude, turns out to derive results that are very similar to the results derived by some of the twentieth century's most respected and innovative scientists.

This trilogy represents serious science folks – the bridge between physics and metaphysics, mind and matter, normal and paranormal. It delivers the first look at a Unified Theory Of Everything – not the way traditional science expected it, but then, if it was the way traditional science expected it, it couldn't actually be a **Big** TOE could it? If you want to see the unified Big Picture, you have to step out of the box that limits your understanding and knowledge. You cannot see it from the vantage point of the old paradigms.

As advertised, *My Big TOE* delivers the integrated results of contemporary modern physics to your doorstep. Why are you so surprised? Did you think *My Big TOE* is not a serious scientific effort? Why not? When you formulate answers to those last two questions, be sure to inspect them for hidden belief traps. A typically trapped mind tends to **believe** that all that goofy metaphysical AUM-stuff necessarily moves the reality model of *My Big TOE* out of the realm of objective science and into the realm of non-provable conjecture. Such a mind **believes** that if AUM is not the **result** of a physical measurement or logically definable by an equation, then AUM falls outside the scope of science.

By now, dear reader, you should immediately recognize the illogic of that particular belief trap. Remember that the logic of the little virtual reality applies only within that little virtual reality. (You may want to re-look at the logic of beginnings discussion in Book 1, Chapter 18.) A virtual reality is, by definition, a closed logical system – its physics is merely a symbolic (mathematical) representation of that logic. The logical causality of the Big Picture cannot be derived from the logical causality of the little picture. The Big Picture cannot logically be contained within, or be a subset of, the little picture. You cannot arrive at the Big Picture if you never leave the little picture.

Einstein and the great quantum theorists got stuck because they were trying to derive the larger reality in terms of the logic (mathematics) of the smaller reality. Little picture logic (the mathematical physics of the space-time rule-set) can tell you only about the little picture. These scientists bumped into the little picture to Big Picture boundary, but could not penetrate that boundary with little picture logic alone. The Big Picture appears to be logically impossible, and thus becomes invisible, when viewed through little picture paradigms.

That *My Big TOE* is presented as a **non-mathematical** extension of the limited science of our virtual physical reality is a logical **requirement**, not a scientific weakness. The logical causality of the Big Picture can be expressed in terms of its own symbolic logic (mathematics), but such constructions are of limited use in thought-space. Before the scientific and mathematical types get lost here, consider the evolutionary purpose of a consciousness system.

Physicists still have a lot to learn about the space-time rule-set that defines our little picture virtual physical reality. Because that rule-set defines a digital virtual simulation within a larger digital reality, little TOEs may need to embrace some of the digital magic we have found so entertaining at the movies and so useful on our desktops. Even so, little picture science and applied mathematics (little picture logic) can never lead directly to a Big TOE. To get there, to experience the Big Picture, you must leave the little picture behind and step through the portal of your subjective (from the little picture view) mind into the larger reality of digital consciousness.

Consider what the little picture scientific establishment will tell you: An AUM-digital-consciousness-system-thing could not possibly be the **larger and more fundamental** reality because it is not **exclusively** manifested as a **physical** substance in the little **virtual** physical reality. Can you imagine a more illogical and irrational belief? You would think that this belief is incredibly stupid if you were not raised to accept that it is unquestionably true.

It should be clear that consciousness is both nonphysical and real – that it is primarily a personal thing and therefore subjective. Furthermore, to separate the external from the internal, the real from the imagined, you should insist on testing and confirming the operational realness and significance of subjective experience by demanding measurable, repeatable, **objective** results before assessing the value of that experience.

We could go on and on exposing this particular PMR *über alles* trap, but we have covered that ground in great detail already. Those who get it do not need to hear it again, and those who do not get it need to wait until they are more open and ready — until they can look beyond the old paradigms and belief systems that have captured and imprisoned their mind. If you are not sure whether you get it or not, finish all three books of the trilogy, give it a rest for a month or so, and then read it again starting with Section 2. Because of *My Big TOE*'s unusualness, and because it is nearly impossible **not** to accept the core beliefs of your culture, it is extremely difficult to get it all the first time through. I know how disconcerting and annoying that thought must be; nonetheless, it is a true statement that applies to most of us.

The wasted potential of a self-limited mind is a sad happening in any reality. A larger consciousness system appears as unsupported conjecture only to those who will not take the time or make the effort to explore (experience) and assess the Big Picture scientifically. Just a little open-minded research and experimentation is all that is required to lend credence to the existence of the larger consciousness system of which we are a part.

Oh good grief! Do you hear all that hysterical shrieking and banging about? That's the sound of the defenders of the scientific and cultural status quo and sacred belief systems building walls instead of bridges. Not to worry, all the folks quoted in this chapter have heard that sound many times before. As always, if the data support it, raise it up, use it, experiment with it, seek new understanding; if it does not, let it self-destruct under the weight of its own failure to produce a greater, more productive paradigm. Fortunately, the truth is not delicate; it can stand up to whatever comes, for however long it takes.

Here, an objective group-proof organized by the establishment (con) or the anti-establishment (pro) has no merit. It is up to you, dear reader, to develop your own proof (knowledge) through your own experience. Let no one succeed in providing the answers for you. At the core, you are consciousness: you have access to all the answers — go find them for yourself and they will make you whole.◀

▶▶ Following the herd, no matter which way it is going, inevitably leads to stasis and dysfunction. Following the herd or conceptual center, allows you to trade the **opportunity** for personal progress and growth for an easy and safe no-brainer glide through existence. Unfortunately, procrastination and immediate gratification team up to produce the worst possible long-term strategy for consciousness evolution.

The herd appears to validate each member with its mutual support mantra of "I'm OK, you're OK, we are all OK as long as we stick together." Fear and ego provide the herd's cohesive forces. Your individual consciousness is a personal thing, not a group thing. If you have the courage to break loose from the herd, don't worry; the larger world is not nearly as scary as it looks. Gather strength and resolve from the fact that, as a solitary seeker, you are following the path of all innovators, discoverers, and creators. Only as an individual do you have the potential to make a significant difference where it counts the most.◀◀

▶ It seems that we have accomplished a great deal with the two basic assumptions (consciousness and evolution) that we started with in Chapter 24 of Book 1. Neither assumption should appear that unusual because we experience a **profound** and **immeasurable** (nonphysical) personal consciousness every day of our lives, and because we have been observing evolution at work in the little picture for a long time. From those simple assumptions we have derived a model of reality, a Big TOE that encompasses all previous knowledge as well as derives new knowledge from a less limiting, more general reality paradigm. Our consciousness is our personal connection to that larger reality – our ticket to view and experience the Big Picture.

Guess what, dear reader? You have successfully been doing Big TOE physics throughout the last four sections of this trilogy. You have, relatively easily and quickly, derived from basic principles the same results that Einstein, Bohm, Bohr, Wigner, Harmon, Fredkin, and other top scientists took several decades and tons of complex mathematical analysis to reach. Furthermore, you have pulled together all of their various pieces of cleverness, discovery, and intuition into one unified whole, something many have tried to accomplish. However, until now, no one had been able to find the perspective wherein they could glimpse the entire Big Picture – a perspective where all the pieces integrate nicely into one. You have now accomplished this long-sought and scientifically remarkable feat. I am very proud of you. I will wait here while you go tell your mother.◀

In the Big Picture, science and philosophy once again become one – two aspects of the same knowledge. We have heard from the scientists; now let's hear what the philosophers have been saying. I will mention only a few of the West's most famous philosophers, all from the group called modern philosophers that arrive at their conclusions through rigorous inductive and deductive reasoning. This is not the king of the opinion hill philosophy that dominated before Descartes, but rather an intensely rational system of logic more closely related to mathematics than persuasive discourse. (Note: Many of the history of philosophy factoids briefly and superficially scattered over the next few pages were cut and pasted directly from Microsoft's Encarta Encyclopedia; how I pitch them is all my own. If you think my history of philosophy facts are not precisely the way you remember them, don't bug me – go see Bill.)

The most famous of all the mutterings of all the philosophers is attributable to Rene Descartes, the initiator of modern philosophy: *"I think, therefore I am."* This simple phrase is a tribute to the primacy of consciousness as a core contributor to any fundamental conception of reality.

Eighteenth-century German philosopher Immanuel Kant held that all that can be known of things in themselves is the way in which they appear in experience – there is no way of knowing what they are substantially in themselves. He also held that the fundamental principles of all science are essentially grounded in the constitution of the mind rather than being derived from the external world. Does this remind you of our discussion of experience in Chapter 66 of Book 2 where we separated the experienceable reality from the un-experienceable reality that lies beyond our limited physical perception?

Gottfried Wilhelm Leibniz was considered a universal genius by his contemporaries and by history. He made brilliant original contributions not only to mathematics and philosophy, but also to theology, law, diplomacy, politics, history, philology, and physics. In the philosophy expounded by Leibniz, the universe is composed of countless conscious centers of spiritual force or energy, known as monads. Each monad represents an individual microcosm, mirroring the universe in varying degrees of perfection and developing independently of all other monads. The universe that these monads constitute is the harmonious result of a divine plan. Do you make the connection between Leibniz's monads and the fundamental dynamic pattern of all existence and creation that forms the basis of the consciousness-evolution fractal we discussed in Chapter 85 of Book 3?

Johann Gottlieb Fichte transformed Kant's critical idealism into absolute idealism by eliminating Kant's things in themselves and making the will the ultimate reality. Fichte maintained that the world is created by an absolute ego, of which the human will is a partial manifestation and which tends toward God-consciousness as an unrealized ideal.

Friedrich Wilhelm Joseph von Schelling went still further in reducing all things to the self-realizing activity of an absolute spirit (individuated unit of consciousness), which he identified with the creative impulse in nature.

One of the most influential philosophical minds of the 19th century was the German philosopher George Wilhelm Friedrich Hegel, whose system of absolute idealism was based on a conception of logic in which conflict and contradiction are regarded as necessary elements of truth, and truth is regarded as a process rather than a fixed state of things. The source of all reality, for Hegel, is an absolute spirit, or cosmic reason, which develops from abstract, undifferentiated being.

The Danish philosopher Søren Kierkegaard attacked the concept that objective reason is the only source of truth. His eloquent defense of feeling and of a subjective approach to the problems of life became one of the main sources of 20th-century existential philosophy.

As you see, modern analytical (logic based, not belief-based) philosophers have been hot on the heels of a larger, more primary reality that is based upon mind, consciousness, or spirit. They have been much closer to Big Truth for centuries than the scientists but didn't know how to integrate their knowledge with the physical world. These philosophers did not fully appreciate the subjective component of reality, or that subjective information requires an objective assessment (tasting the pudding) before its value can be determined.

Without these understandings it was difficult for philosophers, and even more difficult for their followers, to differentiate knowledge from pseudo-knowledge. In an age mesmerized by the results of objective science, narrowly focused critics had little trouble discounting the significance of philosophy's contribution toward understanding a reality defined entirely by physical measurement.

Western philosophers eventually succumbed to both their critics and their culture. They restricted their explorations of reality to remain within the rational boundaries set by the limited understanding of little picture objective causality. They immediately fell from the top of the intellectual relevancy heap (where the title Doctor of Philosophy was the highest of intellectual honors) to the bottom of the heap where philosophers are seen as impractical academicians engaged in mind games that are more or less irrelevant to the real world. Philosophy, when focused on the Big Picture that lies beyond cultural belief traps, can be as potent a tool for discovering Big Truth as science. When it comes to ferreting out Big Truth, neither tool works very well by itself; dramatic success can only occur when science and philosophy work together.

Many philosophers understand the synergy of this connectedness on an intuitive level and carefully track the philosophical implications of science. However, they remain frustrated at being unable to contribute very much to a greater unified understanding of the whole. Today, science leads the search for truth while philosophy provides ad hoc commentary in the margins.

Most scientists think philosophy is a waste of time; fortunately, the best scientists do not share that view. They work with their intuition as well as their high-tech tools and have a larger sense of how their work fits into the whole. Defining that whole is seen to fall within the purview of philosophy. Scientists, the caliber of Albert Einstein (as well as the others mentioned above), realize that science and philosophy are two sides of the same coin, and that one day, when we fully understand the reality of which we are a part, science and philosophy must coalesce into one unified understanding.

By definition, conceptual breakthroughs must necessarily be found outside the boundaries of what is commonly accepted. A profound understanding of reality, a coming together of science and philosophy, can occur only within the Big Picture. When our understanding is complete and whole, all the puzzle pieces fit nicely together. As long as our understanding remains fragmented by our limiting beliefs, competing culturally correct cliques within the little picture will define and therefore limit our vision of reality. The breakthrough solutions that both the scientists and philosophers have been seeking for many decades lie outside the little picture; one cannot get to them without stepping beyond the limits of PMR.

When scientific paradigms (such as the universality of objective causality) become sacred and unquestionable, breakthroughs become impossible. What appears as a weakness in the credibility and scientific stature of *My Big TOE* from the view of the old PMR *über alles* paradigms turns out to be a necessary logical requirement and a condition of credibility from a more expansive, less limited viewpoint.

Appreciating that the dichotomy of modern science and philosophy is an artifact of Big Picture ignorance expressing a limited view will help the reader put *My Big TOE* into a more integrated and unified perspective. To this end, let's round up the concepts we have tossed about in the last several pages and summarize their significance to the bottom line of *My Big TOE*.

As matter and mind split into what appeared to be disconnected opposites, science and philosophy became estranged. Neither could say anything important about the other. Now that *My Big TOE* has shown the mind-matter dichotomy to be a perceptual delusion – all is consciousness – science and philosophy may once again unite as complementary approaches to the same truth. **Good science is good because of the quality of its methodology, rational process, and logic – period**. To further restrict it to a limited little picture causality or local PMR logic is to shut it off from the possibility of explaining its beginnings or seeing the bigger picture.

Scientists require experimentation (a careful objective examination of reality) to separate fact from hypothesis. Within the subjective science of Dr. Harmon, **you** must personally plan and execute your own experiments and derive your own conclusions. Group-proofs, peer review, and majority rule are irrelevant within the mind-space of your personal consciousness. Careful processes and **objective** pudding tasting must be employed before personal science can discover Big Truth.

Similarly, **good philosophy is good because of the quality of its methodology, rational process, and logic – period**. The gap between philosophy and science is not as wide as you might think; one tends more to the possibilities, the other to the actualities. When we explore our reality from a Big Picture perspective, these two approaches to truth become complementary and mutually supportive. It is only when we are far from Big Truth, when we are out of balance – obsessed with the details of an apparently objective physical reality to the exclusion of all else – that science and philosophy seem to exist at opposite poles. It is our inability to see and understand the Big Picture, our lack of inner balance, and our limited PMR paradigms that drive an imaginary wedge between science and philosophy, between mind and matter, between normal and paranormal. The idea that science produces facts while philosophy produces arguments is a simple delusion born of a common prejudice and small view.

Here is an interesting historical comment on the subject. Is not a Ph.D. in Physics, or anything else, a doctor (D) of Philosophy (Ph)? It certainly appears that a long time ago we were less confused on this particular issue than we are now. The scientific attitude that PMR and its little picture logic represent the only possible reality is a common cultural belief that has tricked us into discarding some of the wisdom we had previously acquired. On the other hand, it helped jettison a ton of useless baggage that was holding us back. Obviously, the science-philosophy, mind-matter, normal-paranormal split represented an evolutionary process we sorely needed to go through.

Having made it through our superstitious phase, perhaps now we have grown up enough to simultaneously see a bigger picture and demand objective clarity. Perhaps over the last few hundred years we have learned more than all the kings horses and all the kings men and will be able to put the Big Picture back together again – this time with all the advantages (science and technology) and none of the drawbacks (superstition and belief). That's progress ladies and gentlemen: science eventually saving itself (and everything else) from the ravages of science. No, we are not out of the dark yet, but *My Big TOE* should shed enough light to enable many individuals to make the return trip to wholeness. Hopefully, the glow from this Big TOE will shed enough light to give both science and philosophy a rational first peek over the top of the PMR prison wall. Time will tell.

Thus, *My Big TOE* – obviously a work of metaphysics and philosophy – is by virtue of its methodology, rational process, logic, and careful experimentation, also a work of science. It is in a similar category (reality models) with the atomic shell model, string theory, or any other scientific model for that matter – only much larger in its scope. The data needed to validate this model must come from science, philosophy, and most importantly, from **your** objectively evaluated subjective experience.

The *My Big TOE* trilogy does indeed represent a serious scientific effort that is carefully and objectively investigating the reality that lies both within and without. With this science we have successfully derived, explained, and integrated the results of some of the world's best and most respected physicists. At the same time, we have explained mind, spirit and purpose, uniting the normal and the paranormal under one TOE – one consistent, logical, and rational understanding that explains **all** the data.

▶ This is the perfect place to take another short pause – this time, to round up the objective (shared) side of the Big TOE bottom line. We will focus on the subjective (personal) side of that bottom line a little later.

Even at the introductory level of a 101 survey course, *My Big TOE* provides a sound theoretical basis for understanding many of the scientific, metaphysical, theological, and philosophical enigmas that have been nagging at the minds of scholars and thinkers for decades. Even more importantly, *My Big TOE* provides a logical scientific basis for finally answering some of the most mysterious and pressing **personal** questions that have challenged human understanding since time immemorial – since men and women first stared into a starlit sky and wondered who and why they were.

If you dare to open your mind as well as your eyes and take a hard **objective** look at the world around you, you will find a plethora of solid, respectable, repeatable data points that clearly point to the existence of a larger reality that exists beyond our present understanding. Most of these scientific studies and carefully documented reports of paranormal happenings or other scientific enigmas (such as wave-particle duality or entangled particle pairs) can be adequately explained, as well as given a solid theoretical basis, by applying the reality model presented in *My Big TOE*.

Many individuals have personal experiences that likewise point to the existence of a larger reality. *My Big TOE* provides the logical and rational understanding that is required to pull these experiences out of the closet of "weird things I experienced but don't understand and usually don't talk about," into the liberating light of a scientifically derived Big Picture.

With the understanding provided by *My Big TOE*, much of what we study, observe, and experience that has been beyond explanation and totally without a logical scientific basis, can now be seen as a natural part of the reality we live in, experience, and are a part of. All manner of things mysterious, both scientific and personal, that did not make sense before, make sense now.

How do you determine if a model (theory) is true and accurate? Ask these questions: Does it make sense out of what was previously not understandable? Does it explain **all** the data? Does it provide new direction, new perspectives, and new knowledge in a form that is both useful and profitable? Does it enable significant progress in your effort to understand yourself or the outside world more fully?

You must scientifically evaluate each of these questions relative to your subjective private world, and relative to your objective public world. You are the final judge and it is your responsibility to find and develop the evidence that you need to judge wisely.

Here is the bottom line simply put. If *My Big TOE* delivers useful answers, if it rings true at the deepest level of your knowing, if it is verified by your research, your data, and your experience, if it helps you make sense of your life and your work, then raise it up. Use it as a tool for obtaining greater understanding and knowledge and for converting this new understanding and knowledge into wisdom. Be sure to tell others about the truths that you have discovered.

If, on the other hand, it fails to do these things for you, let it go. Let it stew in its own ineptitude; let it sink under the weight of its unprofitabiliy. Without lifting a finger, the ruthless dynamics of social and intellectual evolution will quickly cast aside any newcomer that rings false. Be sure to tell others about the falseness you have encountered.

In an intelligent and civilized society, ideas are allowed to compete in an open mental space with mutual respect and civility. A theory succeeds or fails by the degree of credibility invested in it by the established community that claims to be the experts, and by those who have applied the theory to good quality data and obtained valid results. In the best of worlds, these two groups should be composed of the same people; too often they are not.

Politics, power, personality, and science usually coalesce into the single abstruse voice of the scientific establishment. (In that last sentence feel free to replace the words "science" and "scientific" with the appropriate names of any group, profession, or organization. The dominating and dumbing influence of politics, power, and personality is as sadly ubiquitous as low quality consciousness.) Whatever the process, the fact remains: **Credibility must always be earned within the context of present understanding**.

Generally, this is a positive thing; social and intellectual viscosity slows the rate of change to ensure that carts stay behind horses – that knowledge does not get too far out ahead of wisdom. The pace of greater change reflects the pace of our greater learning. The world we live in is a direct expression of our collective quality. We are all responsible for the shared reality that we experience – it is our creation.

No idea will take root and blossom before its time, nor be denied when its time has come. To which side of that divide will *My Big TOE* fall? Readers, such as you, will collectively make that determination. *Que será, será.*◄

It would appear that these references and quotes are but the tip of an immense scientific and philosophical iceberg that is in fundamental consonance with *My Big TOE*. Big Truth is the same for everyone for all times; consequently it is not surprising that many of the world's most brilliant minds, both past and present, should have discovered at least some part of it. I am certainly **not** trying to convince you to **believe** this or that by waving about a collection of quotes from a few of the world's most outstanding smart people.

I am equally sure one could find another assortment of smart people who would take the opposite view – one always can. Arguing that my smart people are smarter and more aware than your smart people is not exactly what I had in mind. That a million smart people say "yes" is not a good reason for **you** to say "yes" if you do not have the experience and understanding to support it. Determining truth is not a democratic process ruled by the majority. It is also not something that someone else can do for you. Do not let these smart people sway you to agree with them simply because they are smart. In fact, it is not a given that those I have quoted would support the merit of *My Big TOE*. They may agree or disagree, like it or disown it – that is not the point. These quotes are **not** presented as testimonials – that would be illogical as well as a misrepresentation.

The point here is to allow you to find some safety in numbers, some assurance that the ideas in *My Big TOE* are perhaps not the wild ravings of a single lone lunatic. It is a fact that there have been, and still are, many sober, serious, and very bright folks who have come to supportive and compatible (with *My Big TOE*) conclusions through the application of rigorous logic, mathematics, scientific principles, and careful rational study.

Another point is to demonstrate that perfectly competent, sane, and highly regarded scientists and thinkers can, and often do, hold well justified and thought out concepts that appear to be on or beyond the fringe relative to the normal attitudes and concepts held by the masses – even the masses of scientists and philosophers. All big discoveries (such as the spherical earth, the heliocentric solar system, the atom, quantum mechanics, and relativity) are initially shunned and held to be ridiculous by the great majority of scientists, philosophers, and other keepers of the cultural status quo and sacred belief systems. All of which brings to mind another famous quote from Albert Einstein: *"Great spirits have always encountered violent opposition from mediocre minds."*

And thus it shall ever be: Big ideas, innovation, and creativity always comes from the edge, never from the center. On the other hand, goofiness also comes from the edge, which is why those who are not competent to tell the difference huddle together for safety in the center. This brings us back to the concepts of the fear of not knowing, and finding safety in numbers. Therefore, dear reader, be assured that if you dare to travel near (or even beyond) the edge of acceptability as defined by the center, you may still be traveling in good company.

It is exceedingly rare, if not entirely impossible, for anything of deeper significance or importance to develop from, or within, the great majority living at the center – that is not their function. Nevertheless, the center is critically important because it provides stability and continuity, performs many necessary services, and provides the required infrastructure to maintain the whole. The center and the edge need each other in order to be a successful productive whole; each needs to learn to appreciate the other. The personal bottom line of this chapter is that you may live proudly and productively at the edge, appreciating the center, tolerating the indigenous goofiness found all around, fearlessly thinking Big-thoughts, and tasting the pudding.

It is a more satisfying life when we are living at a place on the continuum – from dead center to ragged edge – that is in consonance with our personality, nature, and capacity. Transitioning your home along that continuum always requires a difficult shift in your local reality.

I have purposely **not** mentioned the many wise persons who have come to conclusions in consonance with *My Big TOE* through means and processes other than Western scientific, mathematical, and rigorously logical methods. From this unmentioned group one could extract a mountain of wise and meaningful words – much of it from direct experience. Indeed, there is wonderful wisdom and insight that flows from all the great religions and spiritual traditions of the world. Though it may be tempting to provide quotes demonstrating a universal spiritual understanding throughout the ages that is supportive of the concepts developed within *My Big TOE*, this exposition does not come from that direction.

It is focused **at** the Western, scientific, logical, process-oriented mind, **by** a Western scientific, logical, process-oriented mind. To optimize its goal of successfully communicating to the growing intellectual products of Western culture, *My Big TOE* has built its logical structure solely upon a Western cultural foundation even if that foundation appears to many to be relatively impoverished in its understanding, acceptance, and experience of spiritual values. Things are often not as they appear; the average quality of consciousness in Western cultures is no less than the quality of consciousness found in other cultures. Tradition, ritual, and acceptance of an extant spiritual reality pay no automatic dividends in individual consciousness quality.

It is not that other than scientific contributions are not important, meaningful and significant, but that you must find those types of inputs on your own, according to your interests. I am not trying to develop a theology or start an argument. All I am doing here is sharing my direct experience and a few of the conclusions and possibilities that can be rationally derived from that experience. If that last sentence seems unlikely and difficult to fathom, it is because you do not share my experience.

The unintended conclusion of the above display of scientific and intellectual support for many of the concepts found within *My Big TOE* may be that the world is full of raving lunatics – that genius and insanity often travel together. You may also conclude that when a collection of sober and clear-headed people of sound intelligence and proven capability say something that you do not understand (sound like lunatics), it may be an error to jump immediately to the conclusion that they obviously do not understand the world as well as you do or that they are either stupid, delusional, or diabolical. Consider the outside possibility that there may be some other reason why you are having difficulty understanding what they are saying.

Unfortunately, open is not a normal state of mind. Admitting ignorance has always been un-cool – something no self-respecting ego would ever do without first being cornered by a mob.

To keep the record straight, lots of goofiness comes from the center as well, but it is a less obvious, more insidious, traditional form of goofiness that both reflects and is supported by the dominant culture (racism, sexism, slavery, witch hunts, religious fundamentalism, stifling conformity and intolerance, irrational fear of change, political correctness, conspicuous consumption, blue laws, grocery store tabloids, pet rocks, and beanie babies come immediately to mind).

You might as well accept it; this is simply the way we humans are. Our state of being accurately reflects the quality of our consciousness and the present extent of our evolution. As long as overgrown egos are the norm and as long as we have AUM derived consciousness at our core, the struggle will continue.

Our struggle is the struggle of an evolving digital consciousness driven by free will, existing in a state of continual becoming, trying to improve itself. We try to grow up, decrease our entropy, actualize the potential we inherit from AUM, and become one with the larger reality – aware in the outermost loop while maintaining individuality. Our struggle to evolve the quality of our consciousness is derived from AUM's struggle to evolve the quality of its consciousness.

It is the struggle of the edge with the middle and the middle with the edge. It is the struggle of fear and ignorance against love and understanding; of ego, wants and needs against compassion, humility and balance; of good against evil; of what you can get versus what you can give. This struggle forces every choice you make, every interaction that takes place within your experience, to fall to one side or the other.

What, dear reader, is **your** part in this epic struggle? Where do you fit in? What is the significance of you? And, if by some odd chance you are not exactly the same person who began reading *My Big TOE* some weeks ago, how long will it take you to return to normal – even if returning to normal is not your **present** stated intent?

For in the end, there is no end – only a beginning.

92

■■■

More Good Company

■■■

In this continuation of the previous chapter, I will not limit our traveling companions to **only** top scientists, mathematicians, and logic driven philosophers, of worldwide fame and stature. In this chapter, I will open the gates a little wider and also include some well known thinkers, artists, writers, political leaders, educators, and a few others – men and women of some renown who have earned reputations for the quality and soundness of their thoughts and deeds – all from our own Western culture.

Anyone I could associate with a known dogma or non-western culture was disqualified from this group, even if the words of these individuals non-dogmatically represented obvious truth and wisdom. Why be that restrictive? The *My Big TOE* trilogy is focused on those immersed within a worldwide Western culture. Unfortunately, most people feel uncomfortable with, and subsequently discount, information coming from a culture other than their own. Most people do not trust information from an unfamiliar source. My restrictiveness is merely an act of straightforward pragmatism – I am simply trying to communicate effectively with the largest number of readers. Please take no offense if I pass over one or more of your favorite wise persons.

A casual inspection of a few Internet quote repositories has produced more than enough material to make the point that many of the thoughts and attitudes expressed in *My Big TOE* are not as far off the beaten path as you might have first thought. By now, this should be an expected result. From whatever direction you approach fundamental truth, the result is the same once differences in personal expression are accounted for.

Actually, the West is not as spiritually bankrupt as it appears; it simply lacks a widely acceptable mode of expression of spiritual values at the cultural level. However, because the quality of consciousness is a personal matter, the lack of a pervasive cultural expression is irrelevant. The lack of a serious pervasive cultural sense of a bigger picture has its benefits as well as drawbacks.

> ▶ I hear some objections to my statement that the West lacks solid spiritual values at the cultural level. There are some spiritual values shallowly embedded within Western culture, but these have mostly been reduced to effete slogans hanging like window dressing on the edge of the dominant religious dogma – they orchestrate little cultural force at a deep level. Why? Because in Western culture there is a huge, nearly unbridgeable gap between the physical and the nonphysical, between being and doing, between I and other, and between the theory and practice of spiritual values.
>
> The large majority of **religious** values have little, if anything, to do with **spiritual** values. This is as true in the East as it is in the West. ◀

This chapter should be fun. Now that the conceptual heavy lifting has been done, I think you will enjoy this relaxed cool-down meander through the intellectual fields of Western sound bite wisdom – a collection of thoughtful gems that bear repeating. I have broken the material into several groups and will say a few words about how each set of quotes relates to *My Big TOE*.

Belief and Self-Imposed Limitations

This set of sound bites speaks to the power of belief to insulate the believer from additional knowledge or finding a larger truth – a point made repetitively in *My Big TOE*. Here, you will also find corroboration that most people have a natural tendency to feel that they know more than they do – that if **they** do not know it, if **they** have not experienced it; it is not likely to be true. These folks are sure that if they and their peer group have no direct experience of a larger reality, the larger reality must be the product of delusion. (The delusional always believe that they are non-delusional and that the non-delusional are delusional – such is the nature of delusion).

Many of Western culture, scientists in particular, are snared in this "arrogance by assumed omniscience" belief trap. It was this same inflated sense of self – that only **other** people could be largely ignorant – that led some of the most respected scientists of the early 1900s to laugh confidently at a young and foolish Albert Einstein for presenting his obviously incorrect theory of relativity. Einstein magnanimously tolerated many arrogant and unkind professional guffaws before getting the last laugh.

As you can see from this first quote, Albert had no use for those caught in the vision limiting belief trap of: "If this doesn't make sense to **me**, then it must be wrong." For those following the well-worn path (steadfastly trod by the large majority in the conceptual center) of common belief masquerading as common knowledge, Dr. Einstein had this to say:

"He who joyfully marches in rank and file has already earned my contempt. He has been given a large brain by mistake, since for him the spinal cord would suffice."

Wow, those are strong words! Quite uncharacteristic of the famously kind and gentle Einstein. Albert had obviously been badly burned a few times by professional and personal know-it-all arrogance.

This lack of appreciation of the extent of your ignorance, and the belief that you already know almost everything important – that what is unknown is small compared to what is known – is usually laid at the feet of common sense. I hear muttering from the intellectuals in the conceptual center – from those who maintain the cultural status quo and staunchly defend our most sacred cultural beliefs:

"Even if I cannot **prove** the absolute truth of our shared cultural assumptions and beliefs, nevertheless, it is only good common sense that **no** reality exists beyond PMR, that we are **not** individuated consciousness experiencing a virtual physical reality, and that a digitally based consciousness is an **obvious** impossibility. Clearly, this Big-TOE-thing is a heap of misguided bullpucky borne of fuzzy wishful thinking and bad science. Why, the logical flaws are so obvious that I do not have to point them out – anyone with half a brain could refute these misguided arguments. Everyone knows that mysticism is dim-witted nonsense that lesser minds use to indulge their inherent weakness."

These attitudes reflect the common sense view of most contemporary westerners, including scientists. Again, it was Albert Einstein who was on the receiving end of a great deal of common sense – enough to prompt him to pen this quip:

"Common sense is the collection of prejudices acquired by age eighteen."

Einstein was not the only one to feel the pressure of cultural conformity; many others also understood the power of belief to create the world in its own image and to reduce thinking to the recitation of current fashion within the politically correct center.

"People see the world not as it is, but as they are."
— AL LEE

"People see only what they are prepared to see."
— RALPH WALDO EMERSON

"If one does not understand a person, one tends to regard him as a fool."
— CARL JUNG

*"The truth which makes men free is for the most part
the truth which men prefer not to hear."*
— HERBERT AGAR (*A Time for Greatness* [1942])

*"The conventional view serves to protect us
from the painful job of thinking."*
— J. K. GALBRAITH

*"Men occasionally stumble over the truth, but most of them pick
themselves up and hurry off as if nothing had happened."*
— WINSTON CHURCHILL

*"To a very large extent men and women are a product
of how they define themselves. As a result of a combination of innate
ideas and the intimate influences of the culture and environment we grow
up in, we come to have beliefs about the nature of being human.
These beliefs penetrate to a very deep level of our psychosomatic systems,
our minds and brains, our nervous systems, our endocrine systems,
and even our blood and sinews. We act, speak, and think
according to these deeply held beliefs and belief systems."*
— JEREMY W. HAYWARD

*"The world we see that seems so insane is the result of a belief system
that is not working. To perceive the world differently, we must be
willing to change our belief system, let the past slip away,
expand our sense of now, and dissolve the fear in our minds."*
— GERALD G. JAMPOLSKY

*"Whatever one believes to be true either is true
or becomes true in one's mind."*
— JOHN C. LILLY

"When we argue for our limitations, we get to keep them."
— PETER MCWILLIAMS

"Nothing is so firmly believed as that which we least know."
— MICHEL EYQUEM DE MONTAIGNE

"Skeptic does not mean him who doubts, but him who investigates or researches, as opposed to him who asserts and thinks that he has found."
— MIGUEL DE UNAMUNO

"Illusions commend themselves to us because they save us pain and allow us to enjoy pleasure instead. We must therefore accept it without complaint when they sometimes collide with a bit of reality against which they are dashed to pieces."
— SIGMUND FREUD

"The pursuit of truth will set you free; even if you never catch up with it."
— CLARENCE DARROW

"People demand freedom of speech as a compensation for the freedom of thought which they seldom use."
— SØREN KIERKEGAARD

"Thoughts have power; thoughts are energy. And you can make your world or break it by your own thinking."
— SUSAN TAYLOR

"There is nothing more frightful than ignorance in action."
— GOETHE

"Nothing is easier than self-deceit. For what each man wishes, that he also believes to be true."
— DEMOSTHENES

"If you don't change your beliefs, your life will be like this forever. Is that good news?"
— DR. ROBERT ANTHONY

Mysticism, Ideas from the Edge

To many in Western culture it is perfectly clear and only common sense that anything connected to metaphysics or mysticism is pure unadulterated nonsense. To the well-educated Westerner, it is obvious that such fools and their delusions have preyed on the gullibility of the uneducated masses since the beginning of time. Let me pass on some advice from the conceptual center:

"An intelligent well-educated person would be well-advised to stay clear of deluded individuals or risk losing their credibility by association. Even if these confused mystics are well meaning, and not simply hustlers and charlatans, their silly jibber jabber flies in the face of science and makes no sense. The paranormal, mysticism, and 'miracles' are mental and emotional pacifiers for the dumb, the uneducated, the emotionally needy, and the gullible."

Do you know anybody who thinks like that? Forty years ago almost everyone centered in Western culture thought like that. Now these cultural and scientific mental straight jackets (beliefs) are worn only by a large majority – the edge has grown ever so slightly amid a greater tolerance of individuality. Einstein evidently knew many individuals who clung to their rational hard-science *über alles* belief systems like a very young child clings to its favorite teddy bear. Much of his writings indicate that he found such people to be extremely limited and boorish – lacking both intelligence and imagination.

"The most beautiful thing we can experience is the mysterious.
It is the source of all true art and science."
— ALBERT EINSTEIN

"There are only two ways to live your life. One is as though nothing is a
miracle. The other is as though everything is a miracle."
— ALBERT EINSTEIN

"The important thing is not to stop questioning.
Curiosity has its own reasons for existing. One cannot help but be in awe
when he contemplates the mysteries of eternity, of life, of the marvelous
structure of reality. It is enough if one tries to comprehend
a little of this mystery every day. Never lose a holy curiosity."
— ALBERT EINSTEIN

"What our eyes behold may well be the text of life but one's meditations
on the text and the disclosures of these meditations are no less
a part of the structure of reality."
— WALLACE STEVENS

"The foundations of a person are not in matter but in spirit."
— RALPH WALDO EMERSON

As I explained in Section 2, logic requires that a successful Big TOE (as opposed to a little PMR-only TOE) must have at least one seemingly mystical assumption. That mystical assumption appears mystical only from the limited perspective of PMR. *My Big TOE* is constructed entirely from two assumptions; 1) the seemingly mystical one: the existence of primordial consciousness (AUO); and 2) the well known and understood Fundamental Process of evolution. From just those two ingredients, all reality flows, including the relatively small but interesting subset of the larger reality we experience as PMR.

"The grand aim of all science is to cover the greatest number
of empirical facts by logical deduction from the
smallest number of hypotheses or axioms."
— ALBERT EINSTEIN

Elegant simplicity is the hallmark of Big Truth. According to Einstein, elegant simplicity is also the grand aim of all science. Combining both ideas logically provides that Big Truth is the grand aim of all science. Tell that to all the little picture scientists who have no interest in Big Truth. In its pursuit of Big Truth, *My Big TOE*, employs the elegant simplicity of a single consciousness-evolution fractal ecosystem to derive All That Is, science and philosophy, point and purpose from two simple assumptions.

The Process and Importance of Growth
Within *My Big TOE*, I describe personal growth – the improvement of the quality of your consciousness – as the point of your existence. It is the evolutionary urge of the Fundamental Process applied to consciousness that pushes us forward to evolve the quality of our being – to lower our entropy. It is growth – spiritual growth – that links us to AUM, defines the

positive direction of progress, and gives us purpose and meaning within the Big Picture. If ever there were a law of being and existence, it would be: Grow or die. Stability lies only in growth, maintaining the status quo is the first sign of impending decay. Growth is the point, the purpose, what makes the experiment run, and the reason for going on. When consciousness and evolution become entangled, modification toward improvement is the result – growth defines the struggle to be. Without growth, without an opportunity to evolve our being, we are nothing.

"The self is only that which is in the process of becoming."
— SØREN KIERKEGAARD

"What is the most rigorous law of our being? Growth. No smallest atom of our moral, mental, or physical structure can stand still a year. It grows – it must grow; nothing can prevent it."
— MARK TWAIN

"Unless you try to do something beyond what you have mastered, you will never grow."
— C.R. LAWTON

"One's mind, once stretched by a new idea, never regains its original dimensions."
— OLIVER WENDELL HOLMES

"Only in growth, reform, and change, paradoxically enough, is true security to be found."
— ANNE MORROW LINDBERGH

Einstein realized that in the atomic age, the spiritual growth of mankind had suddenly become more critical – that the age-old struggle of raising the quality of consciousness would now take on more urgency.

"The release of atomic energy has not created a new problem. It has merely made more urgent the necessity of solving an existing one."
— ALBERT EINSTEIN

Little did he know that mankind would devise several additional ways to destroy life on our planet within a few decades. Given this destructive capacity, future human genetic engineering, psychotropic pharmaceuticals, and the coming Information Age, it has never been more important for mankind to understand the Big Picture before it prematurely terminates this portion of The Great Experiment in Consciousness.

"What lies behind us and what lies before us
are tiny matters compared to what lies within us."
— OLIVER WENDELL HOLMES

Wisdom

Many times within the pages of *My Big TOE*, we have struggled with the problem of separating the truly wise from the delusional fools that would try to appear wise – separating knowledge from pseudo-knowledge, and truth from belief. We concluded that a careful results oriented exploration with plenty of pauses to taste the pudding would be the best process. Wisdom, we said, was derived only from experience and reflects an appreciation and understanding of the Big Picture.

Unfortunately, only the wise can understand and confidently apply wisdom to their reality. Because it takes one to know one, a relatively slow incremental bootstrapping process is required to become wise or to recognize and appreciate the wisdom of others. Because wisdom is non-transferable, only you can decide who are fools and who are wise – and only for yourself. We said that **your** wisdom must flow from **your** experience and that it could not be acquired from psychotropic chemicals, a book, or a guru. We said that wisdom was the natural result of a high quality of consciousness. It seems that others agree with many of these ideas.

"Never mistake knowledge for wisdom. One helps you
make a living; the other helps you make a life."
— SANDRA CAREY

"It requires wisdom to understand wisdom:
the music is nothing if the audience is deaf."
— WALTER LIPPMAN

"Wisdom is not wisdom when it is derived from books alone."
— HORACE

"Try not to become a man of success, but rather
try to become a man of value."
— ALBERT EINSTEIN

> *"We can be knowledgeable with other men's knowledge,*
> *but we cannot be wise with other men's wisdom."*
> — MICHEL DE MONTAIGNE

Fear, Ego, and Delusion

In a long aside in Chapter 42, Book 2, we discuss the interrelationships that exist among fear, ego, intellect, ignorance, and delusion. From this unholy combination it was reasoned that most of our daily problems emerge as self-inflicted wounds. We saw how fear and ignorance prompted a mutual support alliance between the ego and intellect to deny the fear by constructing a personal fantasy (belief system). We watched the pathetic self struggle to maintain the ego's delusion, only to become enslaved by the maintenance requirements of its own deceit. Fear, the opposite of love, was shown to be the root of most everyone's personal problems as well as the main inhibitor of improving the quality of consciousness.

> *"Fear defeats more people than any other one thing in the world."*
> — RALPH WALDO EMERSON

> *"An inflated consciousness is always egocentric and conscious*
> *of nothing but its own existence. It is incapable of learning from the past,*
> *incapable of understanding contemporary events, and incapable of*
> *drawing right conclusions about the future. It is hypnotized by itself*
> *and therefore cannot be argued with. It inevitably dooms itself*
> *to calamities that must strike it dead."*
> — CARL JUNG

> *"To understand the world, one must not be worrying about one's self."*
> — ALBERT EINSTEIN

Truth, Science, and Logic

In Sections 2, 3, and 5 we make a case for expanding the purview of science to include a wider range of phenomena. We explain how the PMR-only view of reality eliminates the possibility of seeing the bigger picture – how it unnecessarily limits the available solution space so that important scientific problems appear to be mysterious and unsolvable. Many for a long time have known the fact that understanding the larger reality requires a vision and perspective that goes beyond our normal sense of PMR. It is no secret that we are intellectually trapped by the limitations of our little picture beliefs about reality; and that there is much more to existence than the local causal physics of PMR.

Likewise, it is clear that rational knowledge, causality, and formal scientific process must rise above its present intellectual little picture entrapment to illuminate the truth that lies beyond simple physical existence. However, it is important to understand that explorations of consciousness also need the rigor of a more general scientific process to separate truth from fiction, the actual from the apparent, the meaningful and significant from the useless and unimportant.

Scientific methodology is not to be abandoned to superstition – in fact, it is exactly the opposite. Scientific methodology needs to rise above the superstition and belief that now cripples its vision in order to separate truth from fiction at a higher level of rational understanding. Personal explorations of consciousness into NPMR are not about otherwise intelligent people (including scientists) becoming mush-heads in PMR – it is more about self-limited PMR boneheads becoming scientists at the next higher level of causality.

"As far as the laws of mathematics refer to reality, they are not certain; and as far as they are certain, they do not refer to reality."
— ALBERT EINSTEIN

"We should take care not to make the intellect our god; it has, of course, powerful muscles, but no personality."
— ALBERT EINSTEIN

"The significant problems we face cannot be solved at the same level of thinking we were at when we created them."
— ALBERT EINSTEIN

"Not everything that can be counted counts, and not everything that counts can be counted."
— ALBERT EINSTEIN

"The further the spiritual evolution of mankind advances, the more certain it seems to me that the path to genuine religiosity does not lie through the fear of life, and the fear of death, and blind faith, but through striving after rational knowledge."
— ALBERT EINSTEIN

*"Every great advance in science
has issued from a new audacity of imagination."*
— JOHN DEWEY, THE QUEST FOR CERTAINTY

*"There is one thing even more vital to science than intelligent methods;
and that is, the sincere desire to find out the truth, whatever it may be."*
— CHARLES SANDERS PIERCE

*"We must learn to tailor our concepts to fit reality,
instead of trying to stuff reality into our concepts."*
— VICTOR DANIELS

Education, Learning and Personal Growth

We said that growth is what life is all about, the point of your existence, the reason for being here, but how does this growth happen? How do you go about growing up? We know that fear and ego block learning and personal growth. Certainly, everyone will agree that growing-up is more a being process than an intellectual process.

Essentially, success is a matter of perspective, attitude, vision, perseverance, and the gumption to get out there and try – and then try again. All you need is a glimpse of the possibilities and the determination to explore at least some of those possibilities. Perhaps, if you are ready, willing, and able, *My Big TOE* will help you see some new tantalizing possibilities. However, supplying the necessary gumption and determination is all up to you – no one can help you with that.

Your intent reflects the quality of your being and drives your choices. It is learning and personal growth that eventually enable you to make the right choices for the right reasons – naturally and intuitively. Every day of every year we are forced to make hundreds of significant choices as we interact with others. We express ourselves with choice. We paint ourselves onto the canvas of the actualized present with brushes of intent and pigments of choice. We juggle a thousand shades of competing need, fear, ego, desire, understanding, compassion, humility, caring and love in a swirling mix of contradiction until it all coalesces into a single result that is actualized into the present irreversible reality by the choices we make. We are the sum total of those choices made over a lifetime, or a series of lifetimes. The substance of our being and the rate of our growth are defined by the quality of our choices.

"One's philosophy is not best expressed in words;
it's expressed in the choices one makes. In the long run,
we shape our lives and we shape ourselves. The process never ends
until we die. And the choices we make are ultimately our responsibility."
— ELEANOR ROOSEVELT

Learning, like growth, is a personal thing – it is generated from the inside out. If we try to educate another by stuffing knowledge in from the outside, like stuffing tissue in a bra, we can achieve only the appearance of education – produce an educated falsie. Our school systems have, during the last century, essentially abandoned education in favor of force-feeding factual information and developing the bare essentials of a functional literacy. Education should never be confused with training. Unfortunately, in our culture most people would be hard pressed to explain the difference.

It is no wonder that the intellectual center is predominately populated by well-educated, well-credentialed, conceptually limited, original-thinking challenged preservers of the status quo. It is also no wonder that so few of us have the inclination and drive to pursue Big Truth on the Path of Knowledge.

"You cannot teach a man anything;
you can only help him find it within himself."
— GALILEO

"Education is what remains after one has
forgotten everything he learned in school."
— ALBERT EINSTEIN

"I find that the great thing in this world is not so much where we stand
as in what direction we are moving; To reach the port of heaven,
we must sail sometimes with the wind and sometimes against it –
but we must sail, and not drift, nor lie at anchor."
— OLIVER WENDELL HOLMES

"Knowing is not enough; we must apply.
Willing is not enough; we must do."
— GOETHE

"Everybody wants to be somebody; nobody wants to grow."
— GOETHE

Growing up is never an easy or quick process. Anxiety and **internal** struggle is always generated when abandoning comfortable and familiar ways of being, thinking, and believing. Diving headlong into the unfamiliar and murky waters of something new and different comes with no guarantee of success. Additionally, there are the **external** struggles with being different – a stranger among your friends – living near the edge in a world dominated by those in the center. Growing up takes great courage as well as great determination.

"Great spirits have always found violent opposition from mediocrities. The latter cannot understand it when a man does not thoughtlessly submit to hereditary prejudices but honestly and courageously uses his intelligence."
— ALBERT EINSTEIN

"What we need is not the will to believe, but the wish to find out."
— BERTRAND RUSSELL

"The journey to happiness involves finding the courage to go down into ourselves and take responsibility for what's there: all of it."
— RICHARD ROHR

"All life is an experiment."
— RALPH WALDO EMERSON

"Happiness is not a reward – it is a consequence. Suffering is not a punishment – it is a result."
— ROBERT GREEN INGERSOLL

Seeing the Big Picture, understanding the larger reality, growing up, increasing the quality or synergy of your consciousness, evolving your consciousness, decreasing the entropy of your consciousness, spiritual growth – or whatever you want to call it – is your responsibility and yours alone. If you are waiting for progress to come to you, or need a growth path that is easy and requires little effort, or cannot stand to be very different from your peers, or require certainty before you can take a step, or are simply stuck in a belief trap of some sort, you are not irretrievably lost. When you are ready, an opportunity to grow will become visible.

Time in PMR is an experience defined by the perception of change. Your deluded perception of a short, somewhat random, existence on planet earth does not limit your responsibility for growth. Every decision, intent, motivation, and action moves you along one way or another in the Big Picture whether you are aware that you are participating in an iterative, repeat it until you get it, consciousness-quality learning lab or not.

Fortunately, the majority of us are growing more or less in the positive direction. The major difference your effort is likely to make is to increase dramatically the rate of your progress and the quality of your existence. That's all. No big deal if you are enjoying your current level of consciousness quality – this is not a timed test. You can accomplish only whatever you are ready for, and capable of. If you decide to kick back and let your opportunities for growth slip by unexplored, one note of caution: Don't let the anti-rats (defined in Chapter 36, Book 2) get you. They primarily prey on the slackers and stragglers. But don't worry; I am sure that won't be a problem for you. Come on along when you are ready.

I have shown you the Big Picture and told you how it works, but I cannot cause you to integrate it into your knowledge-base at a level that is deeper than your intellect.

I have enabled you to see and feel some of the potential that you have as a chip from the old AUM block, but only you can actualize that potential.

As your tour guide, I have pointed out some of the more interesting sights, but you have to get off the intellectual tour bus and experience them on your own.

I have provided a dim glimmer of the possibilities that lie before your being, but you must provide the courage and determination to pursue that vision to some personally profitable conclusion.

I have provided a model of reality to help you interpret, structure, and add rational meaning to your personal experience, but you have to have the experience and then learn from it.

I have pointed out some of the mental and conceptual limitations and constraints that bind you and blind you, but only you can set yourself free and extend your vision.

I have told you some Big Truths, but they must remain vague and powerless until you find them for yourself.

I have explained to you how spirituality, consciousness, love, and paranormal phenomena are interconnected, but I cannot cause you to directly experience any of them.

I have encouraged you with sticks and carrots and pointed you in the right general direction, but I cannot define your personal path much less carry you down it.

I have explained why you are here and what the point and meaning of both your physical and nonphysical existence is, but you must express your own intent and make your own choices in order to make that existence significant and successful.

I have done what I can do. The ball is in your court. How you play it, how the concepts you have found within *My Big TOE* interact with, and ultimately affect, the quality of your being, is entirely up to you. However you currently express your quality and intent – however you make the important choices – if you make them with a long-term vigorous, open-minded, and skeptical commitment to the relentless pursuit of Big Truth, you will eventually succeed. Who you are, what you experience, and the reality you create, are the applied results of whatever Big Truth you have managed to capture and internalize. What could be more straightforward than that? Good consciousness quality is the result of an open mind, a focused steady intent, and the courage to be.

I know what you are thinking: Easy to say, difficult to do. Sometimes growing the quality of your consciousness doesn't seem particularly straightforward or simple because you are exploring unfamiliar territory. The unknown always appears tricky whether it is or not. New paradigms always seem impossible before they become obvious and trivial. If you "just go do it" with a long-term view, seemingly impenetrable barriers will begin to dissolve before your steady unconditional effort. If you never go do it or give up easily because those barriers seem to be hopelessly impenetrable – they will be.

I hope you have had a few valuable insights as well as a few good laughs as we have explored some of the territory that lies beyond the edge of your everyday thoughts. Adventures of mind and spirit are best absorbed in alternating doses of work and play – an interweaving of light and heavy. Learning and growing, once past the boot-camp phase, should be joyful and fun. If you have been swinging a pick in a spiritual salt mine with little to show for it, you are wasting your time on technique, form, and process. That Big Truth yields best to pain, abstention, renunciation, suffering, and sacrifice is pure horsepucky. The Western Puritan ethic and the Western work ethic may combine to produce successful model citizens and productive workers on the outside, but they invariably lead to poverty and frustration on the inside.

Dedication does not mean work, work, work, work until you drop, succeed, quit, or give up the possibility. Dress up the Western work ethic in a monk's robe and you get more anxiety and frustration than spiritual progress. Big Truth is much more likely to yield to loving, lighthearted play than to the stiff ossified narrowness that often defines the mature and dignified Western approach. When happy children laugh and play they live closer to the core of existence – as they grow up and get appropriately serious, they lose their innocence and their intuitive harmony with The One. To be able to freely and seamlessly mix and intermingle knowledge, wisdom, and play is one of the gifts of enlightenment.

The darkness of ignorance surrounds all of us, all the time – we will probably never know all the answers there are to know. Will a bacterium ever become president of the United States? The idea of limitations does not come easy for us Westerners who habitually think in terms of physical objects and processes. Some barriers may need to be gracefully accepted – at least for a while, and maybe forever. Running with the herds of AUMosaurus thundering across the mental plains is simply beyond our capacities and purpose. Regardless, we will and should struggle to understand whatever is within our grasp. If we do not use our full capacity or reach our full potential, shame on us. If we reach limits, let's make sure that they are actual limits, not self-imposed limits. Let's make sure that darkness never becomes a thing in itself, but rather only the absence of light. For no one is ever fearful of the absence of light.

As your awareness evolves from dim to bright, it illuminates more and more of the Big Picture and pushes back the darkness that marks the boundaries of your personal knowing. As a conscious being imbued with the capacities and potential of your source, an ability to expand the light and grow the quality of your consciousness creates the challenge that your existence is all about.

Do not let the magnitude of these Big Picture concepts overwhelm you. We get to where we are going by patiently taking one step at a time. There is no other way – it is the same for everyone. Each of us follows a unique path that is defined as we go, one step at a time, by our free will driven intent. The key is to begin stepping out with a purposeful intent.

If I have succeeded in turning on lights here and there to challenge a few of your personal dark places, I am greatly pleased – but **you** must open your eyes and see what is (and is not) there for yourself. Then you must evaluate what you see – you must separate your personal truth from your personal delusion. The passageway to an improved quality of being lies along the path where personal truth and absolute truth eventually become one.

Recall that absolute truth (Big Truth) is universal and never changes; also that Big Truth is not secretively hidden away where only the chosen few can find it. To the contrary, Big Truth demonstrates itself right in front of your eyes and stares you in the face every day. Its discovery challenges you because your conscious awareness is dimmed and limited by your beliefs to such an extent that you shut yourself off from the experience you need to grow wisdom and more fully understand the bigger picture.

"What one has not experienced, one will never understand in print."
— ISADORA DUNCAN, *My Life*

If you do not yet see light glowing from the possibilities within the bigger consciousness picture, I would hope that you would at least have a dim glimmer of where and how you can find the light-switch on your own in the event you one day get the urge and the courage to explore beyond the tiny little corner of virtual physical existence within which you have walled-off your awareness. Oh, but to lie safe and comfortable in the protective lap of a favorite belief trap – no challenge, no anxiety, no risk, no effort, and no growth – but with lots and lots of reassuring company that occupies itself justifying its beliefs and polishing up the status quo in an endless loop of self-referential ego-boosting blather. That sounds like a comfy and supportive place to live doesn't it? Such an existence – the very definition of normal – goes nowhere significant in the Big Picture, but is safe, easy, reassuring, and often materially rewarding in the little picture.

On the other hand, if you at least dimly grasp that you are a prisoner of your beliefs, confined to a tiny PMR cell of your own making, if you are vaguely aware there is something real beyond the experience of your five physical senses, perhaps now is the time to begin plotting your escape.

Escaping belief traps and finding the necessary mental and emotional freedom to evaluate your experience logically and rationally will become your first-level quest. You will need to be daring and courageous. I am not suggesting an imaginary escape from the supposed real (physical) world – quite the contrary. I propose a real escape from the virtual world of PMR to a larger (superset) world that is closer to The Source and nearer to the core of fundamental existence – one that has an operational effect, and a measurable impact upon PMR – one that fulfills and sustains your deepest purpose as an aware evolving consciousness.

You are what you are – why not become all that you can be? How can you possibly discover what the limits of your being are without reaching beyond where you are now? How far, how consistently, and how steadily will you reach out toward the light of Big Truth? How courageously, carefully, honestly, and scientifically will you explore your personal unknown, your personal ignorance, prejudices, assumptions, fears, and beliefs? Are you ready, willing, and able to challenge the monster that lives under your bed at night? Do you need assurances and expect or require definitive answers to come quickly – or are you prepared to let the process take a lifetime and define its own path?

What's a lifetime worth? What else would you want to use one for? To see how much stuff or power you could accumulate, or how many facts you could figure out or learn? How about to see how much fun you could have, or how much beer you could drink? Perhaps all existence is random and the concepts of consciousness, evolution, entropy reduction, love, personal growth, and profitability are simply delusions because there can be no point or purpose in randomness. Does this feel right to you? What does **your intuition** say about there being significance, pattern and purpose to existence? What is **your** lifetime likely to be worth? (Please, don't bother asking your intellect – it will most likely only confuse you because it will **always** have insufficient data for a logical conclusion.) Even if your intuition is grossly underdeveloped, it is still a better guesser of Big Answers than your intellect. Don't worry; your skepticism will keep you safe until your intuition becomes certifiably reliable.

To help you answer these questions, I have offered you a unified theory of existence and reality, a comprehensive TOE, and a reasonably detailed model of reality-mechanics based upon my own carefully evaluated explorations of the larger reality. The extent to which you have actually absorbed **useful** understanding at a deep level has at least as much, if not more, to do with you, your beliefs, and your experience than it does with the correctness of My Big Picture Theory Of Everything. Think about that; read that last sentence again. Your consciousness, its quality, your awareness, and your reality are intertwined in the Big Picture – their apparent separateness as independent entities is a PMR-based illusion that helps you maintain your focus on the basic things you need to learn first.

> *"Truth can never be told so as to be understood, and not be believed."*
> — WILLIAM BLAKE

If you wish to progress beyond the bottom rung of knowing – which simply is to be a believer – you must develop your own personal understanding and truth. Perhaps *My Big TOE* has provided some of the fertile ground that you need to create and grow **your** Big TOE. Perhaps it has given you some support and encouragement, or merely goaded you to think about and assess your personal truth. If *My Big TOE* has caused you to ponder at least a few big thoughts, then we have had a successful journey together. Whether the thoughts that are rattling around in your head today are in close agreement or strong disagreement with the model I have presented makes little difference. As long as you continue to probe, ponder, and seek Big Truth open-mindedly, you will eventually find it.

Everyone has at least a stub of a Big TOE – whether it is by purposeful construction or by cultural default, whether it is right or wrong, helpful or useless, belief-based and fearful, or love-based and all encompassing. For most, it is a reflection of their core beliefs – their worldview at the bone and sinew level. For others it represents the result of a lifetime of experience and careful intellectual effort, or perhaps a mindless acquiescence to social, religious, scientific, and cultural norms. However well or poorly your stubby (or grand) Big TOE is defined, it is the place where you must start if you wish to grow it beyond its present state. Leave the giant leaps to lottery winners and mythical heroes – consistent plodding in the generally right direction is a better and surer way to get to where you want to go. A personal growth strategy that primarily depends on good luck or some special connection (book, guru, technique, ritual, belief system) for success is little more than a procrastinator's excuse or the common fantasy of those who lack the gumption, self-discipline, or understanding to succeed on their own.

Consciousness, by its nature, is nonphysical (from a PMR point of view) and personal. As an individuated consciousness you create your own local reality that has both shared (objective and PMR-rule-based) and personal (subjective and NPMR-rule-based) components at all levels of awareness.

How closely your personal reality expresses Big Truth determines the quality and the correctness of any Big TOE based upon that personal reality. Each individual must **personally** discover Big Truth. If Big Truth is not discovered (primarily a subjective process) and applied and tested (primarily an objective process) by means of your personal experience, its concept, if known at all, can reside only in your intellect as a relatively powerless theoretical idea.

The personal power and wisdom of a high quality consciousness must be derived through your personal experience of that consciousness. Intellectual transfers of information and factual understanding can do little more than point your intent in the right direction. Nonetheless, the purposeful direction of intent by a free will can be an exceedingly powerful tool. The good news is that profitable growth experience always follows steady, serious, and well-directed intent. The intellect, though poorly understood and badly misused, does have a valued place and function in the evolutionary process of growing consciousness quality. You were born with everything you need to succeed. The only items you must bring to the table are the personal will, drive, courage, and intent to pursue and find Big Truth. If you have those, everything else will fall into place.

Well, that's it ladies and gentlemen, boys and girls, men and women, and the rest of you who fit better under some other category – you are now on your own. Actually, you always have been on your own, but now you should have at least a glimmer of the Big Picture, a better appreciation of the possibilities available to you, and some idea of how to go about actualizing those possibilities if you decide that is what you want to do.

As *My Big TOE*, the book, comes to an end, it is my intent that these quotes and comments have given you some final clarity, direction, and understanding as they wrap-up the **personal** bottom-line of Big-TOEness and provide some assurance that if you do decide to pursue the Path of Knowledge, you will be traveling in good company. Be assured it is not only religious fundamentalists and freelance kooks who are pursuing a bigger picture beyond physical reality. There are plenty of those to be sure, but there are also plenty of seekers with keen analytical minds and careful processes who are making real progress toward actualizing themselves within a larger reality, who are purposely evolving their quality and moving toward regaining their natural capacities as aware individuated portions of AUM consciousness, who are gaining and productively utilizing the awesome power and value of a balanced reduced-entropy consciousness. You can be among them if you care enough to "just do it." You have the capability and potential required for success – you are not excluded by any force or barrier other than what **you** create.

▶ Where one journey ends, another must begin. I won't be traveling with you on your next journey. I have no idea where you are headed, or where you will end up but I do have one little question about how you are going to make the trip. Will you be 1) in charge and driving, 2) along for the ride in the passenger's seat, 3) locked in the trunk, or 4) dragged, kicking and screaming, at the end of a rope?

Bon voyage, amigo! Have a nice trip…. and may you graduate from boot-camp before the rope breaks. ◀

Ah, yes, the fun is never done, but the public unraveling of this particular Big TOE is at its end. Because you have been patient enough to read this strange trilogy all the way to the end, I have come up with a small going away present for you. When and if you decide to go traveling into inner space, I will help you out with all the necessary accommodations – well, that is, as long as you don't require mints placed on your pillow. When and if you begin to understand your existence as consciousness – when you begin exploring the Big Picture that lies on the other side of your mind – when you start experimenting with the larger reality of NPMR and its inhabitants – well…, just tell 'em that ol' Uncle Tom sent you – they'll take good care of you, show you around, and see to it that you stay out of trouble. I'll leave the lights on for you.

http://www.My-Big-TOE.com
http://www.lightningstrikebooks.com